Lecture Notes in Computational Science and Engineering 127

Editors:

More information about this series at http://www.springer.com/series/3527

Xevi Roca • Adrien Loseille
Editors

27th International Meshing Roundtable

Springer

Editors
Xevi Roca
Barcelona Supercomputing Center
Mataró
Barcelona, Spain

Adrien Loseille
Inria
Palaiseau, France

ISSN 1439-7358 ISSN 2197-7100 (electronic)
Lecture Notes in Computational Science and Engineering
ISBN 978-3-030-13994-0 ISBN 978-3-030-13992-6 (eBook)
https://doi.org/10.1007/978-3-030-13992-6

Mathematics Subject Classification (2010): 65M50, 97N80, 97N40

This Springer imprint is published by the registered company Springer Nature Switzerland AG.
The registered company address is: Gewerbestrasse 11, 6330 Cham, Switzerland

Preface

The papers in this volume were peer reviewed and selected for presentation at the 27th International Meshing Roundtable (IMR), held on October 1–5, 2018, in Albuquerque, USA. The International Meshing Roundtable was started by Sandia National Laboratories in 1992 as a small meeting of organizations striving to establish a common focus for research and development in the field of mesh generation. Now after 27 consecutive years, it has become clear that the International Meshing Roundtable has become the recognized international focal point for state-of-the-art meshing research collaboration spanning research and development from universities, commercial companies, and government laboratories.

The 27th International Meshing Roundtable consisted of presentations of peer-reviewed technical papers, research notes, keynote and invited talks, short course presentations, a poster session and competition, a meshing contest, and an open space session. The Program Committee would like to express our appreciation to all who participated in making the International Meshing Roundtable a successful and enriching experience.

The papers in these proceedings present novel contributions that range from the theoretical to practical. The committee selected these papers based on the input from peer reviewers on the perceived quality, originality, and appropriateness to the theme of the International Meshing Roundtable. This year the committee accepted 25 papers. We would like to thank all who submitted papers. We also extend our appreciation to the colleagues who provided reviews of the submitted papers. Their efforts were essential to the process of selecting papers for the International Meshing Roundtable. The names of the reviewers are acknowledged in the following pages.

The conference received travel support from the National Science Foundation (NSF) for student and postdoctoral attendees from the U.S. institutions and additional travel support from CIMNE, csimsoft, MeshGems, Pointwise, Los Alamos

National Laboratory, and Sandia National Laboratory. We deeply acknowledge their support. We extend special thanks to Kathy Loeppky of Sandia National Laboratories for her time and effort to make the 27th International Meshing Roundtable a success.

Barcelona, Spain
Palaiseau, France
November 2018

Xevi Roca
Adrien Loseille

Contents

Part V Parallel and Fast Meshing Methods

Part I
High-Order Adapted Meshes

P^2 Mesh Optimization Operators

Rémi Feuillet, Adrien Loseille, and Frédéric Alauzet

Abstract Curved mesh generation starting from a P^1 mesh relies on mesh deformation and mesh optimization techniques. Mesh optimization techniques consist in locally modifying the mesh in order to improve it with respect to a given quality criterion. This work presents the generalization of two mesh quality-based optimization operators to P^2 meshes. The generalized operators consist in mesh smoothing and generalized swapping. With the use of these operators, P^2 mesh generation starting from a P^1 mesh is more robust and P^2 connectivity-change moving mesh methods for large displacements are now possible.

1 Introduction

In numerical simulation, unstructured meshes are commonly used. More specifically, in Computational Fluid Dynamics (CFD) they are used to help to solve real world problems found in industry and government. In the last decade high-order resolution methods (continuous Galerkin [4], discontinuous Galerkin [10], spectral differences [19], ...) have been used. To preserve the high-order of convergence of these methods, it is required to have a high-order representation of the geometry in the mesh. These meshes are curved in order to fit at best with the boundary of the studied geometric shape. In this context, the generation and the processing of high-order meshes is necessary.

R. Feuillet (✉)
GAMMA3 Team, INRIA Saclay, Palaiseau, France

POEMS Team, CNRS/ENSTA/INRIA, Palaiseau, France
e-mail: remi.feuillet@inria.fr

A. Loseille · F. Alauzet
GAMMA3 Team, INRIA Saclay, Palaiseau, France
e-mail: adrien.loseille@inria.fr; frederic.alauzet@inria.fr

X. Roca, A. Loseille (eds.), *27th International Meshing Roundtable*,
Lecture Notes in Computational Science and Engineering 127,
https://doi.org/10.1007/978-3-030-13992-6_1

To generate high-order meshes, several approaches exist. Some are using a PDE or variational approach to curve a P^1 mesh into a P^k mesh [5, 9, 14], others are based on optimization and smoothing operations and start from a P^1 mesh with a constrained P^k curved boundary in order to generate a suitable P^k mesh [12, 15, 16]. In all these techniques, the key feature is to find the best deformation to apply to the P^1 mesh and to optimize it.

A connectivity-change moving-mesh method [1] that enables closed-advancing boundary layer mesh generation [2] relies on both mesh deformation and mesh optimization techniques. In [1], the motion of vertices is first computed thanks to a linear elasticity model and then the position of these vertices is changed via local mesh optimization operators such as generalized swapping and mesh smoothing. To apply this connectivity-change moving-mesh method to high-order meshes, these two operators need to be generalized to high-order meshes.

In this paper, P^2 mesh quality-based optimization operators are presented. They are a generalization of the P^1 operators and rely on the resolution of a local optimization problem. These two operators can be applied to improve P^2 mesh generation starting from a P^1 and enable high-order connectivity-change moving mesh methods.

The paper is outlined as follow. Section 2 defines what a high-order mesh and sets up validity and quality criteria. Section 3 deals with P^2 mesh optimization. Section 4 shows two applications with examples of P^2 mesh optimization. The first one is an improvement of a P^2 mesh generation technique that curves a P^1 mesh thanks to a high-order linear elasticity solver and the second one describes a P^2 connectivity-change moving mesh method also using a high-order linear elasticity solver. Finally, Sect. 5 deals with conclusions and perspectives driven by this work.

2 High-Order Element, Validity and Quality Criteria

To properly define P^2 quality-based optimization operators, it is fundamental to properly define quality and validity criteria for high-order elements.

In general, a finite element is defined [3] by the triplet $\{K, \Sigma_K, V_h\}$ where K denotes the geometric element (triangle, etc), Σ_K the list of nodes of K, and V_h, the space of the *shape functions*, here it will be the Lagrange polynomial functions (or interpolants). To properly define the geometry and these functions, a reference space (that can also be seen as a parameter space) is defined where all coordinates are between 0 and 1. In this space, the reference element is denoted $\widehat{K}$ and has a fixed (and sometimes uniform) distribution of nodes. The element K, also called physical element, is thus the image of $\widehat{K}$ via a mapping, denoted F_K (see Fig. 1). More specifically, for each point M of K, there is a point $\widehat{M}$ of $\widehat{K}$ such that $M = F_K(\widehat{M})$.

Finally, the position of a point M inside K is defined by

$$M = \sum_{i=1}^{n} \phi_i(\widehat{M}) A_i,$$

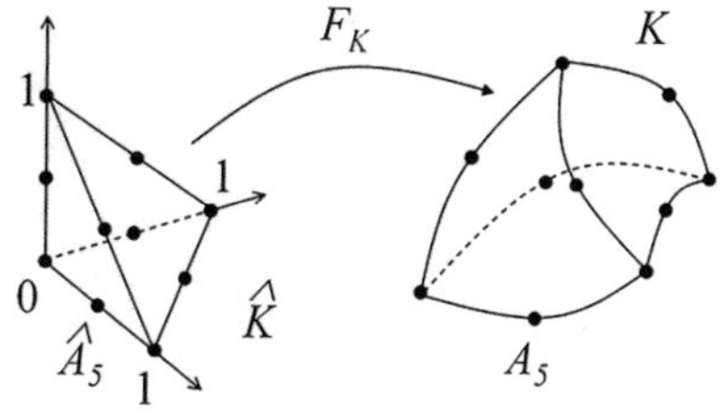

Fig. 1 Mapping F_K from $\widehat{K}$ to K

where n is the number of nodes, $A_i = F_K(\widehat{A}_i)$ with $\widehat{A}_i$ the nodes of the reference element which map to A_i the nodes of the physical element, and ϕ_i are the Lagrange polynomial functions defined such that:

$$\phi_i(\widehat{A}_j) = \delta_{ij} \quad \text{and} \quad \sum_{i=1}^{n} \phi_i = 1.$$

It is important to note that in order to define a complete finite element of degree k on a simplex of dimension d (edge for $d = 1$, triangle for $d = 2$, tetrahedron for $d = 3$), the number of (distinct) nodes needs to be equal to $\frac{\Pi_{j=1}^{d}(k+j)}{d!}$. Also, on a simplex, the reference coordinates $(\widehat{x}, \widehat{y}, \widehat{z})$ can be used to define the simplex barycentric coordinates (u, v, w, t). For instance, for a triangle we have: $u = 1 - \widehat{x} - \widehat{y}$, $v = \widehat{x}$ and $w = \widehat{y}$, and for a tetrahedron $u = 1 - \widehat{x} - \widehat{y} - \widehat{z}$, $v = \widehat{x}$, $w = \widehat{y}$ and $t = \widehat{z}$.

A point M of the simplex K can also be expressed in Bézier form using the Bernstein polynomials B^d_{ijlm} as:

$$M = \sum_{i+j+l+m=k} B^d_{ijlm}(u, v, w, t) P_{ijlm},$$

with $B^d_{ijlm}(u, v, w, t) = \frac{d!}{i!j!l!m!} u^i v^j w^l t^m$. For a triangle, $m = 0$. The points $(P_{ijlm})_{i+j+l+m=k}$, also called $(C_i)_{1 \le i \le n}$ (see Fig. 2) are the Bézier control points and are directly related to the nodes $(A_i)_{1 \le i \le n}$.

For instance, to compute C_5 (P_{1100}) in Fig. 2, we use the formula (see [3] for details):

$$C_5 = \frac{4A_5 - A_1 - A_2}{2}.$$

In general and in the following sections, $(A_i)_{1 \le i \le d+1}$ are called vertices, $(A_i)_{d+2 \le i \le n}$ are called nodes, and $(C_i)_{1 \le i \le n}$ are called control points.

The validity of an element means that the associated mapping F_K is a diffeomorphism. It can be ensured if the minimum of the determinant $\mathcal{J}_K$ of the Jacobian matrix of the mapping F_K is strictly positive everywhere inside the element [7]. In the case of a simplicial element, Jacobian $\mathcal{J}_K$ can be written as polynomial of degree $d \times (k-1)$ in the barycentric coordinates of the simplex, where d is the dimension of

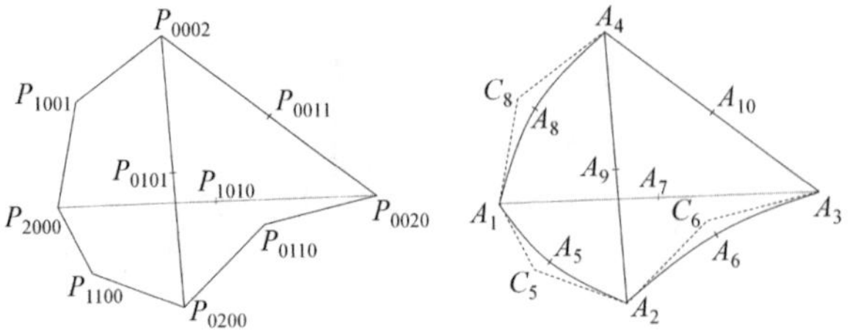

Fig. 2 Correspondence between the control points and the nodes for a P^2 tetrahedron

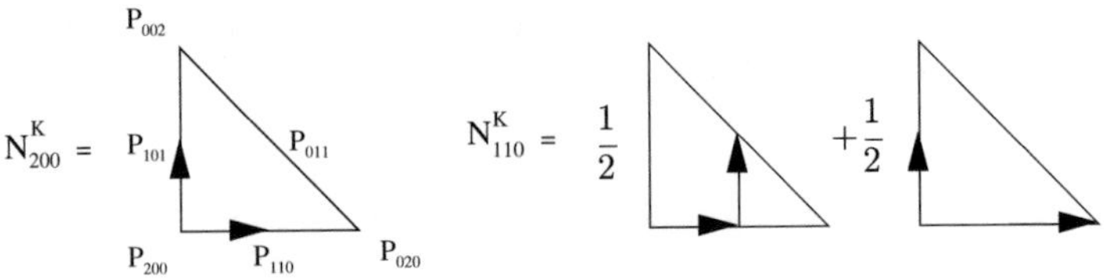

Fig. 3 Vectors involved in the determinant for the computation of a corner (left) and an edge (right) control coefficients of a P^2 triangle

the simplex and k the degree of the simplex. When the element is of degree 1 (e.g. straight-sided), it simply means that the oriented volume/area is strictly positive. The Jacobian can also be expressed in the Bernstein polynomial basis:

$$\mathcal{J}_K(u, v, w, t) = \sum_{i+j+l+m=d(k-1)} B_{ijlm}^{d(k-1)}(u, v, w, t) N_{ijlm}^K,$$

where N_{ijlm}^K are the control coefficient of the Jacobian. These coefficients can be explicitly found and have a geometrical meaning (see [3] for more details). In the case of a P^2-triangle (see Fig. 3), a corner and an edge control coefficients [3] are for example:

$$N_{200}^K = 4\det(\overrightarrow{P_{200}P_{110}}, \overrightarrow{P_{200}P_{101}}), \tag{1}$$

$$N_{110}^K = 2\det(\overrightarrow{P_{200}P_{110}}, \overrightarrow{P_{110}P_{011}}) + 2\det(\overrightarrow{P_{110}P_{020}}, \overrightarrow{P_{200}P_{101}}). \tag{2}$$

Consequently, a sufficient condition to prove that $\mathcal{J}_K$ is strictly positive everywhere is to ensure that all N_{ijlm}^K are strictly positive, but this is a too strong condition. On the contrary, if a N_{ijlm}^K is negative in a corner, it means that $\mathcal{J}_K$ is negative somewhere inside the element as N_{ijlm}^K is the exact value of the Jacobian in this corner [7]. However, if a control coefficient lying on an edge or on a face or in a volume is negative, it does not mean that $\mathcal{J}_K$ is negative somewhere inside the element. We cannot conclude on the positiveness of the Jacobian without any further analysis. In this case, a few iterations of a De Casteljau's algorithm [1, 7] (or a subdivision of the element [11]) are required to have more accurate bound

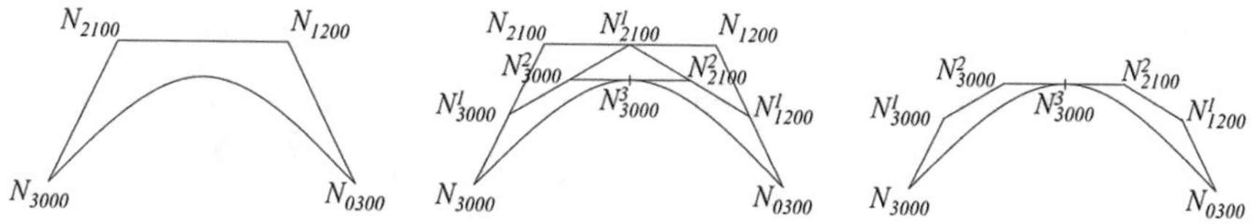

Fig. 4 De Castljau's refinement on two control coefficients of the Jacobian curve on an edge of a P^2 tetrahedron. Left, the initial curve with its control coefficients. Middle, construction of the new control coefficients. Right, the initial curve divided in two curves with their control coefficients

of the Jacobian. The idea of this refinement on an edge/face/volume is to create new control coefficients from the initial ones. These control coefficients can define several curves whose union is the Jacobian curve on the edge/face/volume. This way, a more accurate bound can be found.

For instance, let us consider an edge on a P^2 tetrahedron (see Fig. 4 on the left) with a negative edge control coefficient on it. The De Casteljau's algorithms can be decomposed in three steps (see Fig. 4, middle, K superscripts are omitted for clarity):

$$N^1_{3000} = \frac{N_{3000} + N_{2100}}{2} \quad N^1_{2100} = \frac{N_{2100} + N_{1200}}{2} \quad N^1_{1200} = \frac{N_{1200} + N_{0300}}{2},$$

$$N^2_{3000} = \frac{N^1_{3000} + N^1_{2100}}{2} \quad N^2_{2100} = \frac{N^2_{2100} + N^2_{1200}}{2},$$

$$N^3_{3000} = \frac{N^2_{3000} + N^2_{2100}}{2}.$$

By construction, $N^3_{3000} = \mathcal{J}^K(\frac{1}{2}, \frac{1}{2}, 0, 0)$. So, if $N^3_{3000} \leq 0$ then the element is invalid. Otherwise, $N_{3000}, N^1_{3000}, N^2_{3000}, N^3_{3000}$ (resp. $N^3_{3000}, N^2_{2100}, N^1_{1200}, N_{0300}$) defines a P^3 curve representing the Jacobian between $\mathcal{J}^K(1, 0, 0, 0)$ and $\mathcal{J}^K(\frac{1}{2}, \frac{1}{2}, 0, 0)$ (resp. $\mathcal{J}^K(\frac{1}{2}, \frac{1}{2}, 0, 0)$ and $\mathcal{J}^K(0, 1, 0, 0)$) (see Fig. 4, right). Consequently, the initial analysis can be done on these two sub-curves. If central control coefficients are positive, the Jacobian on the sub-element is valid. Otherwise, De Casteljau's algorithm is reapplied to this subcurve to conclude. If one subcurve gives an invalid Jacobian then the whole curve is invalid.

This way, a recursive method to find the sign of the minimum of the Jacobian on the edge is established.

Once the validity of the element is known, it is interesting to consider a quality criterion for the shape of the element. The chosen one is the one proposed in [8]:

$$Q = \alpha \underbrace{\frac{hS_k}{V_k}}_{①} \underbrace{\frac{\max(V_1, V_k)}{\min(V_1, V_k)}}_{②} \underbrace{\left(\frac{N^K_{max}}{N^K_{min}}\right)^{1/d}}_{③},$$

with:

- d the dimension, S_k the exterior surface of the polyhedron (in 2D, the half perimeter of the polygon) defined by nodes and vertices $(A_i)_{1\leq i\leq n}$
- V_k the volume of the polyhedron (resp. surface of the polygon) defined previously
- h the element's largest edge P^k-length (e.g the length of the union of straight-sided lines defined by the nodes)
- V_1 the volume/surface of the equivalent P^1 element, e.g the element defined by the vertices
- N^K_{min} (resp. N^K_{max}) the smallest (resp. largest) control coefficient of the Jacobian of the element
- α is a normalization factor, dependent of the dimension such that $Q = 1$ for a regular simplex, $\alpha = \frac{\sqrt{3}}{6}$ in 2D and $\alpha = \frac{\sqrt{3}}{36}$ in 3D.

This quality function is actually a product of three terms. (1) is only a generalization of the P^1 quality function and measures the gap to the regular element. (2) measures the distance between the volume of the curved element and the volume of the straight element and ensures the function to be greater than 1 [8]. And, finally (3) gives a measure of the distortion of the element, it can detect if the element is invalid or almost invalid by taking an infinite value. Note that if the element is straight, the standard P^1 quality function [6] is recovered: $Q = Q_{P^1} = \alpha \frac{hS_1}{V_1} = \alpha \frac{h}{\rho}$ where ρ is the inradius of the straight element. Also, this quality function can easily be extended to anisotropic meshes. Based on these definitions, this element-wise quality measure is between 1 and the infinity. The closer the element quality is to 1, the better the quality is.

3 High-Order Mesh Optimization

In the same way as we want to have an optimal P^1 mesh in terms of quality, we want to have an optimal P^k mesh. Several optimization techniques exist to correct an invalid P^k mesh [12, 16] and to optimize the geometrical accuracy [17].

The idea here is to extend two classic mesh quality-based optimization operators [1] to P^2 meshes to increase its quality.

3.1 P^2 Swap Operator

The swap operator (see Fig. 5) locally changes the connectivity of the mesh in order to improve its quality.

In 2D, it consists in flipping an edge shared by two triangles to form two new triangles with the same four vertices (see Fig. 5 left).

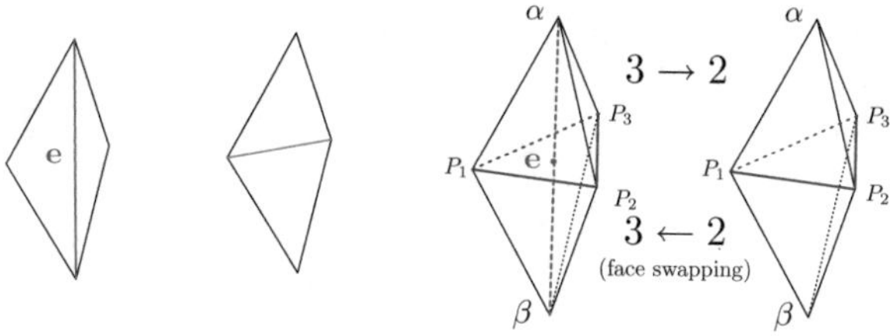

Fig. 5 *Left*, the swap operation in 2D. *Right*, edge swap 3 → 2 and face swap 2 → 3. For all these pictures, shells are in *black*, old edges are in *red*, new edges are in *green*

In 3D, two types of swap exist: face and edge swapping. The face swapping is the extension of the 2D edge swapping, it consists in replacing the common face of two neighboring tetrahedra by the edge linking the opposite vertices to the face of each tetrahedron, also called 2 → 3. The edge swapping is a bit different. First, the shell of the edge to delete (e.g. the set of elements containing this edge) is constructed. From a shell of size n, a non-planar pseudo-polygon formed by n vertices is obtained. The swap consists in deleting the edge, generating a triangulation of the polygon and creating two tetrahedra for each triangle of the triangulation thanks to the two extremities of the former edge. These swaps are designated as $n \rightarrow m$ with $n \geq 3$, where n is the initial number of tetrahedra and $m = 2(n-2)$ is the final number of tetrahedra.

For each possible swapped configuration, if the worst quality of all the elements the shell is improved, the configuration is kept and will be in the new mesh unless another swapped configuration of the shell provides a better quality improvement. When it comes to connectivity-change moving-mesh algorithms in 3D, a small local degradation of the shell's worst quality has been observed as an efficient way to improve the global quality of the mesh.

To generalize it to P^2 meshes, the inner nodes of the edges of the shell have to be taken into account. For instance, for the P^2 case in 2D, there is one node on the swapped edge and if we want the swap to be performed, we need first to find an optimal position for the node in the swapped configuration and then to check if this configuration improves the quality function (see Fig. 6). The key feature is therefore to find a functional whose optimum will give the optimal position for the node in the swapped configuration in term of quality.

In this context, the idea is to find a simple and smooth functional that will be easy to optimize. The quality function is not a good candidate as it is not smooth. We propose to consider the following functional (inspired from the work of [16]):

$$f(\mathbf{X}) = \sum_{K \in \mathcal{S}(e)} \sum_{i+j+k+l=d} \omega_{ijkl} \left(\frac{N_{ijkl}^K(\mathbf{X})}{d! V_1} - 1 \right)^2 ,$$

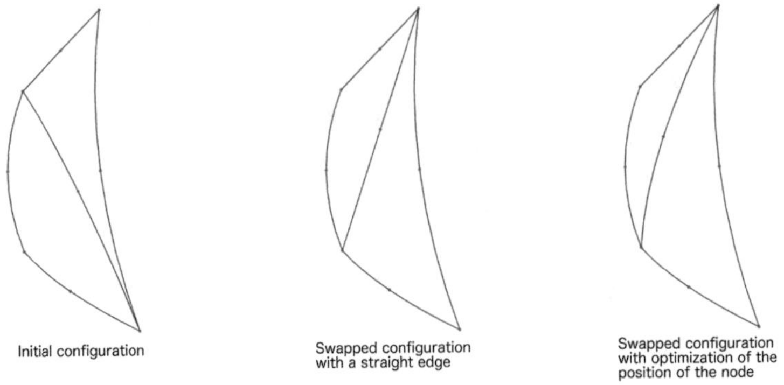

Fig. 6 Three steps of P^2 swap in 2D

where d is the dimension, $\mathbf{X} = [\mathbf{x_1}, \ldots, \mathbf{x_n}]$ is a *set* of n coordinates to be optimized, $\mathbf{X} \rightarrow N_{ijk}^K(\mathbf{X})$ is a function that depends, in the worst case on two variables of $\mathbf{X}$, $\mathcal{S}(e)$ is the shell of the initial edge e (e.g. the set of elements K containing e), ω_{ijkl} is a weight function that measures the importance of each control coefficient, and V_1 is the volume of the straight-sided element deduced from K. Note that $\mathbf{X}$ can either be empty (in which case no optimization is required), or contain more than one node's coordinates as it represents the coordinate of the created edges of the shell.

In 2D, $\mathbf{X}$ is always a singleton and is simply noted $\mathbf{x}$, weights ω_{ijk} $(l = 0)$ are equal to 2 for the corner coefficients and equal to 1 everywhere. In this case, f has the following properties: it is a positive define quadratic form as $\mathbf{x} \rightarrow N_{ijk}^K(\mathbf{x})$ is linear in $\mathbf{x}$, which means that the functional as a unique minimum. Also, on every regular swap configuration, the minimum of f in $\mathbf{x}$ is the same as the minimum of the worst quality of the swapped shell in $\mathbf{x}$. Using the result of the optimization problem in the swap configuration gives therefore a very good approximation of the best configuration that can be obtained. Since the best swap configuration is found, we are able to conclude if this swap will increase the quality or not.

In 3D, weights ω_{ijkl} are equal to 4 for a corner control coefficient, equal to 2 for an edge control coefficient and equal to 1 otherwise. In this work, considered swaps are $2 \rightarrow 3$, $3 \rightarrow 2$ and $4 \rightarrow 4$ which means that the functional of the problem is in the worst case quadratic. This is a consequence of the fact that control coefficients have a linear dependence with respect to a given control coefficient. Also, the problem appears numerically to be definite positive on a regular configuration. Note that these swaps represent $\sim$95% of the swaps performed during a P^1 mesh optimization. For the other swaps ($5 \rightarrow 6$, $6 \rightarrow 8$), the problem begins to be highly costly in term of CPU. The resolution of these optimization problems is performed thanks to a L-BFGS algorithm [13]. Note that even if the quality function is not used for the optimization problem, it is mandatory to keep using it so that it degenerates into classical swap operator when the elements are straight.

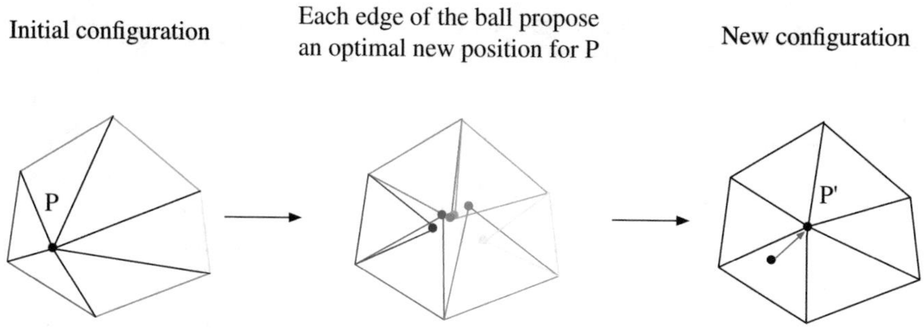

Fig. 7 Laplacian smoothing in two dimensions. Each element of the ball of considered vertex P_i suggests an optimal position for P_i. The resulting new optimal position for P_i is computed as a weighted average of all these proposed locations

3.2 P^2 Mesh Smoothing

Mesh smoothing is a technique that consists in relocating some points inside the mesh to improve the quality of the elements. In P^1, the idea is to relocate each vertex P_i inside its ball of elements (see Fig. 7). For each element K_j in the ball of P_i, the opposite face to P_i denoted by F_j gives an optimal position P_j^{opt}. Then, the vertex is relocated considering a weighted average of the proposed positions. If the proposed new location of the vertex does not improve the ball configuration in term of quality, then a relaxation is performed to check if an improved configuration exists between the original location and the new one. The optimal configuration is computed as follows:

$$P_j^{opt} = G_j + \sqrt{\frac{2}{3}} h_j \frac{\mathbf{n}_j}{||\mathbf{n}_j||},$$

where G_j is the gravity center of F_j, h_j is the average length of the edges of F_j, and $\mathbf{n}_j$ is the outward normal to F_j. The proposed position is then computed with:

$$P_i^{opt} = \frac{\sum_{K_j \in Ball(P_i)} \max(Q(K_j), Q_{\max}) P_j^{opt}}{\sum_{K_j \in Ball(P_i)} \max(Q(K_j), Q_{\max})},$$

where $Q_{\max}$ is a parameter to be defined. Here $Q_{\max} = 10$. Note that if every computed configuration decreases the quality, the smoothing is not performed.

To extend it to P^2 meshes, the edges' node needs to be taken into account. The idea here, is to perform two independent smoothing operations:

- A vertex smoothing

The vertex smoothing is simply a generalization of the P^1 smoothing. The optimal position of the vertex is computed in the same way as in P^1, and the vertex is

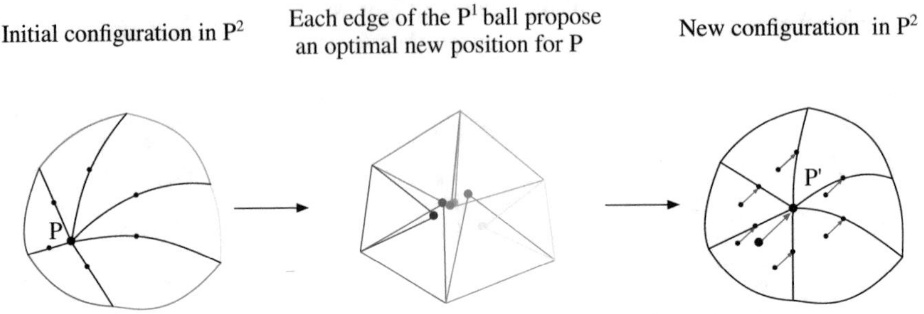

Fig. 8 P^2 laplacian smoothing in two dimensions. Each element of the P^1 ball of considered vertex P suggests an optimal position for P. The resulting new optimal position for P is computed as a weighted average of all these proposed locations. The new position of the nodes of the internal edges of the P^2 ball is then deduced by proportionality

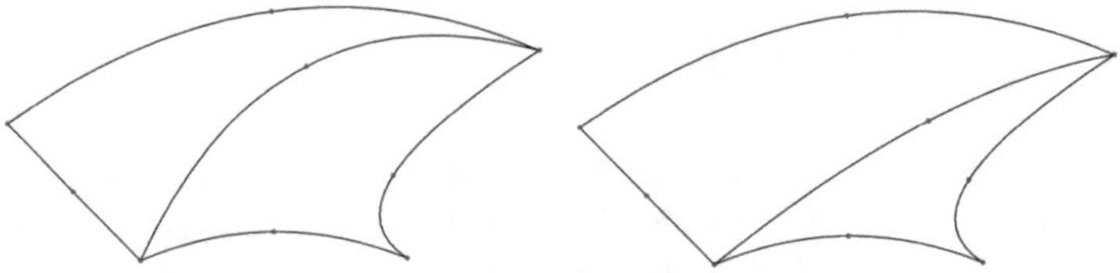

Fig. 9 P^2 node smoothing in two dimensions. The optimal position of the node of the central edge is computed solving an optimization problem. Left, the initial configuration, right, the optimized configuration

located exactly in the same way as before. In order to be consistent with the P^1 vertex smoothing and to keep in the ball straight edges that are initially straight, the displacement of all the inner nodes of the ball cavity is set to half of the value of the displacement of the central vertex (see Fig. 8). In the exact same way as in P^1 if the final configuration does not improve the quality, relaxation is performed the original location and the new one.

- A node smoothing

The optimization of the node position follows the same algorithm as in the P^2 swap operator to find its optimal position. For this purpose, functional f can be re-used to find the optimal node position (see Fig. 9). In this case, there is always only one node coordinates to optimize and consequently the optimization problem is quadratic. In the exact same way as in P^1, if the final configuration does not improve the quality, relaxation is performed the original location and the new one.

4 Applications

4.1 High-Order Mesh Generation by Curving an Initial P^1 Mesh

Most of the techniques to generate an high-order mesh is to start from a P^1 mesh and then to curve it, in a way or another, in order to obtain a P^k mesh [5, 12, 14, 18]. The main reason to use a post-treatment is that all existing P^1 mesh generation algorithms can be reused. It would be harder to implement a directly high-order mesh generator. To curve meshes, used models are numerous: PDE or variational models [5, 9, 14], smoothing and/or optimization procedures [12, 15, 16], ... Our choice here is to use the linear elasticity equation as a model for the motion of the vertices to generate a P^k mesh from a P^1 mesh.

For this purpose, let us consider the linear elasticity equation with Dirichlet boundary conditions:

$$\nabla \cdot (\sigma(\mathcal{E})) = 0\,, \quad \text{with} \;\; \mathcal{E} = \frac{\nabla\xi + {}^T\nabla\xi}{2}\,, \tag{3}$$

where σ, and $\mathcal{E}$ are respectively the Cauchy stress and strain tensors, ξ is the Lagrangian displacement. The Cauchy stress tensor follows the Hooke's law for isotropic homogeneous medium.

Here, Dirichlet boundary conditions represent the gap between the P^k-nodes of the initial straight boundary elements and their position on the *real* boundary. For mesh boundary vertices, the gap is equal to 0.

To compute the gap at the nodes, a continuous representation of the surface mesh is required. It can be either provided by CAD/analytical model or deduced from initial P^1 mesh via a cubic reconstruction technique [20]. Once Dirichlet boundary conditions are set, the high-order finite element linear elasticity code is called. The use of an high-order FE resolution rather than on a subdivided P^1 mesh aims the degrees of freedom to be intrinsically represented. This gives more consistency to the obtained motion.

The elasticity problem using the high-order FEM provides the new position of the internal vertices and nodes. It is then used to generate the high-order mesh by moving the vertices and nodes of the initial straight mesh with the associated values in the elasticity solution vector. If some elements remain invalid after optimization, it is always due to non suitable boundary displacements. In this case, the FEM solution is proportionally reduced in the vicinity (boundary included) of the invalid element until global validity is obtained. The process is summarized by Algorithm 1.

The major fact with this method is that the deformed mesh is only made of isotropic or almost isotropic elements. In this context, the use of the elasticity problem is efficient and always provides a valid mesh. Optimization in the pre-processing makes elements more isotropic and therefore helps curvature process to be more robust whereas optimization in the post-processing improves the quality

Algorithm 1 Mesh curving algorithm

1. Generate a P^1 mesh.
2. Perform P^1 mesh optimization pre-processing: generalized swapping and vertex smoothing.
3. Perform cubic reconstruction of the boundary or use its analytical representation to set Dirichlet boundary conditions for the linear elasticity equation.
4. Solve linear elasticity equation on the P^1 mesh with the FEM at the order of the wanted mesh.
5. Generate the P^k mesh by moving the P^1 mesh with the solution of the linear elasticity.
6. Perform P^2 mesh optimization post-processing: generalized swapping and node/vertex smoothing.
7. Check validity of P^k elements and locally relax the FEM solution if necessary or desired until it is valid.

of the mesh and untangle invalid elements if any. Some results in P^2 are presented below:

4.1.1 NASA RO37 Rotor

This example is a NASA RO37 rotor used for turbomachinery applications. Two different meshes are considered (see Fig. 10).

- A coarse mesh (see Fig. 10, middle) with an initial number of 2434 vertices, 13,326 nodes and 10,970 tetrahedra. The initial mesh average quality is 6.16 with a worst quality of 1682 due to the presence of sharp anisotropic trailing and leading edges on the input wing surface mesh. Curving the mesh without optimization provides an average quality of 5.3 with a worst quality of 1729. Optimization in post and pre processing increase average quality to 2.5 and worst quality to 1346. Note that the number of nodes and tetrahedra is changed to respectively 11,598 and 10,862.
- A fine mesh (see Fig. 10, bottom) with an initial number of 22,145 vertices, 137,393 nodes and 106,562 tetrahedra. The initial mesh average quality is 2.14 with a worst quality of 111. Note that is better than with the coarse mesh as the mesh is finer and therefore deals better with trailing and leading edges. Curving the mesh without optimization provides an average quality of 2.15 with a worst quality of 366. Optimization in post and pre processing increase average quality to 1.7 and worst quality to 27.5. Note that the number of nodes and tetrahedra is changed to respectively 136,924 and 106,093.

In both cases, we clearly observe the benefits of the optimization that improves the average and the worst quality of the final mesh. Note that the worst quality of the final P^2 might be greater than the one of the initial P^1 mesh. This a consequence of the curving process that decreases the quality of straight elements by curving them. We can also observe that the curvature is not propagated a lot in the volume as it is not visible after two or three layers of elements (see Fig. 10, middle and bottom right). This is an illustration of St Venant's principle which states that the elasticity solution can be divided into transmissive effects and local disturbances.

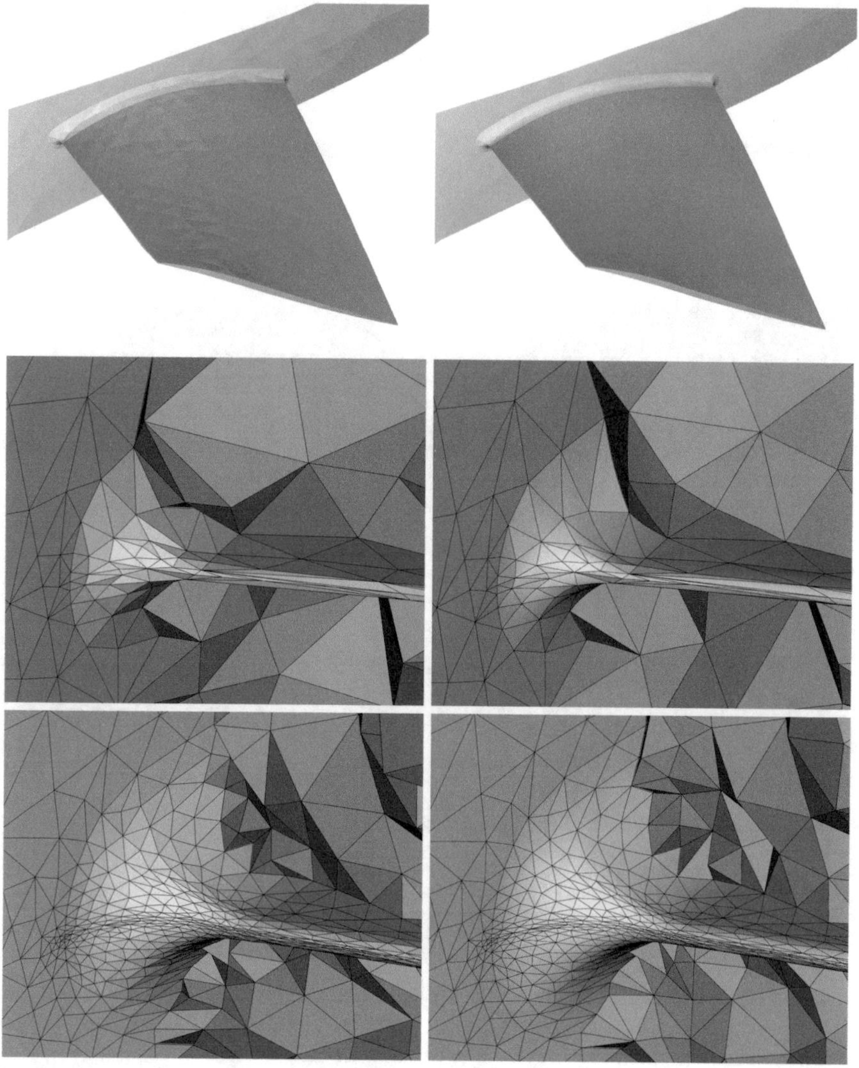

Fig. 10 NASA RO37 rotor turbomachine meshes. From top to bottom, surface, coarse and fine meshes. From right to left, P^1 and P^2 meshes. P^2 meshes are generated with Algorithm 1 using a cubic reconstruction

4.1.2 NASA Common Research Model Aircraft

This example is the NASA Common Research Model, an aircraft model that is massively used in both experimental and numerical simulations in fluid dynamics (see Fig. 11). Again, two meshes are considered:

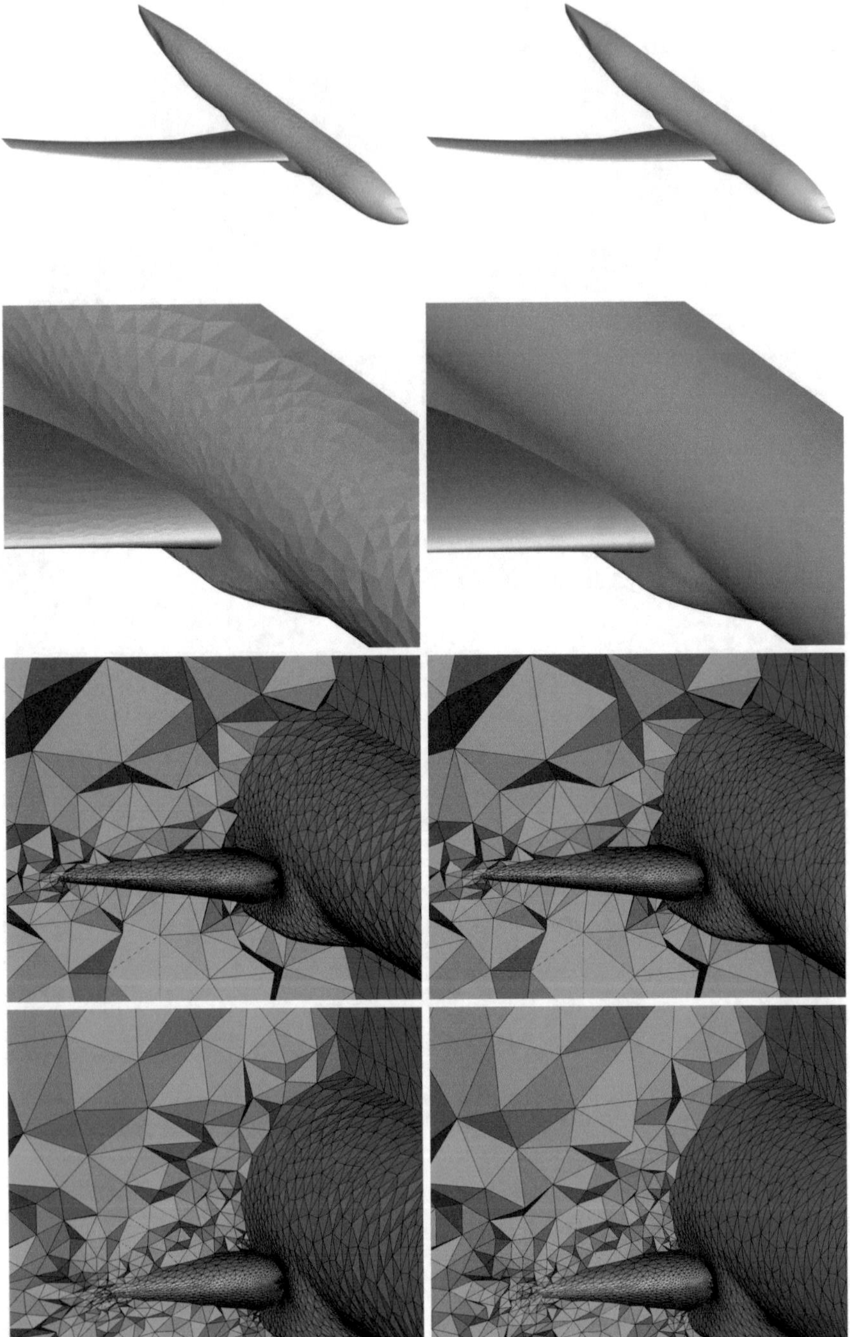

Fig. 11 NASA Common Research Model aircraft meshes. From top to bottom, surface, coarse and fine meshes. From right to left, P^1 and P^2 meshes. P^2 meshes are generated with Algorithm 1 using a cubic reconstruction

- A coarse mesh (see Fig. 11, line 3) with an initial number of 32,479 vertices, 173,468 nodes and 118,012 tetrahedra. The initial mesh average quality is 2.49 with a worst quality of 271 due to the presence of sharp anisotropic trailing and leading edges on the input wing surface mesh. Curving the mesh without optimization provides an invalid configuration with 10 invalid elements. Optimization in post and pre processing increase average quality to 2.13 and worst quality to 271. Note that the number of nodes and tetrahedra is changed to respectively 173,744 and 118,288.
- A fine mesh (see Fig. 11, line 4) with an initial number of 101,422 vertices, 660,071 nodes and 535,672 tetrahedra. The initial mesh average quality is 2.28 with a worst quality of 1314. Curving the mesh without optimization provides an invalid configuration with 4 invalid elements. Optimization in post and pre processing increase average quality to 1.4 and worst quality to 1777. Note that the number of nodes and tetrahedra is changed to respectively 659,545 and 535,146.

In both cases, optimization ensures a valid curved mesh at the end that would not have been obtained without it. In the same way as in the previous example, the curvature is not propagated a lot in the volume as it is not visible after 2 or 3 layers (see Fig. 11, line 3 and 4 right).

4.2 A Moving Mesh Technique for High-Order Elements

In this section, a connectivity-change moving-mesh method inspired from [1] for P^2 meshes is presented. In this case, the initial mesh is a P^2-mesh whose boundary has an initial displacement. Using a linear elasticity analogy, the resolution of the elasticity equation with high-order finite elements gives us a displacement for all the vertices and nodes in the volume. Then the mesh is moved to the new position. The motion of the vertices can be also enhanced by using a local stiffness factor technique [1]. This technique locally multiplies the tensor σ of linear elasticity equation by a factor proportional to $\mathcal{J}_K(\mathbf{x})^{-\chi}$. χ determines the degree by which smaller elements are rendered stiffer than larger ones. We use $\chi = 1$ [1]. Afterwards, connectivity changes are performed on the mesh to improve the *quality* of the elements. It is an efficient way to get rid of any shearing that occurs in the mesh. The high-order moving-mesh algorithm is summarized in Algorithm 2.

The studied case is a moving sphere of radius 0.6 inside a large control volume (see Fig. 12, line 1 left). At each iteration, the sphere is displaced of 0.08 in x direction. The initial mesh (see Fig. 12, line 2) has an average quality of 1.27 and a worst quality of 2.5. Three positions are considered (see Fig. 12, line 1 right) with and without connectivity-change: the initial position, the position after 30 iterations (displacement of 4 radii) and the position after 50 iterations (displacement of 6.7 radii).

When no connectivity-change is done after each iteration, shearing appears in the mesh which constraints the displacement. The high-order linear elasticity resolution

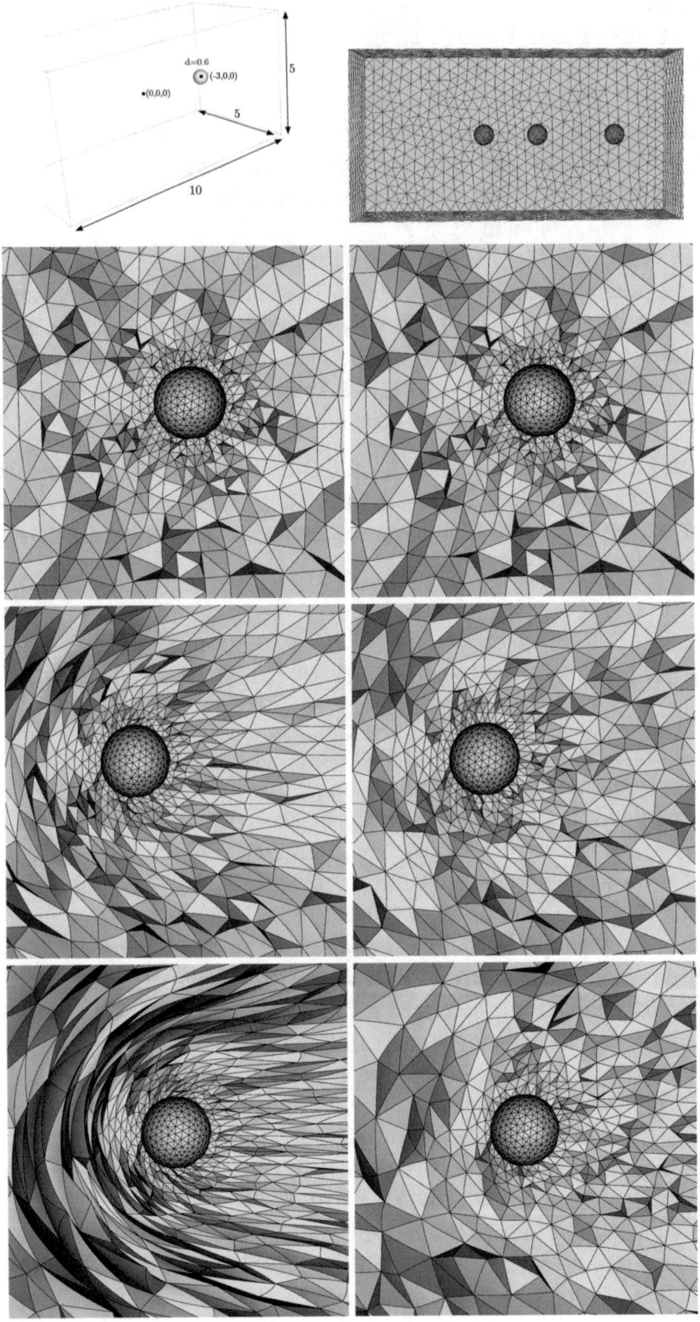

Fig. 12 Case of the moving sphere inside a P^2 mesh. First line, description of the sphere and description of the three considered positions of the moving sphere: the sphere is moving from right to left. And then, from top to bottom, zoom in the vicinity of the sphere for each position. Left, without mesh optimization operators. Right, with mesh optimization operators

Algorithm 2 High-order moving mesh algorithm

1. Mesh deformation algorithm.
   a. Compute body displacement from body translation and rotation data.
   b. Solve linear elasticity equation with the FEM at the order of the mesh.
   c. Perform high-order mesh optimization.
   d. Check validity of the obtained displacement and restart it with a smaller body displacement if necessary/desired until the obtained displacement is valid.
2. Move the mesh.

gives to the elements a curvature that fits to the displacement of the sphere in order to move as far as possible the object when the mesh connectivity is fixed. Indeed, in front of the sphere, the deformed elements fit to the shape of the sphere, whereas in the wake, the curvature of elements is made so that the shearing is reduced. This is a good point but the mesh quality decreases drastically (see Fig. 12, line 3 and 4 left): after 30 iterations, average quality is 1.64 and worst quality is 5.7 and after 50 iterations, 84 elements are invalid. On the contrary, when mesh quality-based optimization operators are considered with the moving mesh algorithm at each iteration, the average and the worst quality do not change a lot (see Fig. 12, line 3 and 4 right): after 30 iterations, average quality is 1.43 and worst quality is 3.1 and after 50 iterations, average quality is 1.48 and worst quality is 3.5.

5 Conclusion and Perspectives

P^2 mesh quality-based optimization operators have been presented. These operators ensure a valid P^2 mesh generation starting from a P^1 mesh and enable to deal with connectivity-change P^2 moving-mesh methods. Note that all these developments are suitable for isotropic meshes but do not work that well on anisotropic meshes and do not work as is with boundary layer meshes. The moving-mesh method gives similar results in term of quality as in P^1 which is promising for future research. The immediate next step will be to generalize it to curved trajectories using [1]. Isotropic degree two meshes were only the first step, further developments will be to generalize it to any higher-order meshes and to deal with anisotropy using metric fields. When it comes to boundary layer meshes, the goal is to extend the closed-advancing boundary layer mesh generation method of [2] to high-order meshes in order to generate directly curved boundary layer meshes. To this end, it is required to:

- Start form an initial high-order mesh that is obtained using the method of Sect. 4.1
- Consider the connectivity-change moving mesh method for high-order mesh presented in Sect. 4.2 with curved trajectories to deform the initial high-order mesh when the boundary layer mesh is inflated inside the domain

- Generate directly high-order elements in the boundary layer when it is inflated using the advancing layer approach presented in [2].

The future work to do is the last item. The advancing layer method will be modified to take into account the high-order boundary layer elements. The new position of the nodes in the boundary layer will be given using the same process as the one for proposing the new position for the vertices. High-order quality functions will be used to check the quality of the boundary layer elements when they are generated.

Note that more accurate normals will be obtained as they will be computed on the high-order mesh instead of a P^1-straight mesh. This is an important point as the quality of the boundary layer is highly dependent on the accuracy of the normal computations [2]. Also, curved boundary layer mesh generation will not need the creation of a new tool and should be more robust and efficient as it directly controls the high-order mesh quality.

Acknowledgements This work was supported by a public grant as part of the *Investissement d'avenir* project, reference ANR-11-LABX-0056-LMH, LabEx LMH.

The authors also would like to thank the reviewers for their fruitful remarks.

References

1. F. Alauzet, A changing-topology moving mesh technique for large displacements. Eng. Comput. **30**(2), 175–200 (2014)
2. F. Alauzet, A. Loseille, D. Marcum, On a robust boundary layer mesh generation process, in *55th AIAA Aerospace Sciences Meeting*, AIAA Paper 2017-0585, Grapevine, TX, USA, 2017
3. H. Borouchaki, P.L. George, *Meshing, Geometric Modeling and Numerical Simulation 1: Form Functions, Triangulations and Geometric Modeling* (Wiley, Hoboken, 2017)
4. P.G. Ciarlet, *The Finite Element Method for Elliptic Problems* (North-Holland, Amsterdam, 1978)
5. M. Fortunato, P.-O. Persson, High-order unstructured curved mesh generation using the Winslow equations. J. Comput. Phys. **307**, 1–14 (2016)
6. P.J. Frey, P.L. George, *Mesh Generation: Application to Finite Elements* (Wiley, New York, 2008)
7. P.L. George, H. Borouchaki, Construction of tetrahedral meshes of degree two. Int. J. Numer. Methods Eng. **90**(9), 1156–1118 (2012)
8. P.L. George, H. Borouchaki, F. Alauzet, P. Laug, A. Loseille, L. Maréchal, *Meshing, Geometric Modeling and Numerical Simulation 2: Metrics, Meshing and Mesh Adaptation* (Wiley, 2019)
9. R. Hartmann, T. Leicht, Generation of unstructured curvilinear grids and high-order discontinuous Galerkin discretization applied to a 3D high-lift configuration. Int. J. Numer. Methods Fluids **82**(6), 316–333 (2016)
10. J.S. Hesthaven, T. Warburton, *Nodal Discontinuous Galerkin Methods: Algorithms, Analysis and Applications*. (Springer, New York, 2008)
11. A. Johnen, J.-F. Remacle, C. Geuzaine, Geometrical validity of curvilinear finite elements. J. Comput. Phys. **233**, 359–372 (2013)
12. S.L. Karman, J.T. Erwin, R.S. Glasby, D. Stefanski, High-order mesh curving using WCN mesh optimization, in *46th AIAA Fluid Dynamics Conference*, AIAA AVIATION Forum. American Institute of Aeronautics and Astronautics, 2016

13. D.C. Liu, J. Nocedal, On the limited memory BFGS method for large scale optimization. Math. Program. **45**(1), 503–528 (1989)
14. D. Moxey, D. Ekelschot, Ü. Keskin, S.J. Sherwin, J. Peirò, High-order curvilinear meshing using a thermo-elastic analogy. Comput. Aided Des. **72**, 130–139 (2016)
15. E. Ruiz-Gironès, X. Roca, J. Sarrate, High-order mesh curving by distortion minimization with boundary nodes free to slide on a 3D CAD representation. Comput. Aided Des. **72**, 52–64 (2016); 23rd International Meshing Roundtable Special Issue: Advances in Mesh Generation
16. T. Toulorge, C. Geuzaine, J.-F. Remacle, J. Lambrechts, Robust untangling of curvilinear meshes. J. Comput. Phys. **254**, 8–26 (2013)
17. T. Toulorge, J. Lambrechts, J.-F. Remacle, Optimizing the geometrical accuracy of curvilinear meshes. J. Comput. Phys. **310**, 361–380 (2016)
18. M. Turner, J. Peirò, D. Moxey, A variational framework for high-order mesh generation. Procedia Eng. **163**(Supplement C), 340–352 (2016); 25th International Meshing Roundtable
19. J. Vanharen, G. Puigt, X. Vasseur, J.-F. Boussuge, P. Sagaut, Revisiting the spectral analysis for high-order spectral discontinuous methods. J. Comput. Phys. **337**, 379–402 (2017)
20. A. Vlachos, P. Jörg, C. Boyd, J.L. Mitchell, Curved PN triangles, in *Proceedings of the 2001 Symposium on Interactive 3D Graphics*, I3D '01 (ACM, New York, 2001), pp. 159–166

Isometric Embedding of Curvilinear Meshes Defined on Riemannian Metric Spaces

Philip Claude Caplan, Robert Haimes, and Xevi Roca

Abstract An algorithm for isometrically embedding curvilinear meshes defined on Riemannian metric spaces into Euclidean spaces of sufficiently high dimension is presented. The method is derived from the Landmark-Isomap algorithm and a previous method for embedding straight-sided meshes. The former is used to decrease the computational complexity of the embedding problem, notably the dense shortest-path problem used to estimate geodesic lengths across the mesh domain as well as the dense eigenvalue decomposition needed to compute the codimension coordinates. A method for defining curvilinear meshes from straight-sided ones in a dimension-independent manner is also discussed. Examples in two- and three-dimensions for both analytic embeddings and analytic metric fields are used to evaluate the method.

1 Introduction

High-order discretisations of partial differential equations demand a curvilinear representation of the mesh geometry [1]. Within an adaptive framework for numerical simulations, the curvilinear metric-conforming mesh can either be generated natively or a posteriori from a straight-sided metric-conforming mesh. Recent interest in using isometric embeddings [3, 5–7, 14], founded on the Nash embedding theorem, offer the potential to generate curvilinear metric-conforming meshes in a native manner [4] by projecting high-order mesh vertices to the embedded mesh during the mesh generation stage. This work answers an important question related to the ability to natively generate curvilinear metric-conforming meshes: how to

P. C. Caplan (✉) · R. Haimes
Massachusetts Institute of Technology, Cambridge, MA, USA
e-mail: pcaplan@mit.edu

X. Roca
Computer Applications in Science and Engineering, Barcelona Supercomputing Center, Barcelona, Spain

X. Roca, A. Loseille (eds.), *27th International Meshing Roundtable*, Lecture Notes in Computational Science and Engineering 127, https://doi.org/10.1007/978-3-030-13992-6_2

embed a curvilinear mesh equipped with a Riemannian metric into a Euclidean space of sufficiently high dimension? In particular, an algorithm for embedding curvilinear meshes given an initial linear mesh and spatially continuous definition of a metric field is introduced. The algorithm is then evaluated by studying the effect of the geometric mesh order and ambient dimension of the Euclidean space on the isometry of the embedded curvilinear mesh. Examples drawn from analytic embeddings and analytic metrics in 2*d* and 3*d* are used to illustrate the method.

2 Methodology

This section describes the methodology used to embed curvilinear meshes, upon which a spatially continuous field of metric tensors is assumed. Before doing so, consider how a curvilinear mesh can be constructed from a linear one in a dimension-independent manner.

2.1 *Construction of Curvilinear Meshes*

In essence, this section describes how any high-order continuous Galerkin field can be defined from a straight-sided mesh but it is important to illustrate the mechanics of this procedure since, to our knowledge, this has not been covered in the literature for higher-dimensional ($d > 3$) simplicial meshes.

Given a straight-sided d-simplicial mesh $\mathscr{T} \in \mathbb{R}^d$, the first step consists of identifying all j-simplices $0 \le j \le d$ in $\mathscr{T}$ along with the set of d-simplices and local facet labels from which each facet is constructed. As an example, a tetrahedral mesh is first decomposed into its vertices, edges, triangles and tetrahedra along with the parent tetrahedra and local indices within each parent tetrahedron that these vertices, edges and triangles are constructed. Note the identification of the 0-simplices and d-simplices is trivial. All other j-simplices are identified by hashing a string representing the sorted integers describing the facet. Another option might be to represent each facet as $\sum_{i=0}^{|f|} f_i n_v^i$ where n_v is the number of vertices in $\mathscr{T}$, however, this risks overflowing the integer representation; as such, the string representation is preferred when hashing the facets upon construction. The labelling of this facet decomposition is illustrated in the sketch of Fig. 1a.

The next step consists of placing equispaced high-order nodes along the *interior* of each j-simplex $j \le d$. Each j-simplex is then fully defined by identifying the local indices of the d-simplex lattice coordinates corresponding to every vertex. Should a boundary representation of the geometry be provided, vertices can be projected to the appropriate entities.

For visualization purposes, the list of triangles and edges can be obtained by extracting the bounding 2- and 1-simplices of the d-triangulation obtained with,

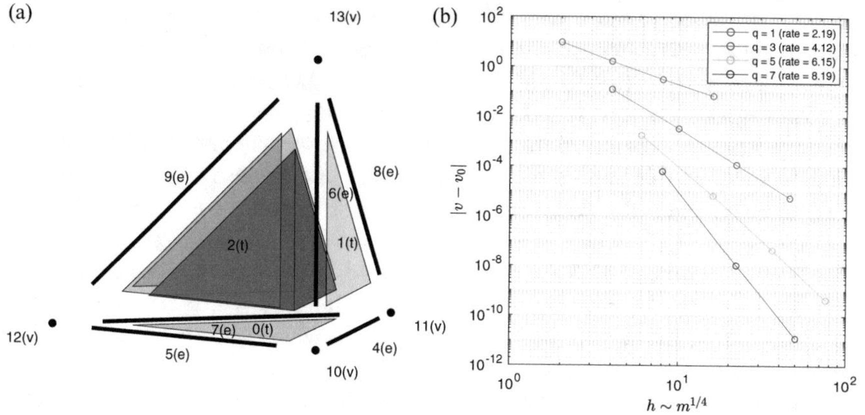

Fig. 1 Demonstration of facet decomposition in $3d$ and verification of curvilinear construction process in $4d$. (**a**) Facet decomposition of the reference tetrahedron. Parentheses indicate either a vertex (v), edge (e) or triangle (t). The only facet not visible is the triangle 1(t). (**b**) Convergence to the analytic volume of a curved domain for a $4d$ curvilinear mesh with various geometric mesh orders

say, a Delaunay triangulation of the lattice coordinates of the curvilinear reference simplex.

As a demonstration of the d-dimensional curvilinear construction process, an initially straight-sided Kuhn-Freudenthal triangulation in $[0, 1]^d$ [13] is augmented to various curvilinear mesh orders. The first two coordinates of the vertices in this curvilinear mesh are then mapped to polar coordinates, $0 \leq r \leq 1$, $0 \leq \theta \leq \pi$ with the remaining dimensions arbitrarily stretched to $l_d = d$. The volume of the mesh is computed by integrating over the elements, using high-order Lagrange basis functions from Burkardt [2] along with Grundmann-Moeller quadrature rules [10]. As seen in Fig. 1b, the volume of a four-dimensional mesh asymptotically approaches the analytic one of $\frac{1}{2}\pi \prod_{i=2}^{d} l_i$ with mesh size h at a rate of h^{q+1} where $h \sim m^{1/d}$ with m the total number of vertices. Standard curved meshing is focused on approximating two-dimensional and three-dimensional geometries. Herein we expose one of the first, if not the first, discussions on setting up curvilinear meshes for $4d$ problems. It is worth noting that the focus is on studying the influence of the metric field (and not the geometry) in the embedding of curvilinear meshes. As such, all geometries studied in the remainder of the work are straight-sided.

2.2 Embedding Algorithm

Here, the mechanics of the algorithm used to embed a curvilinear mesh are presented. The only assumption is that a metric field is defined (or can be evaluated) at the high-order vertex locations. Recent work [4] employs a variant of the Isomap algorithm of Tenenbaum et al. [9] to embed straight-sided meshes along with a Riemannian metric to a higher-dimensional Euclidean space. This approach is advantageous because it reduces the embedding procedure to an eigenvalue decomposition of a centered distance matrix, however, the computational cost of directly applying this approach is restrictive due to the dense eigenvalue problem needed to compute the codimension coordinates, requiring an order of $\mathcal{O}(n^3)$ operations and $\mathcal{O}(n^2)$ where n is the number of vertices in the mesh. Further, the $\mathcal{O}(n^3)$ running time of Dijkstra's algorithm used to estimate the shortest distance between all pairs of vertices in the mesh is prohibitive. Even in the straight-sided case, this cost becomes intractable for large meshes which is further exacerbated by the presence of curvilinear vertices.

Additionally, it is unclear how curvilinear vertices are "connected" in the general case to ensure that the result of the shortest path algorithm is valid. To see this, consider the internal vertex of a $q = 3$ triangle. A triangulation of the reference simplex vertices may yield the connections between all curvilinear vertices to construct the graph for Dijkstra's algorithm but it biases the shortest paths as being the connections within the reference simplex.

For these two reasons, a sparse approach is sought which is equal in cost to that of embedding the straight-sided mesh. The methodology is based on the Landmark-Isomap algorithm of de Silva [8]. Of course, this sparse approach can be used to obtain embeddings of straight-sided meshes at a lower computational cost but that is left for future endeavours since the selection of landmarks is less trivial in that case. Other sparse approaches exist to achieve an isometric embedding at a lower computational cost, such as Locally Linear Embedding [17] or Semidefinite Embedding [19], however, Landmark-Isomap is preferred here due to its isometry preserving properties and its ability to work from the embedding of the straight-sided mesh.

Here, the landmarks are chosen to be the vertices of the straight-sided mesh. As such, the first step consists of embedding the vertices of the straight-sided mesh. The full computational approach to embed a curvilinear mesh is described in Algorithm 1. The number of vertices in the straight sided (q_1) mesh is denoted by n whereas the number of vertices in the high-order curvilinear (q) mesh is denoted by m.

The algorithm up to line 19 essentially computes the embedding of the vertices of the straight-sided mesh. The remainder of the algorithm computes the distance from each curvilinear vertex to each landmark vertex (line 30) which is stored in the matrix $\mathbf{R}$. This is done by first identifying the set of landmark vertices ($\mathcal{N}$) in the set of elements ($\mathbf{K}$) referencing the i-th vertex. The distance from this vertex to all local landmarks in $\mathcal{N}$ is then computed using the provided metric field and stored in $\mathbf{p}$. To estimate the distance to all remaining landmark vertices, the local landmark vertex l corresponding to the minimizer of $||\mathbf{p} + \mathbf{d}(\mathbf{p}, j)||$ is used as the distance from vertex i to landmark j. Lines 33–39 are referred to as the distance-based triangulation portion of the landmark-Isomap algorithm [8]. The embedding coordinates computed in line 40 are modified similar to previous efforts [4] to

<u>function</u> $\mathscr{T}_d^N$ = embed ($\mathscr{T}_d$, $\mathbf{m}$, N_0);

Input : $\mathscr{T}_d$, $\mathbf{m}$, N_0
Output: $\mathscr{T}_d^N$

$E \leftarrow$ `getEdges`($\mathscr{T}_d$);
for $e \in E$ **do**
    $\ell_e =$ `metricLength`(e, $\mathbf{m}$);
    $l_e =$ `euclideanLength`(e);
    $f_e = (\ell_e / l_e)^2$;
end
$f_{\max} \leftarrow \max(\{f_e\})\ \forall e \in E$;
for $e \in E$ **do**
    $\mathbf{G}(e_0, e_1) \leftarrow \sqrt{f_{\max}\ell_e^2 - l_e^2}$; `/* setup the graph adjacency matrix */`
end
$\mathbf{d} =$ `dijkstra`($\mathbf{G}$) ; `/* geodesic length between linear vertices */`
$\boldsymbol{\mu} \leftarrow$ `mean`($\mathbf{d}$) ; `/* row-wise mean of` $\mathbf{d}$ `*/`
$\mathbf{D} = \mathbf{d}^2$; `/* squared geodesic lengths */`
$\mathbf{B} = -\frac{1}{2}\mathbf{JDJ}$; `/* where the centering matrix` $\mathbf{J} = \mathbf{I} - \mathbf{11}^t/n$ `*/`
$(\mathbf{Q}, \boldsymbol{\Lambda}) =$ `eig`($\mathbf{B}$) ; `/* eigendecomposition of` $\mathbf{B}$ `*/`
$\mathbf{V} = \mathbf{Q}_{N_0}$;
$\mathbf{L} = \boldsymbol{\Lambda}_{N_0}$; `/* save` N_0 `largest eigenvalues and eigenvectors */`
$\mathbf{R}_{n,m} = \mathbf{0}$; `/* initial pairwise distance matrix to landmarks */`
for $i = 1, \ldots, m$ **do**
    $\mathbf{K} \leftarrow$ `elementsWithVertex`(i); `/* any` d`-simplex with vertex` i `*/`
    $\mathscr{N} \leftarrow$ `allLinearVertices`($\mathbf{K}$);
    $\mathbf{p} \leftarrow \mathbf{0}$; `/* initial distance from` i `to members of` $\mathscr{N}$ `*/`
    for $j \in \mathscr{N}$ **do**
        $\ell_{i,j} =$ `metricLength`($\{i, j\}$, $\mathbf{m}$) ; `/* distance between` i `and` j `*/`
        $l_{i,j} =$ `euclideanLength`($\{i, j\}$);
        $\mathbf{p}_j = \sqrt{f_{\max}\ell_{i,j}^2 - l_{i,j}^2}$;
    end
    for $j = 1, \ldots, n$ **do**
        $l = \arg\min_{\mathbf{p}} ||\mathbf{p} + \mathbf{d}(\mathbf{p}, j)||$; `/* closest landmark to landmark` j `*/`
        $\mathbf{R}(j, i) = ||\mathbf{p}_l + \mathbf{d}(\mathbf{p}_l, j)||$;
    end
    $\mathbf{L}^{\#} \leftarrow \mathbf{0}$; `/* apply landmark-Isomap to` $\mathbf{D}_{n,m}$ `*/`
    for $i = 1, \ldots, N_0,\ j = 1, \ldots, n$ **do**
        $\mathbf{L}^{\#}(i, j) = \mathbf{V}(j, i)/\sqrt{\mathbf{L}(i, i)}$;
    end
    for $i = 1, \ldots, m$ **do**
        $\mathbf{b} = -\frac{1}{2}\left(\mathbf{R}(:, i)^2 - \boldsymbol{\mu}\right)$;
        $\mathbf{u}_0 = \mathbf{L}^{\#}\mathbf{b}$; `/* codimension coordinates */`
        $\mathbf{u}_i = [\mathbf{x}_i, \mathbf{u}_0/\sqrt{f_{\max}}]$; `/* embedding coordinates by lifting */`
    end
end

Algorithm 1: Embedding algorithm

achieve a one-to-one mapping between the original mesh and the embedded one. Note the same factor $f_{\max}$ is used (computed from the original distance matrix) to modify the local distances between the curvilinear vertices and the landmarks (line 27) since these are shorter than any local distance between landmarks.

As in previous work [4], the algorithm is sensitive to the codimension of the ambient Euclidean space (N_0) which corresponds to the number of eigenvalues and eigenvectors retained from the decomposition of the centered disparity matrix (line 16). The next section studies the effect of N_0 on the isometry achieved by the algorithm.

It is worth mentioning that the current work simply focuses on developing an algorithm to isometrically embed curvilinear meshes with a metric field into a Euclidean space. The current focus is not on the validity of the embedded mesh but is the subject of future work.

3 Numerical Studies

Here the developed algorithm is applied using various analytic metric fields with different geometric orders of the mesh (q) and different embedding dimensions. The procedure used to evaluate the embedding approach is outlined below:

1. Obtain a straight-sided metric-conforming mesh using the software `fefloa` [15, 16] which is based on a unique approach to perform a variety of local mesh operations to achieve a straight-sided metric-conforming mesh,
2. Construct the curvilinear mesh from the straight-sided one using the method of Sect. 2.1,
3. Sweep over embedding dimensions:
    - Embed the curvilinear mesh using the algorithm of Sect. 2.2 into a Euclidean space of dimension $N = d + N_0$,
    - Compute the edge lengths of the embedded mesh. A Lagrange basis for the high-order basis functions is used to compute these lengths. The edge lengths are reported as a histogram in the following section. Since the mesh produced by `fefloa` is metric-conforming, these edges should be of unit length in the embedding space.

3.1 Analytic Embeddings

The first case studied is defined by the embedding into a Euclidean space of unit codimension:

$$\mathbf{u}(\mathbf{x}) = [\mathbf{x}, 5\tanh(10y - 20\cos(\pi x/5) - 50)], \quad \mathbf{x} = [0, 10]^2. \tag{1}$$

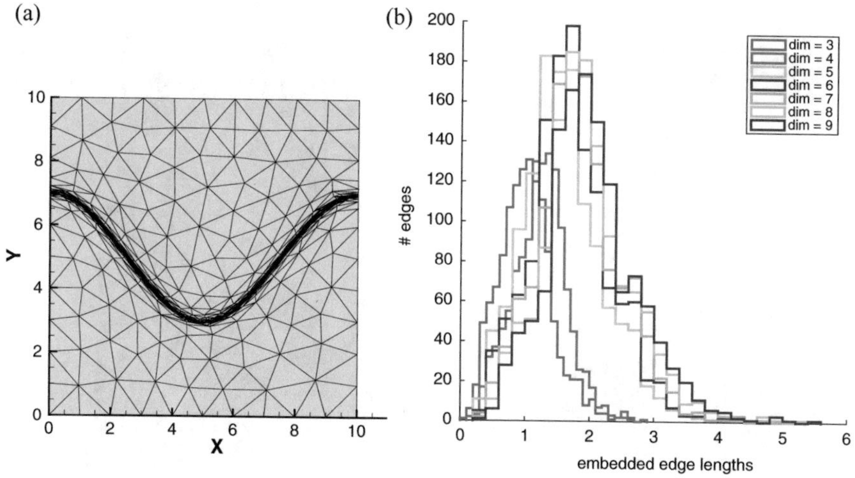

Fig. 2 Edge length histogram for edges produced by embedding the straight-sided mesh conforming to the embedding of Eq. 1 into Euclidean spaces of various dimensions. (**a**) Metric conforming straight-sided mesh for the embedding of Eq. 1. (**b**) Edge length histogram for edges produced by embedding the straight-sided mesh conforming to the embedding of Eq. 1 into Euclidean spaces of various dimensions

The metric is then $\mathbf{m}(\mathbf{x}) = \nabla\mathbf{u}\cdot\nabla\mathbf{u}$. The linear mesh produced by `fefloa` is shown in Fig. 2a and the edge length histograms produced by embedding this straight-sided mesh into Euclidean spaces of various dimensions is shown in Fig. 2b. The best isometry is achieved for a unit codimensional space which is expected since the analytic embedding is defined in three-dimensional space. The edge lengths tend to increase when the dimension of the ambient Euclidean space is increased. The sacrifice in *local* edge length isometry is due to the fact that longer geodesics are better matched by the greater number of eigenvalues taken in the decomposition of the centered disparity matrix (**B** in Algorithm 1).

The embedded linear mesh and curvilinear mesh edges are shown in Fig. 3a and b, respectively. As indicated by the better isometry induced by the greater number of unit edge lengths of the histogram in Fig. 4a, it that seems increasing the geometric order of the meshes improves the representation of the metric in three-dimensional space. However, observe the oscillatory behaviour of these edges in Fig. 3b. Though better isometry seems to be achieved, the resulting embedding may not produce valid elements in the embedding space. Ongoing work consists of evaluating the validity of the embedded mesh with a variant of the algorithm of Johnen et al. in higher-dimensional spaces [11, 12]. Though the best isometry is achieved in three-dimensional space, the edge length histogram produced across various curvilinear mesh orders is given for a five-dimensional embedding in Fig. 4b. As stated earlier, the local edge lengths are longer than they should be, likely a result of better representing the longer geodesics across the domain with the greater number of eigenvalues retained.

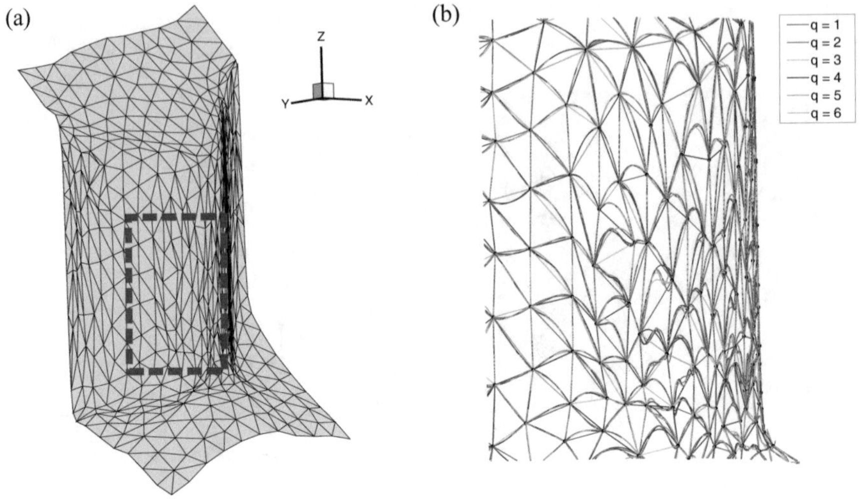

Fig. 3 Straight-sided and curvilinear meshes produced by embedding the mesh conforming to the metric derived from the embedding of Eq. 1. (**a**) Embedded straight-sided mesh produced with the metric field derived from Eq. 1. (**b**) Close-up (of the section in the red box of the left figure) of the curvilinear mesh edges produced by the metric field derived from Eq. 1

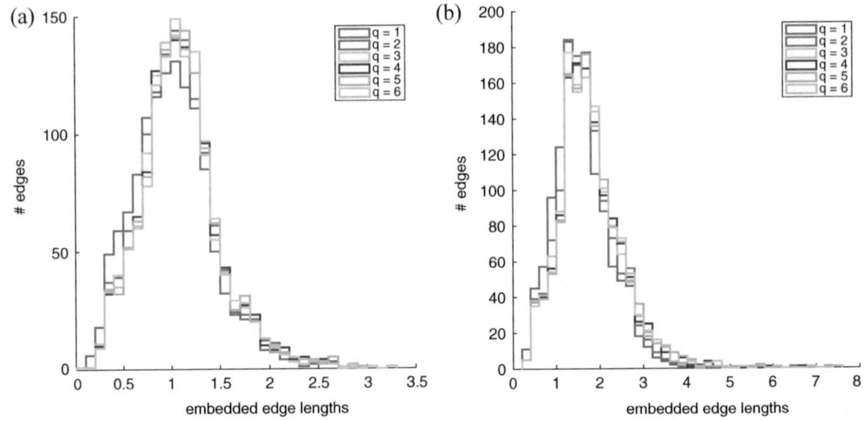

Fig. 4 Edge length histogram for edges produced by embedding curvilinear meshes for the metric conforming to the embedding of Eq. 1 into three- and five-dimensional space. (**a**) Edge length histogram for edges produced by embedding curvilinear meshes for the metric conforming to the embedding of Eq. 1 into three-dimensional space. (**b**) Edge length histogram for edges produced by embedding curvilinear meshes for the metric conforming to the embedding of Eq. 1 into five-dimensional space

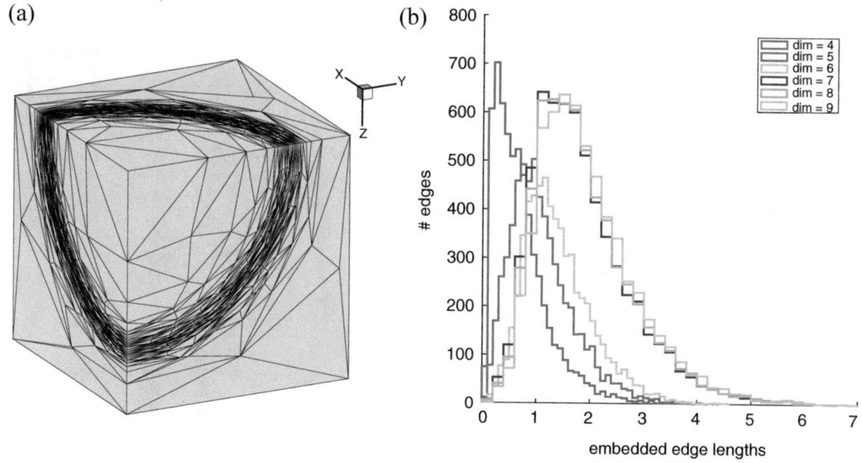

Fig. 5 Edge length histogram for edges produced by embedding the straight-sided mesh conforming to the embedding of Eq. 2 into Euclidean spaces of various dimensions. (**a**) Metric conforming straight-sided mesh for the embedding of Eq. 2. (**b**) Edge length histogram for edges produced by embedding the straight-sided mesh conforming to the embedding of Eq. 2 into Euclidean spaces of various dimensions

A similar procedure is used to study a three-dimensional case, whereby an analytic embedding into four-dimensional space is defined by

$$\mathbf{u}(\mathbf{x}) = \left[\mathbf{x}, 10\tanh(10||\mathbf{x}||^2 - 0.75^2)\right], \quad \mathbf{x} = [0, 1]^3. \tag{2}$$

Unlike its two-dimensional counterpart, this case achieves the best isometry when embedded into a five-dimensional Euclidean space; see Fig. 5b. It is interesting, however, that isometry can be recovered in four dimensions by using a high-order embedding as demonstrated in the edge lengths of Fig. 6a. For higher-dimensional embeddings, however, the lower order embeddings achieve better isometry since the higher-order ones generate longer edge lengths, likely due to approximating the longer geodesics across the domain (Fig. 6b).

3.2 Analytic Metric Fields

Now the method is studied with analytic metric fields. In particular, the metrics defined by

$$\mathbf{m}(\mathbf{x}) = \begin{bmatrix} \frac{1}{(|x-0.5|+0.0025)^2} & 0 & 0 \\ & \frac{1}{(|y-0.5|+0.0025)^2} & 0 \\ & & \frac{1}{(|z-0.5|+0.0025)^2} \end{bmatrix}, \quad \mathbf{x} \in [0, 1]^3 \tag{3}$$

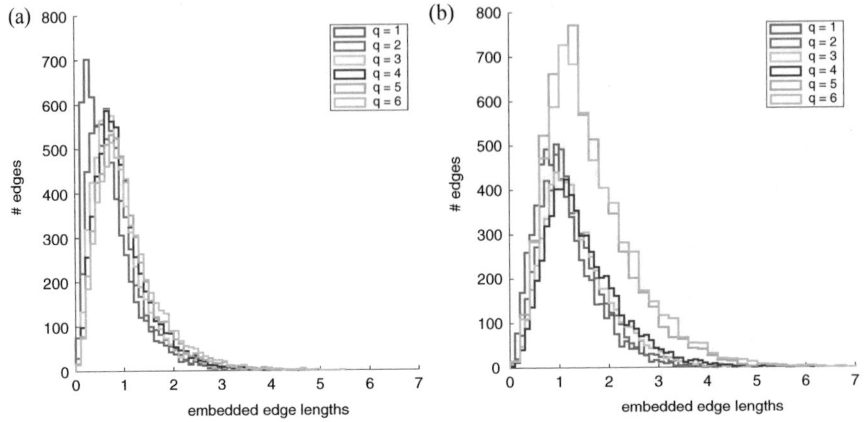

Fig. 6 Edge length histogram for edges produced by embedding curvilinear meshes for the metric conforming to the embedding of Eq. 2 into four- and six-dimensional space. (**a**) Edge length histogram for edges produced by embedding curvilinear meshes for the metric conforming to the embedding of Eq. 2 into four-dimensional space. (**b**) Edge length histogram for edges produced by embedding curvilinear meshes for the metric conforming to the embedding of Eq. 2 into six-dimensional space

and

$$\mathbf{m}(\mathbf{x}) = \begin{bmatrix} \frac{\cos^2\theta}{h_x^2} + \frac{\sin^2\theta}{h_y^2} & (\frac{1}{h_x^2} - \frac{1}{h_y^2})\cos\theta\sin\theta & 0 \\ & \frac{\sin^2\theta}{h_x^2} + \frac{\cos^2\theta}{h_y^2} & 0 \\ & & 4 \end{bmatrix}, \quad \mathbf{x} \in [0, 1]^3 \tag{4}$$

where $h_x = \min(0.005 \cdot 5^a, 0.5)$, $h_y = \min(0.1 \cdot 2^a, 0.5)$ ($a = 10|0.75 - \sqrt{x^2 + y^2}|$ and $\theta = \arctan(x, y)$) [18], are used to create straight-sided metric-conforming meshes with `fefloa` (Figs. 7a and 9a). The embedding algorithm is applied as in the last section, however, the dimension of the ambient Euclidean space cannot be predicted a priori. As such, a sweep over dimensions is used to embed the straight-sided mesh; the histogram of edge lengths resulting from this sweep is shown in Figs. 7b and 9b.

The metric of Eq. 3 appears to be best represented in five- and six-dimensional Euclidean spaces. As such, these spaces are used to now sweep over the geometric orders of the mesh, the histograms of which are shown in Fig. 8a, b. Little difference is observed across the different mesh orders in five-dimensional space but isometry seems to degrade with increased mesh order in six-dimensions since it appears that edges become longer for reasons stated earlier. It is further surprising that only five dimensions are needed to represent this metric field isometrically because of the three directional variations in the metric, implying three codimensions would be needed. Further theoretical work is needed to understand this result.

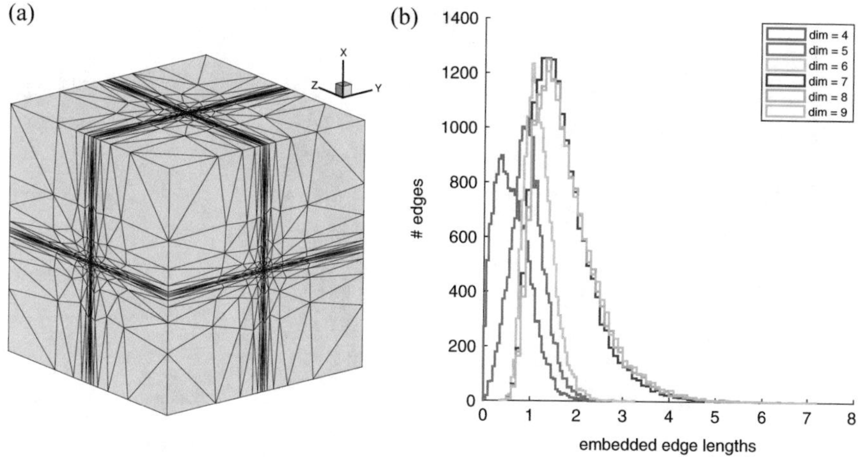

Fig. 7 Edge length histogram for edges produced by embedding the straight-sided mesh conforming to the embedding of Eq. 3 into Euclidean spaces of various dimensions. (**a**) Metric conforming straight-sided mesh for the embedding of Eq. 3. (**b**) Edge length histogram for edges produced by embedding the straight-sided mesh conforming to the embedding of Eq. 3 into Euclidean spaces of various dimensions

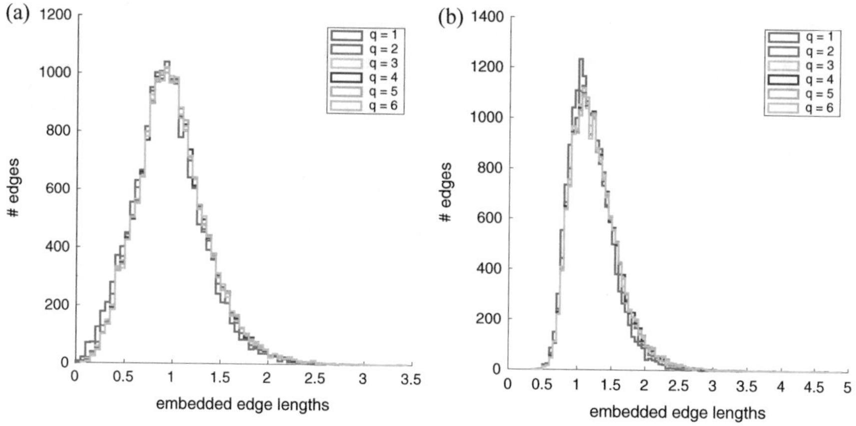

Fig. 8 Edge length histogram for edges produced by embedding curvilinear meshes for the metric conforming to the embedding of Eq. 3 into five- and six-dimensional space. (**a**) Edge length histogram for edges produced by embedding curvilinear meshes for the metric conforming to the embedding of Eq. 3 into five-dimensional space. (**b**) Edge length histogram for edges produced by embedding curvilinear meshes for the metric conforming to the embedding of Eq. 3 into six-dimensional space

The metric of Eq. 4 is best represented in four- and five-dimensional space as observed in the edge length histogram of Fig. 9b. Isometry slightly improves with increasing curvilinear mesh order in four dimensions but little difference is observed in five dimensions (see Fig. 10a, b). This further supports the finding that when the

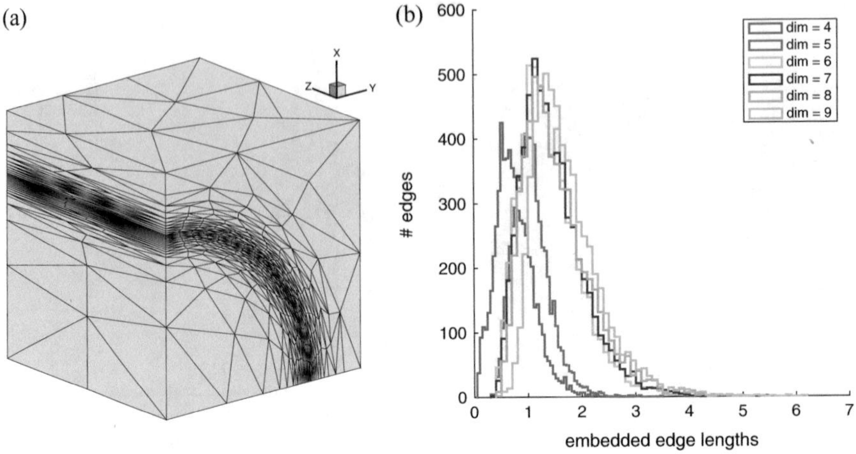

Fig. 9 Edge length histogram for edges produced by embedding the straight-sided mesh conforming to the embedding of Eq. 4 into Euclidean spaces of various dimensions. (**a**) Metric conforming straight-sided mesh for the embedding of Eq. 4. (**b**) Edge length histogram for edges produced by embedding the straight-sided mesh conforming to the embedding of Eq. 4 into Euclidean spaces of various dimensions

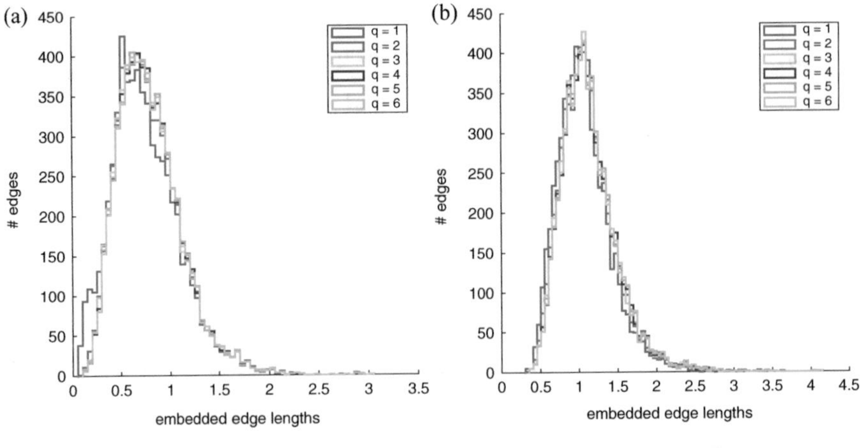

Fig. 10 Edge length histogram for edges produced by embedding curvilinear meshes for the metric conforming to the embedding of Eq. 4 into four- and five-dimensional space. (**a**) Edge length histogram for edges produced by embedding curvilinear meshes for the metric conforming to the embedding of Eq. 4 into four-dimensional space. (**b**) Edge length histogram for edges produced by embedding curvilinear meshes for the metric conforming to the embedding of Eq. 4 into five-dimensional space

ambient dimension of the Euclidean space is not high enough to achieve an isometric embedding of a straight-sided mesh, better isometry can be achieved by embedding a curvilinear representation of the mesh instead. However, when the straight-sided mesh is well represented by an embedding of its vertices into a suitable Euclidean space, then little difference is observed when increasing the geometric order of the embedded mesh.

4 Conclusions and Future Work

This work introduced an algorithm for embedding curvilinear meshes into Euclidean spaces. The first step consisted of defining a curvilinear mesh from a metric-conforming straight-sided one. The next step consisted of embedding the straight sided mesh into Euclidean spaces of various dimensions and analyzing which dimension closely approximates the local geodesic lengths. Curvilinear meshes of varying geometric order were then embedded to the chosen Euclidean spaces. The findings suggest that when the dimension of the Euclidean space is not high enough, then isometry can be improved by increasing the geometric order of the mesh. However, when the embedding of the straight-sided mesh is sufficiently isometric, then little difference is observed in the produced edge lengths across geometric mesh orders.

Future work consists in using the produced embeddings to directly generate curvilinear metric-conforming meshes. A local approach for doing so, similar to the cavity-based approach of Loseille [15, 16] is attractive due to its ability to enlarge initially invalid cavities when inserting high-order vertices.

Acknowledgements The authors wish to thank the members of the Aerospace Computational Design Laboratory at MIT for their encouragement throughout this project and the research staff at the Barcelona Supercomputing Center for hosting the first author while much of Sect. 2.1 was developed. This work was funded by the CAPS project, AFRL Contract FA8050-14-C-2472: “CAPS: Computational Aircraft Prototype Syntheses” with Dean Bryson as the Technical Monitor. This project has received funding from the European Research Council (ERC) under the European Union’s Horizon 2020 research and innovation programme under grant agreement No. 715546. The work of Xevi Roca has been partially supported by the Spanish Ministerio de Economía y Competitividad and the Generalitat de Catalunya under grant agreements RYC-2015-01633 and 2017 SGR 1731, respectively.

References

1. F. Bassi, S. Rebay, High-order accurate discontinuous finite element solution of the 2D Euler equations. J. Comput. Phys. **138**(2), 251–285 (1997)
2. J. Burkardt, The finite element basis for simplices in arbitrary dimensions. Technical Report, Florida State University, Department of Scientific Computing, 2013
3. G.D. Cañas, S.J. Gortler, Surface remeshing in arbitrary codimensions. Vis. Comput. **22**(9), 885–895 (2006)
4. P.C. Caplan, R. Haimes, D.L. Darmofal, M.C. Galbraith, Anisotropic geometry-conforming d-simplicial meshing via isometric embeddings, in *Proceedings of the 26th International Meshing Roundtable* (Springer, Berlin, 2017)

5. F. Dassi, H. Si, S. Perotto, T. Streckenbach, Anisotropic finite element mesh adaptation via higher dimensional embedding. Procedia Eng. **124**, 265–277 (2015)
6. F. Dassi, P. Farrell, H. Si, An anisoptropic surface remeshing strategy combining higher dimensional embedding with radial basis functions. Procedia Eng. **163**, 72–83 (2016); 25th International Meshing Roundtable
7. F. Dassi, S. Perotto, H. Si, T. Streckenbach, A priori anisotropic mesh adaptation driven by a higher dimensional embedding. Comput. Aided Des. **85**, 111–122 (2017); 24th International Meshing Roundtable Special Issue: Advances in Mesh Generation
8. V. de Silva, J.B. Tenenbaum, Sparse multidimensional scaling using landmark points. Technical Report, Stanford University, 2004
9. V. de Silva, J.B. Tenenbaum, J.C. Langford, A global geometric framework for nonlinear dimensionality reduction. Science **290**, 2319–2323 (2000)
10. A. Grundmann, H.M. Moller, Invariant integration formulas for the n-simplex by combinatorial methods. SIAM J. Numer. Anal. **15**, 282–290 (1978)
11. A. Johnen, J.-F. Remacle, C. Geuzaine, Geometrical validity of curvilinear finite elements. J. Comput. Phys. **233**, 359–372 (2013)
12. A. Johnen, C. Geuzaine, T. Toulorge, J.-F. Remacle, Efficient computation of the minimum of shape quality measures on curvilinear finite elements. Procedia Eng. **163**, 328–339 (2016); 25th International Meshing Roundtable
13. H.W. Kuhn, Simplicial approximation of fixed points. Proc. Natl. Acad. Sci. **61**, 1238–1242 (1968)
14. B. Lévy, N. Bonneel, Variational anisotropic surface meshing with Voronoi parallel linear enumeration, in *Proceedings of the International Meshing Roundtable*, 2012
15. A. Loseille, Metric-orthogonal anisotropic mesh generation. Procedia Eng. **82**, 403–415 (2014)
16. A. Loseille, R. Löhner, Cavity-based operators for mesh adaptation, in *51st AIAA Aerospace Sciences Meeting including the New Horizons Forum and Aerospace Exposition.* (American Institute of Aeronautics and Astronautics, Reston, 2013)
17. S.T. Roweis, L.K. Saul, Nonlinear dimensionality reduction by locally linear embedding. Science **290**(5500), 2323–2326 (2000)
18. C. Tsolakis, F. Drakopoulos, N. Chrisochoides, Sequential metric-based adaptive mesh generation, in *2018 Modeling, Simulation, and Visualization Student Capstone Conference*, Suffolk, VA, April 2018
19. K.Q. Weinberger, B.D. Packer, L.K. Saul, Nonlinear dimensionality reduction by semidefinite programming and kernel matrix factorization, in *Proceedings of the Tenth International Workshop on Artificial Intelligence and Statistics*, Barbados, West Indies, ed. by Z. Ghahramani, R. Cowell, 2005

Defining a Stretching and Alignment Aware Quality Measure for Linear and Curved 2D Meshes

Guillermo Aparicio-Estrems, Abel Gargallo-Peiró, and Xevi Roca

Abstract We define a regularized shape distortion (quality) measure for curved high-order 2D elements on a Riemannian plane. To this end, we measure the deviation of a given 2D element, straight-sided or curved, from the stretching and alignment determined by a target metric. The defined distortion (quality) is suitable to check the validity and the quality of straight-sided and curved elements on Riemannian planes determined by constant and point-wise varying metrics. The examples illustrate that the distortion can be minimized to curve (deform) the elements of a given high-order (linear) mesh and try to match with curved (linear) elements the point-wise alignment and stretching of an analytic target metric tensor.

1 Introduction

In the last decades, the utilization of unstructured meshes composed by highly stretched elements and aligned with dominant flow features, such as boundary layers and shock waves, have shown to be very advantageous [1–3]. When compared with uniform refinement or with isotropic meshes with non-uniform sizing, anisotropic meshes lead to a significant reduction on the number of required degrees of freedom to obtain the same approximation accuracy. This allows performing simulations with a significantly reduced, and even unbeatable, computational cost.

The generation of anisotropic meshes requires to determine the location, stretching and alignment of the elements. These features can be prescribed manually with the help of the user interface of a mesh generation environment. They can also be prescribed imposing point-wise varying metric tensors obtained in an automatic and iterative adaption procedure based on error indicators or estimators [4–6]. Then, an anisotropic mesher [6–9] can be used to match the resolution, stretching and alignment determined by the target metric.

G. Aparicio-Estrems · A. Gargallo-Peiró · X. Roca (✉)
Barcelona Supercomputing Center, Barcelona, Spain
e-mail: xevi.roca@bsc.es

X. Roca, A. Loseille (eds.), *27th International Meshing Roundtable*, Lecture Notes in Computational Science and Engineering 127,
https://doi.org/10.1007/978-3-030-13992-6_3

It is standard to use parallelotopes (quadrilaterals and hexahedra) to manually prescribe the alignment and stretching required to capture flow features such as boundary layers. Whereas the flexibility of simplices (triangles and tetrahedra) is the preferred one in automatic adaption iterations. Nevertheless, for both types of elements, the most mature anisotropic mesh generation techniques lead to meshes featuring second order elements such as multi-linear parallelotopes and linear simplices.

The utilization of curved anisotropic meshes composed by third order elements, such as multi-quadratic parallelotopes and quadratic simplices, or piece-wise polynomial elements of higher order has been mainly centered to curve, manually prescribed, straight-sided boundary layer meshes [10–18]. It has not been until recently that the first metric based approaches have been explored to generate anisotropic meshes featuring straight-sided very high-order three dimensional approximations [19], curved quadratic triangles [20], and r-adapted curved high-order 2D elements [21]. However, no specific efforts have been conducted to check the validity and measure the quality of curved high-order anisotropic meshes considering a prescribed metric tensor.

Our main contribution is to define a regularized shape distortion (quality) to measure the deviation of a given linear or high-order 2D element from the stretching and alignment determined by a target metric. The influence of the target metric on the element quality has only been considered in detail for linear elements [22, 23] and not for curved high-order elements [24]. The defined distortion (quality) is suitable to check the validity and the quality of straight-sided and curved elements on Riemannian planes determined by constant and point-wise varying metrics. Furthermore, we illustrate that the distortion can be minimized to curve (deform) the elements of a given high-order (linear) mesh and try to match with curved (linear) elements the point-wise alignment and stretching of an analytic target metric tensor. Specifically, this approach can be used to improve, by curving (deforming) the elements, the alignment and stretching of a mesh obtained with a straight-sided anisotropic mesher.

The rest of the paper is organized as follows. First, in Sect. 2 we introduce the quality measures for high-order isotropic 2D elements. Next, in Sect. 3 we present the new quality measure for high-order anisotropic 2D elements. Following, we present several examples to illustrate the capabilities of the proposed measure, Sect. 4. To finalize, in Sect. 5 we present the main conclusions and sum up the future work to develop.

2 Preliminaries: Measures for High-Order Isotropic 2D Elements

In this section, we review the definition of the Jacobian-based quality measures for linear and high-order isotropic elements. In addition, we introduce the required notation for anisotropic elements.

To define and compute a Jacobian-based measure for linear isotropic triangles [25], three elements are required: the master, the ideal, and the physical, see Fig. 1. The master (E^M) is the element from which the iso-parametric mapping is defined. The ideal element $\left(E^I\right)$ represents the target configuration which, in the isotropic case, is an equilateral (regular) element ($E^{\triangle}$). The physical (E^P) is the element to be measured.

First, the mappings between the ideal and the physical elements through the master element are obtained. By means of these mappings, a mapping between the ideal and physical elements is determined by the composition

$$\boldsymbol{\phi}_E : E^{\triangle} \xrightarrow{\boldsymbol{\phi}_{\triangle}^{-1}} E^M \xrightarrow{\boldsymbol{\phi}_P} E^P.$$

The Jacobian of this affine mapping (denoted by $\mathbf{D}\boldsymbol{\phi}_E$) encodes the deviation of the physical element with respect to the equilateral one. Specifically, the distortion measure η of the physical element is defined as

$$\eta(\mathbf{D}\boldsymbol{\phi}_E) = \frac{\operatorname{tr}\left(\left(\mathbf{D}\boldsymbol{\phi}_E\right)^{\mathrm{T}} \cdot \mathbf{D}\boldsymbol{\phi}_E\right)}{d\left(\det\left(\left(\mathbf{D}\boldsymbol{\phi}_E\right)^{\mathrm{T}} \cdot \mathbf{D}\boldsymbol{\phi}_E\right)\right)^{1/d}} = \frac{\operatorname{tr}\left(\left(\mathbf{D}\boldsymbol{\phi}_E\right)^{\mathrm{T}} \cdot \mathbf{D}\boldsymbol{\phi}_E\right)}{d\left(\det\left(\mathbf{D}\boldsymbol{\phi}_E\right)\right)^{2/d}}. \quad (1)$$

The matrix $\mathbf{D}\boldsymbol{\phi}_E$ is computed for linear triangles as

$$\mathbf{D}\boldsymbol{\phi}_E = \mathbf{D}\boldsymbol{\phi}_P \cdot \mathbf{D}\boldsymbol{\phi}_{\triangle}^{-1} = \begin{pmatrix} x_1 - x_0 & \frac{2x_2 - x_1 - x_0}{\sqrt{3}} \\ y_1 - y_0 & \frac{2y_2 - y_1 - y_0}{\sqrt{3}} \end{pmatrix},$$

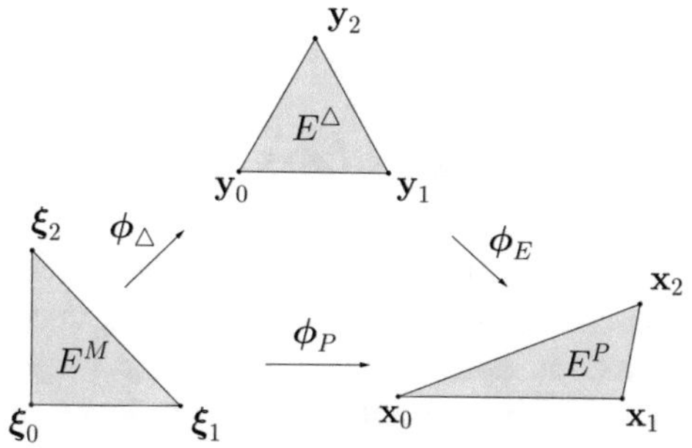

Fig. 1 Mappings between the reference, the ideal and the physical elements in the linear case

where

$$\mathbf{D}\boldsymbol{\phi}_P = \begin{pmatrix} x_1 - x_0 & x_2 - x_0 \\ y_1 - y_0 & y_2 - y_0 \end{pmatrix}, \quad \mathbf{D}\boldsymbol{\phi}_\triangle = \begin{pmatrix} 1 & \frac{1}{2} \\ 0 & \frac{\sqrt{3}}{2} \end{pmatrix}, \tag{2}$$

where $\mathbf{x}_i = (x_i, y_i)$ denotes the coordinates of the physical element E^P, and where the involved matrices have been written for the master element E^M with node coordinates $\{\boldsymbol{\xi}_0 = (-1, -1)\,,\ \boldsymbol{\xi}_1 = (1, -1)\,,\ \boldsymbol{\xi}_2 = (-1, 1)\}$, and the ideal element E^I determined by the nodes $\{\mathbf{u}_0 = \left(-1, -1/\sqrt{3}\right),\ \mathbf{u}_1 = \left(1, -1/\sqrt{3}\right),\ \mathbf{u}_2 = \left(0, 2/\sqrt{3}\right)\}$.

The distortion measure in Eq. (1) quantifies the deviation of the shape of the physical element with respect to the regular shape. The measure gets value 1 when the physical element is the equilateral element. It is important to note that it is invariant under translations, rotations, scalings and symmetries. Moreover, it can be regularized to detect inverted elements. From the distortion measure, the quality measure of an element is defined as

$$q = \frac{1}{\eta}, \tag{3}$$

which takes values in the interval [0, 1], being 0 for degenerated elements and 1 for the ideal element.

For high-order [14, 26, 27] and multi-linear [28] elements with non-constant Jacobian, the distortion measure is reinterpreted as a point-wise measure as

$$\mathcal{N}\boldsymbol{\phi}_E(\mathbf{y}) := \eta(\mathbf{D}\boldsymbol{\phi}_E(\mathbf{y})) \ \ \forall \mathbf{y} \in E^\triangle.$$

Furthermore, the elemental distortion is defined in [24, 26] as

$$\eta_{E^P} := \frac{\left(\int_{E^\triangle} \mathcal{N}\boldsymbol{\phi}_E(\mathbf{y})^2 \, d\mathbf{y}\right)^{1/2}}{\left(\int_{E^\triangle} 1 \, d\mathbf{y}\right)^{1/2}}, \tag{4}$$

and its quality q_{E^P} follows from Eq. (3).

3 Measures for Curved High-Order Anisotropic 2D Elements

In this section, we first present a quality measure for linear triangles equipped with a constant metric, see Sect. 3.1. Next, in Sect. 3.2 we analyze the behavior of the defined measure. Finally, in Sect. 3.3 we present the definition of the new quality measure for anisotropic 2D high-order elements.

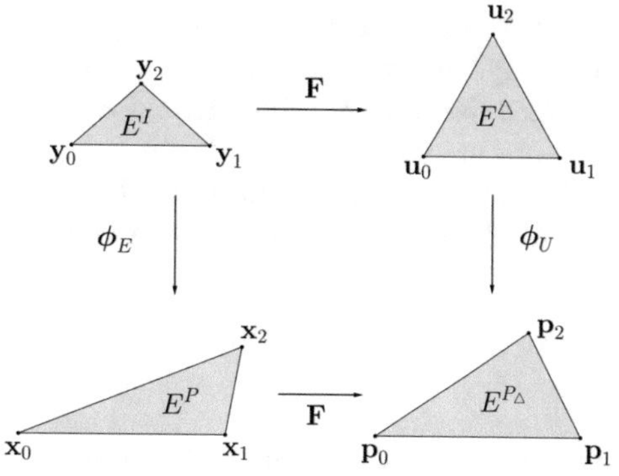

Fig. 2 Mappings between the equilateral, the ideal, the physical and the isotropic physical triangles

3.1 *Linear Elements and Constant Metric*

To define a measure that quantifies the quality of a given element, it is required to define an ideal element that represents the desired configuration, as detailed in Sect. 2. In the isotropic case, with metric $\mathbf{M}$ equal to $\lambda \mathrm{Id}$ for $\lambda > 0$, the ideal triangle is the equilateral triangle $E^{\triangle}$. For non-isotropic metrics, we describe how to obtain the ideal configuration. Then, we measure the distortion of the physical element comparing it to the ideal element.

We define the ideal element as the element with edges of unit length under the desired metric. To compute this configuration, we first decompose $\mathbf{M}$ as:

$$\mathbf{M} = \mathbf{F}^{\mathrm{T}} \cdot \mathbf{F}. \tag{5}$$

Matrix $\mathbf{F}$ can be interpreted as a linear mapping between the space with metric $\mathbf{M}$ and the space with unitary metric. Thus, we define the anisotropic ideal E^I as the preimage by $\mathbf{F}$ of the equilateral triangle, see Fig. 2. In particular, let $\mathbf{u}_i,\ i = 0, 1, 2$ be the nodes of the equilateral triangle $E^{\triangle}$. Then, we define the nodes of the ideal triangle E^I as

$$\mathbf{y}_i = \mathbf{F}^{-1} \cdot \mathbf{u}_i, \quad i = 0, 1, 2.$$

A direct consequence of the above definition is that the ideal triangle has unit edge lengths in the metric sense.

Once the ideal triangle is defined, we measure the deviation between the ideal and physical elements. Similarly to the approaches for a unitary metric, see Sect. 2, in this section we define the distortion between the ideal E^I and physical E^P elements in terms of the mapping between those elements, $\boldsymbol{\phi}_E$.

A priori, we do not know how to compare elements considering the target metric. Nevertheless, we know how to compare elements in the isotropic sense, Sect. 2, and thus we map both elements E^I and E^P to the same Euclidean space using $\mathbf{F}$, see Fig. 2. Then, we compare the image elements $E^{\triangle}$ and $E^{P_{\triangle}}$ using the distortion measure presented in Eq. (1).

Let $E^{P_\triangle}$ be the image of the physical triangle E^P by $\mathbf{F}$. By construction, the image by $\mathbf{F}$ of the ideal triangle is the equilateral triangle. We measure the distortion between the ideal E^I and physical E^P elements in terms of the distortion of the mapping between the $E^\triangle$ and $E^{P_\triangle}$.

Finally, we define the distortion between the physical triangle E^P and the ideal triangle E^I with respect to the desired metric as the distortion of the matrix $\mathbf{D}\boldsymbol{\phi}_U$:

$$\eta_{\mathbf{M}}(\mathbf{D}\boldsymbol{\phi}_E) = \eta_{\mathbf{M}}(\mathbf{F}\cdot\mathbf{D}\boldsymbol{\phi}_U\cdot\mathbf{F}^{-1}) := \eta(\mathbf{D}\boldsymbol{\phi}_U) = \frac{\operatorname{tr}\left(\mathbf{D}\boldsymbol{\phi}_U^{\mathrm{T}}\cdot\mathbf{D}\boldsymbol{\phi}_U\right)}{d\det\left(\mathbf{D}\boldsymbol{\phi}_U^{\mathrm{T}}\cdot\mathbf{D}\boldsymbol{\phi}_U\right)^{1/d}}. \tag{6}$$

Remark 1 The distortion presented in Eq. (6) is well defined because it does not depend on the rotations and symmetries of $E^{P_\triangle}$. That is, all triangles with the same edge lengths have the same quality. We first show the case for rotations.

The rotation of angle θ of $E^{P_\triangle}$ is the triangle $E^{\tilde{P}_\triangle}$ composed by the nodes $\tilde{\mathbf{y}}_i = \mathbf{R}(\theta)\cdot\mathbf{y}_i,\ i=0,1,2$. Then

$$\mathbf{D}\tilde{\boldsymbol{\phi}}_U = \mathbf{R}(\theta)\cdot\mathbf{D}\boldsymbol{\phi}_U,$$

where $\mathbf{D}\tilde{\boldsymbol{\phi}}_U$ is the mapping between the equilateral triangle $E^\triangle$ and $E^{\tilde{P}_\triangle}$. Next,

$$\mathbf{D}\tilde{\boldsymbol{\phi}}_U{}^{\mathrm{T}}\cdot\mathbf{D}\tilde{\boldsymbol{\phi}}_U = \mathbf{D}\boldsymbol{\phi}_U^{\mathrm{T}}\cdot\mathbf{R}(\theta)^{\mathrm{T}}\cdot\mathbf{R}(\theta)\cdot\mathbf{D}\boldsymbol{\phi}_U = \mathbf{D}\boldsymbol{\phi}_U^{\mathrm{T}}\cdot\mathbf{D}\boldsymbol{\phi}_U. \tag{7}$$

By Eqs. (7), (6) and (3) we conclude that the corresponding qualities are equal. The case for symmetries follows analogously since any symmetry $\mathbf{S}$ satisfies that $\mathbf{S}^{\mathrm{T}}\cdot\mathbf{S} = \mathrm{Id}$.

Remark 2 To compute the quality measure in Eq. (6) the explicit decomposition of the metric shown in Eq. (5) is not required. Next, we show how to compute the distortion presented in Eq. (6) without decomposing it using matrix $\mathbf{F}$.

First, in Fig. 3 we include the master element in the diagram of applications illustrated in Fig. 2. Let $\boldsymbol{\phi}_\triangle$ be the mapping between the reference and the equilateral triangle. This mapping is equivalent to the composition of the mappings $\boldsymbol{\phi}_I$ and $\mathbf{F}$, but it can be directly computed from the coordinates of the master and equilateral triangles, as previously done for the isotropic case in Sect. 2.

Taking into account the computation of $\mathbf{D}\boldsymbol{\phi}_\triangle$ in terms of the node coordinates in Eq. (2), the distortion measure $\eta_{\mathbf{M}}(\mathbf{D}\boldsymbol{\phi}_E)$ can be rewritten without requiring to decompose $\mathbf{M}$. We note that, a priori, right-hand side in Eq. (6) depends on $\mathbf{F}$ since

$$\mathbf{D}\boldsymbol{\phi}_U = \mathbf{D}\boldsymbol{\phi}_{P^\triangle}\cdot\mathbf{D}\boldsymbol{\phi}_\triangle^{-1} = \mathbf{F}\cdot\mathbf{D}\boldsymbol{\phi}_P\cdot\mathbf{D}\boldsymbol{\phi}_\triangle^{-1}.$$

Manipulating Eq. (6) one realizes that there is no explicit dependence on $\mathbf{F}$:

$$\mathbf{D}\boldsymbol{\phi}_U^{\mathrm{T}}\cdot\mathbf{D}\boldsymbol{\phi}_U = \left(\mathbf{D}\boldsymbol{\phi}_\triangle\right)^{-\mathrm{T}}\cdot\mathbf{D}\boldsymbol{\phi}_P^{\mathrm{T}}\cdot\mathbf{F}^{\mathrm{T}}\cdot\mathbf{F}\cdot\mathbf{D}\boldsymbol{\phi}_P\cdot\left(\mathbf{D}\boldsymbol{\phi}_\triangle\right)^{-1} \tag{8}$$

$$= \left(\mathbf{D}\boldsymbol{\phi}_\triangle\right)^{-\mathrm{T}}\cdot\mathbf{D}\boldsymbol{\phi}_P^{\mathrm{T}}\cdot\mathbf{M}\cdot\mathbf{D}\boldsymbol{\phi}_P\cdot\left(\mathbf{D}\boldsymbol{\phi}_\triangle\right)^{-1}. \tag{9}$$

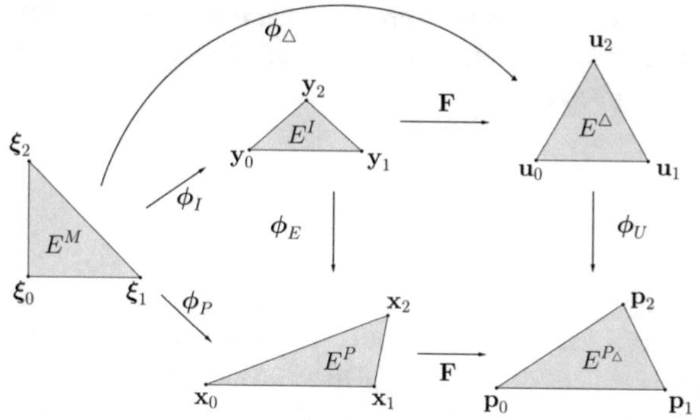

Fig. 3 Mappings between the reference, the equilateral, the ideal, the physical and the isotropic physical triangles

Thus, we obtain an expression for the distortion that does not require to decompose the metric $\mathbf{M}$:

$$\eta_{\mathbf{M}}(\mathbf{D}\boldsymbol{\phi}_E) = \eta(\mathbf{D}\boldsymbol{\phi}_U) = \frac{\operatorname{tr}\left(\left(\mathbf{D}\boldsymbol{\phi}_P \cdot \mathbf{D}\boldsymbol{\phi}_\triangle^{-1}\right)^{\mathrm{T}} \cdot \mathbf{M} \cdot \mathbf{D}\boldsymbol{\phi}_P \cdot \mathbf{D}\boldsymbol{\phi}_\triangle^{-1}\right)}{d\left(\det\left(\left(\mathbf{D}\boldsymbol{\phi}_P \cdot \mathbf{D}\boldsymbol{\phi}_\triangle^{-1}\right)^{\mathrm{T}} \cdot \mathbf{M} \cdot \mathbf{D}\boldsymbol{\phi}_P \cdot \mathbf{D}\boldsymbol{\phi}_\triangle^{-1}\right)\right)^{1/d}}. \tag{10}$$

3.2 *Behavior of the Metric Distortion Measure*

In this section, we illustrate the behavior of the shape quality measure corresponding to the distortion measure presented in Eq. (6) for linear anisotropic triangles equipped with a constant metric. We first show the level curves of the quality measure of a triangle when we fix two nodes and we let the third node to move in $\mathbb{R}^2$. Second, we analyze the behavior of the measure with respect to the alignment of the element with the metric.

3.2.1 Level Curves of the Shape Quality Measure

To show the behavior of the level curves of the shape quality measure we consider two cases, the Euclidean or isotropic case when $\mathbf{M} = \mathrm{Id}$ and the anisotropic case when $\mathbf{M}$ has two different eigenvalues.

We apply two tests to a triangle, one for each metric. We illustrate the behavior by plotting the level sets in terms of a free node of the triangle. We consider the anisotropic metric given by

$$\mathbf{M} = \begin{pmatrix} 1 & 0 \\ 0 & \frac{1}{h^2} \end{pmatrix}, \quad h = 1/3. \tag{11}$$

This metric is aligned with the canonical axes and features a stretching ratio of 1 against 3. Specifically, it is devised to ensure that vectors $(1, 0)$ and $(0, h)$ have unit length. The ideal element E^I is expected to be an element of height h and base 1. In each test, we consider a free node, keeping the rest of nodes fixed at their original location, and we compute the quality of the element in terms of the location of this node. The free node considered is the vertex node $\mathbf{x}_2$.

In Fig. 4, we show the contour plots of the quality for each test when the free node is allowed to move in a region of R^2. The locus of the points where the element has positive Jacobian, feasible region, is independent of the metric and corresponds to the half-plane $y > 0$. As expected, for each metric the optimal node location is different. Furthermore, we can observe that the level sets and the height of the ideal triangle corresponding to the metric of Eq. (11) are more stretched than in the isotropic case. Similarly, the level sets of the quality measure become more stretched as the anisotropy of the metric increases.

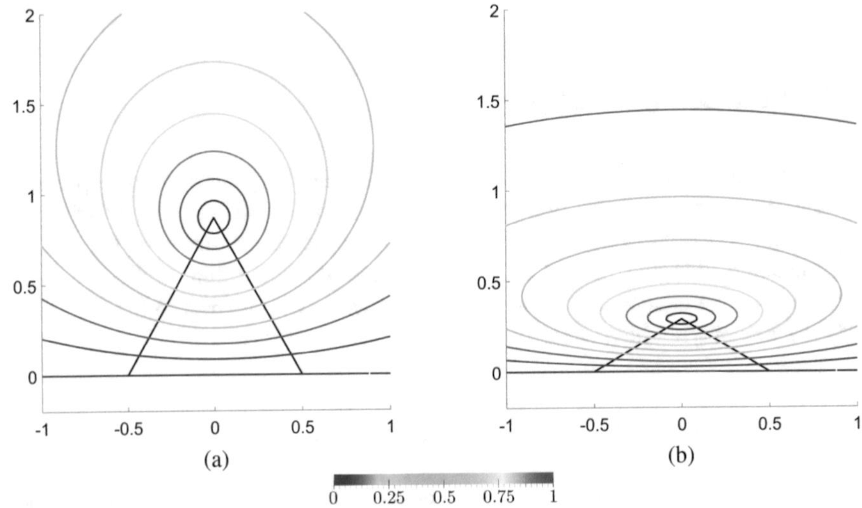

Fig. 4 Level sets for the shape quality measure: **(a)** isotropic and **(b)** anisotropic with metric presented in Eq. (11)

3.2.2 Alignment Dependence of the Shape Quality Measure

In the second test, we illustrate how the quality measure depends on the alignment between the anisotropy axes and the element. We compute the quality measure of a sequence of physical elements generated rotating the ideal element. We consider the metric presented in Eq. (11).

Let $\mathbf{R}(\theta)$ be the rotation at the origin of angle $\theta \in [0, 2\pi)$ which is given by

$$\mathbf{R}(\theta) = \begin{pmatrix} \cos\theta & -\sin\theta \\ \sin\theta & \cos\theta \end{pmatrix}.$$

We define the physical element as the ideal element rotated θ radians, with nodes $\mathbf{x}_i = \mathbf{R}(\theta) \cdot \mathbf{y}_i,\ i = 0, 1, 2$. For each θ we compute the quality of the corresponding physical element.

In Fig. 5, we plot the quality of each physical element with respect to the angle of the rotation applied to the ideal element to generate it. The angle of rotation θ is represented (in radians) in the x-axis and the quality measure is represented in the y-axis. The cases $\theta = 0, \pi/2, \pi, 3\pi/2$ and 2π are marked with a black dot and the corresponding rotations of the ideal element are shown in Fig. 5a, b, c, d and e, respectively. A rotation of the unit circle in the Euclidean space is mapped as the same ellipse in the metric space, see Fig. 5a–e. We highlight that independently of the applied rotation, the ellipse remains constant. An element with quality one must have the nodes on the ideal ellipse.

In the isotropic case, rotations of the equilateral triangle have quality 1. In the anisotropic case, when two axes correspond to different eigenvalues of the metric,

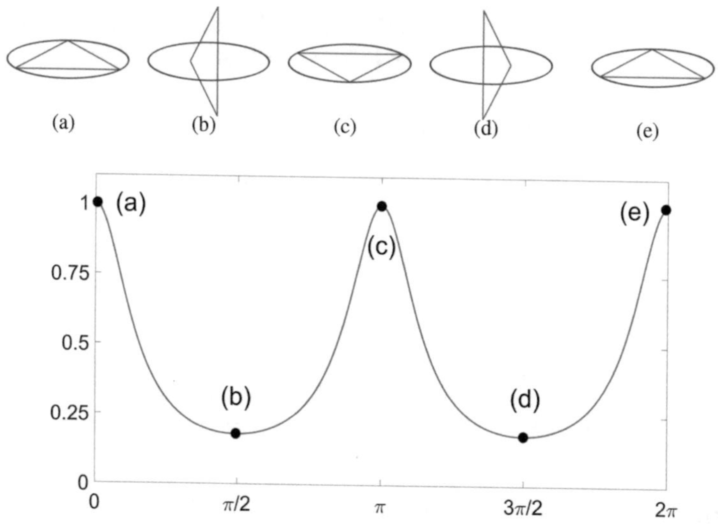

Fig. 5 Shape quality measure of physical elements which are rotations of the ideal element

we observe that the quality oscillates having two maxima and two minima in $[0, 2\pi)$. The maxima are obtained in $\theta = 0$ and $\theta = \pi$ and the minima at $\theta = \frac{\pi}{2}$ and $\theta = \frac{3\pi}{2}$. When $\theta = 0$ the rotation $\mathbf{R}(\theta)$ is the identity and $E^P = E^I$. When $\theta = \frac{\pi}{2}$ then the axes are interchanged (up to sign) and the quality at $\theta = \frac{\pi}{2}$ attains a minimum. The minima are attained when both axes are interchanged (up to sign) and the maxima are attained when the axes coincide with the eigenvectors of the metric (up to sign).

3.3 Measures for High-Order 2D Elements on Varying Metric

In Sect. 3.1, we have presented the distortion measure for linear elements equipped with a constant metric. For high-order elements, the Jacobian of the mapping is not constant. In this section, we describe the analogous formulation for high-order elements and for linear elements equipped with a non-constant metric field.

The point-wise distortion measure for an element E^P equipped with a metric $\mathbf{M}$, at a point $\mathbf{u} \in E^{\triangle}$ is defined as

$$\mathscr{N}\boldsymbol{\phi}_U(\mathbf{u}) := \eta(\mathbf{D}\boldsymbol{\phi}_U(\mathbf{u})).$$

Following Eq. (4), the distortion measure for an element E^P equipped with a metric $\mathbf{M}$ is defined as

$$\eta_{(E^P,\mathbf{M})} = \frac{\left(\int_{E^{\triangle}} \left(\mathscr{N}\boldsymbol{\phi}_U(\mathbf{u})\right)^2 \, d\mathbf{u}\right)^{1/2}}{\left(\int_{E^{\triangle}} 1 \, d\mathbf{u}\right)^{1/2}}. \tag{12}$$

Equation (12) can be written in terms of $\boldsymbol{\xi}$ on the master element. That is, the Jacobian of the map $\boldsymbol{\phi}_U$ can be written in terms of $\boldsymbol{\xi}$ as:

$$\mathbf{D}\boldsymbol{\phi}_U(\boldsymbol{\phi}_{\triangle}(\boldsymbol{\xi})) = \mathbf{F}(\boldsymbol{\phi}_P(\boldsymbol{\xi})) \cdot \mathbf{D}\boldsymbol{\phi}_P(\boldsymbol{\xi}) \cdot \left(\mathbf{D}\boldsymbol{\phi}_{\triangle}(\boldsymbol{\xi})\right)^{-1},$$

where

$$\mathbf{M}(\boldsymbol{\phi}_P(\boldsymbol{\xi})) = \mathbf{F}(\boldsymbol{\phi}_P(\boldsymbol{\xi}))^{\mathrm{T}} \cdot \mathbf{F}(\boldsymbol{\phi}_P(\boldsymbol{\xi})).$$

Then, Eq. (12) reads

$$\eta_{(E^P,\mathbf{M})} = \frac{\left(\int_{E^M} \left(\mathscr{N}\boldsymbol{\phi}_U(\boldsymbol{\phi}_{\triangle}(\boldsymbol{\xi}))\right)^2 \, |\det \mathbf{D}\boldsymbol{\phi}_{\triangle}(\boldsymbol{\xi})| \, d\boldsymbol{\xi}\right)^{1/2}}{\left(\int_{E^M} |\det \mathbf{D}\boldsymbol{\phi}_{\triangle}(\boldsymbol{\xi})| \, d\boldsymbol{\xi}\right)^{1/2}}. \tag{13}$$

Similarly to Eq. (10), the decomposition of the metric is not required:

$$\mathbf{D}\boldsymbol{\phi}_U(\boldsymbol{\phi}_{\triangle}(\boldsymbol{\xi}))^{\mathrm{T}} \cdot \mathbf{D}\boldsymbol{\phi}_U(\boldsymbol{\phi}_{\triangle}(\boldsymbol{\xi})) = \mathbf{A}(\boldsymbol{\xi})^{\mathrm{T}} \cdot \mathbf{M}(\boldsymbol{\phi}_P(\boldsymbol{\xi})) \cdot \mathbf{A}(\boldsymbol{\xi}),$$

where

$$\mathbf{A}(\boldsymbol{\xi}) := \mathbf{D}\boldsymbol{\phi}_P(\boldsymbol{\xi}) \cdot \left(\mathbf{D}\boldsymbol{\phi}_\triangle(\boldsymbol{\xi})\right)^{-1}.$$

Using the above equation we obtain the final expression on each point $\boldsymbol{\xi}$ of the master element:

$$\mathcal{N}\boldsymbol{\phi}_U(\boldsymbol{\phi}_\triangle(\boldsymbol{\xi})) = \frac{\operatorname{tr}\left(\mathbf{A}(\boldsymbol{\xi})^{\mathrm{T}} \cdot \mathbf{M}(\boldsymbol{\phi}_P(\boldsymbol{\xi})) \cdot \mathbf{A}(\boldsymbol{\xi})\right)}{d\left(\det\left(\mathbf{A}(\boldsymbol{\xi})^{\mathrm{T}} \cdot \mathbf{M}(\boldsymbol{\phi}_P(\boldsymbol{\xi})) \cdot \mathbf{A}(\boldsymbol{\xi})\right)\right)^{1/d}}. \tag{14}$$

In order to detect inverted elements [14, 29–31] we regularize the determinant in the denominator of Eq. (14) to

$$\sigma_0 = \frac{1}{2}(\sigma + |\sigma|),$$

where

$$\sigma = \det \mathbf{D}\boldsymbol{\phi}_P(\boldsymbol{\xi}) \det \mathbf{D}\boldsymbol{\phi}_\triangle(\boldsymbol{\xi})^{-1} \sqrt{\det \mathbf{M}(\boldsymbol{\phi}_P(\boldsymbol{\xi}))}.$$

Then, the point-wise regularized distortion measure of a physical element E^P is defined as

$$\mathcal{N}_0\boldsymbol{\phi}_U(\mathbf{u}) := \eta_0(\mathbf{D}\boldsymbol{\phi}_U(\mathbf{u})) := \frac{\operatorname{tr}(\mathbf{D}\boldsymbol{\phi}_U^{\mathrm{T}} \cdot \mathbf{D}\boldsymbol{\phi}_U)}{d\sigma_0^{2/d}}.$$

Finally, the regularization of the elemental distortion given in Eq. (12) is given by

$$\eta_{0,(E^P,\mathbf{M})} := \frac{\left(\int_{E^\triangle} \left(\mathcal{N}_0\boldsymbol{\phi}_U(\mathbf{u})\right)^2 \, d\mathbf{u}\right)^{1/2}}{\left(\int_{E^\triangle} 1 \, d\mathbf{u}\right)^{1/2}}$$

and its corresponding quality

$$q_{0,(E^P,\mathbf{M})} = \frac{1}{\eta_{0,(E^P,\mathbf{M})}}. \tag{15}$$

4 Results

In this section, we present several examples to illustrate the main features of the proposed quality measure. Each example features an analytic and point-wise varying metric with continuous first derivatives. For each case, we generate an initial mesh and we measure its quality according to the new measure for anisotropic elements. Next, we optimize the location of the nodes to minimize the element distortion using the framework presented in [14, 32].

The formulation to obtain the ideal element described in Sect. 3.1 can be extended to quadrilaterals in a straight-forward manner by considering the unit square $E^{\square}$ as the ideal element in the Euclidean case [28]. To illustrate the applicability of the presented measures for both triangles and quadrilaterals, each example is presented for an initial mesh featuring triangles, and an initial mesh featuring quadrilaterals.

In each case, we present a table summarizing the element quality statistics. Specifically, we show the minimum quality, the maximum quality, the mean quality and the standard deviation of the initial and optimized meshes. We highlight that in all cases, the optimized mesh increases the minimum element quality and it does not include any inverted element. The meshes resulting after the optimization are composed by elements as aligned and stretched as possible to match the target metric tensor. In all figures, the meshes are colored according to the elemental quality, see Eq. (15).

As a proof of concept, a mesh optimizer has been developed in MATLAB using the Optimization Toolbox, the PDE Toolbox and the Symbolic Math Toolbox. The MATLAB prototyping code is sequential (one execution thread), corresponds to the implementation of the method presented in this work. In all the examples, the optimization is reduced to find a minimum of a nonlinear unconstrained multi-variable function, where a trust-region algorithm is used. The stopping condition is set to reach a relative residual smaller than 10^{-8}.

4.1 Boundary Layer Mesh: Triangles and Quadrilaterals

In this example, we illustrate how our approach deals with boundary layers. We choose a metric with a geometric grow of the size in the y direction. We consider a rectangular domain $\Omega = [0, 1] \times [0, y_{\max}]$ with the metric

$$\mathbf{M}(x, y) := \begin{pmatrix} 1 & 0 \\ 0 & 1/h(y)^2 \end{pmatrix},$$

where

$$h(y) := \begin{cases} h_{\min} + \gamma y & \text{for } y < y_0 \\ 1 - (1 - h_0)e^{-r(y-y_0)} & \text{for } y \geq y_0 \end{cases}, \quad (x, y) \in \Omega,$$

being $\gamma = 1.3$ the growing ratio, $h_{\min} = 10^{-3}$ the minimum size in the y direction, and $y_{\max} = 5(1 - h_{\min})/\gamma \approx 3.8423$ the top height of the domain. The parameters $h_0 = 1 - 10^{-5}$ and $r = \gamma/(1 - h_0) = 1.3 \cdot 10^5$ are chosen to determine a differentiable metric and $y_0 = (h_0 - h_{\min})/\gamma \approx 0.7685$ is the height of the transition between the anisotropic and the isotropic regions determined by the metric. The metric attains the highest level of anisotropy at $y = 0$, with a maximum quantity $1 : 1/h_{\min}$. When $y \geq y_0$ the metric is almost isotropic.

We generate two isotropic meshes, one featuring triangles and one featuring quadrilaterals. The unstructured triangular linear mesh is composed by 230 nodes and 394 elements. The structured quadrilateral linear mesh is composed by 208 nodes and 175 elements. The initial isotropic meshes are illustrated in Fig. 6a, c, coloring the elements according to the quality for anisotropic meshes presented in Eq. (15). We observe that the elements of minimal quality (in blue) lie in the region of higher anisotropy.

For each mesh we optimize the location of the nodes to minimize the distortion, letting the inner nodes move in $\mathbb{R}^2$ and the boundary nodes slide on their corresponding boundaries. The optimized meshes are shown in Fig. 6b, d. The quality statistics of both the initial and optimized meshes are shown in Table 1. The minimum is improved after the optimization process and the standard deviation of the element qualities is reduced. We observe that in the optimized meshes the elements lying in the region $y \geq y_0$ have been enlarged vertically to allow the elements located in the region $y < y_0$ to be compressed vertically.

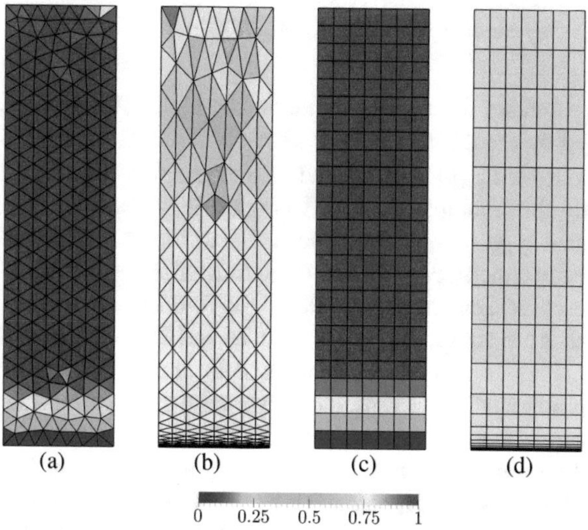

Fig. 6 Meshes for a rectangle domain featuring a vertical boundary layer metric: (**a**) initial triangular mesh, (**b**) optimized triangular mesh, (**c**) initial quadrilateral mesh, and (**d**) optimized quadrilateral mesh

Table 1 Quality statistics for boundary layer case

Figure	Mesh	Element type	Min	Max	Mean	Std. dev.
6a	Initial	Triangles	0.0068	1.0000	0.9255	0.2076
6b	Optimized	Triangles	0.5754	0.9046	0.7411	0.0459
6c	Initial	Quadrilaterals	0.0168	0.9733	0.9241	0.2150
6d	Optimized	Quadrilaterals	0.7097	0.7229	0.7182	0.0063

4.2 *Curved Anisotropy: Curved Quartic Quadrilaterals and Triangles*

In this example, we illustrate the behavior for high-order elements of the proposed quality measure and of the optimization framework [14, 32] applied to the new developed quality measure. We consider the quadrilateral domain $\Omega := [0, 1]^2$ and the metric $\mathbf{M} := \nabla\boldsymbol{\varphi}^{\mathrm{T}} \cdot \nabla\boldsymbol{\varphi}$ induced by the surface $\boldsymbol{\varphi}$:

$$\boldsymbol{\varphi}(\mathbf{x}) := (x, y, z(x, y)), \quad z(x, y) := \tanh\left(f \cdot h(x, y)\right),$$

where

$$h(x, y) = \frac{1}{\sqrt{4\pi^2 + 100}} \cdot (10y - \cos(2\pi x) - 5), \quad (x, y) \in \Omega, \quad f = 5.$$

The metric is extracted from [33] and attains the highest level of anisotropy at the curve described by the points $(x, y) \in \Omega$ such that $h(x, y) = 0$. The anisotropy ratio of this metric is $1 : \sqrt{1 + |\nabla z|^2}$ and its maximum is given by $1 : \sqrt{1 + f^2}$.

We illustrate this example both for triangles, Fig. 7, and quadrilaterals, Fig. 8. For each element type, we generate two different initial straight-sided high-order meshes: an isotropic mesh, Figs. 7a and 8a, and an anisotropic mesh, Figs. 7c and 8c. The quadrilateral meshes are composed by 1681 nodes and 100 elements and the triangular meshes are composed by 1905 nodes and 228 elements. The anisotropic initial meshes, Figs. 7c and 8c, are generated by means of optimizing the linear isotropic mesh and next, increasing the polynomial degree of the linear optimized mesh. We highlight that the two straight-sided anisotropic meshes are better initial configurations to curve the high-order mesh according to the metric. Although they are not optimal meshes of polynomial degree 4, they have been obtained from an optimal linear mesh. All the meshes are colored in the figures according to the elemental quality measure presented in Eq. (15).

The optimized meshes for the triangle case are illustrated in Fig. 7b, d, and for the quadrilateral case are illustrated in Fig. 8b, d. In the optimized meshes the elements away from the anisotropic region are enlarged vertically whereas the elements lying in the anisotropic region are compressed. For the triangle and quadrilateral cases, the two meshes optimized from the two different initial configurations are equal up to the chosen optimization tolerance. The quality statistics of both the initial and

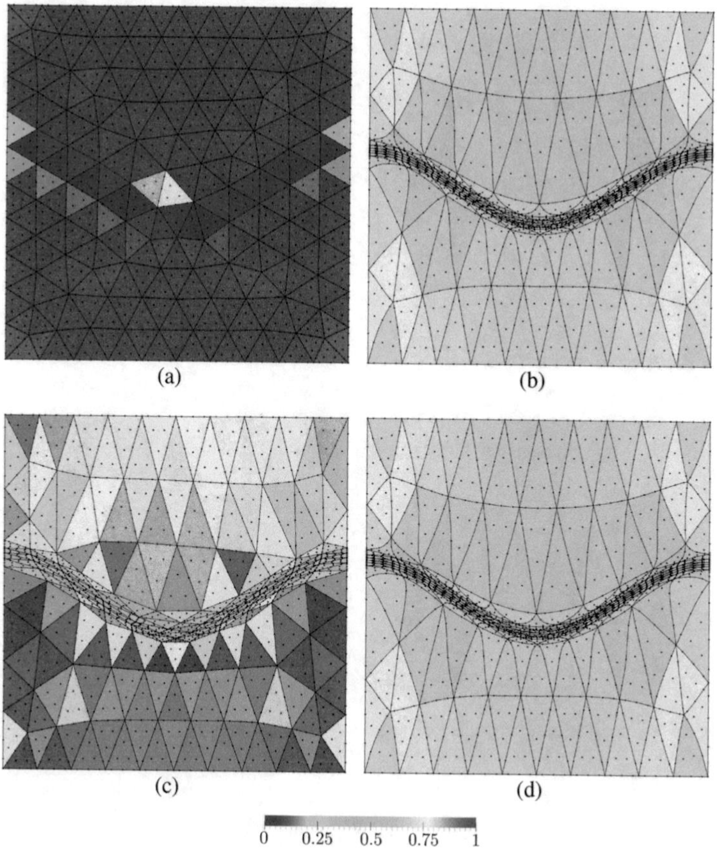

Fig. 7 Triangle meshes of polynomial degree 4: (**a**) initial straight-sided isotropic mesh, (**b**) optimized mesh from initial configuration (**a**), (**c**) initial straight-sided anisotropic mesh, (**d**) optimized mesh from initial configuration (**c**)

optimized meshes are shown in Table 2. In all the optimized meshes the minimum is improved and the standard deviation of the element qualities is reduced when compared with the initial configuration.

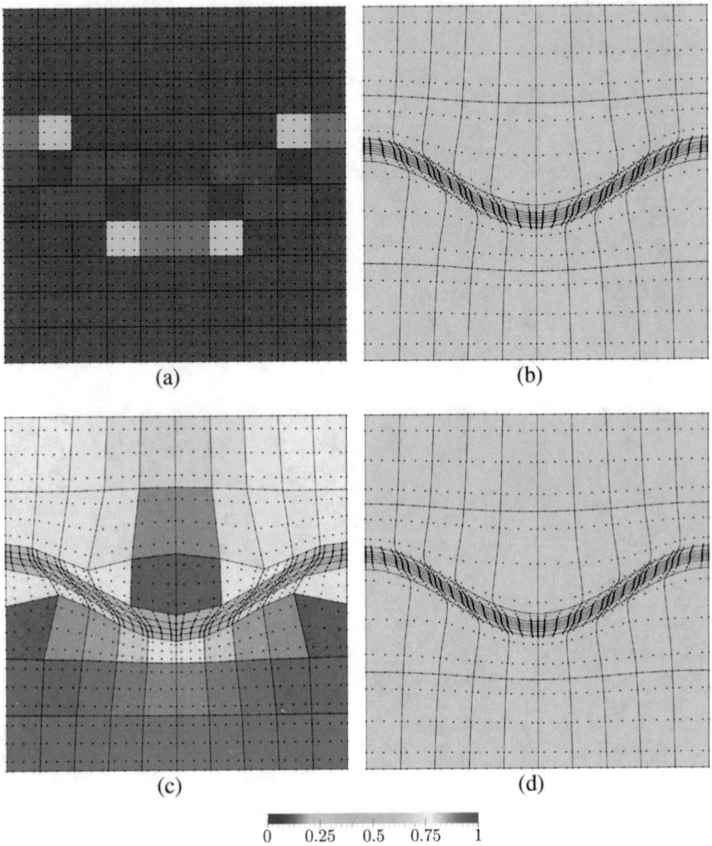

Fig. 8 Quadrilateral meshes of polynomial degree 4: (**a**) initial straight-sided isotropic mesh, (**b**) optimized mesh from initial configuration (**a**), (**c**) initial straight-sided anisotropic mesh, (**d**) optimized mesh from initial configuration (**c**)

Table 2 Quality Statistics of the meshes of polynomial degree 4 for curved anisotropy

Figure	Mesh	Element type	Min	Max	Mean	Std. dev.
7a	Initial	Triangles	0.0569	1.0000	0.8849	0.2824
7b	Optimized	Triangles	0.5381	0.7626	0.6570	0.0435
7c	Initial	Triangles	0.1726	0.9905	0.6739	0.2308
7d	Optimized	Triangles	0.5381	0.7626	0.6570	0.0435
8a	Initial	Quadrilaterals	0.0631	1.0000	0.8408	0.3324
8b	Optimized	Quadrilaterals	0.6384	0.6696	0.6506	0.0130
8c	Initial	Quadrilaterals	0.1852	0.9405	0.6631	0.2417
8d	Optimized	Quadrilaterals	0.6384	0.6696	0.6506	0.0130

5 Concluding Remarks and Future Work

In this work, we have presented a new definition of distortion (quality) measures for linear and high-order planar anisotropic meshes equipped with a point-wise metric. The proposed quality measures the alignment and stretching of the elements according to the given metric. In addition, it is valid for any interpolation degree and allow to detect the validity of a high-order element equipped with a metric. To assess the reliability of the technique, we have first analyzed the behavior of the measure for linear triangles equipped with a constant metric. The tests show that for a given metric the obtained quality measure detects invalid and low-quality configurations and, the alignments and stretching described by the metric.

The defined distortion measure is applied to curve linear meshes to improve the node configuration according to the desired metric. To perform the distortion optimization we use the framework for high-order optimization presented in [14, 32]. The numerical examples show optimized meshes with an improved alignment and stretching according to the metric. This improvement leads in all the cases to an increase of the minimum elemental mesh quality and a reduction of the standard deviation between the different element qualities.

Our long term goal is to extend the quality measure for 3D anisotropic meshes. In addition, the quality measure developed in this work is devised to quantify the alignment and the stretching of the mesh according to the target metric. Thus, we would like to extend the proposed measure to also quantify the mesh sizing.

Acknowledgements This project has received funding from the European Research Council (ERC) under the European Union's Horizon 2020 research and innovation programme under grant agreement No 715546. This work has also received funding from the Generalitat de Catalunya under grant number 2017 SGR 1731. The work of X. Roca has been partially supported by the Spanish Ministerio de Economía y Competitividad under the personal grant agreement RYC-2015-01633.

References

1. F. Alauzet, A. Loseille, A decade of progress on anisotropic mesh adaptation for computational fluid dynamics. Comput. Aided Des. **72**(1), 13–39 (2016)
2. C. Gruau, T. Coupez, 3d Tetrahedral, unstructured and anisotropic mesh generation with adaptation to natural and multidomain metric. Comput. Methods Appl. Mech. Eng. **194**(48–49), 4951–4976 (2005)
3. P. Frey, F. Alauzet, Anisotropic mesh adaptation for cfd computations. Comput. Methods Appl. Mech. Eng. **194**(48–49), 5068–5082 (2005)
4. M. Yano, D.L. Darmofal, An optimization-based framework for anisotropic simplex mesh adaptation. J. Comput. Phys. **231**(22), 7626–7649 (2012)
5. A. Loseille, F. Alauzet, Continuous mesh framework part i: well-posed continuous interpolation error. SIAM J. Numer. Anal. **49**(1), 38–60 (2011)

6. T. Coupez, L. Silva, E. Hachem, Implicit boundary and adaptive anisotropic meshing, in *New Challenges in Grid Generation and Adaptivity for Scientific Computing* (Springer, Cham, 2015), pp. 1–18
7. F. Hecht, Bamg: bidimensional anisotropic mesh generator. User Guide. INRIA, Rocquencourt (1998)
8. A. Loseille, R. Lohner, Cavity-based operators for mesh adaptation, in *51st AIAA Aerospace Sciences Meeting including the New Horizons Forum and Aerospace Exposition*, 2013, p. 152
9. A. Loseille, Metric-orthogonal anisotropic mesh generation. Procedia Eng. **82**, 403–415 (2014)
10. O. Sahni, X.J. Luo, K.E. Jansen, M.S. Shephard, Curved boundary layer meshing for adaptive viscous flow simulations. Finite Elem. Anal. Des. **46**(1), 132–139 (2010)
11. P.-O. Persson, J. Peraire, Curved mesh generation and mesh refinement using lagrangian solid mechanics, in *Proceeding of 47th AIAA*, 2009
12. T. Toulorge, C. Geuzaine, J.-F. Remacle, J. Lambrechts, Robust untangling of curvilinear meshes. J. Comput. Phys. **254**, 8–26 (2013)
13. A. Gargallo-Peiró, X. Roca, J. Peraire, J. Sarrate, Inserting curved boundary layers for viscous flow simulation with high-order tetrahedra, in *Research Notes, 22nd International Meshing Roundtable* (Springer International Publishing, Cham, 2013)
14. A. Gargallo-Peiró, X. Roca, J. Peraire, J. Sarrate, Optimization of a regularized distortion measure to generate curved high-order unstructured tetrahedral meshes. Int. J. Numer. Methods Eng. **103**, 342–363 (2015)
15. D. Moxey, M.D. Green, S.J. Sherwin, J. Peiró, An isoparametric approach to high-order curvilinear boundary-layer meshing. Comput. Methods Appl. Mech. Eng. **283**, 636–650 (2015)
16. M. Fortunato, P-O. Persson, High-order unstructured curved mesh generation using the winslow equations. J. Comput. Phys. **307**, 1–14 (2016)
17. A. Gargallo-Peiró, G. Houzeaux, X. Roca, Subdividing triangular and quadrilateral meshes in parallel to approximate curved geometries. Procedia Eng. **203**, 310–322 (2017)
18. D. Moxey, D. Ekelschot, Ü. Keskin, S.J. Sherwin, J. Peiró, High-order curvilinear meshing using a thermo-elastic analogy. Comput. Aided Des. **72**, 130–139 (2016)
19. O. Coulaud, A. Loseille, Very high order anisotropic metric-based mesh adaptation in 3d. Procedia Eng. **163**, 353–365 (2016); Proceedings of the 25th International Meshing Roundtable
20. T. Coupez, On a basis framework for high order anisotropic mesh adaptation, in *Research Note of the 26th International Meshing Roundtable* (2017)
21. J. Marcon, M. Turner, D. Moxey, S.J. Sherwin, J. Peiró, A variational approach to high-order r-adaptation. in *IMR26*, 2017
22. W. Huang, Y. Wang, Anisotropic mesh quality measures and adaptation for polygonal meshes (2015), https://arxiv.org/abs/1507.08243
23. W. Huang, Measuring mesh qualities and applications to variational mesh adaptation. SIAM J. Sci. Comput. **26**(5), 1643–1666 (2005)
24. X. Roca, A. Gargallo-Peiró, J. Sarrate, Defining quality measures for high-order planar triangles and curved mesh generation, in *Proceedings of 20th International Meshing Roundtable* (Springer International Publishing, Cham, 2012), pp. 365–383
25. P.M. Knupp, Algebraic mesh quality metrics. SIAM J. Numer. Anal. **23**(1), 193–218 (2001)
26. A. Gargallo-Peiró, X. Roca, J. Peraire, J. Sarrate, Distortion and quality measures for validating and generating high-order tetrahedral meshes. Eng. Comput. **31**, 423–437 (2015)
27. A. Gargallo-Peiró, X. Roca, J. Peraire, J. Sarrate, A distortion measure to validate and generate curved high-order meshes on CAD surfaces with independence of parameterization. Int. J. Numer. Methods Eng. **106**(13), 1100–1130 (2015)
28. A. Gargallo-Peiró, E. Ruiz-Gironés, X. Roca, J. Sarrate, On curving high-order hexahedral meshes, in *24th International Meshing Roundtable (IMR24), October 11–14, 2014, Austin, TX* (Elsevier, Amsterdam, 2015), pp. 1–5
29. L.V. Branets, V.A. Garanzha, Distortion measure of trilinear mapping. application to 3-d grid generation. Numer. Linear Algebra Appl. **9**(6–7), 511–526 (2002)
30. E.J. López, N.M. Nigro, M.A. Storti, Simultaneous untangling and smoothing of moving grids. Int. J. Numer. Methods Eng. **76**(7), 994–1019 (2008)

31. J.M. Escobar, E. Rodríguez, R. Montenegro, G. Montero, J.M. González-Yuste. Simultaneous untangling and smoothing of tetrahedral meshes. Comput. Methods Appl. Mech. Eng. **192**(25), 2775–2787 (2003)
32. A. Gargallo-Peiró, Validation and generation of curved meshes for high-order unstructured methods. PhD thesis, Universitat Politècnica de Catalunya, 2014
33. P.C. Caplan, R. Haimes, D.L. Darmofal, M.C. Galbraith, Anisotropic geometry-conforming d-simplicial meshing via isometric embeddings. Procedia Eng. **203**, 141–153 (2017); 26th International Meshing Roundtable.

Curvilinear Mesh Adaptation

Ruili Zhang, Amaury Johnen, and Jean-François Remacle

Abstract This paper aims at addressing the following issue. Assume a unit square: $\Omega = \{(x^1, x^2) \in [0, 1] \times [0, 1]\}$ and a Riemannian metric $g_{ij}(x^1, x^2)$ defined on U. Assume a mesh $\mathcal{T}$ of U that consist in non overlapping valid quadratic triangles that are potentially curved. Is it possible to build a unit quadratic mesh of U i.e. a mesh that has quasi-unit curvilinear edges and quasi-unit curvilinear triangles? This paper aims at providing an embryo of solution to the problem of curvilinear mesh adaptation. The method that is proposed is based on standard differential geometry concepts. At first, the concept of geodesics in Riemannian spaces is quickly presented: the geodesic between two points as well as the unit geodesic starting at a given point with a given direction are the two main tools that allow us to address our issue. Our mesh generation procedure is done in two steps. At first, points are distributed in the unit square U in a frontal fashion, ensuring that two points are never too close to each other in the geodesic sense. Then, a simple isotropic Delaunay triangulation of those points is created. Curvilinear edge swaps as then performed in order to build the unit mesh. Notions of curvilinear mesh quality is defined as well that allow to drive the edge swapping procedure. Examples of curvilinear unit meshes are finally presented.

1 Introduction

There is a growing consensus that state of the art Finite Volume and Finite Element technologies require, and will continue to require too extensive computational resources to provide the necessary resolution, even at the rate with which computational power increases. The requirement for high resolution naturally leads us to consider methods with higher order of grid convergence than the classical (formal)

R. Zhang · A. Johnen · J.-F. Remacle (✉)
Université catholique de Louvain, Louvain-la-Neuve, Belgium
e-mail: ruili.zhang@uclouvain.be; amaury.johnen@uclouvain.be; jean-francois.remacle@uclouvain.be

X. Roca, A. Loseille (eds.), *27th International Meshing Roundtable*, Lecture Notes in Computational Science and Engineering 127,
https://doi.org/10.1007/978-3-030-13992-6_4

2nd order provided by most industrial grade codes. This indicates that higher-order discretization methods will replace at some point the finite volume/element solvers of today, at least for part of their applications. The development of high-order numerical technologies for CFD is underway for many years now. For example, Discontinuous Galerkin methods (DGM) have been largely studied in the literature, initially in a quite theoretical context [4], and now in the application point of view [9]. In many contributions, it is shown that the accuracy of the method strongly depends of the accuracy of the geometrical discretization [3]. In other words, the following question is raised: yes we have the high order methods, but how do we get the meshes?

Several research teams are now actively working in the domain of curvilinear meshing. This new subject is considered as crucial for the future of CFD [13] and large fundings have been given to some brilliant researchers to allow innovation in the domain (our colleague Xevi Roca has recently obtained an ERC starting grant on the subject).

A good research project should ideally be summarized as a simple yet fundamental question. It is very much the case here. Assume a unit square

$$\Omega = \{(x^1, x^2) \in [0, 1] \times [0, 1]\}$$

and a smooth function $f(x^1, x^2)$ defined on the square. Consider a mesh $\mathcal{T}$ made of P^2 triangles that exactly covers the square. How can we compute the mesh $\mathcal{T}$ that minimizes the discretization error $\|\Pi f - f\|_{\Omega}$. Here, Π is the so-called Clément interpolation of f on the mesh [5]. This problem is the problem of curvilinear mesh adaptation. The solution of that problem requires to address three main open questions:

1. What is the geometrical structure of the discretization error in the P^2 case?
2. How can we relate this structure with the geometry/shape of a P^2 triangle?
3. How can we build a mesh made of optimal P^2 triangles?

The first question is related to error estimation and we will not deal with it in this paper.

In this first attempt, we will start with a simpler statement. A Riemannian metric field $g_{ij}(x^1, x^2)$ is defined on the unit square. This metric field is supposed to be the result of the error estimation. Our aim is thus to build a unit P^2 mesh with respect to that metric. A discrete mesh $\mathcal{T}$ of a domain Ω is a unit mesh with respect to Riemannian metric space $\mathbf{g}(x^1, x^2)$ if all its elements are quasi-unit. More specifically, a curvilinear triangle t defined by its list of edges e_i, $i = 1, 2, 3$ is said to be quasi-unit if all its adimensional edges lengths $\mathcal{L}_{e_i} \in [0.7, 1.4]$.[1] Generating unit straight-sided meshes is a problem that has been largely studied, both in the theoretical point of view and on the application point of view [6]. Here, our aim

[1] This range is not arbitrary. When a long edge of size 1.4 is split, it should not become a short edge. Other authors choose $[\sqrt{2}/2, \sqrt{2}]$.

is to allow edges to become curved, leading to unit meshes that would potentially contain way less triangles.

The paper is structured as follows. Our mesh generation technique essentially relies on the computation of the shortest parabola between two points and on a unit-size parabola starting in a given direction. In Sect. 2, standard notions of geodesics in Riemann spaces are briefly exposed. Algorithms that compute geodesic parabolas are explained as well.

The mesh generation approach that we advocate is in two steps. We first generate the points in a frontal fashion [1]. In that process, we ensure that (1) two points $\mathbf{x}_i$ and $\mathbf{x}_j$ are never too close to each other and (2) that there exist four points $\mathbf{x}_{ij},\ j = 1,\ldots,4$ in the vicinity of each point $\mathbf{x}_i$ that are not too far to $\mathbf{x}_i$ i.e. that can form edges in the prescribed range [0.7, 1.4].

Then, points are connected in a very standard "isotropic" fashion. The mesh is subsequently modified using curvilinear edge swaps in order to form the desired unit mesh. A curvilinear mesh quality criterion is proposed that allow to drive the edge swapping process.

In Sect. 5, some unit meshes are presented that adapt to analytical metric fields.

In what follows, we illustrate concepts of unit circle and geodesics using the following *toy metric tensor*:

$$\mathbf{g}(x^1,x^2) = \begin{pmatrix} g_{11} & g_{12} \\ g_{12} & g_{22} \end{pmatrix} = \begin{pmatrix} \cos\theta & \sin\theta \\ -\sin\theta & \cos\theta \end{pmatrix} \begin{pmatrix} \frac{1}{l_{\min}^2} & 0 \\ 0 & \frac{1}{l_{\max}^2} \end{pmatrix} \begin{pmatrix} \cos\theta & -\sin\theta \\ \sin\theta & \cos\theta \end{pmatrix} \quad (1)$$

with

$$\mathbf{x} = \{x^1, x^2\},\quad r = \|\mathbf{x}\|,\quad \theta = \arctan(x^2/x^1),$$

$$l_{\min} = \epsilon + l_{\max}(1 - \exp(-((r - r_0)/h)^2).$$

2 Geodesics

In a Riemannian space, the length of curve $\mathcal{C}$ is computed as

$$\mathcal{L}_{\mathcal{C}} = \int_{\mathcal{C}} \sqrt{g_{ij}dx^i dx^j}$$

The geodesic between two points $\mathbf{x}_1$ and $\mathbf{x}_s$ is the shortest path $\mathcal{C}$ between those two points. It is possible to compute geodesics by solving a set of coupled ordinary differential equation (ODE). Defining the so-called Christoffel symbols

$$\Gamma^i{}_{kl} = \tfrac{1}{2} g_{im}^{-1}\left(\frac{\partial g_{mk}}{\partial x^l} + \frac{\partial g_{ml}}{\partial x^k} - \frac{\partial g_{kl}}{\partial x^m}\right) = \tfrac{1}{2} g_{im}^{-1}(g_{mk,l} + g_{ml,k} - g_{kl,m}),$$

the ODE's of geodesics are written:

$$\frac{d^2x^i}{dt^2} + \Gamma^i_{jk}\frac{dx^j}{dt}\frac{dx^k}{dt} = 0. \tag{2}$$

2.1 Geodesics and Unit Circle

Assume a point $\mathbf{x} = \{x^1, x^2\}$ and an initial velocity $\dot{\mathbf{x}} = \{\cos(\alpha), \sin(\alpha)\}$. Equation (2) allows to compute geodesic $\mathcal{C}(\alpha)$ which is the geodesic passing by $\mathbf{x}$ and which tangent vector at $\mathbf{x}$ is $\dot{\mathbf{x}}$. In this work, a simple RK2 scheme is used to integrate Eq. (2) explicitly.

The unit circle centered at $\mathbf{x}$ is the set of end-points of all geodesics $\mathcal{C}(\alpha)$ with $\mathcal{L}_{\mathcal{C}(\alpha)} = 1$ starting at point $\mathbf{x}$. Figure 1 shows unit circles with different centers for the toy metric (1).

The tangent plane assumption that is usually made in anisotropic meshing theory [6] leads to unit circles that are ellipsis and where geodesic remain straight lines. Here, geodesics have a *banana shape* that differes very much with an ellipsis. On Fig. 2, geodesics corresponding to the principal directions of the metric at point

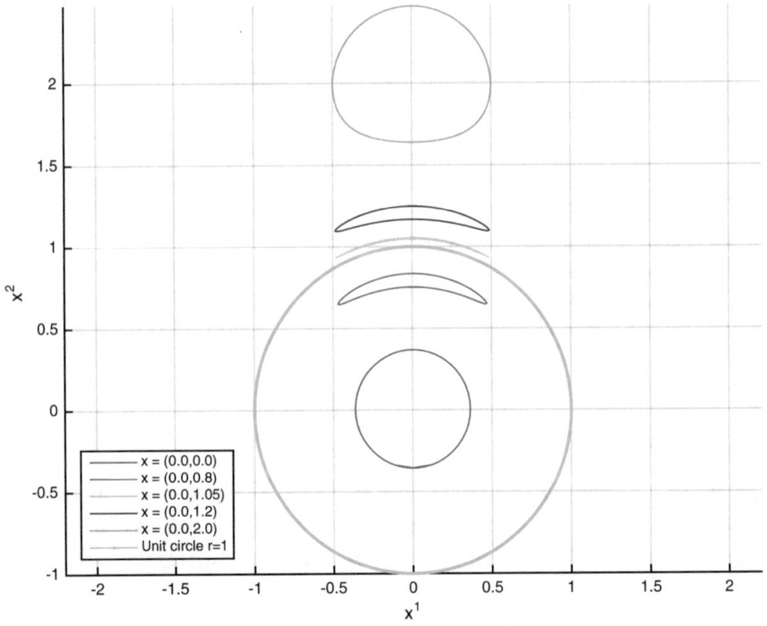

Fig. 1 Unit circles at different centers for the toy metric (1)

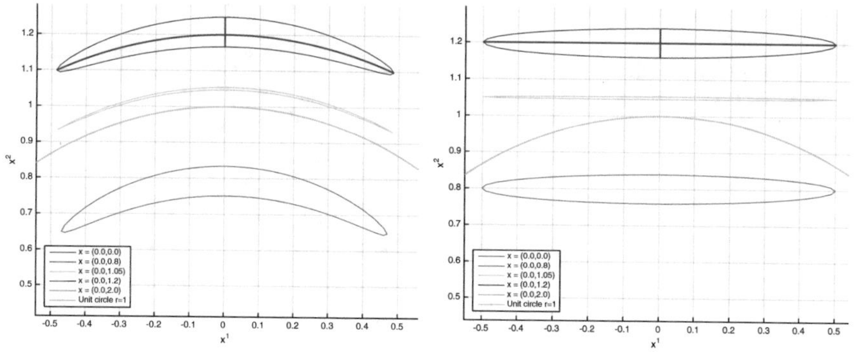

Fig. 2 Unit circles at different centers for the toy metric (1). Left figure shows circles computed using the exact geodesics while right figure assumes a constant metric (tangent plane approximation)

$\{x^1, x^2\} = \{0, 1.2\}$ are drawn, both for true geodesics (left) and in the case of the tangent plane approximation (right).

2.2 *Geodesic Curve Between Two Points*

Shooting a geodesic from a point $\mathbf{x}$ with velocity $\dot{\mathbf{x}}$ can be solved by integrating the geodesic ODE (2) explicitly in t. Now, consider two points $\mathbf{x}_1$ and $\mathbf{x}_2$. If our aim is to find a geodesic between those points, we need to integrate the geodesic ODE (2) implicitly. In this work, we choose to simplify that procedure. Quadratic meshes are considered in this paper, which means that “mesh geodesics” are parabola. In order to simplify our formulation even more, we assume that the mid point $\mathbf{x}_{12}$ on the geodesic parabola $\mathcal{C}_{12}$ between $\mathbf{x}_1$ and $\mathbf{x}_2$ is located on the orthogonal bissector of segment $\mathbf{x}_1\mathbf{x}_2$ as:

$$\mathbf{x}_{12} = \frac{1}{2}(\mathbf{x}_1 + \mathbf{x}_2) + \alpha(\mathbf{x}_2 - \mathbf{x}_1) \times \mathbf{e}_3 \ , \quad \alpha \in R.$$

Parametric equation of this geodesic parabola is given by:

$$\begin{aligned} \mathcal{C}_{12} \equiv \mathbf{x}(t, \alpha) &= (1-t)(1-2t)\mathbf{x}_1 + t(2t-1)\mathbf{x}_2 + 4t(1-t)\mathbf{x}_3(\alpha) \\ &= \mathbf{x}_1 + t(\mathbf{x}_2 - \mathbf{x}_1) + 4t(1-t)\alpha(\mathbf{x}_2 - \mathbf{x}_1) \times \mathbf{e}_3. \end{aligned}$$

Tangent vector at t is computed as,

$$\dot{\mathbf{x}}(t, \alpha) = (\mathbf{x}_2 - \mathbf{x}_1) + (4 - 8t)\alpha(\mathbf{x}_2 - \mathbf{x}_1) \times \mathbf{e}_3.$$

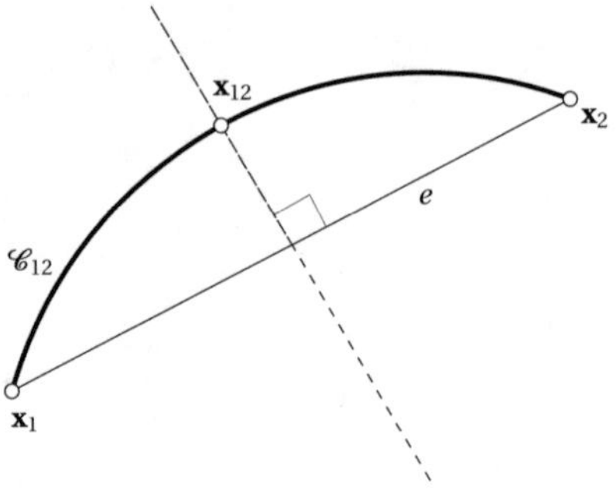

Fig. 3 Midpoint $\mathbf{x}_{12}$ of a parabola situated on the orthogonal bissector of the straight line $\mathbf{x}_1\mathbf{x}_2$

So, point $\mathbf{x}_{12}$ is computed by minimizing the length of that parabola

$$\mathbf{x}_{12} = arg \min_{\alpha} \mathcal{L}_{\mathcal{C}_{12}} = \int_0^1 \sqrt{\dot{x}^i \dot{x}^j \, g_{ij}(x^i, x^j)} \, dt \tag{3}$$

using a golden section algorithm (Fig. 3).

3 Generation of Points

Assume a $1D$ mesh of the unit square that is compatible with the metric field $g_{ij}(\mathbf{x})$ i.e. where every boundary mesh edges is quasi-unit. The main idea here is to proceed as we did for generating hex dominant meshes [1]. The point sampling algorithm is presented in Algorithm 1.

Algorithm 1 Point sampling for the generation of a unit curvilinear mesh

Input: A LIFO queue Q is initialized containing all mesh vertices of the 1D mesh and a metric field $g_{ij}(\mathbf{x})$.
Output: A list L of accepted vertices
while Q is not empty **do**
  $\mathbf{x} \leftarrow Q$: pop vertex $\mathbf{x}$ at the begin of the queue
  Compute $\mathbf{g}(\mathbf{x})$ as well as its eigenvectors $\mathbf{v}_1$ and $\mathbf{v}_2$ at point $\mathbf{x}$
  Four tentative points $\mathbf{x}_1, \mathbf{x}_2, \mathbf{x}_3, \mathbf{x}_4$ are computed at a geodesic distance equal to 1 in the four directions $\mathbf{v}_1, -\mathbf{v}_1, \mathbf{v}_2, -\mathbf{v}_2$ solving Eq. (2).
  for $i = 1, \ldots, 4$ **do**
    if $\mathbf{x}_i$ is not too close to any accepted point in L **then**
      Add $\mathbf{x}_i$ at the end of the queue Q
    end if
  end for
  $L \leftarrow L + \mathbf{x}$: add $\mathbf{x}$ in the list L of accepted vertices
end while

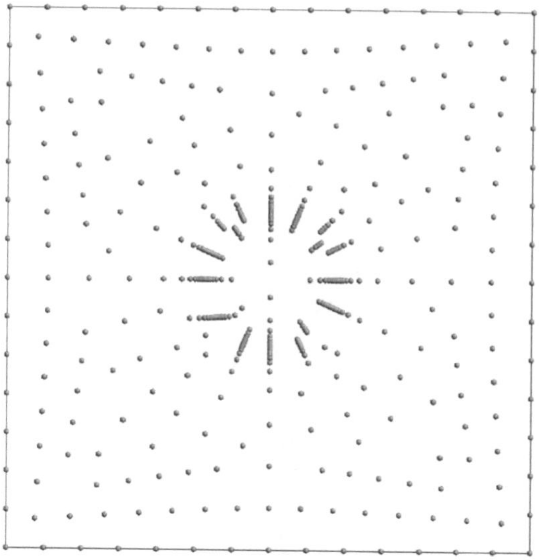

Fig. 4 Sampling of points using toy metric (1) with parameters $\epsilon = 0.01$, $h = 1/\sqrt{10}$, $r_0 = 0.5$ and $l_{\max} = 0.3$. The square is of size 4×4 and is centered at $(x^1, x^2) = (0, 0)$

Algorithm 1 ensures that there exists no point in the mesh that are too close to another while, on the other hand, ensuring that there exist 4 points that are sufficiently close to any point of the mesh. Principal directions of the metric field $\mathbf{v}_1$ and $\mathbf{v}_2$ are used as a "direction field". This is an arbitrary choice. Yet, it has the advantage in most cases to generate meshes that are more structured.

Ensuring that two points are not too close is done using a RTree [2] spatial search structure. The distance between two points is computed as the shortest parabola in the given metric (see Eq. (3)). Our sampling algorithm applied to the toy metric (1) provides the set of points of Fig. 4.

4 Generation of Triangles

The set of points optimally sampled are then triangulated using an off the shelf constrained Delaunay triangulator such as Gmsh [7] or Triangle [11]. We see on Fig. 5 that isotropic straight sided elements are not suited for the proposed metric. Here, local mesh modifications [10] will be used to align the mesh with the desired metric. We do not move the points that are optimally sampled. Only edge swaps will be performed, yet in a non usual fashion.

High order points are initially placed on every edge of the straight sided mesh using Eq. (3). Assume two triangles $t_1(\mathbf{x}_1, \mathbf{x}_2, \mathbf{x}_4)$ and $t_2(\mathbf{x}_2, \mathbf{x}_1, \mathbf{x}_3)$ that share an edge e (see Fig. 6). Triangles t_1 and t_2 are possibly curvilinear (as in the figure) and we aim at evaluating the opportunity of replacing edge e by edge e' (edge e' is the geodesic between $\mathbf{x}_3$ and $\mathbf{x}_4$). Two indicators will help us to decide whether an edge swap should be performed:

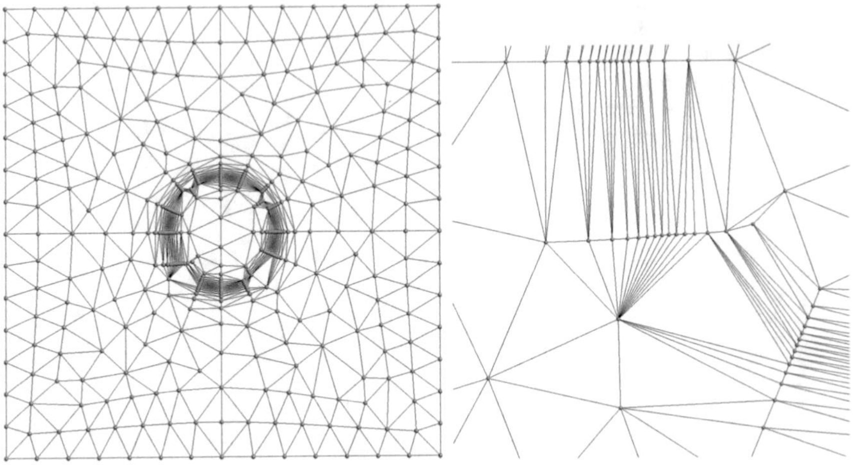

Fig. 5 Constrained Delaunay mesh constructed using sampled points of Fig. 4. The triangulation is straight sided. It has been done using no specific metric and is thus clearly not adapted

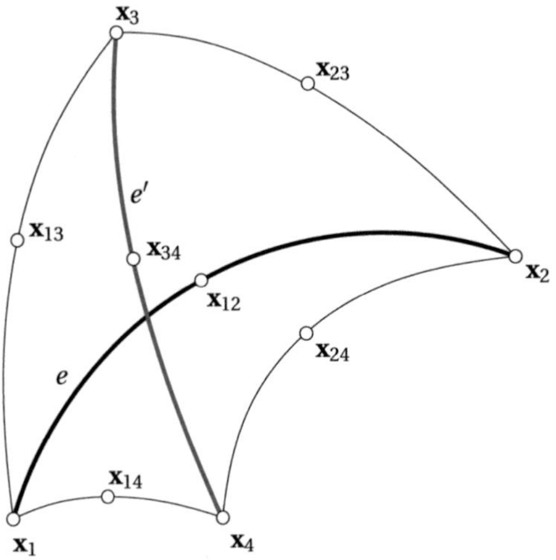

Fig. 6 Curvilinear edge swap

- The new curvilinear triangles $t_1'(\mathbf{x}_4, \mathbf{x}_3, \mathbf{x}_2)$ and $t_1'(\mathbf{x}_3, \mathbf{x}_4, \mathbf{x}_1)$ have to be both valid. The validity criterion that is used is based on robust estimates that have been developed in [8]. In short, for t_1', determinants of jacobians J_4,J_3,J_2,J_{43},J_{32},J_{24} are computed at its 6 nodes. A sufficient condition for triangle t_1' to be valid is

$$J_4 > 0\,,\ J_3 > 0\,,\ J_2 > 0\,,\ 4J_{43} > J_3 + J_4\,,\ 4J_{32} > J_3 + J_2\,,\ 4J_{24} > J_2 + J_4.$$

- The quality of the mesh has to be improved by the swap:

$$\min(q_{\mathbf{g}}(t_1), q_{\mathbf{g}}(t_2)) < \min(q_{\mathbf{g}}(t_1'), q_{\mathbf{g}}(t_2'))$$

  where $q_{\mathbf{g}}(t)$ is a curvilinear quality measure of triangle t with respect to metric field $\mathbf{g}$.

The quality measure that is used here is a direct extension to standard quality measures defined in [12]. We define

$$q_{\mathbf{g}}(t) = \frac{12}{\sqrt{3}} \frac{\int_t \sqrt{\det \mathbf{g}}\, d\mathbf{x}}{\mathcal{L}_{e_1}^2 + \mathcal{L}_{e_2}^2 + \mathcal{L}_{e_3}^2} \tag{4}$$

where e_1, e_2 and e_3 are the three edges of t, $\mathcal{L}_e$ is the length of e with respect to the metric. Note that triangle inequality is not necessary verified in Riemannian metrics i.e. $\mathcal{L}_{e_1} \leq \mathcal{L}_{e_2} + \mathcal{L}_{e_3}$ is not necessary true. In consequence, quality measure $q_{\mathbf{g}}(t)$ may be larger than one. Edges are swapped until a stable configuration is found.

5 Examples

5.1 *Unit Mesh for the Toy Metric*

Figure 7 present meshes for the toy metric (1). All triangles are valid by construction.

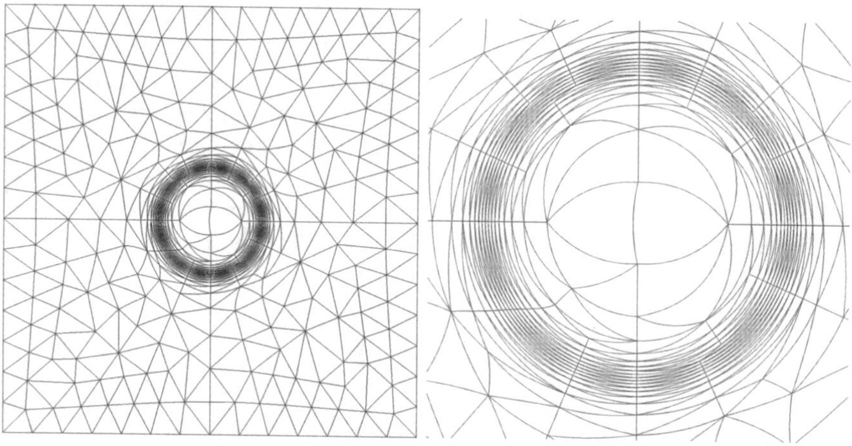

Fig. 7 Curvilinear mesh of the unit square using the toy metric

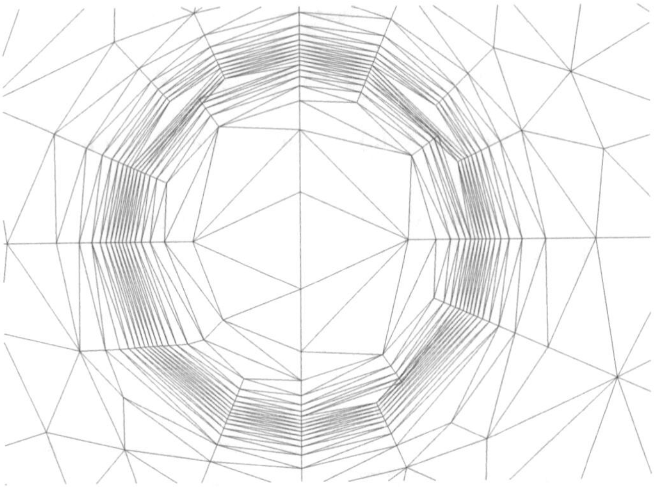

Fig. 8 This figure depicts the corresponding P^1 straight sided version of the curvilinear mesh of Fig. 7. A large amount of the P^1 triangles are invalid while every single P^2 triangle of Fig. 7 is valid

Note here that the corresponding P^1 mesh of our P^2 mesh is totally invalid. It is indeed not possible to generate a P^1 mesh and curving it afterwards without doing curvilinear local mesh modifications (see Fig. 8).

In the sampling process, points are placed along true geodesics while edges of the mesh are parabola. Parabola that have the same endpoints as true unit geodesics could potentially be longer than 1. Even though the number of long edges that are the consequence of this approximation is quite small, this discrepancy could potentially become annoying. We have addressed that issue by reducing the size of geodesics with the aim at producing parabolas that are of the right unit size. With this fix, edge lengths are in the range [0.701, 1.66] which is very close to the optimal range (see Fig. 9). Note that no short edges can exist in the mesh by construction. Long edges are due to the inability of the swapping process to connect points that are close enough without generating invalid P^2 triangles. In further work, other mesh optimizations will be put into place that could enhance even further the quality of the P^2 meshes. Quality measures (4) are also depicted in Fig. 9.

5.2 *Intersection of Three Toy Metrics*

This example consist in placing three toy metrics M_1, M_2 and M_3 in the 4×4 square, centered at different locations with different mesh sizes and intersecting them [6]:

$$M = M_1 \cap M_2 \cap M_3.$$

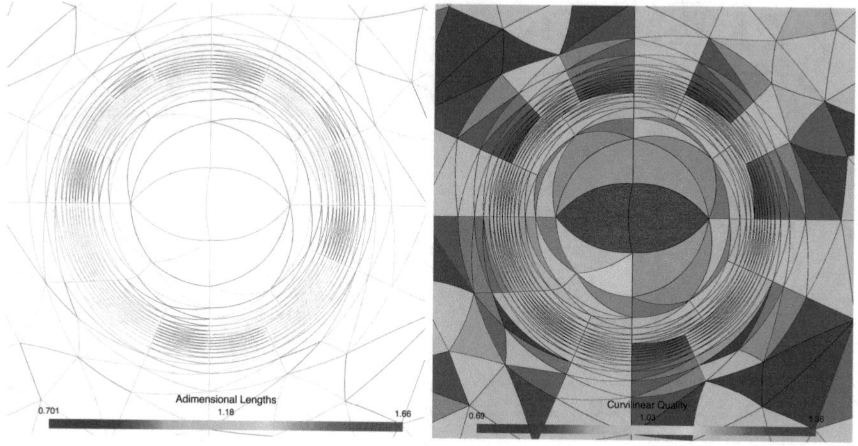

Fig. 9 Left figure shows adimensional lengths of edges of the mesh for the toy metric. Right figure present P^2 triangle quality measures (4)

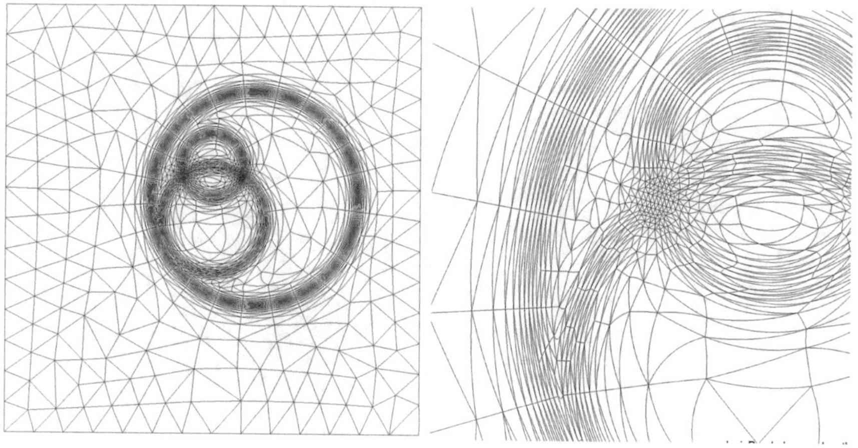

Fig. 10 Curvilinear mesh of the unit square using the intersection of three toy metrics

Meshes are presented in Fig. 10. A total of 1270 mesh vertices were inserted in the unit square. Then, 840 curvilinear swaps were performed to produce the final mesh. Edges of the mesh have sizes that are in the range [0.7, 1.8].

5.3 Other Analytical Metrics

We have used our technique to adapt to iso zero of two functions (Figs. 11 and 12). Our procedure seems to remain stable and robust for thicker and thiner adaptations.

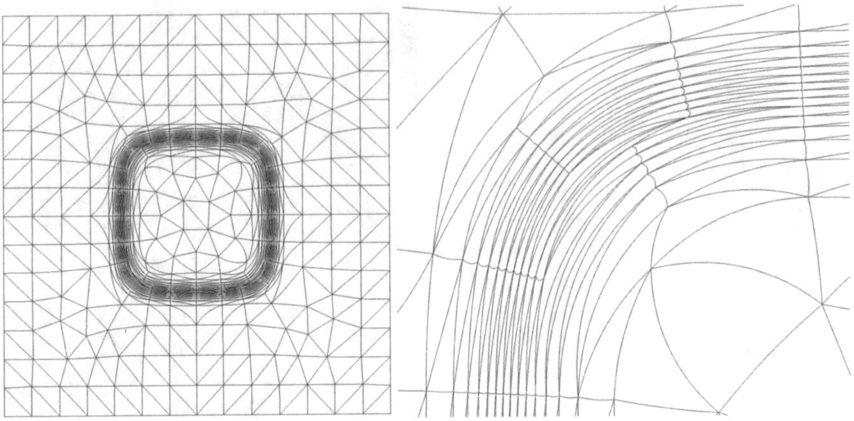

Fig. 11 Curvilinear mesh adapted to capture $(x^1)^4 + (x^2)^4 = R^4$

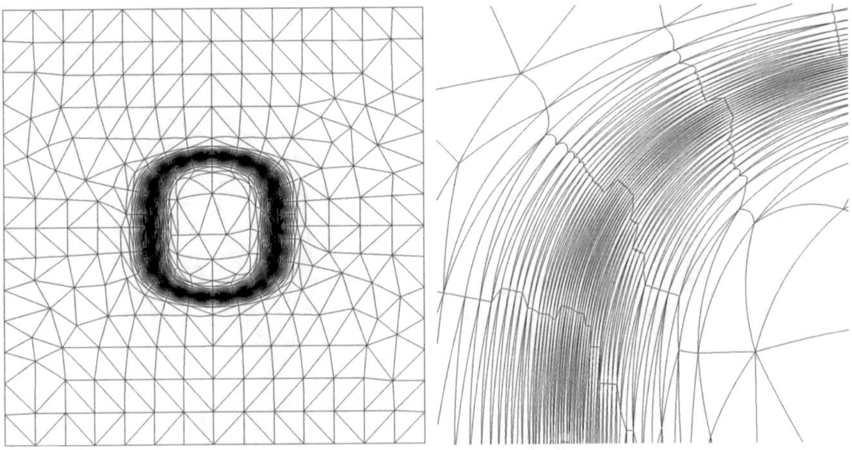

Fig. 12 Curvilinear mesh adapted to capture $(x^1)^2 + 2(x^2)^4 = R^4$

6 Conclusions

In this paper, a new methodology for generating unit curvilinear meshes has been proposed. The method guarantees two important properties in the final mesh:

1. Generated meshes are valid. This important property is due to the fact that P2 meshes are valid at any point of the algorithm. The initial mesh is curved along geodesic. A backtracking step is applied to ensure that every triangle of the mesh is valid. Then, edge swaps are only applied if elements are valid. Note that the validity criterion that is used is robust.

2. No short edges will be exist in the mesh. A spatial search procedure is used for ensuring that any point that is inserted is not to close in the sense of geodesics than any other point.

This work is now being extended to true adaptation i.e. adapting a mesh to a given function $f(x^1, x^2)$. Even though metrics are still the right tool for driving mesh adaptation at higher orders, basing **g** on hessians of f is not correct anymore for higher orders of approximation. Our future work will be to build metric fields that are suited for high order.

Acknowledgements This research is supported by the European Research Council (project HEXTREME, ERC-2015-AdG-694020) and by the Fond de la Recherche Scientifique de Belgique (F.R.S.-FNRS).

References

1. T. C. Baudouin, J.-F. Remacle, E. Marchandise, F. Henrotte, C. Geuzaine, A frontal approach to hex-dominant mesh generation. Adv. Model. Simul. Eng. Sci. **1**(1), 8 (2014)
2. N. Beckmann, H.-P. Kriegel, R. Schneider, B. Seeger, The r*-tree: an efficient and robust access method for points and rectangles, in *ACM Sigmod Record*, vol. 19 (ACM, New York, 1990), pp. 322–331
3. P.-E. Bernard, J.-F. Remacle, V. Legat, Boundary discretization for high-order discontinuous galerkin computations of tidal flows around shallow water islands. Int. J. Numer. Methods Fluids **59**(5), 535–557 (2009)
4. B. Cockburn, C.-W. Shu, TVB Runge-Kutta local projection discontinuous galerkin finite element method for conservation laws. II. general framework. Math. Comput. **52**(186), 411–435 (1989)
5. A. Ern, J.-L. Guermond, *Theory and Practice of Finite Elements*, vol. 159 (Springer Science & Business Media, Heidelberg, 2013)
6. P.-J. Frey, F. Alauzet, Anisotropic mesh adaptation for CFD computations. Comput. Methods Appl. Mech. Eng. **194**(48–49), 5068–5082 (2005)
7. C. Geuzaine, J.-F. Remacle, Gmsh: A 3-D finite element mesh generator with built-in pre-and post-processing facilities. Int. J. Numer. Methods Eng. **79**(11), 1309–1331 (2009)
8. A. Johnen, J.-F. Remacle, C. Geuzaine, Geometrical validity of curvilinear finite elements. J. Comput. Phys. **233**, 359–372 (2013)
9. N. Kroll, The adigma project, in *ADIGMA-A European Initiative on the Development of Adaptive Higher-Order Variational Methods for Aerospace Applications* (Springer, Heidelberg, 2010), pp. 1–9
10. X. Li, M. S. Shephard, M. W. Beall, 3D anisotropic mesh adaptation by mesh modification. Comput. Methods Appl. Mech. Eng. **194**(48–49), 4915–4950 (2005)
11. J. R. Shewchuk, Triangle: engineering a 2D quality mesh generator and delaunay triangulator, in *Applied Computational Geometry Towards Geometric Engineering* (Springer, Berlin, 1996), pp. 203–222
12. J. Shewchuk, What is a good linear finite element? Interpolation, conditioning, anisotropy, and quality measures (preprint). Univ. Calif. Berkeley **73**, 137 (2002)
13. J. Slotnick, A. Khodadoust, J. Alonso, D. Darmofal, W. Gropp, E. Lurie, D. Mavriplis, *CFD Vision 2030 Study: A Path to Revolutionary Computational Aerosciences* (National Aeronautics and Space Administration, Langley Research Center, Hampton, 2014)

Part II
Mesh and Geometry Blocks, Hex Mesh Generation

A 44-Element Mesh of Schneiders' Pyramid

Kilian Verhetsel, Jeanne Pellerin, and Jean-François Remacle

Abstract This paper shows that constraint programming techniques can successfully be used to solve challenging hex-meshing problems. Schneiders' pyramid is a square-based pyramid whose facets are subdivided into three or four quadrangles by adding vertices at edge midpoints and facet centroids. In this paper, we prove that Schneiders' pyramid has no hexahedral meshes with fewer than 18 interior vertices and 17 hexahedra, and introduce a valid mesh with 44 hexahedra. We also construct the smallest known mesh of the octagonal spindle, with 40 hexahedra and 42 interior vertices. These results were obtained through a general purpose algorithm that computes the hexahedral meshes conformal to a given quadrilateral surface boundary. The lower bound for Schneiders'pyramid is obtained by exhaustively listing the hexahedral meshes with up to 17 interior vertices and which have the same boundary as the pyramid. Our 44-element mesh is obtained by modifying a prior solution with 88 hexahedra. The number of elements was reduced using an algorithm which locally simplifies groups of hexahedra. Given the boundary of such a group, our algorithm is used to find a mesh of its interior that has fewer elements than the initial subdivision. The resulting mesh is untangled to obtain a valid hexahedral mesh.

1 Introduction

From the finite element practitioners point of view, hexahedral meshes have several advantages over tetrahedral meshes. However, there is no algorithm to generate a hexahedral mesh conformal to a given quadrilateral boundary. The state-of-the-art hexahedral meshing methods fail on small polyhedra such as the octogonal spindle and Schneiders' pyramid (Fig. 1). Schneiders' pyramid is a square-based pyramid

K. Verhetsel (✉) · J. Pellerin · J.-F. Remacle
Université catholique de Louvain, Louvain-la-Neuve, Belgium
e-mail: kilian.verhetsel@uclouvain.be; jeanne.pellerin@uclouvain.be; jean-francois.remacle@uclouvain.be

X. Roca, A. Loseille (eds.), *27th International Meshing Roundtable*, Lecture Notes in Computational Science and Engineering 127,
https://doi.org/10.1007/978-3-030-13992-6_5

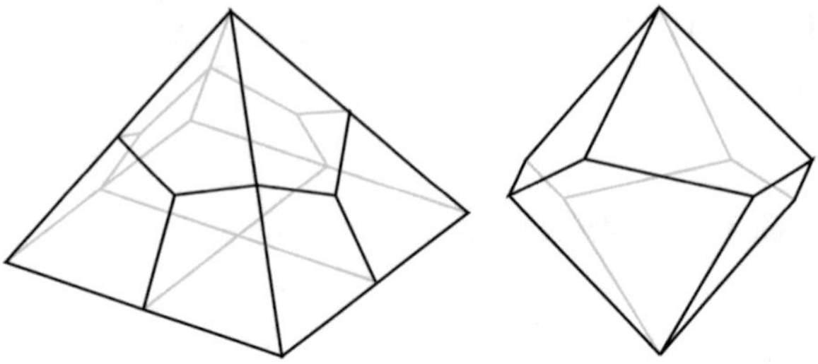

Fig. 1 Left: Schneiders'pyramid. Right: the octogonal spindle

with eight additional vertices at edge midpoints and five at face midpoints, and with its triangular and quadrangular faces split into three and four quadrangles respectively. The octogonal spindle, or tetragonal trapezohedron, can be used to construct Schneiders' pyramid by adding four hexahedra to form the pyramid base. The problem of meshing this pyramid with hexahedra was introduced by Schneiders [13] as an example of a boundary mesh for which no hexahedral mesh was known.

The question of the existence of a solution was settled by Mitchell [10] who proved that all quadrilateral surface meshes of the sphere with an even number of quadrilateral facets do have a hexahedral mesh. The algorithm deduced from the proof of this important theoretical result, as well as those of [5] and [6], generates too many hexahedra to be practical. Carbonera and Shepherd [4] constructs 5396 times more hexahedra than there are quadrilateral facets. In 2002, [18] introduced the hexhoop template family and constructed a hexahedral mesh of Schneiders' pyramid with 118 hexahedra. Later on, they improved their solution, building an 88-element mesh [19]. Very recently, a 36-element mesh was constructed by finding a sequence of flipping operations to transform the cube into Schneiders' pyramid, interpreting each operation as the insertion of a hexahedron [17].

In this paper, we propose a backtracking algorithm to enumerate the combinatorial meshes of the interior of a given quadrilateral surface (Sect. 2). Our first contribution is to prove that there is no hexahedral mesh of Schneiders' pyramid with strictly fewer than 12 interior vertices. Using the same approach, we also prove that there is no hexahedral mesh of the octagonal spindle with strictly fewer than 21 interior vertices.

The second contribution of this paper is an algorithm allowing the construction of a new hexahedral mesh of Schneiders' pyramid and the smallest known mesh of the octagonal spindle (Fig. 2). This construction uses a modified version of the backtracking algorithm to simplify the 88-element solution of [19] and reduce the number of hexahedra to 66 by locally simplifying groups of hexahedra (Sect. 3). The realized operations may be viewed as a general form of cube flips [2]. They substitute a set of hexahedra by another set without changing their boundary. However, instead of having a predefined set of flips, as do other local operations on hexahedral meshes [15], our algorithm automatically operates generically on any group of hexahedra. This group of hexahedra is replaced by fewer hexahedra using a combinatorial approach. The resulting mesh is untangled to obtain a valid

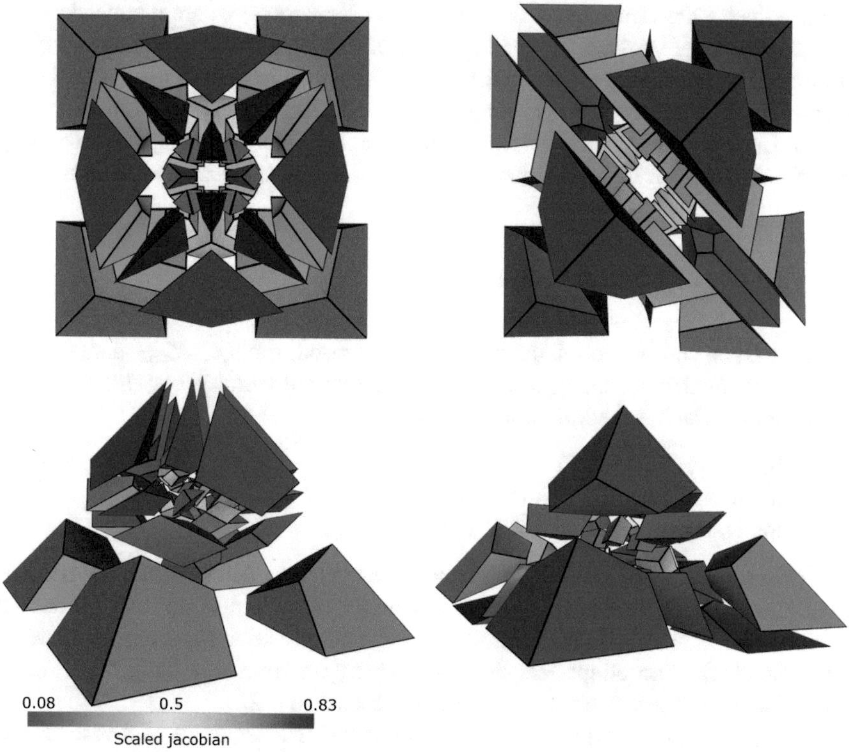

Fig. 2 Comparison of our 44-element mesh of Schneiders' pyramid (left) with the smallest known 36-element solution (right). Both admit two planar symmetries

hexahedral mesh. Furthermore, we used a sheet extraction procedure to construct a 40-element mesh of the octagonal spindle, and a 44-element mesh of the pyramid. [3, 9]

The C implementation of our algorithms and the resulting meshes can all be downloaded from https://www.hextreme.eu/.

2 Enumerating Combinatorial Hexahedral Meshes

In this section, we describe an algorithm that lists all possible hexahedral meshes with a prescribed boundary. We use this algorithm to determine lower bounds for the number of vertices and hexahedra needed to mesh the octagonal spindle and Schneiders' pyramid. It is also the key to the local mesh simplification algorithm we propose in Sect. 3.

When discussing the existence of hexahedral meshes or when enumerating those of the interior of a given quadrilateral mesh, we first ignore geometric issues and consider *combinatorial hexahedral meshes*. In a combinatorial hexahedral mesh, the hexahedra are represented as sequences of 8 integers, where distinct integers represent distinct vertices. A set of hexahedra defines a valid combinatorial mesh if all pairs of hexahedra are *compatible*: their intersection must be a shared combinatorial face (i.e. one of their 8 vertices, 12 edges, or 6 quadrangular facets) or be empty. Each quadrangle is also required to either be on the boundary (i.e. in exactly one hexahedron), or in the interior of the mesh (i.e. in exactly two hexahedra).

2.1 Backtrack Search Algorithm

Given ∂H, a combinatorial quad-mesh of a closed surface, H_{max} a maximum number of hexahedra, and V_{max} a maximum number of vertices, our algorithm lists all combinatorial hexahedral meshes H such that:

- the boundary of H is ∂H,
- the number of hexahedra $|H|$ is at most $H_{\max}$,
- the total number of vertices in H is at most $V_{\max}$.

This problem we are solving has similarities with problems commonly encountered in *constraint programming*: (1) efficiently filtering a large set of potential solutions and (2) managing solutions having multiple equivalent representations. Our implementation adopts concepts and strategies from this field. For a more general study of these problems, we refer the reader to [12].

The hexahedra are built one at a time by choosing a sequence of 8 vertices. At each step, all possible candidates for one of the 8 vertices are considered and the algorithm branches for each possibility. Each branch corresponds to the addition of a vertex to the current hexahedron. When a complete solution is determined, or when the search fails (no available candidates to complete a hexahedron), the algorithm backtracks to the previous choice. This process is repeated until all possibilities have been explored. Algorithm 1 corresponds to the exploration of a search tree (Fig. 3) where each branching node represents the choice of a point, and the leaves represent either solutions or failure points where the algorithm backtracks. The search tree has an exponential size in the maximum number of hexahedra in a solution. This high complexity is managed by pruning branches that cannot contain a solution and by using efficient implementations of all performed operations.

2.2 Search Space Reduction Strategies

In this section, we describe the key points of our implementation of Algorithm 1, all of which aim at reducing the search space explored by the algorithm:

- the order in which the hexahedra are constructed is crucial—we use an advancing-front strategy and start the construction of hexahedra from the boundary;

Algorithm 1 Recursive enumeration of the hexahedral meshes of the interior of ∂H

Input: ∂H, the boundary; S, a partial solution; $C = (C_1, \ldots, C_8)$, the sets of candidate vertices for the current hexahedron

if the boundary of S is ∂H **then**
  Print solution S
else if $|S| = H_{\max}$ **then**
  Backtrack
else
  $C \leftarrow$ FILTER-CANDIDATES($\partial H, S, C$)
  if $|C_1| = \cdots = |C_8| = 1$ **then**
    $S' \leftarrow S \cup \{(v_1, v_2, v_3, v_4, v_5, v_6, v_7, v_8)\}$
    SEARCH($\partial H, S'$, INITIALIZE-CANDIDATES(S'))
  else if $\min_{i \in \{1,\ldots,8\}} |C_i| = 0$ **then**
    Backtrack
  else
    $i \leftarrow$ PICK-HEX-VERTEX(C)
    for each $v \in C_i$ **do**
      $C' \leftarrow C$
      $C'_i \leftarrow \{v\}$
      SEARCH($\partial H, S, C'$)
    end for
  end if
end if

- an efficient filtering algorithm that eliminates candidate vertices that would create incompatible combinatorial hexahedra in the solution;
- a method to manage the high number of symmetries of this problem;
- the order in which the current hexahedron vertices are selected.

Advancing-Front Construction While the hexahedra of a combinatorial mesh can be arbitrarily reordered, constructing them in a specific order makes the algorithm significantly faster. We use a classical advancing front generation strategy and require the hexahedron under construction to share a face with a front of quadrangles. There are then only four vertices needed to complete a hexahedron. The quadrangle front is constituted of the interior facets that are in only one hexahedron, or of boundary facets that are in no hexahedra. At the root of the search tree, it is set to be the boundary ∂H. An interior facet is added to the front after its first appearance in the mesh. The facet is removed from the front when it is added to the partial solution. When the front becomes empty, the boundary of the solution matches the input (Fig. 4).

Filtering Out Candidate Vertices For each of the eight vertices of the hexahedron under construction, we store a set of candidate vertices that could be part of the solution. Some of these vertices would make the current hexahedron incompatible with some already existing hexahedra. Therefore when initiating the construction of a hexahedron, or when adding a vertex to a hexahedron, vertices that cannot be added without creating incompatibilities between the current hexahedron and the

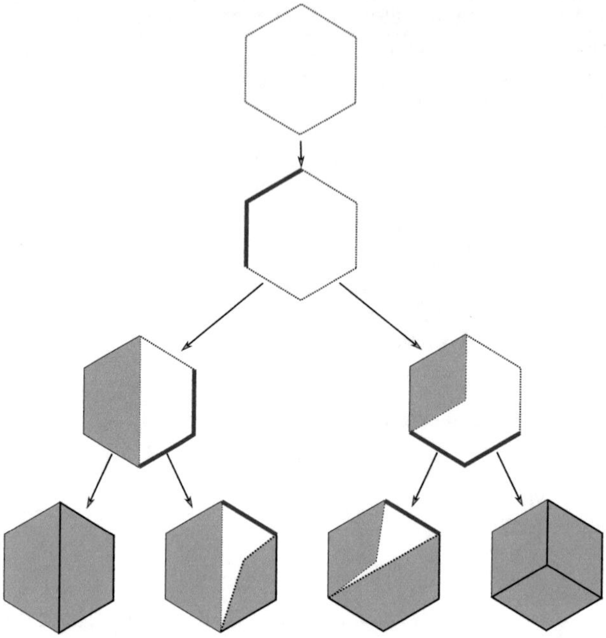

Fig. 3 Searching all quadrilateral meshes of a polygon with up to one interior point. The search tree leaves are either valid solutions, or correspond to detected failure points where Algorithm 1 backtracks

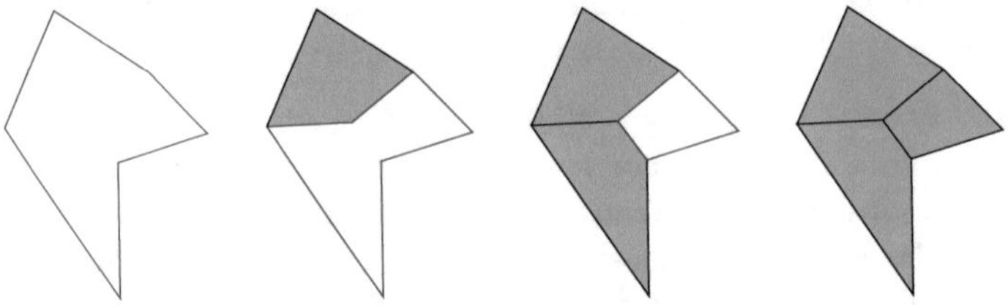

Fig. 4 Each new element must share a face with the front of boundary faces (red)

already built hexahedra are filtered out. The following rules are used to eliminate candidates:

1. an edge cannot match the diagonal of an existing quadrangle, or an interior diagonal of an existing hexahedron;
2. conversely an interior diagonal cannot match the diagonal of an existing quadrangle, or an existing edge, or an interior diagonal of an existing hexahedron;
3. a facet diagonal cannot match an existing hexahedron edge, or an existing hexahedron interior diagonal;
4. if one facet diagonal matches an existing quadrangle diagonal, so must the second one;
5. all eight vertices must be different.

Algorithm 2 INITIALIZE-CANDIDATES(S): compute the sets candidate vertices

Input: S, a set of hexahedra.
Output: $C = (C_1, \ldots, C_8)$, the sets of candidate vertices for the next hexahedron.

Let (v_1, v_2, v_3, v_4) be some quadrangle that needs to occur in the mesh.
for each $i \in \{1, \ldots, 4\}$ **do**
  $C_i \leftarrow \{v_i\}$
end for
for each $i \in \{1, \ldots, 4\}$ **do**
  $C_{4+i} \leftarrow$ ALLOWED-NEIGHBORS$(v_i) \setminus \{v_1, v_2, v_3, v_4\}$
  for each $j \in \{1, \ldots, 4\}$ **do**
    if $i \neq j$ **then**
      $C_{4+i} \leftarrow C_{4+i} \setminus$ KNOWN-DIAGONALS$(v_j) \setminus$ KNOWN-NEIGHBORS(v_j)
    end if
    if $i = j + 2 \mod 4$ **then**
      $C_{4+i} \leftarrow C_{4+i} \setminus \{v_k \mid (v_j, v_k)$ is the diagonal of a quadrangle$\}$
    end if
  end for
end for
return $(C_1, \ldots, C_8)$

Our implementation tracks three sets of vertices for each vertex v. These sets are used to build the candidate set for each vertex of a new hexahedron (Algorithm 2). These sets are updated whenever an edge, quadrangle or interior diagonal is added to the mesh, and they are:

- ALLOWED-NEIGHBORS(v), the set of vertices u such that an edge (u, v) could be added to the mesh without creating incompatibilities with existing hexahedra or facets of the boundary;
- KNOWN-NEIGHBORS(v), the set of vertices that are adjacent to v in existing hexahedra or in quadrangles contained in the boundary;
- KNOWN-DIAGONALS(v), the set of all vertices u such that (u, v) is one of the four interior diagonals of a hexahedron.

Because the execution time of the search algorithm blows up as the number of vertices increases, the number of vertices each set contains is always small, making them good candidates for being represented as bit-sets.

Symmetry Breaking Combinatorial meshes are characterized by their large number of symmetries, a major challenge when operating on combinatorial hexahedral meshes. Indeed, a combinatorial hexahedral mesh has many equivalent representations:

1. interior vertices can be relabelled (Fig. 5)—for boundary vertices, the algorithm uses the same labels as the input;
2. the hexahedra of the solution can be constructed in a different order (Fig. 6);
3. for a given hexahedron, written as an ordered sequence of 8 vertices, there are $1680 = 8!/24$ ways to reorder these vertices while leaving the hexahedron unchanged (Fig. 7).

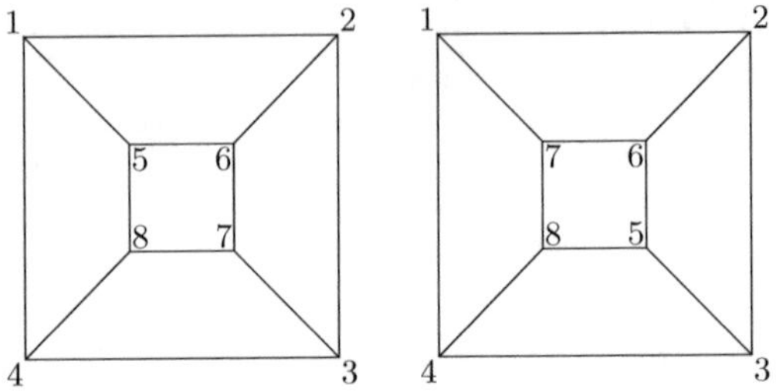

Fig. 5 Two of the 4! ways to label the 4 interior vertices of this mesh

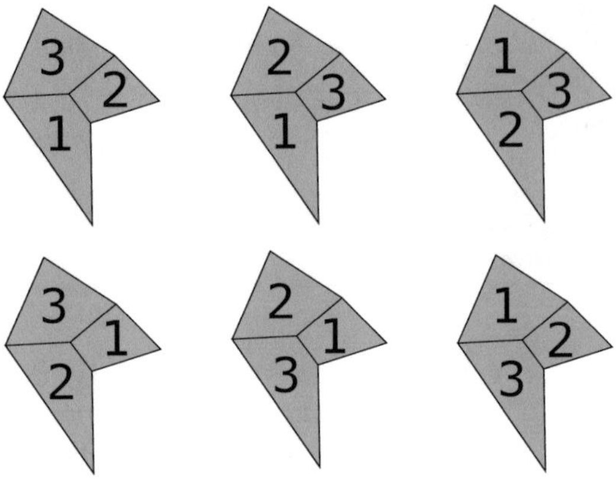

Fig. 6 The 3! different ways to number the elements of a 3-element mesh

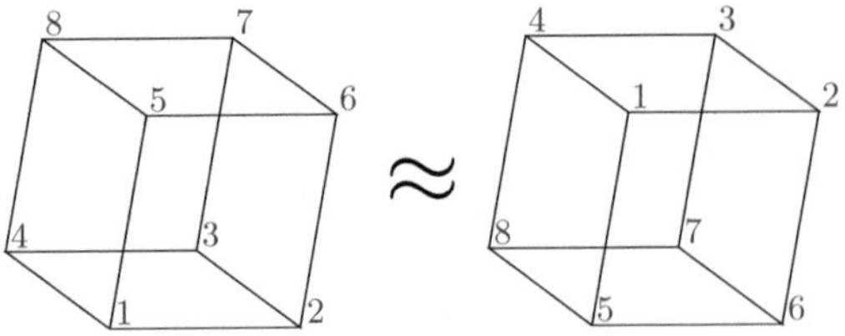

Fig. 7 Two combinatorially equivalent hexahedra

The advancing front strategy defines the order in which the solution hexahedra are constructed (symmetry 2). This also uniquely determines the order of vertices in a hexahedron (symmetry 3). To prevent the relabelling of interior vertices (symmetry 1), we add *value precedence* constraints to our problem [8]. A solution H found by the algorithm can be written as an array of $8|H|$ integers, writing down the vertices of each hexahedron in the order in which they were constructed by the algorithm. In an array, x *precedes* y when the first occurrence of x is before the first occurrence of y. Enforcing a total precedence order on interior vertices, we guarantee that only one of their permutations is a solution.

Optimization of Hexahedron Construction The efficiency of Algorithm 1 depends on the size of the search tree needed to explore all possibilities. A cheap approach to reduce the number of nodes in the search tree is to choose the vertex with the smallest set of candidate vertices when deciding which vertex to branch on [1]. This does not affect the correctness of the algorithm, as long as a vertex with more than one candidate is selected.

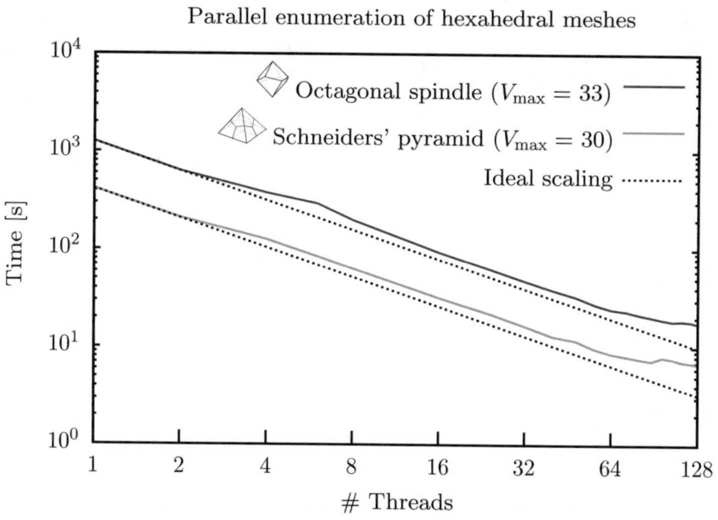

Fig. 8 Time to explore a search tree in parallel on a machine with two AMD EPYC 7551 CPUs (32 cores each, 2 threads per core). Using 64 threads, the speed-up is of 48 for Schneiders' pyramid and 52 for the octagonal spindle

2.3 *Parallel Search*

The exploration a search tree can be parallelized in a natural way by exploring different subtrees in parallel, making the algorithm much faster on parallel architectures (Fig. 8). We use an approach similar to the embarrassingly parallel search of [11]. The main challenge to overcome is that some subtrees are multiple orders of magnitude larger than other ones without any possibility to determine it ahead of time.

We solve this issue by attributing many subtrees to each worker thread, so that all threads must on average perform the same amount of work (we used 4096 subtrees per thread). At the start of the search, the tree is explored in a breadth-first manner until a layer with enough subproblems is reached. The nodes of this layer are then explored in parallel by independent worker threads using Algorithm 1.

2.4 *Lower Bounds for Hex-Meshing Problems*

Using Algorithm 1, we computed lower bounds for the number of vertices and hexahedra required to mesh Schneiders' pyramid and the octagonal spindle (Fig. 1). The algorithm is run multiple times, and we increment either V_{max} or H_{max} between each run. At each step, we verify that no solution was found by the algorithm.

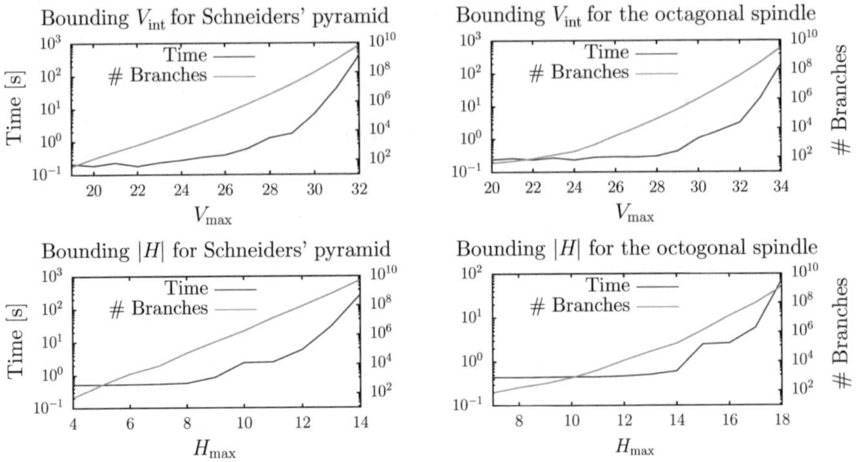

Fig. 9 The time to prove lower bounds for the number of interior vertices V_{int} and the number of hexahedra H required to mesh a polyhedron increases exponentially. This is due to the exponential size of the search tree explored by the algorithm

The time required to compute these bounds increases exponentially as the bounds become tighter (Fig. 9).

Theorem 1 *Any hexahedral mesh of Schneiders' pyramid has at least 18 interior vertices and 17 hexahedra.*

Theorem 2 *Any hexahedral mesh of the octagonal spindle has at least 29 interior vertices and 21 hexahedra.*

3 Simplifying Hexahedral Meshes

The algorithm described in the previous section can be used to find the smallest hexahedral mesh with a given boundary. In this section, we use this algorithm to accomplish our goal of computing upper bounds for the number of hexahedra required to mesh Schneiders' pyramid. From the 88-element solution of [19], we locally simplify the mesh. By simplification we mean decreasing the number of hexahedra (Fig. 10). The realized operations may be viewed as a generalized form of cube flips [2] that substitute a set of hexahedra by another set without changing their boundary. However, instead of a finite set of transformations, the algorithm introduced in this section automatically determines them at execution time.

Globally minimizing the number of hexahedra in the mesh is a computationally demanding task. Our algorithm therefore selects a small subset of the mesh, or *cavity*, and focuses on modifying the connectivity of the mesh only within this cavity. Our hexahedral mesh simplification algorithm is based on Algorithm 1.

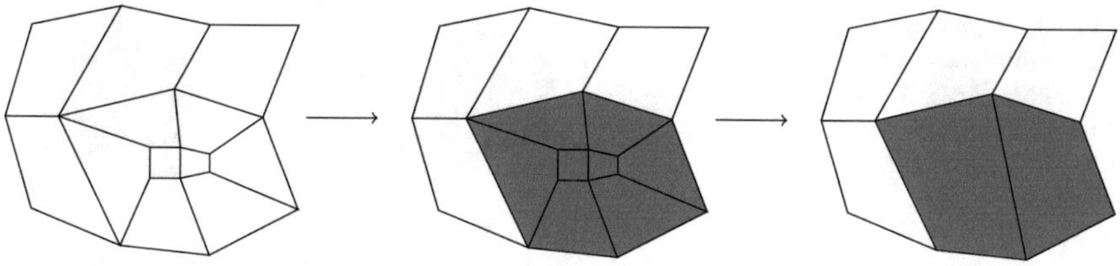

Fig. 10 The number of elements in a mesh can be reduced by operating locally on a cavity

Algorithm 3 Cavity selection algorithm

Input: H, the mesh; n, the size of the cavity
Output: A cavity $\mathscr{C}$ of n elements
$h \leftarrow$ a random element of H
$\mathscr{C} \leftarrow \{h\}$
while $|\mathscr{C}| \neq n$ **do**
  $h \leftarrow$ a random element of $H \setminus \mathscr{C}$ sharing a facet with a hexahedron in $\mathscr{C}$
  $\mathscr{C} \leftarrow \mathscr{C} \cup \{h\}$
end while
return $\mathscr{C}$

From a geometric hexahedral mesh it outputs a geometric hexahedral mesh whose boundary is strictly identical and which has fewer elements.

The mesh simplification procedure has three main steps:

1. the selection of a cavity, the group of hexahedra to simplify, $\mathscr{C}$;
2. finding the smallest hexahedral mesh $\mathscr{C}_{\text{min}}$ compatible with the cavity boundary $\partial\mathscr{C}$ and replacing the cavity with this smaller mesh;
3. untangling the hexahedra to determine valid coordinates for the mesh vertices.

1. Cavity Selection The cavity selection algorithm is a greedy algorithm that starts from a random element of the input hexahedral mesh (Algorithm 3). When the target size, in terms of number of hexahedra is reached, this process stops. The choice of a target cavity size is a trade-off between the cost of finding the hexahedral meshes of the cavity and the likelihood that the mesh can be simplified by remeshing the cavity. Cavities with many hexahedra are more likely to accept smaller meshes, but the cost of finding the smallest hexahedral mesh $\mathscr{C}_{\text{min}}$ increases exponentially with the number of hexahedra in the cavity. In practice, we start by considering relatively small cavities containing up to 10 hexahedra, and increase this limit when no improvement is possible. We require the cavity to contain at least 4 interior vertices. Indeed, when there are no interior vertices (e.g. with a stack of hexahedra), it is not possible to remove any hexahedra. As the number of interior vertices increases, so does the likelihood that the cavity can be simplified.

2. Cavity Remeshing To find a smaller mesh of the boundary of a cavity $\mathscr{C}$, we first solve the combinatorial problem, i.e. we find the smallest combinatorial hexahedral

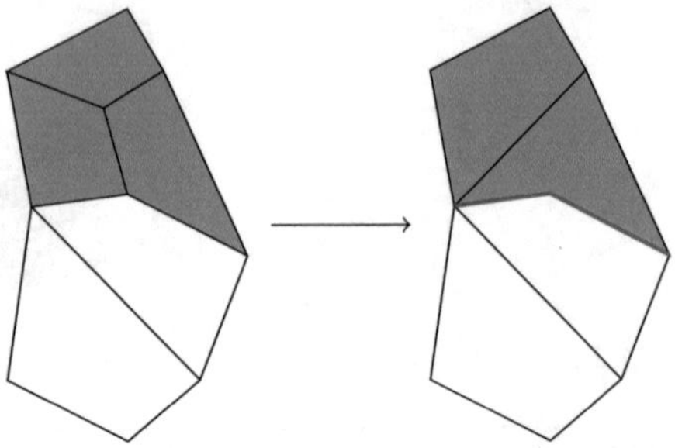

Fig. 11 Replacing the cavity with a valid mesh sharing the same boundary still produces an invalid mesh by creating two quadrangles sharing two edges

mesh of $\partial\mathscr{C}$, and then solve the geometric problem of finding valid coordinates for the modified mesh vertices.

The combinatorial problem of finding the smallest mesh of $\partial\mathscr{C}$ is but an application of Algorithm 1 which enumerates all combinatorial meshes of a given surface. The maximum number of hexahedra H_{max} of the solution is set to a smaller value than $|\mathscr{C}|$. Changing the parity of a hexahedral mesh is known to be a difficult operation [14], so we set H_{max} to $|\mathscr{C}| - 2$. We also set the limit to the number of interior vertices V_{max} to one less than the number of interior vertices in $\mathscr{C}$ to accelerate the search.

There is a subtle but important difference between meshing a cavity in an existing mesh and meshing a stand-alone polyhedron: the hexahedra inside the cavity must be compatible with the other elements of the input mesh. An example where new elements from a cavity are not compatible with elements adjacent to the cavity is given in Fig. 11. A 3-element cavity is replaced by 2 quadrangles, but one of these two quadrangles shares two edges with an existing element, which is an invalid configuration. To guarantee that the algorithm does not break the mesh validity, the data structures used to filter out inadequate vertex candidates (Sect. 2.2) are modified to take into account the hexahedra that are not part of the cavity.

3. Untangling The previous step of the algorithm found a new connectivity for the mesh. The simplified mesh obtained by using this result is not valid in general because the interiors of hexahedra may intersect (Fig. 12). To obtain a valid geometric mesh we use the untangling algorithm described in [16]. The vertices are iteratively moved until all hexahedra in the mesh are valid. If the untangling fails, connectivity changes are undone. The validity of the final mesh is evaluated with the method proposed by Johnen et al. [7].

A 66-Element Mesh of Schneiders' Pyramid

We applied our algorithm to Yamakawa's mesh of Schneiders' pyramid and obtained a valid hexahedral mesh with 66 hexahedra and 63 interior vertices (Fig. 2). Table 1 shows the sizes of the different cavities simplified by our algorithm. It takes a few minutes for our algorithm to reduce the number of hexahedra in the mesh from 88 down to 66 in our final mesh. Figure 13 shows the changes to the connectivity of the mesh performed in two different iterations of the algorithm. The vertices

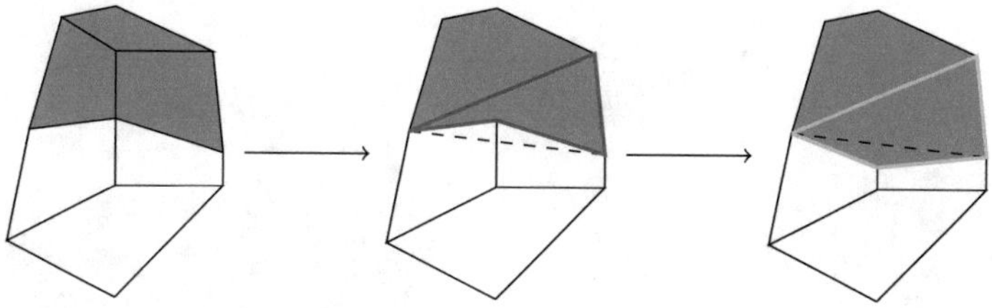

Fig. 12 A valid change to the connectivity of the mesh can create a geometrically invalid mesh, fixed by moving the vertices

Table 1 Cavity remeshing operations performed by our hex-mesh simplification algorithm on Yamakawa's 88-element mesh of Schneiders' pyramid [19]

Initial mesh		Initial cavity			Remeshed cavity		New mesh	
#hex	#vert.	#hex	#vert.	#bd. facets	#hex	#vert.	#hex	#vert.
88	105	8	23	18	6	21	86	103
86	103	8	23	18	6	21	84	101
84	101	8	23	18	6	21	82	99
82	99	14	33	24	8	27	76	93
76	93	6	16	10	2	12	72	89
72	89	18	40	30	12	32	66	81

had to be moved to obtain a valid mesh, but the combinatorial boundary remains the same. For example, for the second pair of cavities in the figure, the same 30 facets can be seen before and after the remeshing operation: there is a central facet, surrounded by a ring of five quadrangles, followed by three rings of six quadrangles, followed by one more ring of five quadrangles surrounding a single face. We also determined a combinatorial mesh with 64 hexahedra and 59 interior vertices, on which the untangling failed.

A 44-Element Mesh of Schneiders' Pyramid

We used the 72-element mesh described in the previous paragraph to create a 40-element mesh of the octagonal spindle. Indeed, such a mesh can be constructed by extracting one of the two sheets of the dual of our 72-element solution using the method described in [3]. The mesh resulting from this operation is the smallest known mesh of the octagonal spindle, with 40 hexahedra and 42 interior vertices. Our 44-element mesh of the pyramid mesh was obtained by adding 4 hexahedra to our mesh of the spindle (Fig. 2).

All the meshes discussed in this section are available online at https://www.hextreme.eu/.

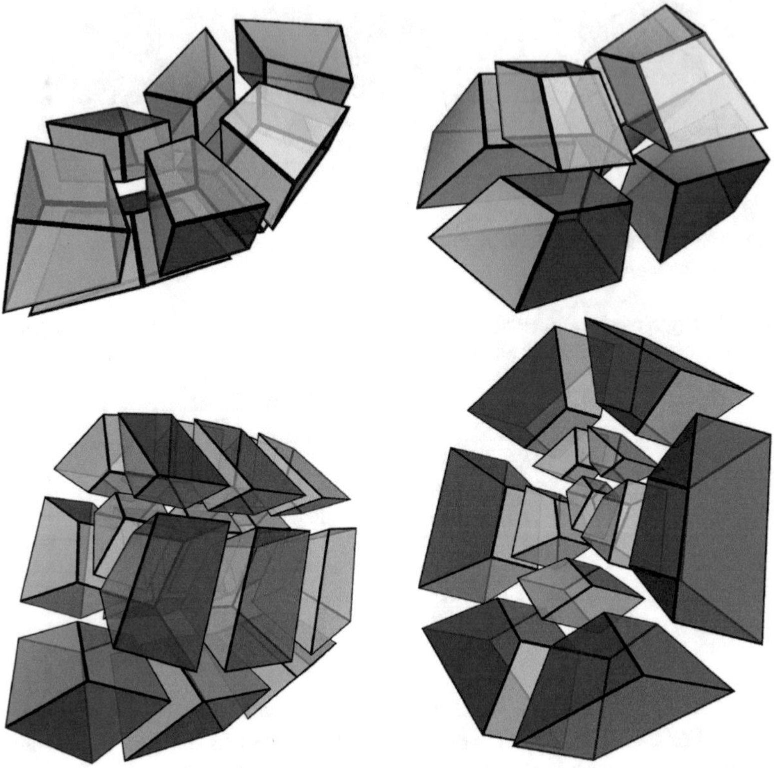

Fig. 13 (top) Removal of two hexahedra from Schneiders' pyramid; (bottom) removal of six hexahedra. The initial cavity (left) and the remeshed cavity (right) have the same combinatorial boundary (top: 18 facets; bottom: 30 facets). Colors highlight the correspondence between faces

4 Conclusion

The main contribution of this paper is an algorithm to prove new bounds for Schneiders' pyramid and the octagonal spindle hex-meshing problems. The relatively large number of vertices required to mesh the pyramid implies that subdividing pyramids into hexahedra to create all-hexahedral meshes will necessarily create many additional hexahedra. This makes it likely that some of the hexahedra created in this manner will be invalid. Because these algorithms are general, they could be used to evaluate the viability of other subdivision schemes to create all-hexahedral meshes.

One limitation of the hex-mesh simplification algorithm described in this paper is that its execution becomes expensive, because finding the smallest hexahedral mesh of a cavity becomes exponentially more time consuming as its size increases. A cheaper algorithm could be designed by finding a small set of local operations and an algorithm to choose which of them to perform in order to reduce the size of the mesh.

Acknowledgements This research is supported by the European Research Council (project HEXTREME, ERC-2015-AdG-694020). Computational resources have been provided by the supercomputing facilities of the Université catholique de Louvain (CISM/UCL) and the Consortium des Équipements de Calcul Intensif en Fédération Wallonie Bruxelles (CÉCI) funded by the Fond de la Recherche Scientifique de Belgique (F.R.S.-FNRS) under convention 2.5020.11.

References

1. J.C. Beck, P. Prosser, R.J. Wallace, Trying again to fail-first, in *International Workshop on Constraint Solving and Constraint Logic Programming* (Springer, Berlin, 2004), pp. 41–55
2. M. Bern, D. Eppstein, J. Erickson, Flipping cubical meshes. Eng. Comput. **18**(3), 173–187 (2002)
3. M. J. Borden, S. E. Benzley, J. F. Shepherd, Hexahedral sheet extraction, in *IMR* (2002), pp. 147–152
4. C. D. Carbonera, J. F. Shepherd, A constructive approach to constrained hexahedral mesh generation. Eng. Comput. **26**(4), 341–350 (2010)
5. D. Eppstein, Linear complexity hexahedral mesh generation. Comput. Geom. **12**(1–2), 3–16 (1999)
6. J. Erickson, Efficiently hex-meshing things with topology. Discrete Comput. Geom. **52**(3), 427–449 (2014)
7. A. Johnen, J.-C. Weill, J.-F. Remacle, Robust and efficient validation of the linear hexahedral element. Procedia Eng. **203**, 271–283 (2017)
8. Y. C. Law, J. HM. Lee, Global constraints for integer and set value precedence, in *International Conference on Principles and Practice of Constraint Programming* (Springer, Berlin, 2004), pp. 362–376
9. F. Ledoux, J. Shepherd, Topological modifications of hexahedral meshes via sheet operations: a theoretical study. Eng. Comput. **26**(4), 433–447 (2010)
10. S. A. Mitchell, A characterization of the quadrilateral meshes of a surface which admit a compatible hexahedral mesh of the enclosed volume, in *Annual Symposium on Theoretical Aspects of Computer Science* (Springer, Berlin, 1996), pp. 465–476
11. J.-C. Régin, M. Rezgui, A. Malapert, Embarrassingly parallel search, in *International Conference on Principles and Practice of Constraint Programming* (Springer, Berlin, 2013), pp. 596–610
12. F. Rossi, P. V. Beek, T. Walsh, *Handbook of Constraint Programming* (Elsevier, Amsterdam, 2006), p. 978
13. R. Schneiders, A grid-based algorithm for the generation of hexahedral element meshes. Eng. Comput. **12**(3–4), 168–177 (1996)
14. A. Schwartz, G. M. Ziegler, Construction techniques for cubical complexes, odd cubical 4-polytopes, and prescribed dual manifolds. Exp. Math. **13**(4), 385–413 (2004)
15. T. J. Tautges, S. E. Knoop, Topology modification of hexahedral meshes using atomic dual-based operations. Algorithms **11**, 12 (2003)
16. T. Toulorge, C. Geuzaine, J.-F. Remacle, J. Lambrechts, Robust untangling of curvilinear meshes. J. Comput. Phys. **254**, 8–26 (2013)
17. S. Xiang, J. Liu, A 36-element solution to schneiders' pyramid hex-meshing problem and a parity-changing template for hex-mesh revision. Preprint arXiv:1807.09415 (2018)
18. S. Yamakawa, K. Shimada, HEXHOOP: modular templates for converting a hex-dominant mesh to an all-hex mesh. Eng. comput. **18**(3), 211–228 (2002)
19. S. Yamakawa, K. Shimada, 88-element solution to schneiders' pyramid hex-meshing problem. Int. J. Numer. Methods Biomed. Eng. **26**(12), 1700–1712 (2010)

Representing Three-Dimensional Cross Fields Using Fourth Order Tensors

Alexandre Chemin, François Henrotte, Jean-François Remacle, and Jean Van Schaftingen

Abstract This paper presents a new way of describing cross fields based on fourth order tensors. We prove that the new formulation is laying in a linear space in R^9. The algebraic structure of the tensors and their projections on SO(3) are presented. The relationship of the new formulation with spherical harmonics is exposed. This paper is quite theoretical. Due to pages limitation, few practical aspects related to the computations of cross fields are exposed. Nevertheless, a global smoothing algorithm is briefly presented and computation of cross fields are finally depicted.

1 Introduction

We call a cross f a set of six distinct unit vectors mutually orthogonal or opposite to each other (Fig. 1).This geometric object composed of vectors lives in the tangent space of Euclidean spaces E^3. A cross field $F = \{x \in \Omega \subset E^3 \mapsto f(x)\}$, now, is a function that associates a cross $f(x)$ to each point of a subset Ω of E^3. Cross fields are auxiliary in 3D mesh generation to define local preferred orientations for hexahedral meshes, or for the computation of the polycube decomposition of a solid. Automatic polycube decomposition is a necessary step for multiblock or isogeometric meshing of 3D domains.

Let the Euclidean space E^3 be equipped with a Cartesian coordinate system $\{x_1, x_2, x_3\}$. The six vectors $(\pm 1, 0, 0)$, $(0, \pm 1, 0)$ and $(0, 0, \pm 1)$ form a cross, which we call the reference cross f_{ref}. Crosses being rigid objects, their orientation in space can be *identified* by a rotation respective to f_{ref}, that is a member of SO(3) represented by, e.g., the Euler angles α, β and γ, (Fig. 1). This *representation* of f is however not unique due to the symmetries of the cross, which are fully characterized by regarding the cross as set of six points at the summits of an octahedron. The

A. Chemin · F. Henrotte · J.-F. Remacle (✉) · J. V. Schaftingen
Université catholique de Louvain, Louvain-la-Neuve, Belgium
e-mail: Alexandre.Chemin@uclouvain.be; francois.henrotte@uclouvain.be; jean-francois.remacle@uclouvain.be; jean.vanschaftingen@uclouvain.be

X. Roca, A. Loseille (eds.), *27th International Meshing Roundtable*, Lecture Notes in Computational Science and Engineering 127,
https://doi.org/10.1007/978-3-030-13992-6_6

Fig. 1 3D crosses representation. Left image shows the reference cross f_{ref} and right image shows a cross f that is a rotation of the reference cross

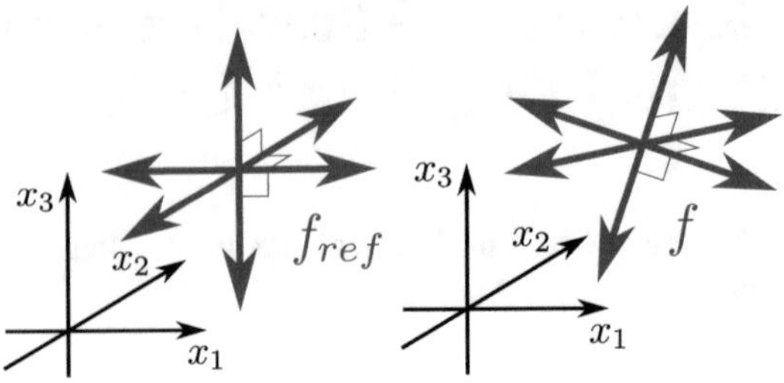

symmetry group of this point set has 24 elements, which are the 24 rotations that apply the cross onto itself, and is called the group of cube rotation O. We call attitude of the cross f its orientation in space up to the symmetries of the cross, and we have

$$f \in \mathrm{SO}(3)/\mathrm{O}.$$

The expansion of a discretized field F into coefficients and shape functions in finite element analysis is by definition a linear combination, leading then by orthogonalization in this linear space to a linear system of equations to solve. It is hence necessary in this finite element context to have a representation of the point value of the discretized field $F(x)$ in a linear space, i.e., a space containing the linear combinations (here with real coefficients) of all its members. This is however in general not the case with fields taking their values in non-trivial group manifolds like, e.g., SO(3)/O.

The sketch of the solution to this problem can be illustrated with a simple 2D example. Consider the unit circle S^1 and two points $e^{\imath\theta_1}$ and $e^{\imath\theta_2}$ on this manifold. Clearly, linear combinations

$$a\, e^{\imath\theta_1} + b\, e^{\imath\theta_2} \quad , \quad a, b \in \mathrm{R}$$

do not all belong to S^1. In order to have a practical representation of the elements of S^1 in a linear space amenable to finite element analysis, one has to expand S^1 to the enclosing complex plane, $\mathrm{C} \supset S^1$, which is a linear space. The finite element problem can so be formulated in terms of complex valued unknowns that are afterwards projected back into S^1 by means of a projection operator, e.g.,

$$\Pi : \mathrm{C} \mapsto S^1 \quad , \quad x + \imath y \mapsto e^{\imath\, \mathrm{atan2}(y,x)}.$$

A similar approach is followed in this paper for the 3D finite element smoothing of cross attitudes belonging to the group manifold SO(3)/O. The approach rely on a new way of representing 3D cross fields as a particular class of fourth order tensors, themselves in close relations to fourth degree homogeneous polynomials of the Cartesian coordinates. 3D cross field representations based on tensors have been used for 3D solid texturing and hex-dominant meshing [3–6], but none of them was addressing symmetry issues or projections. The use of fourth order tensors allows to build a 9-dimensional linear space $\mathcal{A}$, containing SO(3)/O as a subset, together with a projection operator

$$\Pi : \mathcal{A} \mapsto \mathrm{SO}(3)/\mathrm{O}.$$

The approach leads eventually to a very efficient smoother for cross fields, one order of magnitude faster than state-of-the art implementations. The proposed representation also allows easy computation of the distance between a finite element computed cross f, and its projection back into SO(3)/O. This distance indicates the presence of singular lines and singular points in the cross field in a straightforward fashion.

The paper is organized as follows. The fourth order tensor representation for crosses is first introduced, and the useful mathematical properties of this tensor space are then derived. The projection method is then presented and results obtained with a naive 3D crossfield smoothing on some benchmarks problem are finally discussed.

2 Cross Representation with Fourth Order Tensors

2.1 *The Reference Cross f_{ref}*

Point groups, like O, are isometries leaving at least one point of space, the center, invariant. As such, they have very convenient and useful representations on the sphere, and hence also in terms of spherical harmonics. In [1, 2], spherical harmonics of degree four are proposed as a polynomial basis to represent 3D cross fields. They exhibit the required octahedral symmetry and span a linear polynomial space $\mathcal{H}_4$ of dimension nine. The projection operator

$$\Pi : \mathcal{H}_4 \mapsto \mathrm{SO}(3)/\mathrm{O},$$

however, is tedious as it relies on a complex minimization process that is not ensured to converge to the true projection. Moreover, the differential properties of spherical harmonics (they are eigenfunctions of the laplacian operator) are of no use to the purpose of cross representation.

The idea promoted in this paper is thus also to work with polynomials whose isovalues exhibit the sought octahedral symmetry but, instead of expanding them in a spherical harmonics basis, they are represented as explicit rotations of a reference polynomial f_{ref}. With $x \in E^3$ and (x_1, x_2, x_3) its coordinates in the Cartesian coordinate system associated to E^3,

$$f_{ref}(x_1, x_2, x_3) = \|x\|_4^4 \equiv x_1^4 + x_2^4 + x_3^4, \tag{1}$$

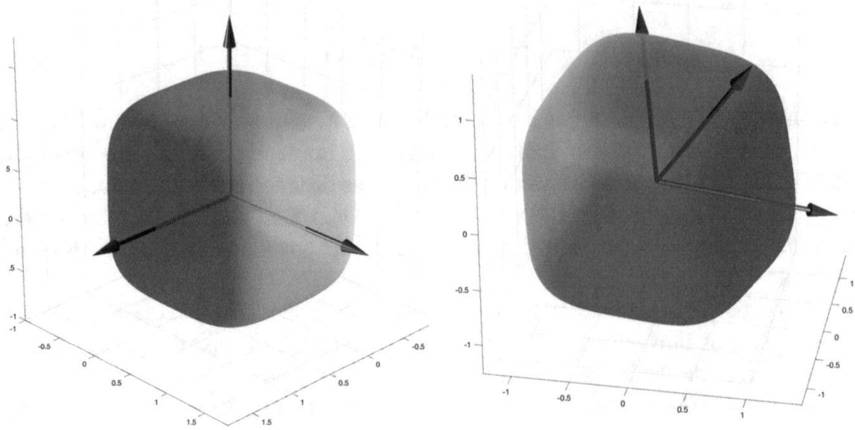

Fig. 2 Representation of the reference cross as a dice-shaped polynomial isovalue surface (left), and of a general cross attitude as a rotation of the latter (right)

whose isovalue $f_{ref} = 1$ is the dice-shaped surface depicted in Fig. 2 (left). fourth order is the lowest polynomial order exhibiting distinctive octahedral symmetry, which is by the way rather natural in a Cartesian coordinate system, as it simply amounts to the invariance against any argument inversion and/or permutation:

$$f_{ref}(x_1, x_2, x_3) = f_{ref}(-x_1, x_2, x_3) = f_{ref}(x_2, -x_1, x_3) = \dots$$

AC: pourquoi on montre ici l'invariance par reflexion alors que dans tout ce qu'on fait avant on ne parle que des isometries positives? Du coup je pense qu'il vaudrait mieux parler de $O(3)/O_h$ plutot que $SO(3)/O$.

In tensor notations, we have

$$f_{ref}(x_1, x_2, x_3) = \tilde{A}_{ijkl}\, x_i x_j x_k x_l$$

assuming Einstein's implicit summation over repeated indices. As a polynomial is characterized by its coefficients, not by the power terms which act as a basis, the fourth order tensor

$$\tilde{A}_{ijkl} = \sum_{q=1}^{3} \delta_{iq}\delta_{jq}\delta_{kq}\delta_{lq}. \tag{2}$$

is another full-fledged representation of the reference cross f_{ref}. It has only three non-zero components

$$\tilde{\mathbb{A}}_{1111} = \tilde{\mathbb{A}}_{2222} = \tilde{\mathbb{A}}_{3333} = 1.$$

2.2 Rotation of the Reference Cross

The reference cross (1) exhibits octahedral symmetry and rotations, which are isometries, preserve this symmetry. It can therefore be stated that the space of all possible cross attitudes in E^3 is the set

$$f(x_1, x_2, x_3) = f_{ref}(R_{1i}x_i, R_{2j}x_j, R_{3k}x_k) \quad , \quad R_{ij} \in \mathrm{SO}(3), \tag{3}$$

whose corresponding tensor representation reads

$$\begin{aligned} \mathbb{A}_{ijkl} = R_{im}R_{jn}R_{ko}R_{lp}\,\tilde{\mathbb{A}}_{mnop} &= \textstyle\sum_{q=1}^{3} R_{im}R_{jn}R_{ko}R_{lp}\,\delta_{mq}\delta_{nq}\delta_{oq}\delta_{pq} \\ &= \textstyle\sum_{q=1}^{3} R_{iq}R_{jq}R_{kq}R_{lq}. \end{aligned} \tag{4}$$

This tensor, noted $\mathbb{A}$, represents a general *attitude* of the cross in E^3, and the isovalue of the associated polynomial

$$f(x_1, x_2, x_3) = \mathbb{A}_{ijkl}\, x_i x_j x_k x_l = 1$$

is a rotation of the axis-aligned dice-shaped surface f_{ref}, Fig. 2 (right).

Rotation matrices play a pivotal role in these definitions. For convenience, let us define the following nomenclature:

- The indices 1, 2, 3 refer to the angles α, β and γ, i.e. the angles corresponding to the first, second and third elemental rotations, respectively.
- The matrices X, Y, Z represent the elemental rotations about the axes x_1, x_2, x_3 of the Cartesian reference cross in $\mathbb{R}^3$ (e.g., Y_1 represents a rotation about x_2 by an angle α).
- The shorthands s and c represent sine and cosine (e.g., s_1 represents the sine of α).

We have for example the rotation matrix in $\mathbb{R}^3$

$$R = Z_1X_2Z_3 = \begin{pmatrix} c_1c_3 - c_2s_1s_3 & -c_1s_3 - c_3c_2s_1 & s_2s_1 \\ c_1c_2s_3 + c_3s_1 & c_1c_2c_3 - s_1s_3 & -c_1s_2 \\ s_3s_2 & c_3s_2 & c_2 \end{pmatrix}, \tag{5}$$

which non-linearly depends on only three degrees of freedom: the angles α, β and γ.

2.3 Algebraic Structure of $\mathbb{A}_{ijkl}$

A 4th order tensor in E^3 has at most $3^4 = 81$ independent components. The specific algebraic structure of (4) makes it so, however, that the tensor space of interest for cross fields is much smaller than that, and can be characterized as a linear space $\mathcal{A}$ of dimension nine, convenient for finite element interpolation, together with a non-linear projection operator

$$\Pi : \mathcal{A} \mapsto SO(3)/O \tag{6}$$

from the 9-dimensional linear space onto a 3-dimensional nonlinear manifold.

The demonstration of this algebraic structure is in several steps. First, the number of independent components of $\mathbb{A}$ cannot be larger than the dimension of the space of homogeneous polynomials of order four with three variables, i.e., $\binom{4+3-1}{4} = 15$. This is a mere consequence of the fact that the products of coordinates, as the product of real numbers, obviously commute, $x_i x_j = x_j x_i$, and that all terms associated with components of $\mathbb{A}$ whose indice sets are permutations of each other eventually contribute to the same term in the polynomial. In mathematical terms, it amounts to require the tensor $\mathbb{A}$ be *fully symmetric*, a condition usually written

$$\mathbb{A}_{ijkl} = \mathbb{A}_{(ijkl)}$$

with

$$\mathbb{A}_{(ijkl)} = \frac{1}{24}\left(\mathbb{A}_{ijkl} + \mathbb{A}_{jikl} + \ldots\right)$$

where the 24 permutations of the set $ijkl$ are enumerated at the right-hand side.

The tensors (4) have however deeper structures, related with the unitary property

$$R^t R = I \quad , \quad R_{ik} R_{jk} = \delta_{ij}$$

of rotations matrices. The so called "partial traces"[1]

$$\mathbb{A}_{iikl} = \sum_{q=1}^{3} R_{iq} R_{iq} R_{kq} R_{lq} = \sum_{q=1}^{3} \delta_{qq} R_{kq} R_{lq} = R_{kq} R_{lq} = \delta_{kl} \tag{7}$$

leads to the 6 additional relationships

$$\mathbb{A}_{1111} + \mathbb{A}_{2211} + \mathbb{A}_{3311} = 1,$$

$$\mathbb{A}_{1122} + \mathbb{A}_{2222} + \mathbb{A}_{3322} = 1,$$

[1] Make sure to clearly distinguish sums over two repeated indices, which are implict in our notation, and sums over four repeated indices, which are explicitly written.

$$\mathbb{A}_{1133} + \mathbb{A}_{2233} + \mathbb{A}_{3333} = 1,$$
$$\mathbb{A}_{1112} + \mathbb{A}_{2212} + \mathbb{A}_{3312} = 0,$$
$$\mathbb{A}_{1113} + \mathbb{A}_{2213} + \mathbb{A}_{3313} = 0,$$
$$\mathbb{A}_{1123} + \mathbb{A}_{2223} + \mathbb{A}_{3323} = 0.$$

Other partial traces would give linearly dependent relationships, due to the full symmetry of the tensor mentioned above.

It is important to note that partial traces are conserved under affine combination of tensors. Tensors in $\mathbb{A}$ form thus a $15 - 6 = 9$ dimensional linear space, noted $\mathcal{A}$, convenient for finite element interpolation. Interestingly enough, this dimension is also that of the space of 4th order spherical harmonics, used by some authors to represent crosses [2].

2.4 $\mathbb{A}$ *is a Projector*

Let X be the set of symmetric 2^d order tensors in E^3. This is a linear space, and any tensor $\mathbf{d} \in X$ can be expanded in terms of rank one basis tensors

$$\mathbf{d} = d_{mn}\, e_m \otimes e_n \quad , \quad d_{mn} = d_{nm}$$

where e_m, $m, 1, 2, 3$, are the orthormal basis vectors of the Cartesian coordinate system. Alternatively, basis tensors rotated by a matrix $R \in SO(3)$ can be used as well,

$$\mathbf{d} = d'_{mn}\, r_m \otimes r_n \quad , \quad d'_{mn} = d'_{nm} \tag{8}$$

with now

$$r_m = R(e_m) \quad , \quad (r_m)_i = R_{ij}\delta_{jm} = R_{im}$$

the mth column vector of the rotation matrix R. The polynomials we are using in this paper to represent cross attitudes are built from special tensors $B \in X$ for which one simply has $B_{mn} = x_m x_n$.

Once the space X is appropriately characterized, the tensor $\mathbb{A}$ defined by (4) can be regarded as a linear application

$$\mathbb{A} : X \mapsto X,$$

and it is readily shown that it is a projection operator:

$$\mathbb{A}^2 = \mathbb{A} : \mathbb{A} \equiv \mathbb{A}_{ijmn}\mathbb{A}_{mnkl}$$
$$= \sum_{q=1}^{3}\sum_{s=1}^{3} R_{iq}R_{jq}R_{mq}R_{nq}R_{ns}R_{ms}R_{ks}R_{ls}$$

$$= \sum_{q=1}^{3} \sum_{s=1}^{3} R_{iq} R_{jq} R_{ks} R_{ls} \underbrace{R_{mq} R_{ms}}_{\delta_{qs}} \underbrace{R_{nq} R_{ns}}_{\delta_{qs}}$$

$$= \sum_{q=1}^{3} R_{iq} R_{jq} R_{kq} R_{lq} = \mathbb{A}_{ijkl} = \mathbb{A}, \tag{9}$$

To avoid any confusion later in the article, an eigenvector $\mathbf{d} \in X$ of a fourth order tensor $\mathbb{A} \in \mathcal{A}$ will be called an *eigentensor* and an eigenvector $r \in E^3$ of a 2^d order tensor $B \in X$ will be called an *eigenvector.*

To characterize this projection, the image by $\mathbb{A}$ of the basis tensors $r_m \otimes r_n$ in (8) is evaluated. One has

$$\begin{aligned}(\mathbb{A} : r_m \otimes r_n)\Big|_{ij} &= \mathbb{A}_{ijkl}\; R_{km} R_{ln} \\ &= \sum_{q=1}^{3} R_{iq} R_{jq} R_{kq} R_{lq} R_{km} R_{ln} \\ &= \sum_{q=1}^{3} R_{iq} R_{jq} \delta_{qm} \delta_{qn},\end{aligned}$$

from where follows

$$\mathbb{A} : (r_m \otimes r_m) = r_m \otimes r_m \quad m = 1, 2, 3 \text{ (no sum)} \tag{10}$$

$$\mathbb{A} : (r_m \otimes r_n) \quad = 0 \qquad \text{if } m \neq n. \tag{11}$$

As expected for a projector, eigen values are either 0 or 1. The eigenspace corresponding to the eigenvalues $\lambda_1 = \lambda_2 = \lambda_3 = 1$ is

$$\text{range}\, \mathbb{A} = \text{span}\, (r_1 \otimes r_1, r_2 \otimes r_2, r_3 \otimes r_3) \subset X$$

whereas that corresponding to $\lambda_4 = \lambda_5 = \lambda_6 = 0$ is the kernel space

$$\ker \mathbb{A} = \text{span}\, (r_1 \otimes r_2 + r_2 \otimes r_1, r_2 \otimes r_3 + r_3 \otimes r_2, r_3 \otimes r_1 + r_1 \otimes r_3)\,.$$

Obviously, the eigentensors $\mathbf{d}^j$ are both symmetric ($\mathbf{d}^j_{mn} = \mathbf{d}^j_{nm}$), and orthonormal to each other ($\mathbf{d}^i : \mathbf{d}^j = \delta_{ij}$) under the Frobenius norm $\| \mathbf{d} \|_F^2 = \mathbf{d} : \mathbf{d} = d_{mn} d_{nm}$.

3 Three Main Results

The three main results of the paper are now presented in this section. We first prove that any tensor $\mathbb{A} \in \mathcal{A}$ (a fully symmetric tensor obeying (7)) corresponds to a cross attitude (i.e., a rotation of the reference cross) if it is a projector $\mathbb{A}^2 = \mathbb{A}$ onto a three-dimensional subspace of X. Then, we show how to project a tensor $\mathbb{A} \in \mathcal{A}$ that is not a projector onto another tensor in $\mathcal{A}$ verifying $\mathbb{A}^2 = \mathbb{A}$ with three non-zero eigenvalues. Finally, we show the direct relationship between spherical harmonics and our representation on terms of fourth order tensors.

3.1 Sufficiency

Theorem 1 *A tensor $\mathbb{A} \in \mathcal{A}$, (fully symmetric fourth order tensor obeying the partial trace condition(7)) that is also a projector on a 3-dimensional subspace of X corresponds to a cross attitude (i.e., to a rotation of the reference cross)*

Proof If $\mathbb{A}$ is a projector onto a 3-dimensional subspace of X, there exist three orthonormal symmetric second order tensors $\mathbf{d}^a, \mathbf{d}^b, \mathbf{d}^c \in X$ such that

$$\begin{aligned} \mathbf{d}^a \otimes \mathbf{d}^a + \mathbf{d}^b \otimes \mathbf{d}^b + \mathbf{d}^c \otimes \mathbf{d}^c &= \mathbb{A} \\ \mathbf{d}^l : \mathbf{d}^m &= \delta_{lm} \; l, m = a, b, c \\ \mathbf{d}^l_{ij} &= \mathbf{d}^l_{ji} \; l = a, b, c, \; i, j = 1, 2, 3. \end{aligned} \tag{12}$$

Note that there is no implicit summation on upper indices in this proof. The key point of the proof is to show that the eigentensors $\mathbf{d}^a$, $\mathbf{d}^b$ and $\mathbf{d}^c$ commute with each other. If this is the case, they are joint diagonalizable and share therefore the same set of eigenvectors. It is then easy to see that $\mathbb{A}$ is the fourth order tensor representation of a cross.

Let

$$[\mathbf{d}^l, \mathbf{d}^m] = \mathbf{d}^l \cdot \mathbf{d}^m - \mathbf{d}^m \cdot \mathbf{d}^l$$

be the commutator of $\mathbf{d}^l$ and $\mathbf{d}^m$, of which we have to prove the Frobenius norm is zero,

$$\| [\mathbf{d}^l, \mathbf{d}^m] \|_F^2 = [\mathbf{d}^l, \mathbf{d}^m] : [\mathbf{d}^l, \mathbf{d}^m] = 0.$$

One first notes that

$$\begin{aligned} (\mathbf{d}^a \cdot \mathbf{d}^b) : (\mathbf{d}^e \cdot \mathbf{d}^f) &= d^a_{ik} d^b_{kj} d^e_{il} d^f_{lj} = \mathrm{tr}\,(\mathbf{d}^a \mathbf{d}^b \mathbf{d}^f \mathbf{d}^e) \\ &= (\mathbf{d}^b \cdot \mathbf{d}^a) : (\mathbf{d}^f \cdot \mathbf{d}^e) \end{aligned}$$

$$= (\mathbf{d}^a \cdot \mathbf{d}^e) : (\mathbf{d}^b \cdot \mathbf{d}^f)$$
$$= (\mathbf{d}^e \cdot \mathbf{d}^a) : (\mathbf{d}^f \cdot \mathbf{d}^b) \tag{13}$$

exploiting the symmetry of the individual $\mathbf{d}^l$ tensors, and all possible reorganizations of the matrix products. As the contraction operator : is also symmmetric, there are thus eight equivalent argument permutations (out of 24) for such scalar quantities. With this, one shows that

$$\begin{aligned}\| [\mathbf{d}^a, \mathbf{d}^b] \|_F^2 &= (\mathbf{d}^a \cdot \mathbf{d}^b - \mathbf{d}^b \cdot \mathbf{d}^a) : (\mathbf{d}^a \cdot \mathbf{d}^b - \mathbf{d}^b \cdot \mathbf{d}^a) \\ &= (\mathbf{d}^a \cdot \mathbf{d}^b) : (\mathbf{d}^a \cdot \mathbf{d}^b) - (\mathbf{d}^a \cdot \mathbf{d}^b) : (\mathbf{d}^b \cdot \mathbf{d}^a) - \\ &\quad (\mathbf{d}^b \cdot \mathbf{d}^a) : (\mathbf{d}^a \cdot \mathbf{d}^b) + (\mathbf{d}^b \cdot \mathbf{d}^a) : (\mathbf{d}^b \cdot \mathbf{d}^a) \\ &= 2(\mathbf{d}^a \cdot \mathbf{d}^a) : (\mathbf{d}^b \cdot \mathbf{d}^b) - 2(\mathbf{d}^a \cdot \mathbf{d}^b) : (\mathbf{d}^b \cdot \mathbf{d}^a) \end{aligned} \tag{14}$$

the last two terms being not reducible to each other by the permutation rules given above. As

$$(\mathbf{d}^a \cdot \mathbf{d}^b) : (\mathbf{d}^e \cdot \mathbf{d}^f) = \mathrm{tr}\,(\mathbf{d}^a \mathbf{d}^b \mathbf{d}^f \mathbf{d}^e),$$

the identity (14) can also be interpreted as

$$\| [\mathbf{d}^a, \mathbf{d}^b] \|_F^2 = \mathrm{tr}\,(\mathbf{d}^a \mathbf{d}^a \mathbf{d}^b \mathbf{d}^b - \mathbf{d}^a \mathbf{d}^b \mathbf{d}^a \mathbf{d}^b) = \mathrm{tr}\,\left((\mathbf{d}^a)^2 (\mathbf{d}^b)^2 - (\mathbf{d}^a \mathbf{d}^b)^2\right). \tag{15}$$

The identity tensor I being in the range of $\mathbb{A}$, it is an eigen tensor of $\mathbb{A}$, one has thus

$$\delta_{ij} = A_{ijkl}\delta_{kl} = A_{iklj}\delta_{kl}$$

where the full symmetry of $\mathbb{A}$ has been used. This reads, without components,

$$I = \mathbf{d}^a \mathbf{d}^a + \mathbf{d}^b \mathbf{d}^b + \mathbf{d}^c \mathbf{d}^c = (\mathbf{d}^a)^2 + (\mathbf{d}^b)^2 + (\mathbf{d}^c)^2,$$

wherefrom directly follows

$$(\mathbf{d}^a)^2 = (\mathbf{d}^a)^4 + (\mathbf{d}^a)^2 (\mathbf{d}^b)^2 + (\mathbf{d}^a)^2 (\mathbf{d}^c)^2. \tag{16}$$

On the other hand, using now the fact that $\mathbf{d}^a$ is an eigen tensor of $\mathbb{A}$, one has

$$d^a_{ij} = A_{ijkl} d^a_{kl} = d^a_{ij} d^a_{kl} d^a_{kl} + d^b_{ij} d^b_{kl} d^a_{kl} + d^c_{ij} d^c_{kl} d^a_{kl}$$

and, using again the full symmetry of $\mathbb{A}$

$$d^a_{ij} = d^a_{ik} d^a_{lj} d^a_{kl} + d^b_{ik} d^b_{lj} d^a_{kl} + d^c_{ik} d^c_{lj} d^a_{kl}$$

so that

$$\mathbf{d}^a = (\mathbf{d}^a)^3 + \mathbf{d}_b\mathbf{d}_a\mathbf{d}_b + \mathbf{d}_c\mathbf{d}_a\mathbf{d}_c$$

and premultiplying with $\mathbf{d}^a$

$$(\mathbf{d}^a)^2 = (\mathbf{d}^a)^4 + (\mathbf{d}_a\mathbf{d}_b)^2 + (\mathbf{d}_a\mathbf{d}_c)^2. \tag{17}$$

Substraction of (17) and (16) yields

$$0 = (\mathbf{d}_a\mathbf{d}_b)^2 + (\mathbf{d}_a\mathbf{d}_c)^2 - (\mathbf{d}^a)^2(\mathbf{d}^b)^2 + (\mathbf{d}^a)^2(\mathbf{d}^c)^2,$$

the trace of which gives, using (15),

$$0 = \| [\mathbf{d}^a, \mathbf{d}^b] \|_F^2 + \| [\mathbf{d}^a, \mathbf{d}^c] \|_F^2.$$

As this is a sum of positive terms, both terms are zero, and we have proven that $\mathbf{d}^a$ commutes with $\mathbf{d}^b$ and $\mathbf{d}^c$.

As $(\mathbf{d}^a, \mathbf{d}^b, \mathbf{d}^c)$ are symmetric and commute, there exist an othonormal basis $(r^1, r^2, r^3) \in (\mathbb{R}^3)^3$ such as:

$$\begin{aligned}
\mathbf{d}^a &= \alpha_1 r^1 \otimes r^1 + \alpha_2 r^2 \otimes r^2 + \alpha_3 r^3 \otimes r^3 \;\; \alpha_1, \alpha_2, \alpha_3 \in \mathbb{R} \\
\mathbf{d}^b &= \beta_1 r^1 \otimes r^1 + \beta_2 r^2 \otimes r^2 + \beta_3 r^3 \otimes r^3 \;\; \beta_1, \beta_2, \beta_3 \in \mathbb{R} \\
\mathbf{d}^c &= \gamma_1 r^1 \otimes r^1 + \gamma_2 r^2 \otimes r^2 + \gamma_3 r^3 \otimes r^3 \;\; \gamma_1, \gamma_2, \gamma_3 \in \mathbb{R}
\end{aligned} \tag{18}$$

We will now show that $r_i \otimes r_i, i \in \{1, 2, 3\}$ are eigentensors of $\mathbb{A}$. First, we know that $\mathbf{d}^l$ are orthogonal and of norm one. So, $((\alpha_1, \alpha_2, \alpha_3), (\beta_1, \beta_2, \beta_3), (\gamma_1, \gamma_2, \gamma_3))$ forms an orthonormal basis of $\mathbb{R}^3$.

Therefore, it exists a unique vector $v \in \mathbb{R}^3$ such as:

$$\begin{pmatrix} \alpha_1 & \beta_1 & \gamma_1 \\ \alpha_2 & \beta_2 & \gamma_2 \\ \alpha_3 & \beta_3 & \gamma_3 \end{pmatrix} v = \begin{pmatrix} 1 \\ 0 \\ 0 \end{pmatrix} \tag{19}$$

and we have

$$v_1\mathbf{d}^a + v_2\mathbf{d}^b + v_3\mathbf{d}^c = r_1 \otimes r_1 \tag{20}$$

As, $\mathbf{d}^l$ are eigentensors of $\mathbb{A}$ associated to eigenvalue 1,

$$\begin{aligned}
\mathbb{A} : (r_1 \otimes r_1) &= \mathbb{A} : (v_1\mathbf{d}^a + v_2\mathbf{d}^b + v_3\mathbf{d}^c) \\
&= v_1\mathbf{d}^a + v_2\mathbf{d}^b + v_3\mathbf{d}^c \\
&= (r_1 \otimes r_1)
\end{aligned} \tag{21}$$

Consequently, $r_1 \otimes r_1$ is an eigentensor of $\mathbb{A}$ assiociated to eigenvalue 1. We can show in the same way that $(r_2 \otimes r_2)$ and $(r_3 \otimes r_3)$ are also eigentensors of $\mathbb{A}$ assiociated to eigenvalue 1.

Thus, as $\mathbb{A}$ is a projector with only three non zero eigenvalues, we finally have:

$$\mathbb{A} = r_1 \otimes r_1 \otimes r_1 \otimes r_1 + r_2 \otimes r_2 \otimes r_2 \otimes r_2 + r_3 \otimes r_3 \otimes r_3 \otimes r_3 \tag{22}$$

Therefore, $\mathbb{A}$ is the representation of the cross with orthogonal directions (r_1, r_2, r_3).

3.2 Recovery

The representation that is advocated here relies heavily on the computation of eigentensors of fourth order tensors. Disappointingly, numerical tools for linear algebra are designed to manipulate vectors and matrices. Hopefully, it is possible to represent symmetric fourth order tensors as matrices.

A fourth order tensor $\mathbb{A}$ endowed with minor symmetry conditions $\mathbb{A}_{ijkl} = \mathbb{A}_{jikl} = \mathbb{A}_{ijlk}$ has 36 independant components. It is useful to write it in the so called Mandel notation as the following matrix 6×6 matrix:

$$A = \begin{pmatrix} \mathbb{A}_{1111} & \mathbb{A}_{1122} & \mathbb{A}_{1133} & \sqrt{2}\mathbb{A}_{1123} & \sqrt{2}\mathbb{A}_{1113} & \sqrt{2}\mathbb{A}_{1112} \\ \mathbb{A}_{2211} & \mathbb{A}_{2222} & \mathbb{A}_{2233} & \sqrt{2}\mathbb{A}_{2223} & \sqrt{2}\mathbb{A}_{2213} & \sqrt{2}\mathbb{A}_{2212} \\ \mathbb{A}_{3311} & \mathbb{A}_{3322} & \mathbb{A}_{3333} & \sqrt{2}\mathbb{A}_{3323} & \sqrt{2}\mathbb{A}_{3313} & \sqrt{2}\mathbb{A}_{3312} \\ \sqrt{2}\mathbb{A}_{2311} & \sqrt{2}\mathbb{A}_{2322} & \sqrt{2}\mathbb{A}_{2333} & 2\mathbb{A}_{2323} & 2\mathbb{A}_{2313} & 2\mathbb{A}_{2312} \\ \sqrt{2}\mathbb{A}_{1311} & \sqrt{2}\mathbb{A}_{1322} & \sqrt{2}\mathbb{A}_{1333} & 2\mathbb{A}_{1323} & 2\mathbb{A}_{1313} & 2\mathbb{A}_{1312} \\ \sqrt{2}\mathbb{A}_{1211} & \sqrt{2}\mathbb{A}_{1222} & \sqrt{2}\mathbb{A}_{1233} & 2\mathbb{A}_{1223} & 2\mathbb{A}_{1213} & 2\mathbb{A}_{1212} \end{pmatrix}. \tag{23}$$

Major symmetry conditions $\mathbb{A}_{ijkl} = \mathbb{A}_{klij}$ ensure that A is symmetric. Factors two and $\sqrt{2}$ in (23) allow to write the cross representation as the following usual quadratic form:

$$(x \otimes x)^t \; A \; (x \otimes x) = 1. \tag{24}$$

with

$$x \otimes x = \left(x_1^2 \;\; x_2^2 \;\; x_3^2 \;\; \sqrt{2}x_2x_3 \;\; \sqrt{2}x_1x_3 \;\; \sqrt{2}x_1x_2\right)^t.$$

Let us now compute Mandel's representation of the reference cross $\tilde{\mathbb{A}}$ (see (1)):

$$\tilde{A} = \begin{pmatrix} 1&0&0&0&0&0 \\ 0&1&0&0&0&0 \\ 0&0&1&0&0&0 \\ 0&0&0&0&0&0 \\ 0&0&0&0&0&0 \\ 0&0&0&0&0&0 \end{pmatrix} \tag{25}$$

In a previous section, we have shown that only nine scalar parameters $(a_1, \ldots, a_9)$ are required to represent $\mathbb{A}$. Taking into account symmetries and partial traces, we can write

$$A=\begin{pmatrix} a_1 & & & & & \\ \frac{1}{2}(1+a_3-a_2-a_1) & a_2 & & & SYM & \\ \frac{1}{2}(1+a_2-a_3-a_1) & \frac{1}{2}(1+a_1-a_2-a_3) & a_3 & & & \\ -\sqrt{2}(a_4+a_5) & \sqrt{2}a_4 & \sqrt{2}a_5 & 1+a_1-a_3-a_2 & & \\ \sqrt{2}a_6 & -\sqrt{2}(a_6+a_7) & \sqrt{2}a_7 & -2(a_8+a_9) & 1+a_2-a_3-a_1 & \\ \sqrt{2}a_8 & \sqrt{2}a_9 & -\sqrt{2}(a_8+a_9) & -2(a_6+a_7) & -2(a_4+a_5) & 1+a_3-a_2-a_1 \end{pmatrix} \tag{26}$$

with the following correspondances between the $\mathbb{A}_{ijkl}$'s and the a_i's:

$$a_1=\mathbb{A}_{1111}\,, a_2=\mathbb{A}_{2222},\ a_3=\mathbb{A}_{3333},$$

$$a_4=\mathbb{A}_{2322}\,, a_5=\mathbb{A}_{2333},\ a_6=\mathbb{A}_{1311},$$

$$a_7=\mathbb{A}_{1333}\,, a_8=\mathbb{A}_{1211},\ a_9=\mathbb{A}_{1222}.$$

Mandel's notation allows to write tensor contractions as matrix products. For example, $\mathbb{A}:\mathbb{A}=\mathbb{A}$ (see (9)) is written using Mandel's notation as $A\cdot A=A$.

Eigenvectors of A are the eigentensors of $\mathbb{A}$. Their six components are the six independent entries of eigentensors $\mathbf{d}^k$ that are symmetric second order tensors. The two following MATLAB functions allow to transform fourth order tensors $\mathbb{A}$ into Mandel's form and transform eigenvectors of A into second order tensors. We also see factors of $\sqrt{2}$ that accounts for the symmetry of A.

```
function D = Vec6ToTens2 (v)
    s = 2^(1./2.);
    D = [
       v(1)  , v(6)/s , v(5)/s ;
       v(6)/s , v(2)  , v(4)/s ;
       v(5)/s , v(4)/s , v(3)  ;
    ];
    end
```

```
function a = Tens4ToMat6 (A)
s = 2^(1./2.);
a = [
  A(1,1,1,1)  , A(1,1,2,2)  , A(1,1,3,3)  ,
   s*A(1,1,2,3), s*A(1,1,1,3), s*A(1,1,1,2)  ;
  A(2,2,1,1)  , A(2,2,2,2)  , A(2,2,3,3)  ,
    s*A(2,2,2,3), s*A(2,2,1,3), s*A(2,2,1,2)  ;
  A(3,3,1,1)  , A(3,3,2,2)  , A(3,3,3,3)  ,
    s*A(3,3,2,3), s*A(3,3,1,3), s*A(3,3,1,2)  ;
  s*A(2,3,1,1)  , s*A(2,3,2,2)  , s*A(2,3,3,3)  ,
    2*A(2,3,2,3), 2*A(2,3,1,3), 2*A(2,3,1,2)  ;
  s*A(1,3,1,1)  , s*A(1,3,2,2)  , s*A(1,3,3,3)  ,
    2*A(1,3,2,3), 2*A(1,3,1,3), 2*A(1,3,1,2)  ;
  s*A(1,2,1,1)  , s*A(1,2,2,2)  , s*A(1,2,3,3)  ,
    2*A(1,2,2,3), 2*A(1,2,1,3), 2*A(1,2,1,2)  ;
```

```
];
end
```

Note that those MATLAB routines are made for testing and that 3D large codes will only manipulate the nine nodal unknowns a_i.

Computing 3D cross fields implies to propagate tensors that have known values on the boundary of a 3D domain inside the domain. Assume a tensor $\mathbb{A}$ that has the right structure and that is such that $\mathbb{A} : \mathbb{A} = \mathbb{A}$. With such properties, we know that $\mathbb{A}$ is a rotation of $\tilde{\mathbb{A}}$. The first important issue is about backtracking R from $\mathbb{A}$ i.e. find the three orthonormal column vectors r^q of R that form $\mathbb{A}$ through Eq. (4).

An eigentensor $\mathbf{d}^n$ of $\mathbb{A}$ that is associated with eigenvalue one is the linear combination

$$\mathbf{d}^n_{ij} = \sum_{q=1}^{3} c_{nq} r^q_i r^q_j.$$

We have

$$\mathbf{d}^n_{im} r^k_m = c_{nk} r^k_i$$

which means that the eigenvectors of $\mathbf{d}^n$ are indeed the r^k. One issue here could be that $\mathbf{d}^n$ is not of full rank. Yet, the sum

$$\mathbf{d} = \sum_{k=1}^{3} \mathbf{d}^k$$

is of full rank. Eigenvectors of $\mathbf{d}$ are the wanted three directions. Assume a representation $\mathbb{A}$ in of the form (26) and let us recover rotation matrix R. The following MATLAB code recovers the rotation matrix R starting from a tensor $\mathbb{A}$ that is a rotation.

```
function R = Tens4ToRotation (A)
a     = Tens4ToMat6 (A); % transform A into
  its matrix form
[V,D] = eig (a)    ; % compute eigenspace
[X,I] = sort(diag(D)); % sort eigenvalues
% compute the sum of eigentensors of A associated
% to eigenvalues equal to 1
V2      = Vec6ToTens2 (V(:,I(4))+V(:,I(5))
 +V(:,I(6)));
[R,d2] = eig(V2); % get rotation matrix R
end
```

This code has been tested to thousands of random rotations, giving the right answer in a 100% robust fashion.

The aim of our work is to build smooth cross fields in general 3D domains. For that, we will solve a boundary value problem for the nine linearly independant components $(a_1, \ldots, a_9)$ of the tensor representation. Consider two representations X and Y with their representation vectors $(x_1, \ldots, x_9)$ and $(y_1, \ldots, y_9)$ Any smoothing procedure computes (weighted) averages of such representations. For example, representation vector

$$\frac{1}{2}(x_1 + y_1, \ldots, x_9 + y_9)$$

allows to build Mandel's representation $Z = \frac{1}{2}(X + Y)$ that as the same structure as matrix (26).

Assume a cross attitude $A(\alpha, \beta, \gamma)$ that depends on Euler angles α, β and γ. The projection of Z into the space of rotations of the reference cross is defined as the cross attitude A that verifies

$$A = \min_{\alpha,\beta,\gamma} \|A(\alpha, \beta, \gamma) - Z\|.$$

The following function

```
function P = projection (A)
    b_guess = [0 0 0];
    [b_guess(1) b_guess(2) b_guess(3)] =
     EulerAngles(Tens42Rotation (A));
    vA = Tensor4ToMat6 (A);
    fun = @(x) norm(Tensor4ToMat6(makeTensor
     (makeEulerRotation (x(1),x(2),x(3))))-vA) ;
    b_min = fminsearch(fun, b_guess);
    P = makeTensor (makeEulerRotation (b_min(1),
      b_min(2),b_min(3)));
end
```

allows to compute such a projection. In that function, we choose as an initial guess for Euler angles the value computed by `Tens4ToRotation` which uses the eigenspace of A relative to its three largest eigenvalues. Figure 3 shows that this initial guess is indeed a very good approximation of the projection. In reality, it is such a good approximation that it can be used as is without doing the exact minimization.

Let's define $\Pi_1 A$ is the exact projection of A on the space of SO(3) / O and $\Pi_2 A$ is the approximation of Π_1 computed using function `Tens42Rotation`. On Fig. 3 is displayed with blue dots, for 4000 random tensor A, $\Pi_2 A$ regarding $\Pi_1 A$. The red line represents the function $\Pi_2 X = \Pi_1 X$, $X \in \mathcal{A}$. The less accurate the projection approximation Π_2 will be for a tensor A, the further the blue dot representing $\Pi_2 A$ regarding $\Pi_1 A$ will be from the red line. We see that the approximation Π_2 is always very good with respect to the projection Π_1, while being extremely simple and fast to compute.

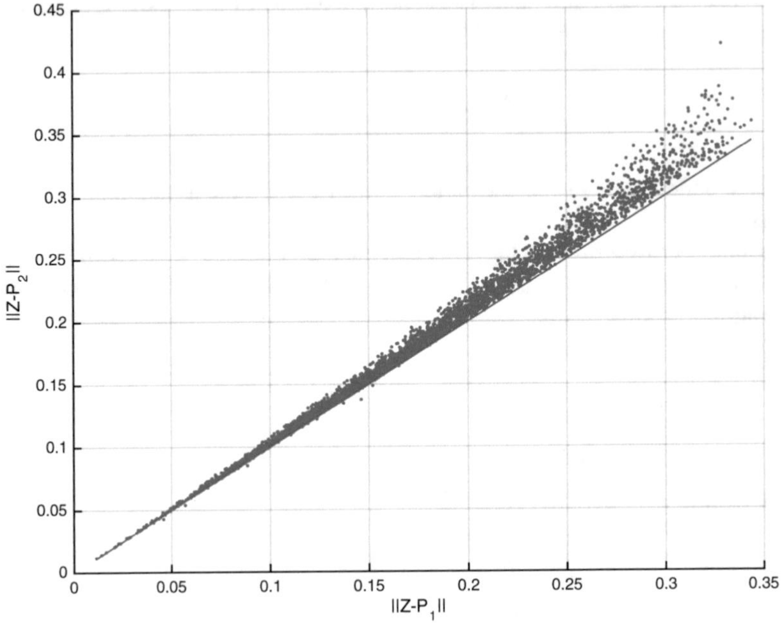

Fig. 3 Projection of 4000 random tensors Z. Π_1 is the true projection while Π_2 is the approximation computed using function `Tens42Rotation`

3.3 Relation with Spherical Harmonics

Harmonic polynomials $h(x)$ are polynomials that are such $\nabla^2 h = 0$. Consider the rotated diced cube polynomial representation

$$\alpha(x) = \sum_{q=1}^{3} (r^q \cdot x)^4$$

We have

$$\nabla^2 \alpha = \sum_{j=1}^{3} \frac{\partial^2 \alpha}{\partial x_j^2} = 12 \sum_{q=1}^{2} (r^q \cdot x)^2 (r_j^q)^2$$
$$= 12 \sum_{q=1}^{3} \left[(r^q \cdot x)^2 \left((r_1^q)^2 + (r_2^q)^2 + (r_2^q)^2 \right) \right] = 12 \sum_{q=1}^{3} (r^q \cdot x)^2.$$

The equation $\sum_{q=1}^{3} (r^q \cdot x)^2$ is the one of the unit sphere that is invariant by rotation. Thus,

$$\nabla^2 \alpha = 12\,|x|^2\,.$$

Representation polynomial $\alpha(x)$ is thus not harmonic. Yet, acknowledging that

$$\nabla^2 |x|^4 = 20|x|^2,$$

we can define the following projection operator of diced cubes onto harmonic polynomials

$$P_{\mathcal{H}_4}(\alpha(x)) = \alpha(x) - \frac{3}{5}|x|^4.$$

Operator $P_{\mathcal{H}_4}$ essentially remove three fifth of a sphere to the diced cube so that the representation retains its symmetry properties while becoming itself harmonic. Let us show that $P_{\mathcal{H}_4}$ is an orthogonal projector with respect to a norm that is related to spherical harmonics. Consider the unit sphere S^2 and compute

$$\int_{S^2} h(x)\left[P_{\mathcal{H}_4}(\alpha(x)) - \alpha(x)\right] dx = -\frac{3}{5}\int_{S^2} h(x)|x|^4 dx = -\frac{3}{5}\int_{S^2} h(x)dx.$$

Harmonic functions are endowed with the mean value property which states that the average of $h(x)$ over any sphere centered at c is equal to $h(c)$. So,

$$\int_{S^2} h(x)\left[P_{\mathcal{H}_4}(\alpha(x)) - \alpha(x)\right] dx = -\frac{3}{5}\int_{S^2} h(0)dx.$$

Harmonic polynomials are homogeneous so $h(0) = 0$ and operator $P_{\mathcal{H}_4}$ is an orthogonal projector onto fourth order spherical harmonics.

As an example, consider our reference diced cube that is represented by $\alpha(x) = x_1^4 + x_2^4 + x_3^4$. Its projection onto $\mathcal{H}_4$ is

$$P_{\mathcal{H}_4}(x_1^4 + x_2^4 + x_3^4) = \frac{2}{5}(x_1^4 + x_2^4 + x_3^4 - 3(x_1^2x_2^2 + x_1^2x_3^2 + x_2^2x_3^2))$$

This is indeed interesting to see that, for $x \in S^2$, we have

$$P_{\mathcal{H}_4}(x_1^4 + x_2^4 + x_3^4) = \sqrt{\frac{12\pi}{7}\frac{16}{3}}\left(\sqrt{\frac{7}{12}}Y_{4,0} + \sqrt{\frac{5}{12}}Y_{4,4}\right)$$

where $Y_{4,j}$, $j = -4, \ldots, 4$ are the orthonormalized real spherical harmonics. In [2], authors define their reference frame as

$$\tilde{F} = \sqrt{\frac{7}{12}} Y_{4,0} + \sqrt{\frac{5}{12}} Y_{4,4}$$

which is to a constant the orthogonal projection of our reference frame onto spherical harmonics.

4 Practical Computations

Assume a domain Ω with its non smooth boundary Γ that may contain sharp edges and corners. Our aim is to find a crossfield F that is smooth, and such as for all $x \in \Gamma$, $F(x)$ has one direction aligned to the boundary normal $n(x)$. Here, a simple smoothing procedure that consists in locally averaging cross attitudes at every vertex of a mesh that covers Ω is proposed. The issue of boundary conditions is not treated here.

Let $\mathbf{a}_i \in \mathrm{R}^9$ the representation vector at vertex i. The energy function that is considered is pretty standard

$$E = \frac{1}{2} \sum_{ij} \|A_i - A_j\|_F^2 \tag{27}$$

where $\sum_{ij}$ is the sum over all edges of the mesh and $\|\cdot\|_F$ is the Frobenius norm. The energy is minimized in an explicit fashion. The global smoothing algorithm follows the steps:

- Solution is initialized on the whole domain,
- tensor representations $\mathbf{a}_i \in \mathbb{R}^9$ are averaged at every vertex of the mesh,
- then projected on SO(3) in the approximate fashion developped above.

The algorith is stopped when the global energy E has decreased by a factor of 10^4. We have generated three uniform meshes of a unit sphere with different resolutions. Results are presented in Fig. 4. The iterations were started with every node assigned to the reference frame aligned with the axis. Crosses with values of η in the range $\eta \in [0.3, 0.5]$ are drawn on the figures. Figures show the usual polycube decomposition of the sphere with 12 singular lines made of "cylinders" that form an internal topological cube plus eight singular lines connecting the corners of the topological cube to the surface. Refining the mesh allows to produce more detailed representation of the decomposition. Our method is significantly faster than the ones using spherical harmonics thanks to the efficient projection operator that only requires to compute eigenvectors and eigenvalues of 3×3 and 6×6 symmetric matrices.

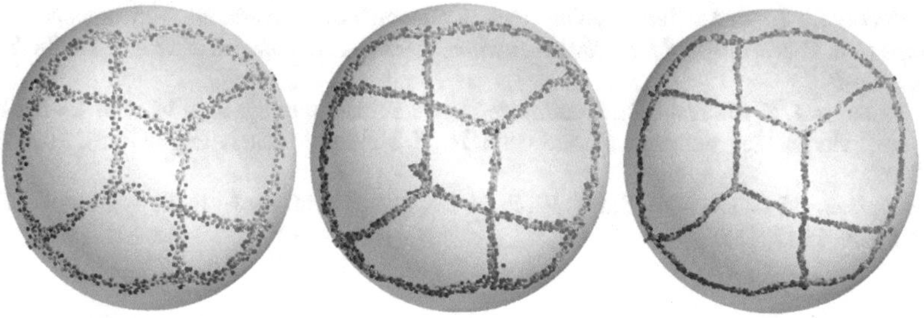

Fig. 4 Computation of cross field on a sphere meshed with tetrahedra. The three meshes used contain respectively 447,405, 2,124,801 and 6,128,555 tetrahedra. Resolution time for reducing the residual from 1 to 10^{-5} was respectively 3 s, 34 s and 81 s

5 Conclusion

The method to represent and compute crossfields on 3D domains offers a lot of advantages.At first, the new formulation is, to our opinion, way easier to understand geometrically: rotations of tensors, recovery procedures and projections have a clear geometrical representation. Then, we have shown that there exists a one-to-one relationship between our representation and fourth order spherical harmonics. It should be possible to build a 9×9 matrix that allows to change of base. The fourth order tensor representation used allows to approximate in a very efficient way projections on the crosses space $\mathcal{F}$. The direct consequence is a fast resolution of the smoothing problem.

We are aware that this paper is quite theoretical: way more practical results about this new representation are in our hands: detection of singularities, boundary conditions, norms. . . Due to page limitations, we have deliberatlely made the choice to present basic results. More practical aspects of that new representation as well as computations of cross fields on complex geometries will appear in forthcoming articles.

Acknowledgements This research is supported by the European Research Council (project HEXTREME, ERC-2015-AdG-694020) and by the Fond de la Recherche Scientifique de Belgique (F.R.S.-FNRS).

References

1. J. Huang, Y. Tong, H. Wei, H. Bao, Boundary aligned smooth 3D cross-frame field, in *ACM Transactions on Graphics (TOG)*, vol. 30 (ACM, New York, 2011), p. 143
2. N. Ray, D. Sokolov, B. Lévy, Practical 3D frame field generation. ACM Trans. Graph. (TOG) **35**(6), 233 (2016)
3. K. Takayama, M. Okabe, T. Ijiri, T. Igarashi, Lapped solid textures: filling a model with anisotropic textures, in *ACM Transactions on Graphics (TOG)*, vol. 27 (ACM, New York, 2008), p. 53

4. V. Vyas, K. Shimada, Tensor-guided hex-dominant mesh generation with targeted all-hex regions, in *Proceedings of the 18th International Meshing Roundtable* (Springer, Berlin, 2009), pp. 377–396
5. S. Yamakawa, K. Shimada, Fully-automated hex-dominant mesh generation with directionality control via packing rectangular solid cells. Int. J. Numer. Methods Eng. **57**(15), 2099–2129 (2003)
6. G.-X. Zhang, S.-P. Du, Y.-K. Lai, T. Ni, S.-M. Hu, Sketch guided solid texturing. Graph. Model. **73**(3), 59–73 (2011)

Medial Axis Based Bead Feature Recognition for Automotive Body Panel Meshing

Jonathan E. Makem, Harold J. Fogg, and Nilanjan Mukherjee

Abstract As a feature sensitive meshing investigation, this paper focuses on *beads* which are tangent continuous, high curvature, raised surfaces meant to stiffen and enhance the durability and specific strength of automotive body panels. An improvised and enhanced medial axis based strategy is proposed for identifying three broad types of bead features. Appropriate boundary discretisation, inclusion of zero medial vertex case for annulus identification, medial axis topology modifications to eliminate undesirable pathologies, T-junction squaring with cubic filtering smoothing highlight some of the improvisations to the medial axis technology employed. *Ridge curves* representing the crest lines of the bead are extracted and inserted on the face. A combination of multi-blocking, clamping and face node-loop insertion, followed by boundary connection strategies are used to generate high fidelity, feature sensitive, quasi-structured meshes.

1 Introduction

Beads constitute the most important feature in contemporary automotive vehicle bodies. These are tangent continuous, medium to high curvature feature faces, usually single or two-looped, raised above the flat panel zones to provide structural stiffness, enhance durability and optimise panel weight, as shown in Fig. 1. Beads are manufactured by a deep-drawing process and their shapes and distribution are determined based on structural stress, NVH behaviour, crash response etc.

Beads are usually defined by many geometry parameters like curvature, form geometry, arrangement, height etc. It is imperative that the finite element mesh model capture these characteristics with high fidelity to ensure desired accuracy

J. E. Makem (✉) · H. J. Fogg
Francis House, Cambridge, UK
e-mail: jonathan.makem@siemens.com

N. Mukherjee
Siemens, Milford, OH, USA

X. Roca, A. Loseille (eds.), *27th International Meshing Roundtable*,
Lecture Notes in Computational Science and Engineering 127,
https://doi.org/10.1007/978-3-030-13992-6_7

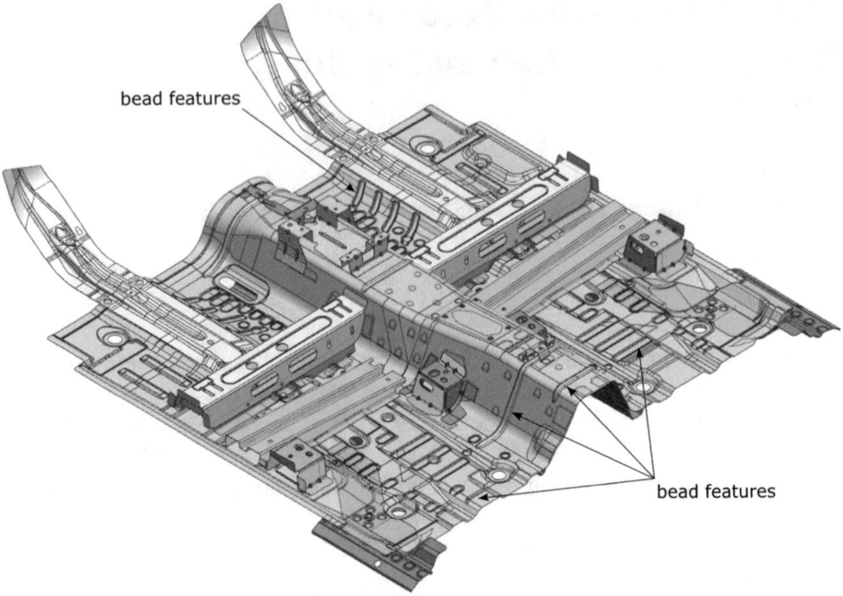

Fig. 1 Automotive body panel assembly, courtesy of GM-Opel [1]

in the plethora of analyses performed on these panels. In the various analyses on the body panel different qualities and behaviours are targeted. For example, the structural engineer focuses on panel strength, the crash analyst on shock absorption, local buckling strength and crack propagation and the NVH engineer attempts to recover panel stiffness and lower levels of air and structure-borne noise and vibration leading to more efficient and lighter automotive body panels. Bead identification and modelling play a key role in meeting these goals.

This paper describes a medial axis based strategy for bead identification and classification. An efficient and robust framework for computing the medial axis from a 2-D Delaunay tessellation is proposed. Simple strategies are incorporated for robustly optimising the medial axis topology and for improving its geometric shape. This enables its use for robust feature recognition. Based on the bead feature type and the ridge curve data generated to represent the crest line of the feature, the surface mesher employs feature specific methods to generate a desired mesh whose local parameters can be userdriven.

2 Previous Work

By far the most extensively used and practical form of medial axis computation is derived from a discretised point set of the boundary. These boundary-sampling based approaches generally approximate the medial axis from the Voronoi diagram or its dual, the Delaunay tessellation of the point set. Amenta et al. [2] proposed a "Power Crust" method to extract the medial axis from the Voronoi vertices of a sample point set. An inverse transform is then applied to reconstruct the original shape. Dey et al. [3] describe an algorithm which generates a sub-plot of the Voronoi diagram that approximates the medial axis from the Hausdorff distance. However, a major limitation with the approach is that the sample points are only generated on smooth surfaces. Li et al. [4] presented a "Q-MAT" technique which generates a piecewise linear approximation of the medial axis derived from an initial sample mesh using quadratic error minimisation for axis refinement. Its main drawback is its inability to preserve very sharp features on the boundary. Sun et al. [5] propose a more efficient approximation error estimator that evaluates the 1-sided Hausdorff distance from the input shape to the boundary represented by the medial axis. The smoothness of the final medial mesh is questionable and the authors suggest further refinement by expensive re-meshing.

There are several alternatives to using the aforementioned geometric approximation error to govern the axis simplification process. Foskey et al. [6] observe that angle-based filtration often causes significant changes in the topology of the medial axis. Chassard et al. [7] devised the λ-medial axis which utilises the circumradius of the closest boundary point to a medial edge point. If the circumradius of a medial edge point is smaller than the λ tolerance, it is removed. Nevertheless, when the feature size changes dramatically, the approach is not consistent as small values of λ will not successfully prune medial points on larger features. Miklos et al. [8] developed the Scaled Axis Transform (SAT) where all medial discs are scaled by a factor, $s > 1.0$. Any scaled disc which overlaps another is removed.

Aside from the above-mentioned boundary sampling techniques, a different method and indeed one of the earliest approaches [9] to computing a medial axis is referred to as "thinning". Given a voxel-based representation of the initial figure, the premise of this approach is to incrementally remove voxels from the boundary until enough voxels have been sufficiently removed to reveal the medial axis. In any case, an inherent limitation with thinning is that the final axis is normally off-center and generally exhibits a non-smooth, noisy definition [10]. A more advanced approach which yields a better solution is to use a distance transform [11, 12]. However, these techniques have only been applied to relatively basic geometries.

Theoretically, it is possible to derive an exact definition of the medial axis for shapes defined by semi-algebraic sets, each set the solution to a finite system of algebraic equations and inequalities. Attali et al. [13] observes however, that even for simple planar shapes bounded by primitive curves the algebraic complexities in the axis computation are severe. Aichholzer et al. [14] implemented an approach that approximates the shape boundary to a series of bi-arcs. An algorithm is then applied to shapes which are bounded by circular arcs to generate the medial axis. But the approach was limited to single loop faces. Buchegger et al. [15] improved this work by introducing a regularisation method aimed at simplifying the medial axis by smoothing the boundary of the domain and reducing the number of local curvature extrema while maintaining a low approximation error.

Peng et al. [16] developed a locus method associated with the moving Frenet Frame [17] for generating the medial axis on B-rep models to simplify fillet features. A series of insertion points are established within close proximity of the fillet and are incrementally adjusted along a direction derived from the Frenet Frame formula. The algorithm finally converges when the point is equidistant from opposing boundary edges. The method is applied only to rectangular fillet faces.

3 Medial Axis Generation

Applications in the fields of motion planning [18, 19], feature recognition [20], surface reconstruction [21, 22] shape analysis [23] and (quasi-)structured meshing [24–26], to name a few, all utilise the medial axis to a high degree. Therefore, an extensively used tool such as this which is heavily relied on requires an efficient and reliable computation. Moreover, because the medial axis is inherently unstable, especially on geometries of industrial complexity, regularisation techniques must be employed to counteract this tendency.

3.1 Boundary Discretisation

An appropriately sized Delaunay mesh is first generated in 2-D. A reliable flattening algorithm [27] facilitates the creation of accurate medial axes on non-planar faces. The circumcentres of triangles of the Delaunay mesh are approximately located on the medial axis. For regions which are densely tessellated, circumcentres of adjacent triangles may be super-imposed which subsequently requires a complex system of collapse and update operations to produce a viable medial topology. This highlights the fundamental problem of adequate boundary discretisation. Dense point sampling on the boundary produces a more accurate approximation of the medial axis at the expense of more triangles which will also have a downstream impact on the computational efficiency of the topology march. Conversely, a sparse discretisation on the boundary will result in a coarse tessellation and a poorly defined medial object. Hence, the problem of producing an appropriate boundary sampling remains non-trivial as any improvement in accuracy by modifying the sample size will be negated by an increase in computational cost.

In industry, particularly in Computer Aided Engineering (CAE) applications, complex CAD geometries are often represented by facetted models which don't have a precise mathematical definition. Consequently, algebraic methods [14] for computing an adequate discretisation are not a feasible solution. Alternatively, as the underlying polygon faceting has already been sufficiently sized within acceptable bounds to respect the original parent CAD geometry, it is an acceptable compromise to use this information to determine an appropriate sample size for the Delaunay

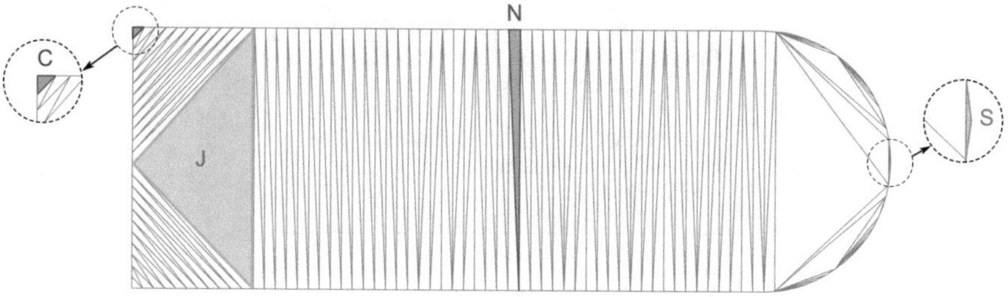

Fig. 2 2-D Delaunay mesh with triangle types identified

Table 1 Delaunay triangle types

Type(t)	Adj(t).size()	Other cond.
J	3	
N	2	
C	1	θ € θ_{threch}
S	1	θ Σ θ_{threch}

mesh. This avoids more expensive alternatives which seek to minimise geometric approximation errors by creating multiple medial objects [4, 5].

3.2 Delaunay Triangle Classification

Before the medial axis can be computed, the 3-D surface is flattened [27] and a constrained Delaunay mesh [28] is generated in parametric space. The triangles, t, of the Delaunay mesh are labelled according to the numbers of their adjacent triangles, $adj(t)$, as listed in Table 1, and depicted in Fig. 2.

The types C and S are distinguished from each other by the corner angle between their two free edges, θ. If it is less than a threshold, θ_{threch} (e.g 160°), it is classified as C and the boundary is treated as a tangent discontinuity and a medial edge will connect to the vertex between the two free edges. If it is more than the threshold angle the boundary is treated as being tangent continuous.

3.3 Building the Medial Axis Topology

The medial axis topology is derived from the Delaunay triangles and their assigned types. Each medial vertex corresponds to a single triangle of type J, C or S. Each medial edge corresponds to sequences of triangles beginning and ending at J, C and S triangles and running along adjacent N triangles. Pseudo code of the algorithms for establishing the medial vertices and medial edge topologies are given in Algorithms 1 and 2. (C++ STL Containers Library terminology is used [29].)

Algorithm 1. – Establish Medial Vertices

Output: List of Medial Vertex triangles,
mediaVertexTs.

```
medialVertexTs = [ ] // a sequence container
for t in Delaunay Mesh:
   if type(t) != N:
      medialVertexTs.push_back(t)
```

Algorithm 2. – Establish Medial Edges

Output: List of medial edge triangle sequences,
medialEdgeTs.

```
medialEdgeTs = [ ]
for mvT in medialVertexTs:
   mark(mvT)
   for adjT0 in adj(mvT):
      if hasMark(adjT0):
         continue
   meTs = [ ]
   meTs.push_back(mvT)
   mark(meTs.back())
   while type(meTs.back() == N:
      for adjT1 in adj(meTs.back()):
         if (adjT1 != *(meTs.end() -2):
            meTs.push_back(adjT1)
            mark(meTs.back())
   medialEdgeTs.push_back(meTs)
```

A special case is where there are only *N* and no *J* triangles and therefore there are no medial vertices. The only possibility is that the face is an annulus and the medial axis forms a single closed loop.

Due to the empty circumcircle property of the constrained Delaunay mesh, the circumcentres of the triangles tend to the medial axis as the discretisation of the boundary goes to zero. Thus, the circumcentres of the triangles are used as approximate positions for the medial axis.

3.4 Medial Edge Classification

By classifying medial edges based on the types of the associated Delaunay triangles of the end medial vertices, there are six possible types: *S-S*, *S-C*, *S-J*, *C-C*, *C-J* and *J-J*. Closed medial edges without medial vertices make a seventh type called *N*. These are illustrated in Fig. 3.

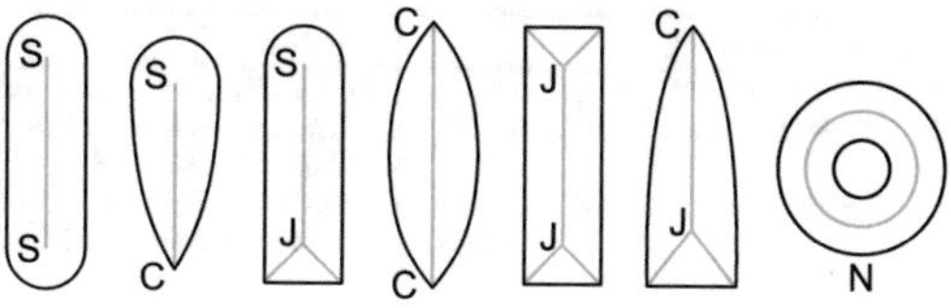

Fig. 3 Medial edge types

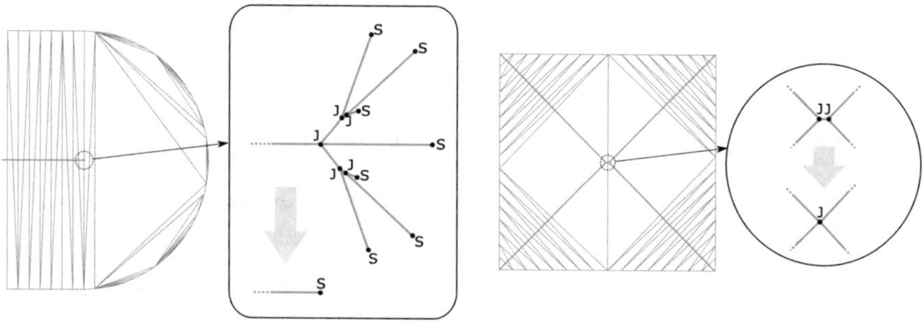

Fig. 4 Examples of medial axis topology clean up

3.5 Medial Axis Topology Processing

The initially generated medial axis may have unwanted artefacts as consequences of the finite discretisations of the boundaries and the subtle particulars of the Delaunay mesh. Typically, these are medial edges with lengths below the discretisation size of the boundaries and they most commonly occur in finite contact zones. A topology clean up process as outlined in Alg. 3 is used to modify the topology to remove these medial edges. Examples are shown in Fig. 4.

3.6 Medial Axis Geometry Processing

The initial geometric representations of medial edges are piecewise-linear curves through the Delaunay triangle circumcentres. To improve their smoothness two iterations of cubic smoothing filtering are applied [30]. This involves optimising the position of every vertex, v_i, in the piecewise-linear curve in turn except for the end vertices. A cubic polynomial is fitted to the surrounding vertices, $\boldsymbol{v}_{curr} = (v_{i-2}, v_{i-1}, v_{i+1}, v_{i+2})$, in a local coordinate system and then the position of v is adjusted to lie on the curve, as illustrated in Fig. 5a. The local coordinate system is chosen to have its origin at the average position in $\boldsymbol{v}_{curr}$ and its x-axis aligned with the line through $\boldsymbol{v}_{curr}[0]$ and $\boldsymbol{v}_{curr}[-1]$ (i.e. the first and last elements in $\boldsymbol{v}_{curr}$). If v_i is adjacent to an end vertex then v_{i-2} or v_{i+2} is left out of $\boldsymbol{v}_{curr}$ and a least norm solution is found. An example showing the smoothed result is given in Fig. 5b.

Algorithm 3. – Topology Clean Up	**Notes**

mvNearbyData = [ ] /* sequence container of tuples:
(medial vertices, distance, type) */
for mv **in** medialVertices:
nearbyMvs, culumlativeDis = *findNearbyMedialVertices*(mv, tolDis)
mvNearbyData.*push_back*((nearbyMvs, culumlativeDis, type(mv)))
 	
sortedIndices = (0, 1, ... medialVertices.size() - 1)
sort(sortedIndices.begin(), sortedIndices.end(),
lambda[mvNearbyData](i0, i1):
return lexiographical_greater(
mvNearbyData[i0], mvNearbyData[i1])
 	
for i **in** sortedIndices: // visit all medial vertices in sorted order
mv0 = medialVertices[i]
if *hasMark*(mv0)
continue
mark(mv0)
nearbyMvs0 = mvNearbyMvs[i][0]
for mv1 **in** nearbyMvs0:
mark(mv1)
replace(mv1, mv0)
	findNearbyMedialVertices returns:
	- the nearby medial vertices within inputted traversal distance tolerance and
	- the cumulative traversal distances from the medial vertex to its nearby medial vertices
	
	lexiographical_greater performs a sequence of greater_than comparisons of the elements of the tuples:
	if mvNearbyData[i0][0].size() != mvNearbyData[i1][0].size():
	return mvNearbyData[i0][0].size() > mvNearbyData[i1][0].size())
	else if mvNearbyData[i0][1] != mvNearbyData[i1][0]:
	return mvNearbyData[i0][1] > mvNearbyData[i1][1]
	else if mvNearbyData[i0][2] == J **xor** mvNearbyData[i0][2] == J
	return mvNearbyData[i0][2] == J
	else
	return false
	
	replace(mv1, mv0) replaces all instances of mv1 with mv0 in the medial axis data model

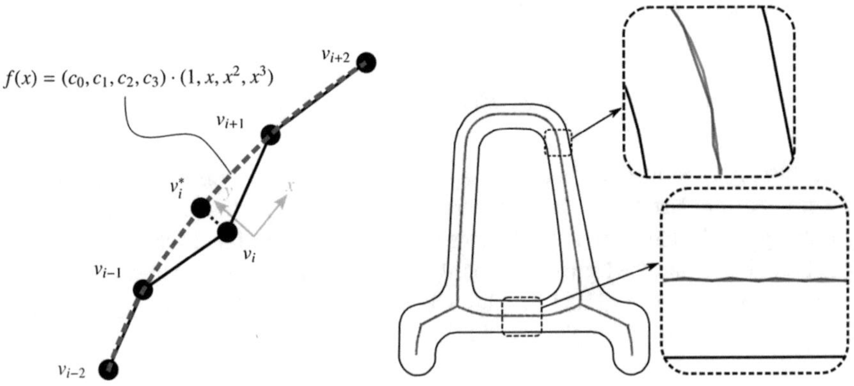

(a) (b)

Fig. 5 (**a**) Cubic smoothing filtering of the vertex v_i by fitting a cubic polynomial to the surrounding vertices v_{i-2}, v_{i-1}, v_{i+1}, v_{i+2}. (**b**) An example showing the original (in blue) and optimised (in red) polyline medial edges after 2 iterations of cubic smoothing filtering

3.7 *Medial Object Data Model*

The medial object is the data model for organising the medial axis points. In 2-D it involves medial edges that join at medial vertices. Additionally, thickness, subtended angle, touching point and length information are associated with the medial object entities, as shown in Table 2.

Table 2 Medial Object Data Model

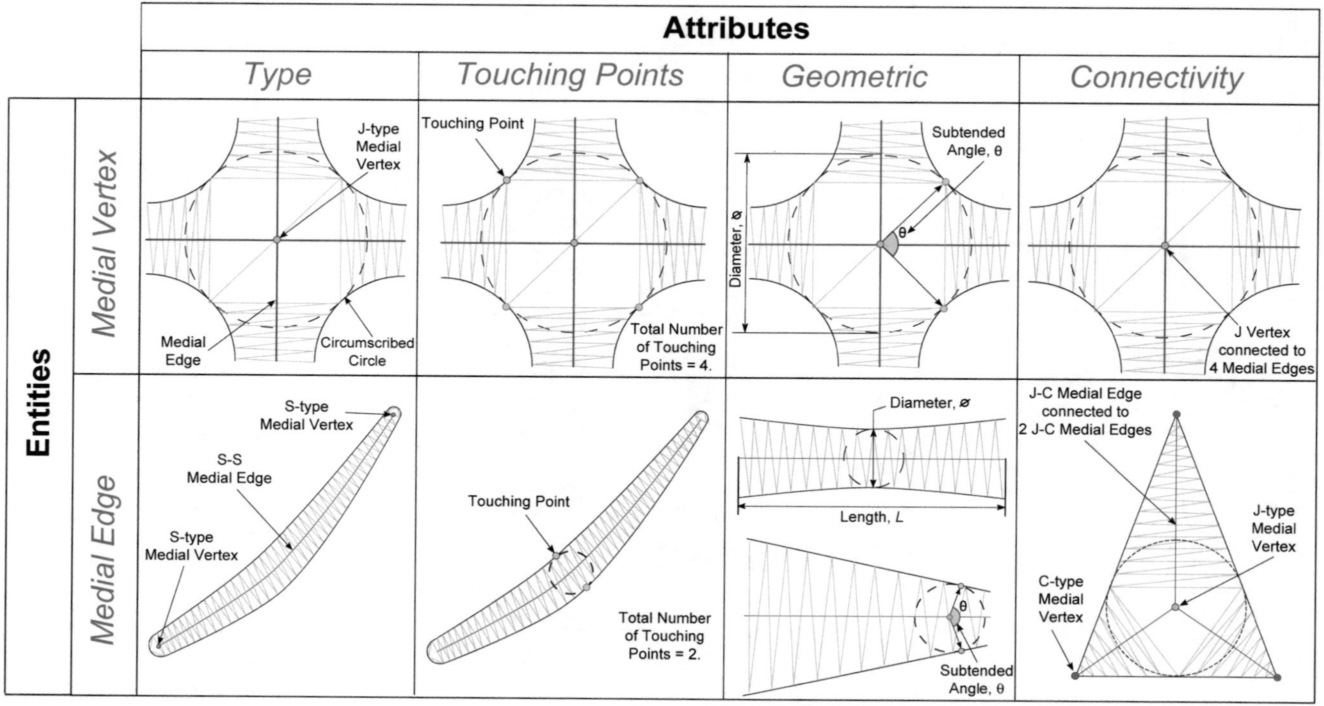

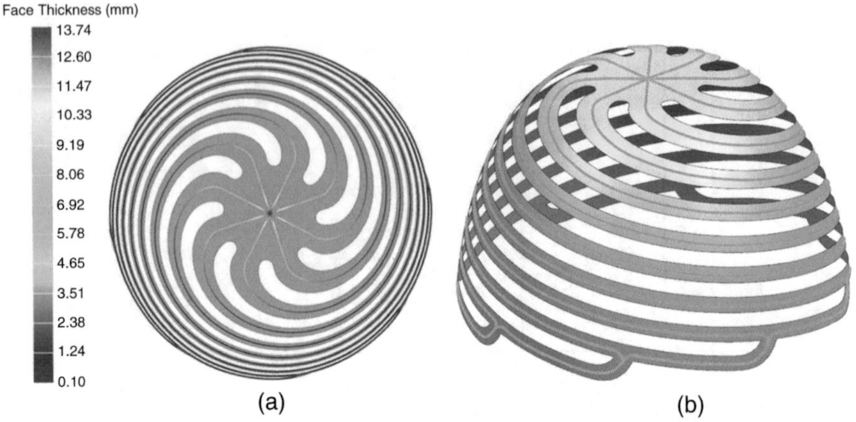

Fig. 6 A spiral bowl with its medial axis (**a**) in flattened 2-D space with face thickness contour plot and (**b**) transformed back to 3-D space

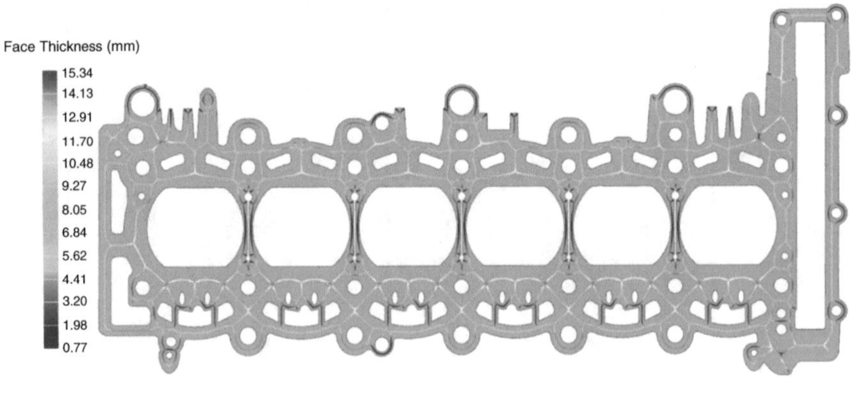

Fig. 7 Medial axis of an engine cylinder head gasket with a thickness contour plot

3.8 Examples

The developed medial object generation strategy is generic and can be used for any face, not just beads or feature faces. Examples of computed Medial Axes on complex geometries are shown in Figs. 6, 7, 8 and 9.

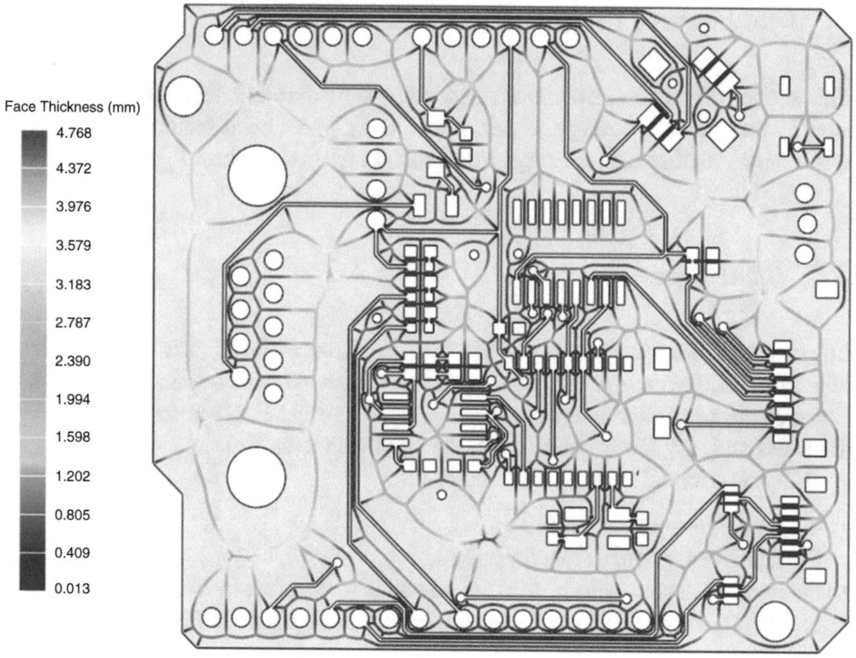

Fig. 8 Medial axis of a printed circuit board with a thickness contour plot

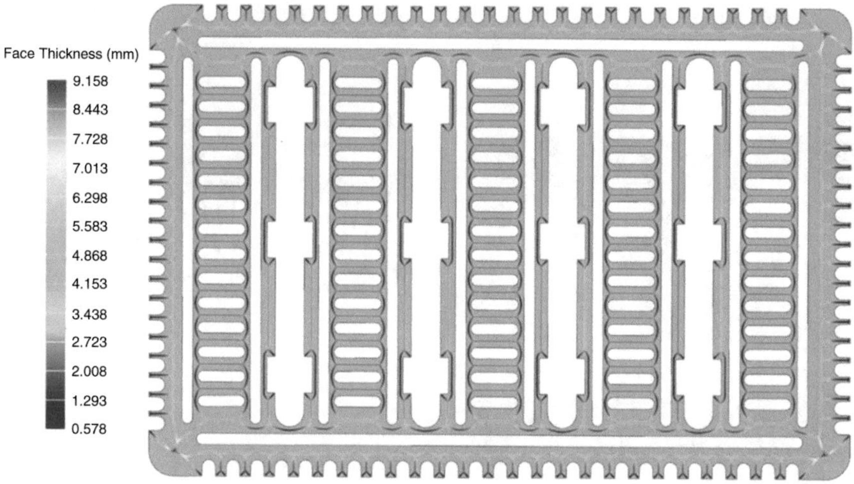

Fig. 9 Medial axis of a complex automotive component with a thickness contour plot

4 Bead Feature Identification and Processing

This section describes how the aforementioned medial axis technology is used to identify and process a range of automotive body panel bead features, and thereby enables the creation of feature-specific quasi-structured meshes.

4.1 Bead Types

In general, a bead feature is defined as a high or medium curvature (single or two-loop) face which is long and slender with a tangent continuous boundary. Consequently, the medial object can be used to identify such features using the type, thickness and length attributes of the edge and vertex entities.

A face is deemed to be a bead feature if:

- the boundary is G1 smooth,
- the face has a single or two loops and
- the average aspect ratio of the medial edges (computed by averaging the length to thickness ratio of its medial edges) is greater than a predefined threshold.

Beads can be of many shapes and based on the mesh preferred they can be further sub-categorised into two broad variants:

- Generic-beads which have more than 1 medial edge (Fig. 10a, b) and
- I-beads which have only 1 medial S-S edge (Fig. 10c, d)

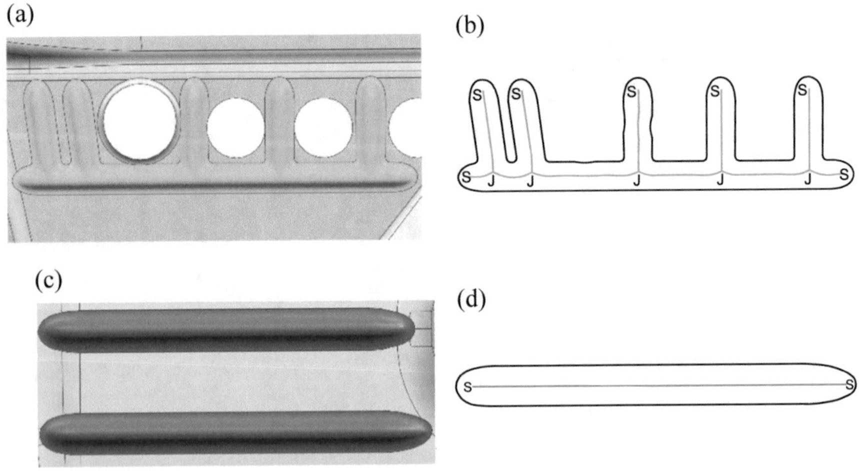

Fig. 10 Bead features. (**a**) Generic Bead. (**b**) Medial Object (All J-J and J-S Edges). (**c**) I-Bead. (**d**) Medial Object (One S-S Edge)

The Generic-bead can be further sub-classified into X, Y, J, G, L etc. shapes as templatised meshes are required on them. Once these features have been identified a crucial metric to consider while preparing them for meshing is the Minimum Element Length (*MEL*). The *MEL* is critical for crash analyses solution stability and efficiency. It is the lowest element size below which no element edge length must fall.

4.2 I-Beads

To facilitate the generation of a structured mesh the I-bead is partitioned or multiblocked into three regions. A central rectangular portion and a tip at each end. Two pairs of medial axis touching points are used to define virtual vertices and virtual edges that establishes the multi-block topology, as shown in Fig. 11. These touching points are located at positions along the *S-S* medial edge where

- the face thickness (diameter of the inscribed circle) is greater than $2 \times MEL$ and
- $r + d \; \Sigma \; 4 \times MEL$, where r is the radius of the inscribed circle at the end *S*-type medial vertex and d is distance along the medial axis from the end medial vertex.

Once the virtual edges have been established the multi-block topology is formed using an approach by Mukherjee and Makem [31].

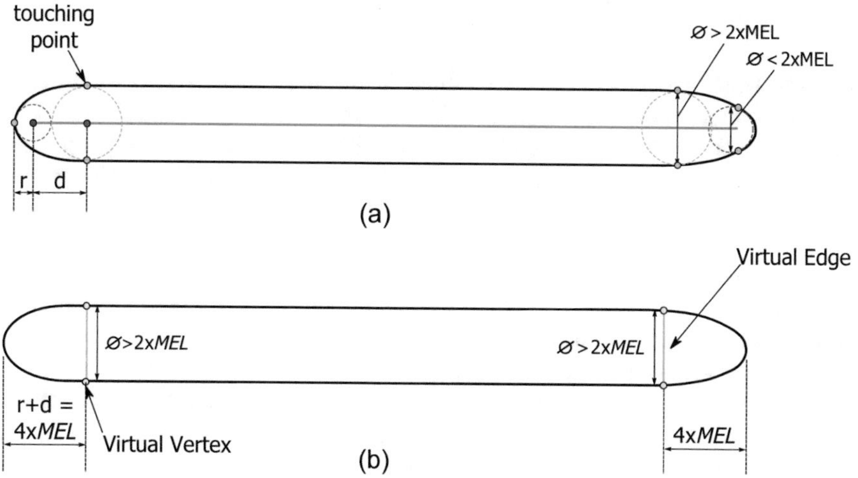

Fig. 11 Multi-blocking of I-bead feature (**a**) determining the split lines (**b**) forming the block topology

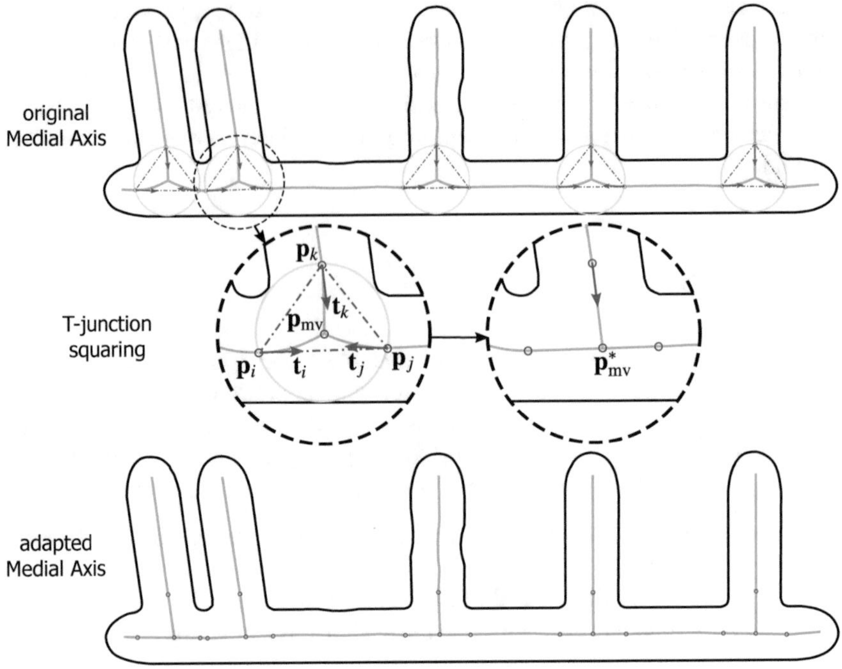

Fig. 12 T-junction squaring of the medial axis on a generic bead

4.3 Generic Beads

Generic beads are processed for meshing in manner different than for I-beads. Instead of multi-blocking the face, the medial axis is used as a ridge curve which the mesh is forced to respect. Typically, near medial vertices, medial edges possess a degree of curvature which is not conducive for ridge curve use. Thus, a T-junction squaring process is performed as illustrated in Fig. 12.

Medial vertices with three medial edges are candidates for T-junction squaring. First, positions on the medial edges at a distance of the inscribed circle radius from the medial vertex are approximated. If the medial edge ends before that distance the position of the medial vertex on the other end is used. Next, local tangents of the medial edges are evaluated using a modified version of the cubic filtering smoothing method. Two conditions are required to carry out the squaring. These are (referring to Fig. 12):

1. $\angle(\boldsymbol{t}_\mathrm{i}, \boldsymbol{t}_\mathrm{j})\Sigma\ \theta_{\mathrm{threch}}$ where $i, j \in \{1,2,3\}, i\ \mathrm{G}\ j\ \ldots$ (165° is used for θ_{threch}), 2. $\angle((\boldsymbol{p}_\mathrm{i} - \boldsymbol{p}_\mathrm{k}), \boldsymbol{t}_\mathrm{k})$€ $\angle\ ((\boldsymbol{p}_\mathrm{i} - \boldsymbol{p}_\mathrm{k}), (\boldsymbol{p}_\mathrm{j} - \boldsymbol{p}_\mathrm{k}))$

The medial vertex location $\boldsymbol{p}_{\mathrm{mv}}$ is moved to $\boldsymbol{p}_*$, the intersection of the line through $\boldsymbol{p}_\mathrm{k}$ in the direction of $\boldsymbol{t}_\mathrm{k}$ and the line through $\boldsymbol{p}_\mathrm{i}$ and $\boldsymbol{p}_\mathrm{j}$. The medial edges are trimmed from their respective points and extended to the new medial vertex.

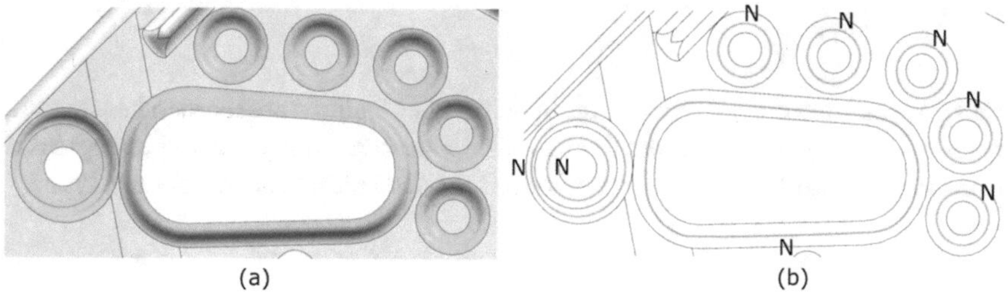

Fig. 13 A typical body panel displaying (**a**) annular beads with (**b**) associated medial objects (single N-type medial edges per face)

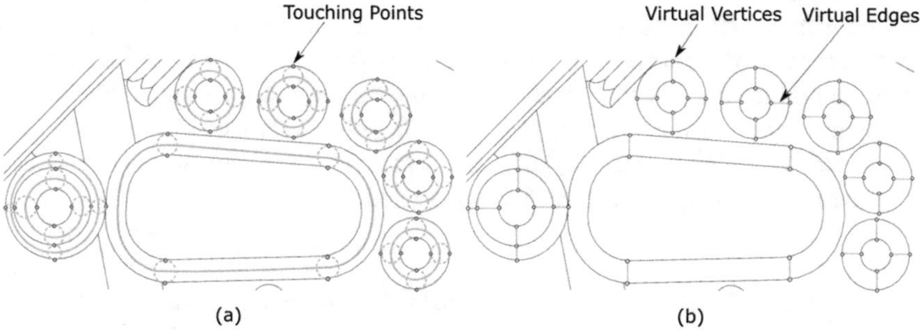

Fig. 14 Multi-blocking of annular beads (**a**) determining the split lines (**b**) forming the block topology

4.4 Annular Beads

Annular beads are another common type of feature specific to body panels where a structured mesh must be applied. Unlike the I and Generic beads, these beads are two-loop faces with G1 boundaries and one loop is a distance offset of the other. The medial axis can identify any face as an annulus if it has a single N-type medial edge, as shown in Fig. 13.

After the annuli have been identified the touching points of the inscribed circle of the medial axis are used to define virtual vertices at a parametric position of $t = 0.0$, 0.25, 0.5 and 0.75 along the medial edge. In a similar fashion to the I-bead, these vertices are linked to form virtual edges which are used to decompose the face into a series of 4-sided, map-meshable blocks [31], as shown in Fig. 14.

5 Meshing

Being special features, the beads, along with other features are meshed before other faces of the body panel in order to allow for lesser constraints and higher degrees of freedom. The meshing methods employed for beads are a function of their type.

5.1 *I-Beads*

The I-Bead is multi-blocked using a procedure developed recently [32] into a rectangular mid-section and two semi-oval end caps as shown in Fig. 11. The mid-section is mapped meshed with an even number of elements along the thickness, so as to capture the crest of the feature. The end caps are meshed using a templatised clamping procedure as described in Fig. 15a. The sub-face is decomposed into 4 sub-areas, each of which is map-meshed. The element count scheme used is described by letters m,n,p and q obeying the following relations

$m \,\%\, 2 = 0,$

$q = m - 2.$

The final mesh is the result of variational [33] and optimisation smoothing [34, 35] of the mesh assembly from the three sub-sections, as shown in Fig. 15b.

5.2 *Generic Beads*

For the generic bead, ridge curves computed from the medial object data are handed on to the mesher. The mesher constructs a scar-loop (self-retracing) from the discrete 3D point data and inserts it into the face as an artificial face-loop (Fig. 16a). The mesher modifies the face topology with this injected, retraceable inner loop.

The face loops are discretised and the inner loop is connected (Fig. 16b) with the outer at locations (in red) based on proximity determined from the underlying voxel model the 2d flattened geometry is overlaid on. The face area, therefore, is

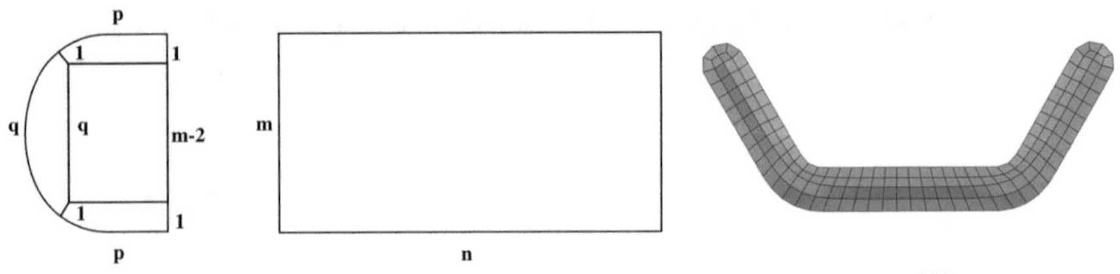

(a) (b)

Fig. 15 Meshing I-beads showing element count distribution (**a**) and the final mesh (**b**)

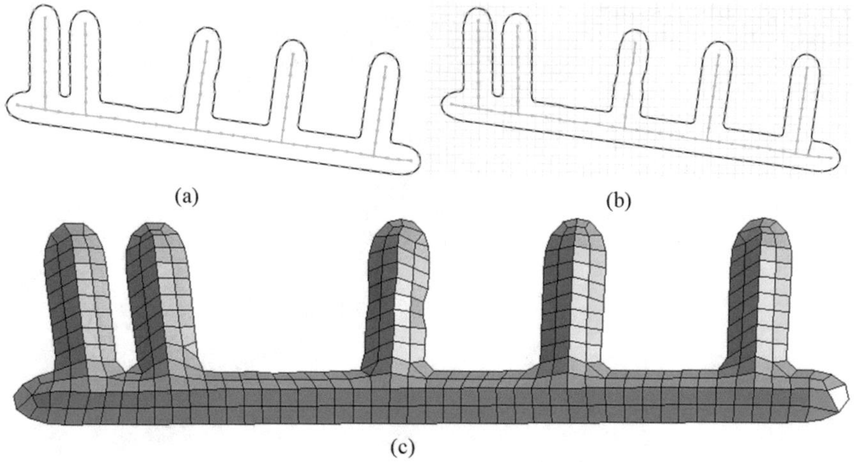

Fig. 16 Meshing a generic comb-shaped bead with ridge curves, where (**a**) the ridge curve is face-inserted, (**b**) discretised face node loops are joined and meshed (**c**)

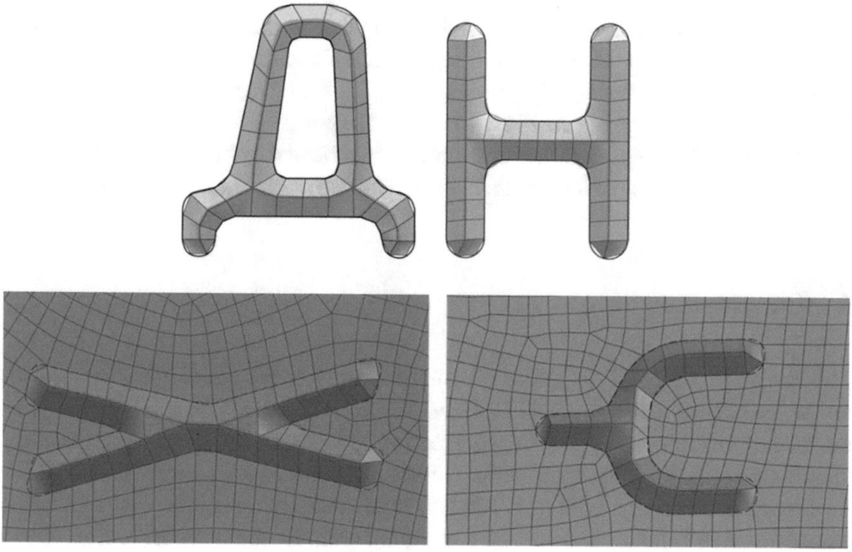

Fig. 17 Example meshes on generic beads using ridge curves

reduced to a single self-touching node loop which is meshed using a subdivision meshing procedure [27]. This approach ensures that the generic bead feature line is embedded on the face as a meshing constraint and is accordingly honoured by the mesh (Fig. 16c). The mesh on the comb-shaped bead (Fig. 10a) is unstructured, but the crest line (ridge curve) of the bead feature is honoured by the mesh. Further examples are shown in Fig. 17.

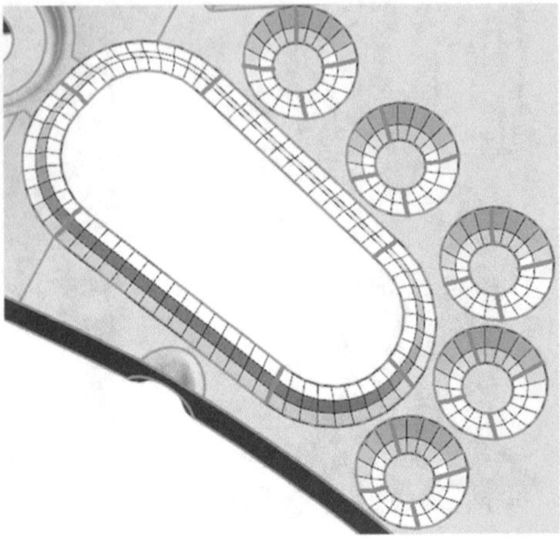

Fig. 18 Examples of meshes on annular beads

5.3 Annular Beads

Perfectly structured meshes are desired on annular beads. To ensure this, annular bead faces are multi-blocked into 4–6 segments and each annular segment or virtual face is then mapped meshed. Figure 18 shows such a cluster of concentric annular beads (from the examples in Figs. 13 and 14) split into 4–6 sub-faces (mesher-native virtual edges representing the "splits" are indicated by blue lines) and mapped meshed with usually an even number of elements around the circumference.

6 Conclusions

This paper presents an assembly of unique strategies for identifying, processing and subsequently meshing bead features in automotive body panels. State-of-the-art medial axis technology is improvised for appropriate boundary discretisation, annulus identification by zero medial vertex instances and topology modifications to eliminate unwanted artefacts. Medial edge geometry is improved with cubic smoothing and T-junction squaring employing careful vertex repositioning. Three broad types of bead features, namely I-shaped, generic and annular beads are identified and processed. While multi-blocking strategies are used for meshing the I-shaped and annular beads, a ridge curve, representing the feature centre or crest line, is extracted for generic beads. Feature-specific surface meshing methods involving clamping templates and face node loop insertion followed by boundary connection are used to mesh the beads. The results show high quality structured meshes honouring crest lines.

References

1. http://www.opel.com, Opel international – Product & Company Information, News, Experience, Excitement (2018). [online] Available at: https://www.opel.com/. Accessed 13 Jun 2018
2. N. Amenta, S. Choi, R.K. Kolluri, The power crust. In Proc. ACM Solid Modeling, 2001, pp. 249–260
3. P. Li, B. Wang, F. Sun, X. Guo, C. Zhang, W. Wenping, Q-Mat: computing medial axis transform by quadratic error minimization. ACM Transactions on Graphics **35**, 1–16 (2015)
4. F. Sun, Y.K. Choi, Y. Yu, W. Wang, Medial meshes – a compact and accurate representation of medial axis transform. IEEE Transactions on Visualization and Computer Graphics **22**, 1278–1290 (2016)
5. M. Foskey, M.C. Lin, D. Manocha, Efficient computation of a simplified medial axis. Journal of computing and information science in Engineering **4**, 274–284 (2003)
6. J. Chaussard, M. Couprie, H. Talbot, A discrete λ-medial axis. Proc. 15th IAPR Int. Conf. Discrete Geometry Comput. Imagery, 2009, pp. 421–433
7. B. Miklos, J. Giesen, M. Pauly, Discrete scale axis representations for 3D Geometry. ACM Trans Graph. **29**, 1–10 (2010)
8. A. Tagliasacchi, T. Delame, M. Spagnuolo, N. Amenta, A. Telea, 3D skeletons: a state- of-the-art report. Eurographics **35**, 1–24 (2016)
9. P.K. Saha, G. Borgefors, G.S. Di Baja, A survey on skeletilization algorithms and their applications. Pattern Recognition Letters **76**, 3–12 (2016)
10. C. Arcelli, G. Sanniti, L. Serino, Distance driven skeletalization in voxel images. IEEE TPAMI **33**(4), 709–720 (2011)
11. H. Xia, P.G. Tucker, Fast equal and biased distance fields for medial axis transform with meshing in mind. Journal of Applied Mathematical Modelling **35**, 5804–5819 (2011)
12. D. Attali, J.D. Boissonmat, H. Edelsbrunner Stability and computation of medial axes – a state-of-the-art-review. Mathematical Foundations of Scientific Visualization, Computer Graphics, and Massive Data Exploration, 2009, pp. 109–125
13. O. Aichholzer, W. Aigner, F. Aurenhammer, T. Hackl, B. Juttler, M. Rabl, Medial axis computation for planar free-form shapes. Journal of Computer Aided Design **41**, 339–349 (2009)
14. F. Buchegger, B. Juttler, M. Kapl, Total curvature variation fairing for medial axis regularization. Graphical Models **76**, 633–647 (2014)
15. J. Peng, H. Wang, J. Li, C. Song, Generation method and application of product-oriented medial axis. 6th International Conference on Logistic, Informatics and Service Science, vol. 16, 2016, pp. 160–174
16. L. Cao, L. Liu, Computation of the medial axis and offset curves of curved boundaries in the planar domain. Computer Aided Design **40**, 465–475 (2008)
17. D. Ding, Z. Pan, D. Cuiuri, H. Li, N. Larkin, Adaptive path planning for wire-feed additive manufacturing using medial axis transformation. Journal of Cleaner Production **133**, 942–952 (2016)
18. B. Durix, G. Morin, S. Chambon, Skeleton-based multiview reconstruction. IEEE International Conference on Image Processing, 2016, pp. 4947–4051
19. J. Yuan, A.M. Cheriyadat, Image feature based GPS trace filtering for road network generation and road segmentation. Machine and Vision Applications **27**, 1–12 (2016)
20. A. Tagliasacchi, T. Delame, M. Spagnuolo, N. Amenta, A. Telea, 3D skeletons: a state- of-the-art report. Eurographics **35**, 1–26 (2016)
21. Z. Yasseen, A. Verroust-Blonder, A. Nasri, Shape matching by part alignment using extended chordal axis transform. Pattern Recognition **57**, 115–135 (2016)
22. T. Tam, C. Armstrong, 2D finite element mesh generation by medial axis subdivision. Advances in Engineering Software **13**, 313–324 (1991)
23. H.J. Fogg, C.G. Armstrong, T.T. Robinson, Enhanced medial-axis-based block-structured meshing in 2D. Journal of Computer Aided Design **72**, 87–101 (2016)

24. J.E. Makem, C.G. Armstrong, T.T. Robinson, Automatic decomposition and efficient semi-structured meshing of complex solids. Engineering with Computers **30**, 345–361 (2014). https://doi.org/10.1007/s00366-012-0302-x
25. K. Beatty, N. Mukherjee, Flattening 3D triangulations for quality surface mesh generation. Proc. 17th Int. Meshing Roundtable, 2008, pp. 125–139
26. J.R. Shewchuk, Triangle: engineering a 2D quality mesh generator and delaunay triangulator. Applied computational geometry towards geometric engineering, 1996, pp. 203–222
27. C++ STL Containers Library. Available at: https://en.cppreference.com/w/cpp/container
28. G.P. Bonneau, S. Hahmann. Smooth polylines on polygon meshes. Geometric modeling for scientific visualization, 2004, pp. 69–84
29. N. Mukherjee, J.E. Makem, A Cartesian slab based multiblocking strategy for irregular cylindrical surfaces. Proc. 26th Int. Meshing Roundtable, 2017
30. J.E. Makem, N. Mukherjee, Mesh generation system and method, Patent Application WO2017040006A1, Siemens PLM Software Inc., 2015-09-01. https://patents.google.com/patent/WO2017040006A1/en
31. N. Mukherjee, A hybrid, variational 3D smoother for orphaned shell meshes, Proc. 11th Int. Meshing Roundtable, 2002, pp. 379–390
32. J.E. Makem, N. Mukherjee, System and method for element quality improvement in 3d quadrilateral-dominant surface meshes. Patent Application WO2018080527A1, Siemens PLM Software Inc., 2016-10-31. https://patents.google.com/patent/WO2018080527A1/en
33. N. Mukherjee, J.E. Makem, Fogg, H.J. A 3D constrained optimisation smoother to post-process quadrilateral meshes for body-in-white. Proc. 25th Int. Meshing Roundtable, 2016, pp. 262–275
34. T.K. Dey, W. Zhao, Approximating the medial axis from the Voronoi diagram with a convergence guarantee. Algorithmica **38**, 179–200 (2004)
35. W. Van Toll, A. Cook., M.J. Van Kreveld, R. Geraerts. The explicit corridor map: using the medial axis for real-time path planning and crowd simulation. Proceedings of the 32nd International Symposium on Computational Geometry, vol. 51. 2016, pp. 72–75

An Angular Method with Position Control for Block Mesh Squareness Improvement

Jin Yao and Douglas Stillman

Abstract We optimize a target function defined by angular properties with a position control term for a basic stencil with a block-structured mesh, to improve element squareness and mesh spacing in 2D and 3D. Comparison with the condition number method shows that besides a similar mesh quality regarding orthogonality can be achieved, the new method converges faster, provides a more uniform global mesh spacing, and is more perturbation resistant.

1 Introduction

Mesh orthogonality, if achieved, reduces computational errors by eliminating the cross-terms in the truncation error. Therefore, a more accurate result can be obtained in a numerical simulation [1, 2]. There are various ways to approach mesh orthogonality and the condition-number [3] mesh smoothing is among the most widely employed in practice. It reduces a target function defined by the ratio between the sum of element edge lengths squared and a power of the element-volume consistent with the dimension of length squared, defined for a corner for each corner in an element. When orthogonality is achieved, the volume of a quad (or hex) element takes its maximum possible value and the target function is minimized. In general, the condition-number method often provides good element shapes and it also works in the case of an unstructured mesh.

The condition-number method produces squarish elements when the boundary of a mesh is consistent with orthogonality. However, it also provides relatively small element sizes around a reduced-connectivity point. Sometimes it pulls the mesh surfaces toward a concave boundary and causes thin or even flipped elements. These behaviors are not desired and may limit the time steps in a simulation with the Courant-Friedrichs-Levy (CFL) condition to control instability, or even crush the

J. Yao (✉) · D. Stillman
Lawrence Livermore National Laboratory, Livermore, CA, USA
e-mail: yao2@llnl.gov; stillman2@llnl.gov

X. Roca, A. Loseille (eds.), *27th International Meshing Roundtable*, Lecture Notes in Computational Science and Engineering 127,
https://doi.org/10.1007/978-3-030-13992-6_8

run. We have also observed a relatively slow convergence with the condition-number method, especially when the initial mesh is twisted.

A relatively fast mesh-smoothing approach is proposed in this article. The new method is almost parallel to the condition-number method in the two-dimensional case, except with a different target function to minimize. In three-dimensions the new method sums up target functions similar to its 2D ones, defined on the three logically 2D stencils shared by a given center node. Therefore, the new method is easier to implement than the condition-number method in 3D.

Besides providing a comparable element squareness as the condition-number method does, the new method converges much faster, has a better ability to resist perturbations, and provides globally more uniform mesh sizes.

2 An Angle-Based Quartic Target Function

In (Fig. 1) a basic two-dimensional stencil is shown on the left. The red node at the center is to move and one may attempt to make the sum of squared angle cosines of the α, β angles as small as possible for orthogonality. In the ideal case, the four α angles (with the center node as the tip) and the eight β angles (with middle nodes on each wall of a regular stencil as tips) would all be $\pi/2$ and the sum of their cosine squares is *zero*. However, with an initially much twisted mesh a Newton's method for minimizing the sum of cosine squared often diverges or finds undesired solutions.

To simplify the problem, we consider that a mesh would have nearly straight mesh lines after smoothing. Thus, we modify the basic stencil in (Fig. 1, left) by

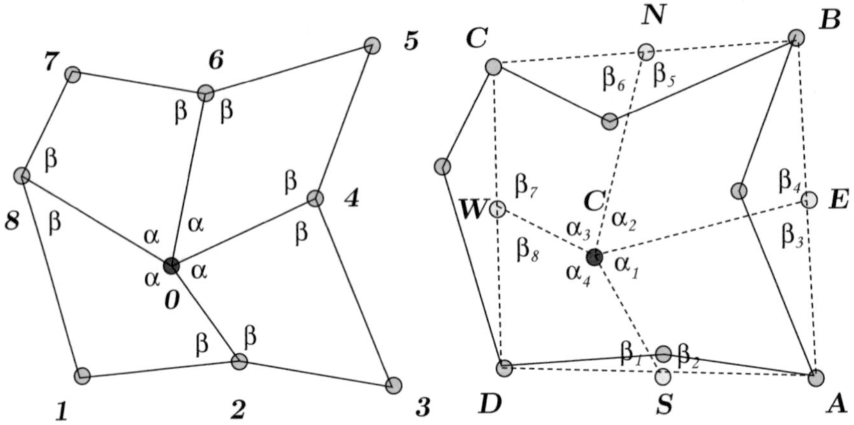

Fig. 1 Left: A normal stencil in 2D with 9 nodes; right: a simple stencil modified from the left. Yellow points S(south), N(north), W(west), and E(east) are mid-points on faces of the simplified stencil ABCD (a quad defined by dashed lines)

ignoring the middle nodes on the four walls of the stencil. To provide some position control we take the four midpoints on the faces of the resulting quad and call them S(south), E(east), N(north), and W(west) (Fig. 1, right). Then, we link the node at the center node C to S, E, N, and W (as desired node positions). Then 12 angles are formed with $\alpha_i\,(i = 1, 2, 3, 4)$ stand for the corners with C as their tip and $\beta_i\,(i = 1, 2, 3, \ldots 8)$ with the *four* mid-points (S, E, N, and W) on faces as corner tips. Since in the ideal case with orthogonality all the α and β are right angles. We propose a target function

$$T = \frac{1}{2}\left(\sum_{i=1}^{4}\cos^2(\alpha_i) + \sum_{i=1}^{8}\cos^2(\beta_i)\right). \qquad (1)$$

For each angle involved in the above target function, the square of its cosine is computed by the square of the *inner-product* of the two vectors defined by a corner tip and the two closest nodes in the stencil that define the corner, divided by the product of the length squared of the two vectors. For examples (in Fig. 1)

$$\cos^2(\alpha_1) \equiv \frac{(\mathbf{CS}\cdot\mathbf{CE})^2}{|CS|^2\cdot|CE|^2},$$

and

$$\cos^2(\beta_1) \equiv \frac{(\mathbf{SC}\cdot\mathbf{SD})^2}{|SC|^2\cdot|SD|^2}.$$

The target-function defined above, when being minimized by a Newton's method, still sometimes diverges or finds spurious roots. To achieve stability, we simplify the target function further by fixing the denominators in the Newton's iterations. This is to say the leg-lengths in a smoothing iteration are taken to their values of the previous iteration. As the smoothing converges, the length of a given leg gradually reaches its limiting value. Thus, when the proposed method converges, the above approximation of cosine with the target function in (Eq. (1)) shall not change the solution, however, it simplifies the algebra quite a bit.

Finally, we are left with a positively definite quartic function (Eq. (1)). A minimizer must exist because the target function is non-negative. In the case of the target function equaling zero, all angles involved become right-angles and a perfect orthogonality is achieved. The proposed target function has smooth derivatives. In addition, the denominator with each term in the target function shall always be finite unless a pair of corner nodes exactly overlap. Therefore, the proposed algorithm is rather stable, behaves well with the Newton's method.

2.1 A Local Optimization with the Newton's Method

In a smoothing step of the proposed method we loop over all internal nodes (boundary nodes are assumed fixed) and for each internal node, the simplified target function can be written as

$$T = \frac{1}{2}\sum_{i=1}^{12} c_i \left((x - a_{1i})(x - a_{2i}) + (y - b_{1i})(y - b_{2i})\right)^2 . \tag{2}$$

The subscript 'i' stands for the contribution of angle 'i' and there are 12 angles in total. However, when a corner-node is also a reduced connectivity point in a regular stencil, because a mesh-line is not expected to be straightened there, we ignore contribution of a corner with a reduced-connectivity at the tip. The constants $a_{1i}, a_{2i}, b_{1i}, b_{2i}$, and c_i are all in terms of coordinates of the simplified stencil (Fig. 1, right). The iterator 'i' counts α angles then β ones. For example the first term in the sum would be

$$\cos^2(\alpha_1) = \frac{((x - x_S)(x - x_E) + (y - x_S)(y - y_E))^2}{[(x_C - x_S)^2 + (y_C - y_S)^2][(x_C - x_E)^2 + (y_C - y_E)^2]}$$

which is in the format of (Eq. (2)) with

$$a_{11} = x_S, a_{21} = x_E;\ b_{11} = y_S, b_{21} = y_E;\ \text{and}$$

$$c_1 = \omega_i[(x_C - x_S)^2 + (y_C - y_S)^2]^{-1}[(x_C - x_E)^2 + (y_C - y_E)^2]^{-1}.$$

In the above expression x, y are the coordinates to be updated of the center node C. x_C, y_C are the coordinates of C at the last smoothing iteration, similar with x_E, y_E and x_S, y_S. ω_i is a numerical weight. In this study we have taken $\omega_i = 1$ unless in the case of a reduced connectivity corner-tip at node 'i', ω_i is taken to 0.

At a minimizer one must have $(\partial T/\partial x) = 0$ and $(\partial T/\partial y) = 0$ where

$$\frac{\partial T}{\partial x} = \sum_{i=1}^{12} c_i(2x + a_{1i} + a_{2i})\left((x + a_{1i})(x + a_{2i}) + (y + b_{1i})(y + b_{2i})\right)$$

$$\frac{\partial T}{\partial y} = \sum_{i=1}^{12} c_i(2y + b_{1i} + b_{2i})\left((x + a_{1i})(x + a_{2i}) + (y + b_{1i})(y + b_{2i})\right) \tag{3}$$

With the Newton's method for optimization the second derivatives of the target function are also needed that

$$\frac{\partial^2 T}{\partial x^2} = \sum_{i=1}^{12} c_i[(2x + a_{1i} + a_{2i})^2 + 2\,((x + a_{1i})(x + a_{2i}) + (y + b_{1i})(y + b_{2i}))],$$

$$\frac{\partial^2 T}{\partial y^2} = \sum_{i=1}^{12} c_i[(2y + b_{1i} + b_{2i})^2 + 2\,((x + a_{1i})(x + a_{2i}) + (y + b_{1i})(y + b_{2i}))],$$

$$\frac{\partial^2 T}{\partial xy} = \sum_{i=1}^{12} c_i[(2x + a_{1i} + a_{2i})(2y + b_{1i} + b_{2i})]. \quad (4)$$

The position of the minimizer is not related to the original position of a given node. To start with, the initial guess of the minimizer is taken to the geometrical center of a quad formed by S, E, N, and W

$$x_0 = \frac{1}{4}(x_S + x_E + x_N + x_W),$$

$$y_0 = \frac{1}{4}(y_S + y_E + y_N + y_W).$$

This can be justified by considering that when mesh orthogonality and even mesh-spacing are achieved, this initial guess would be exactly the solution point.

A single Newton's iteration is then carried out with

$$x_1 = x_0 - \left[\frac{\partial^2 T}{\partial x \partial y}\frac{\partial T}{\partial y} - \frac{\partial^2 T}{\partial y^2}\frac{\partial T}{\partial x}\right] \Big/ \left[\frac{\partial^2 T}{\partial x^2}\frac{\partial^2 T}{\partial y^2} - (\frac{\partial^2 T}{\partial x \partial y})^2\right]$$

$$y_1 = y_0 - \left[\frac{\partial^2 T}{\partial x \partial y}\frac{\partial T}{\partial x} - \frac{\partial^2 T}{\partial x^2}\frac{\partial T}{\partial y}\right] \Big/ \left[\frac{\partial^2 T}{\partial x^2}\frac{\partial^2 T}{\partial y^2} - (\frac{\partial^2 T}{\partial x \partial y})^2\right]. \quad (5)$$

We perform the above Newton's iteration (Eq. (5)) only once in a smoothing step for each internal node to obtain an improved position (x_1, y_1). Then a loop over all the internal nodes moves each one to its improved location. There may be nodes at singular connectivity points that do not own normal stencils. In this case we simply take the geometrical average of the neighbor nodes directly linked to the given node by single legs. We stop the smoothing when the global L_2 difference between the results of two consecutive smoothing steps is smaller than a preset threshold, or a preset limit of the number of iterations is reached.

2.2 An Additional Position Control Term

Minimizing the target function defined in the last section usually gives good mesh quality that is comparable to the condition-number method with fewer steps and better mesh-spacing. However, when the element aspect ratio is big, the contribution from corners that have short sides sometimes cannot balance contributions from other corners. As a result, the minimizer maybe located exterior to the stencil while the mesh squareness is reasonably achieved locally. To avoid this situation, we add a position control term to the target function proposed in (Eq. (1)) defined by the distances from the center node to the mid-face points squared

$$U = \frac{1}{2}[(x - x_S)^2 + (y - y_S)^2 + (x - x_E)^2 + (y - y_E)^2 + (x - x_N)^2 \\ +(y - y_N)^2 + (x - x_W)^2 + (y - y_W)^2], \tag{6}$$

and the target function to minimize then becomes

$$T + \sigma U. \tag{7}$$

The factor σ is proportional to the aspect ratio defined by $|NS|^2/|WE|^2$ or its reverse, whichever is bigger. If minimizing T alone takes a node out of its stencil, the term σU tends to move the node back in the stencil. There are other possibilities for a position control function. We use the current one because it works well in all the cases we have tested. σ is certainly an adjustable parameter and can be taken to zero in suitable cases.

It should be pointed out that simply taking the geometric average of S, E, N and W (the initial guess chosen above) also smooths the mesh (and it can be seen as a Laplacian smoothing [4]), but does not provide a more uniform mesh spacing. An example about this is showed in a later section.

2.3 Mesh-Quality Measurements in Two-Dimensions

To quantitatively compare the mesh-qualities obtained from different smoothing methods in two-dimensions, we employ three measurements (mesh-metric) with a given mesh. The first measurement is the *size-uniformity* which is the deviation of mesh size relative to it of a 'perfect' element. In this study, the mesh-size is defined by the area of an element divided by its shortest edge-length, and the ideal mesh size is the squared root of the average element area. In the ideal situation, the element-size deviation is 0 with a perfectly regular mesh. The range of *size-uniformity* is $[0, \infty)$. The second measurement is the *squareness* defined as the average of squared cosine of the four inner angles of an element and is ranged in $[0, 1]$. The third measurement is the *condition-number*, the average of element edge-length squared divided by the area of element, ranged $[1, \infty)$. It worth to notify that the *condition-number* is not a sharp indicator of element squareness for its dependence on the aspect-ratio.

3 The Three-Dimensional Case

In three-dimensions the target function is the sum of three target functions defined on logically orthogonal two-dimensional stencils.

3.1 A Simplified 3D Stencil with Direction-Nodes

In three-dimensions, a given regular interior node (i.e., not at a singular point) is shared by three logically two-dimensional regular stencils logically perpendicular to each other, one in each logical direction. To achieve three-dimensional mesh orthogonality, it is necessary that the two-dimensional mesh orthogonality is obtained with each of the logically 2D spatial normal stencil.

In each logical direction, we define a pair of direction-nodes (similar to the S, E, N, W nodes in the 2D case). Each direction node is the geometrical average of the corner nodes of a corresponding stencil wall to allow each pair of logically perpendicular 2D stencils sharing the same direction node (in order to achieve a even mesh-spacing). The three modified logically 2D stencil are shown in (Fig. 2).

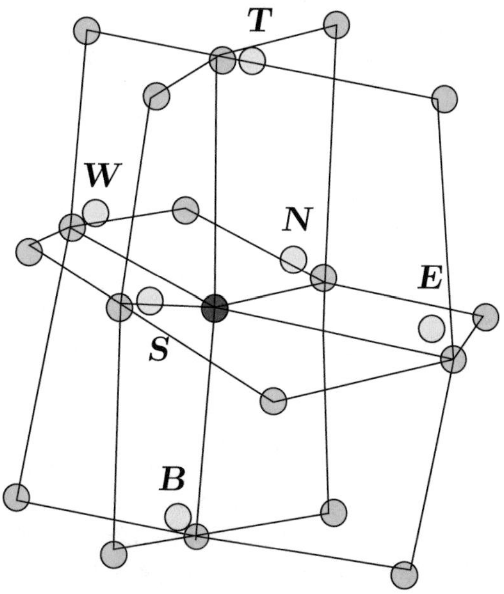

Fig. 2 A regular node in three-dimensions is shared by three logically two-dimensional regular stencils. The original nodes of these stencils are marked blue. The yellow nodes B(bottom), T(top), W(west), E(east), S(south), and N(north) are 'direction-nodes' with each one marks a logical direction and on a wall of the given stencil, is defined by the geometrical average of the four corner nodes of the corresponding stencil-wall. Corner nodes of the 3D stencil are not shown in the figure but are utilized in defining the direction-nodes

3.2 The Target Function in Three-Dimensions

We propose to sum the three two-dimensional target functions associated with a given regular node for a three-dimensional target function to minimize. The two vectors involved in an inner product (to compute cosine of an angle) are both three-dimensional in this case. The target function has *three* independent variables x, y, z instead of only x and y in the previously described two-dimensional case.

Similar to the 2D case, we pick the initial guess as the geometrical average of the positions of the direction-nodes. A Newton's iterative scheme can again be used to reduce the quartic target function. Although it is a common practice, we still write down the standard procedure for minimizing a general function $F(x, y, z)$ for reference. Let the initial guess of the minimizer be (x_0, y_0, z_0). Because the condition $\nabla F = 0$ must be satisfied at a minimizer, we expand the gradient of F around (x_0, y_0, z_0) and keep only the linear terms, then, solve the following 3 by 3 linear system

$$\begin{bmatrix} \partial^2 F/\partial x^2 & \partial^2 F/\partial x \partial y & \partial^2 F/\partial x \partial z \\ \partial^2 F/\partial x \partial y & \partial^2 F/\partial y^2 & \partial^2 F/\partial y \partial z \\ \partial^2 F/\partial x \partial z & \partial^2 F/\partial y \partial z & \partial^2 F/\partial z^2 \end{bmatrix} \cdot \begin{bmatrix} \delta x \\ \delta y \\ \delta z \end{bmatrix} = - \begin{bmatrix} \partial F/\partial x \\ \partial F/\partial y \\ \partial F/\partial z \end{bmatrix}. \tag{8}$$

Then the location of the previous guess gets updated by $x_1 = x_0 + \delta x$, $y_1 = y_0 + \delta y$, and $z_1 = z_0 + \delta z$. The above Newton's scheme is performed for a single time in each smoothing step. The three-dimensional target function we take is written as the follows

$$F = (T_I + \sigma_I U_I) + (T_J + \sigma_J U_J) + (T_K + \sigma_K U_K), \tag{9}$$

where I, J, and K stand for the three logical directions, T, σ, and U have the same definitions as in the two-dimensional case except that the position of a point has *three* components x, y, and z in this case.

4 Comparison with the Condition-Number Method

The proposed method can somehow be seen as a condition-number method with its target function replaced by Eq. (1) or (9). Therefore, it should be easy to modify an existing condition number implementation to perform the proposed algorithm.

In all the numerical problems we have tested, the proposed method provides a faster convergence and a globally more uniform mesh with better spacing around reduced connectivity points. The mesh-squareness provided by the proposed method is at least comparable to the condition number method in all the cases.

In the following examples we compare the results obtained with the proposed method and the condition-number method. Some results obtained with an angle-based method [5], an equi-potential method [6], and an equal-distance method [7, 8] are also provided for references.

The treatment for nodes at reduced connectivity points is the same with all the methods, by taking the geometrical center of the triangle formed by the nodes directly linked to a reduced-connectivity point.

4.1 A Deformed Butterfly Mesh

A two-dimensional five-block butterfly configuration is contained in a perfect 6 by 6 square. Each block is assigned an algebraic mesh by a bi-linear mapping from an ideal mesh in the parametric space to the physical space.

In the first case the center block is twisted counter-clock-wise by $(\pi/6)$ (Fig. 3, left). Each block is assigned a 15 by 15 mesh. Figure 4 shows the comparison of smoothing results between the proposed method and other well known methods.

The proposed method has converged before 320 iterations is reached with a relative tolerance of 10^{-3}, as well as the Winslow-Crowley method. The angle-based method converges slower and the condition-number method converges the slowest. Finally, we take 6400 iterations with each method for comparisons between the converged results. The angle-based method and the condition-number resulted the best squareness and condition number with marginal differences, but with the worst size-uniformity measurements. The proposed method arrived at a comparable squareness, however, with the best mesh-size uniformity and the fastest convergence.

Table 1 shows the comparison of mesh metric averaged between methods. Clearly, with a fixed number of smoothing steps, the proposed method produced symmetry and the largest mesh size around a reduced connectivity point most

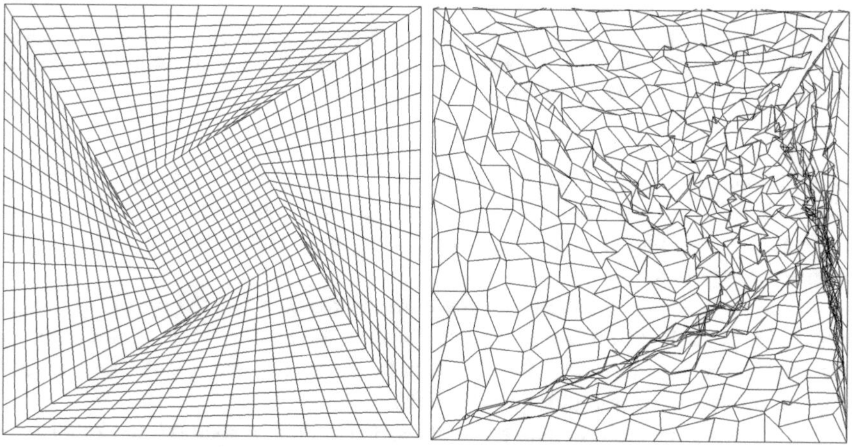

Fig. 3 Left: a butterfly mesh on a 6 by 6 square with the center block twisted counter-close-wise by $(\pi/6)$. Right: the left figure with the center block shifted by 3 units in the x-direction with a random perturbation added

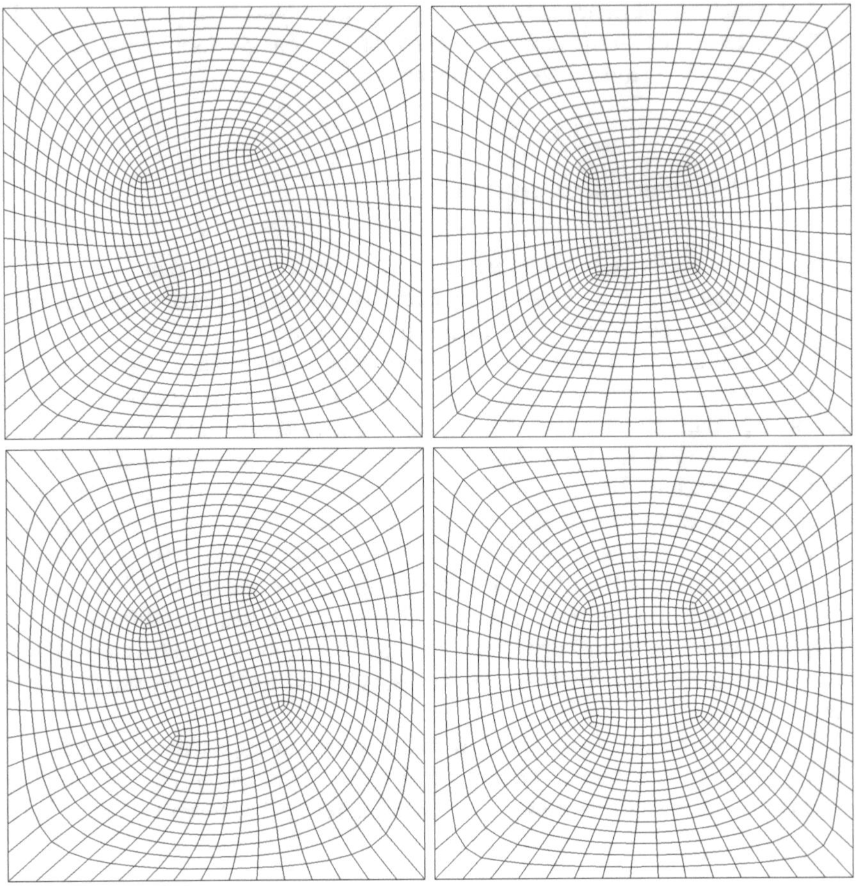

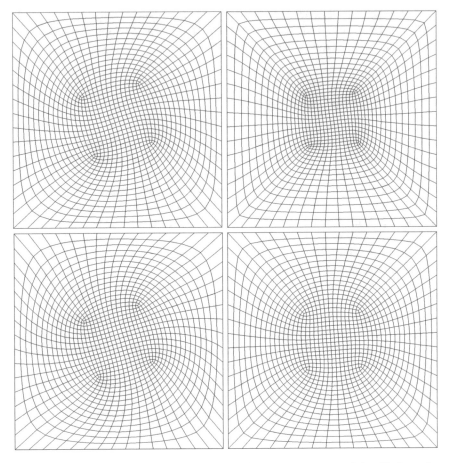

Fig. 4 The initial mesh in (Fig. 3, left) smoothed with 160 iterations: upper left: with the angle-based method; upper right: with the Winslow-Crowley method; lower left: with the condition-number method; lower right: with the proposed method

quickly. The condition number method and the angle-based method converge slower. The Winslow-Crowley method, although converges not as slow (but slower than the proposed method), produces undesirably small elements around the reduced connectivity points.

In the second case the center of the butterfly mesh is twisted by $(\pi/3)$ then is shifted in the x-direction with a distance of three units. A random perturbation of the maximum displacement of $(1/2)$ in both the x- and the y-directions is then applied for every interior node (Fig. 3, right).

Figure 5 shows the comparison between the proposed method and other well known methods with the initial configuration in (Fig. 3, right). In this case the condition-number method did no converge, perhaps caused by zero or negative element areas from the perturbation added to the deformation of mesh. The proposed

Table 1 Mesh metrics against the number of iterations regarding the initial mesh in Fig. 3, left

Iters	ANG	EQP	CND	ORT
10	(0.449, 0.193, 1.280)	(0.458, 0.171, 1.242)	(0.429, 0.177, 1.235)	(0.440, 0.130, 1.185)
20	(0.440, 0.167, 1.233)	(0.459, 0.143, 1.194)	(0.422, 0.150, 1.191)	(0.443, 0.095, 1.142)
40	(0.442, 0.134, 1.184)	(0.464, 0.108, 1.141)	(0.423, 0.122, 1.152)	(0.452, 0.066, 1.108)
80	(0.456, 0.096, 1.136)	(0.484, 0.075, 1.096)	(0.444, 0.094, 1.119)	(0.469, 0.045, 1.085)
160	(0.479, 0.061, 1.096)	(0.506, 0.064, 1.079)	(0.485, 0.066, 1.085)	(0.480, 0.035, 1.074)
320	(0.512, 0.036, 1.065)	(0.514, 0.078, 1.089)	(0.544, 0.042, 1.053)	(0.482, 0.033, 1.072)
∞	(0.657, 0.021, 1.030)	(0.514, 0.080, 1.090)	(0.677, 0.022, 1.029)	(0.482, 0.033, 1.072)

'CND' stands for condition number; 'OTR' for the proposed method; 'ANG' for angle-based; and 'EQP' for equi-potential (Winslow-Crowley). The first number in the metrics is the uniformity measure (the standard deviation of relative mesh-sizes), the second number is the squareness (average cosine squared of angles), the last one is the condition-number. The ideal metrics would be (0, 0, 1) for a perfect square mesh

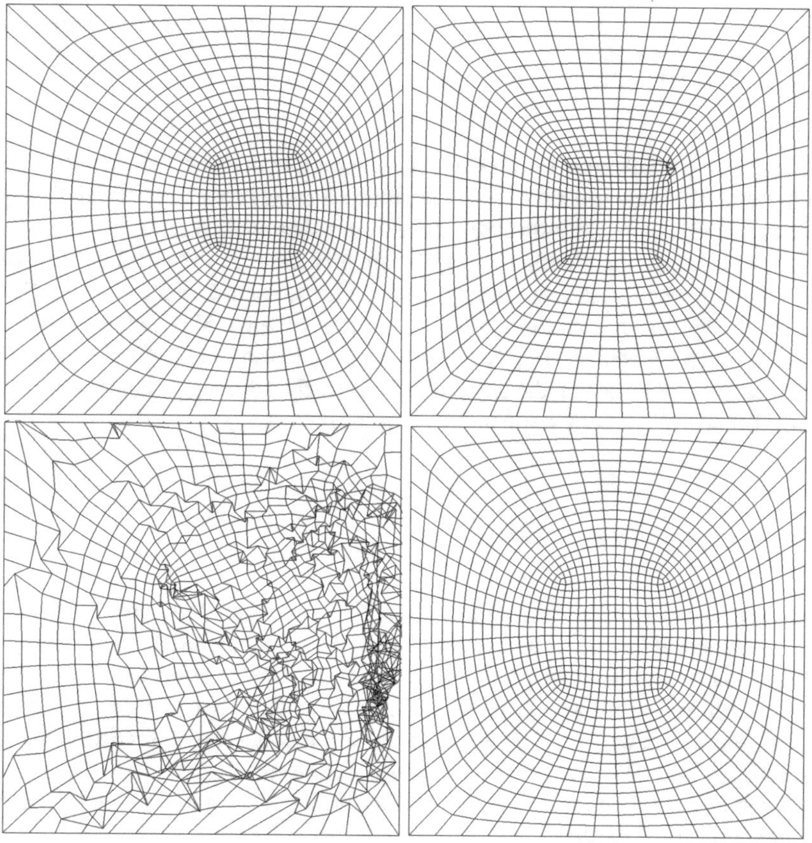

Fig. 5 The initial mesh in (Fig. 3, right) smoothed with 800 iterations: upper left: with the angle-based method; upper right: with the Winslow-Crowley method; lower left: with the condition-number method; lower right: with the proposed method

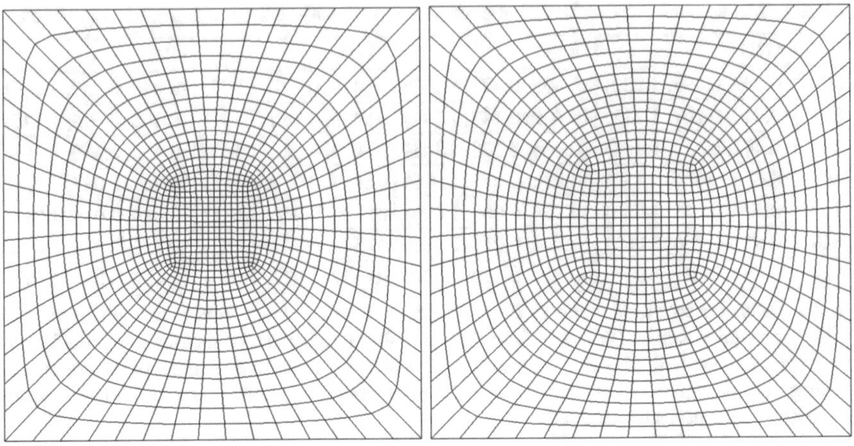

Fig. 6 The initial mesh in (Fig. 3, right) smoothed with 2000 iterations for convergence: left: without the Newton's step for minimizing the target function (taking the geometric average of S, E, N, and W in Fig. 1); right: with the Newton's step (the proposed method)

method not only converges the fastest, but also provides the best mesh quality among all the methods. It clearly has a better capability to resist perturbations than the equipotential method, certainly also the condition-number method.

To demonstrate the contribution of the angular terms in the target function, we compare the results between simply taking the geometric average of S, E, N, and W in Fig. 1 (the initial guess of the solution of smoothing), against with the Newton's step for minimizing the target function. Figure 6 show their difference. The result with minimizing the target function clearly has a bigger center box which is consistent with globally more uniform mesh sizes.

4.2 A Checkerboard Mode

The proposed target function takes information from only the corner nodes of a regular stencil. If the nodes involved are on a perfect square mesh, the center node shall not move at all with the proposed method. Let's take a perfect two-dimensional lattice with each node at the position (i, j) and color its nodes in such a way that all the nodes with $(i + j)$ equal to an odd number are colored red, the rest are colored blue. We give all the red nodes a uniform displacement. Clearly, the proposed method shall not do any smoothing, because the blue nodes and red nodes are both on perfect meshes.

We argue that the above *checkerboard* mode is unrealistic because the boundary of mesh has to be zigzag to support such a configuration. However, we still provide an example with an interior *checkerboard* configuration to examine the effectiveness of the proposed method.

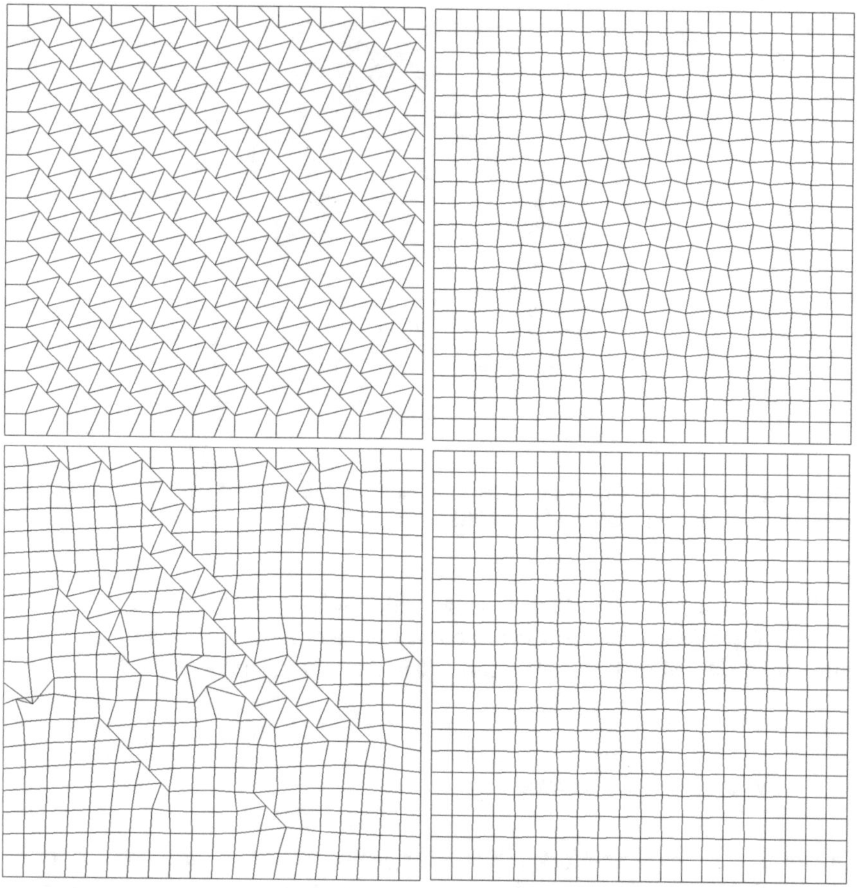

Fig. 7 A mesh with an interior *checkerboard* configuration (upper-left) is smoothed with 100 iterations by the equi-potential method (upper-right); the condition-number method(lower-left); and the proposed method (lower-right)

Figure 7 shows the smoothing of a 20 by 20 perfect mesh with each node originally at an integer lattice point (i, j) (with the lower left corner at $(-10, -10)$). To setup the checkerboard mode, each node (i, j) with $(i + j)$ being an odd number is moved to the position $(i + 0.6, j + 0.4)$, except for a node on the boundary. With 100 iterations, the proposed method has nearly removed the checkerboard mode while the equal-potential method still displays zigzag patterns in the center (comparable to the result of the proposed method with 50 iterations). The condition number method suffers with numerical difficulty of dividing a vanishingly small volume and we choose not to move a node in this situation for the code to keep running and the result is less than satisfactory.

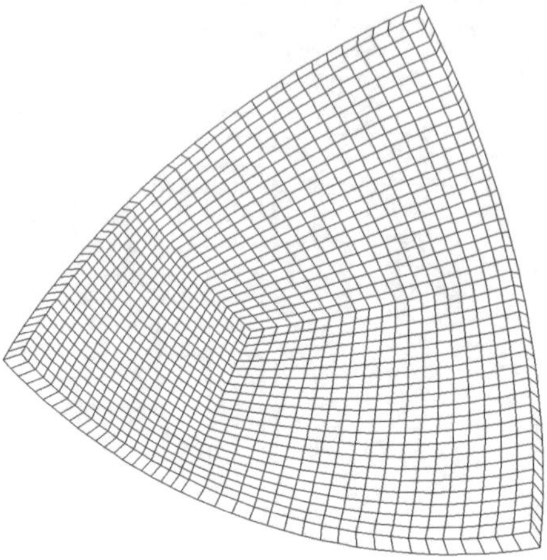

Fig. 8 The initial three-block mesh on the inner side of a $1/8$ spherical shell. The smaller block has element dimensions nearly half of the dimensions of elements in rest of the blocks

This checkerboard mode test shows that the proposed method is not only numerically more stable than the condition-number method, but also propagates boundary information faster, even faster than the equi-potential method.

4.3 On a Curved Surface

Figure 8 shows a three-block initial mesh which is on the inner side of a one-eighth spherical shell with a radius of 15 units centered at the origin. We perform smoothing with four approaches: the condition number [3]; the equipotential [6]; the equal-distance [7]; and the proposed method for orthogonality and even mesh spacing. Each smoothing method is applied for 500 times. The stencil associated with each node is projected on a plane tangent to the sphere $r = 15$ at the given point, in order to perform a two-dimensional smoothing, then the improved node position is projected back to the original surface. The results of smoothing are shown in Fig. 9. The original equi-potential method (upper left figure) does a fair job for mesh squareness, however it also produces relatively small mesh-sizes around the reduced-connectivity point. The equal-distance method (upper-right figure), which is aimed at producing a globally uniform mesh size, produces the largest mesh sizes around the reduced-connectivity point. It does not produce the best squareness of elements because orthogonality is not a concern. The condition-number method (lower left figure) provides mesh-squareness, but generates small mesh-sizes around the reduced-connectivity point. The proposed orthogonality method also provides mesh-squareness. Nevertheless, its smallest mesh-size is larger than produced by the condition-number method. This is consistent with previous results.

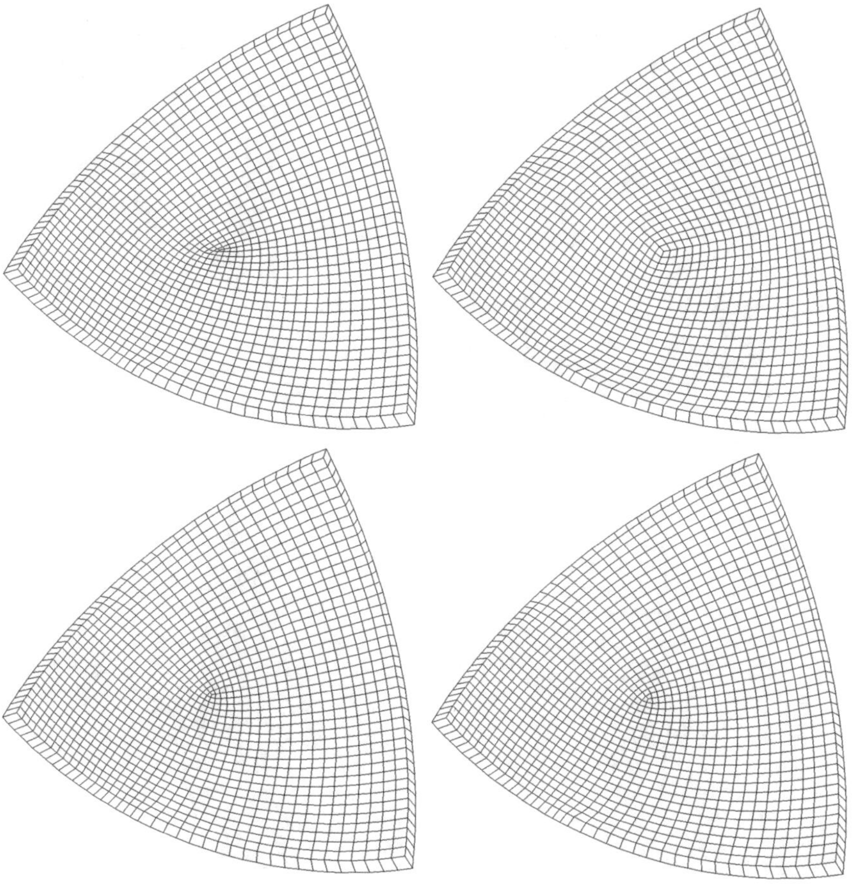

Fig. 9 The initial mesh in Fig. 8 smoothed with 500 iterations: upper left: with the original equi-potential method; upper right: with the equal-distance method. lower left: with the condition-number method; lower right: with the proposed method

4.4 A Three-Dimensional Shell

In Fig. 10 (upper-left), a *one-eighth* spherical shell is meshed with three blocks and this is essentially the region in the previous test with a thickness of 5 units in the radial direction. The inner radius is 10 units. The condition number method generates thin elements near the inner surface and this is can be seen with slicing the mesh after smoothing with the $x = 0$ plane (Fig. 10, upper right). The angle-based method does a better job (Fig. 10 lower left), but the proposed method behaves the best (Fig. 10, lower right). This example clearly demonstrates that the proposed method is able to generate mesh surfaces that resist the attraction of a concave boundary, and provide a better global mesh-quality than the condition-number method.

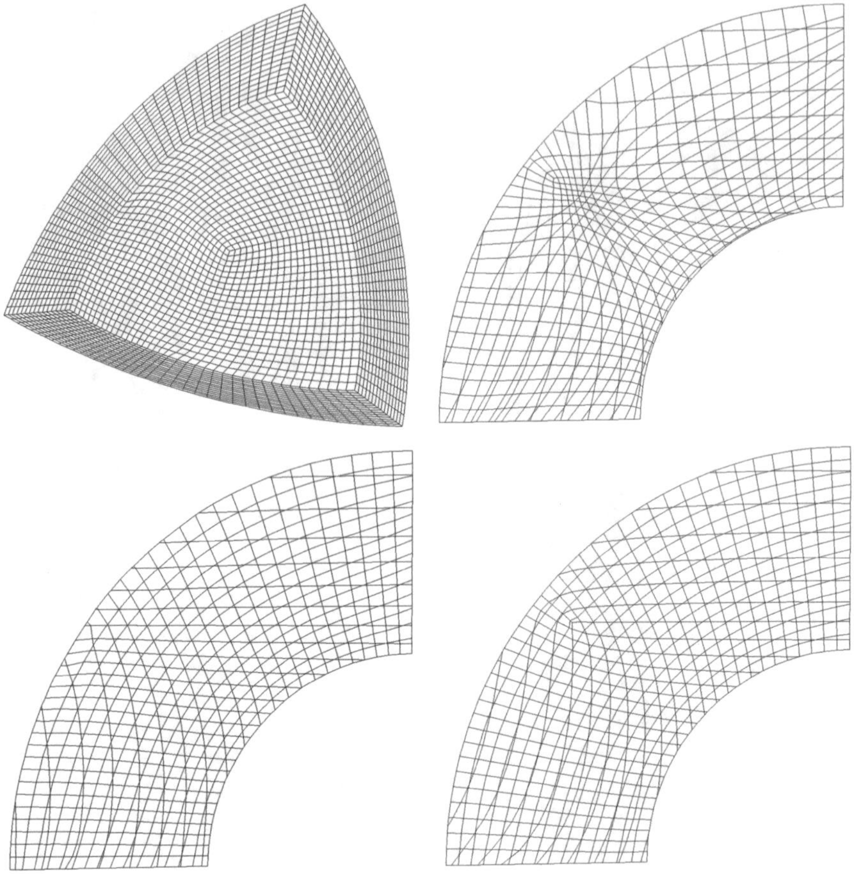

Fig. 10 A 3D shell-mesh is smoothed with 800 iterations for convergence with four methods: upper left: the equal-distance method; upper right: condition-number method; lower left: the angle-based method; lower right: the proposed method. The surface mesh with the equal-distance smoothing is shown for providing a $3D$ view of the region. The meshes smoothed with the other methods are sliced with the y–z plane passing the origin

4.5 A Three-Dimensional Block Mesh with Center Twisted

We show here the comparison between various smoothing methods with three mesh-quality metrics for another three-dimensional test problem. This time the metrics are chosen to be: (a) the relative minimum element size; (b) the smallest two-dimensional angle; and (c) the aspect ratio of an element.

The minimum element size is computed with the volume of a given element divided by the maximum face area of the same element. The smallest angle is computed by finding the largest inner product of a pair of element-edges that share the common tip, divided by the products of the two edge-lengths, then taking its

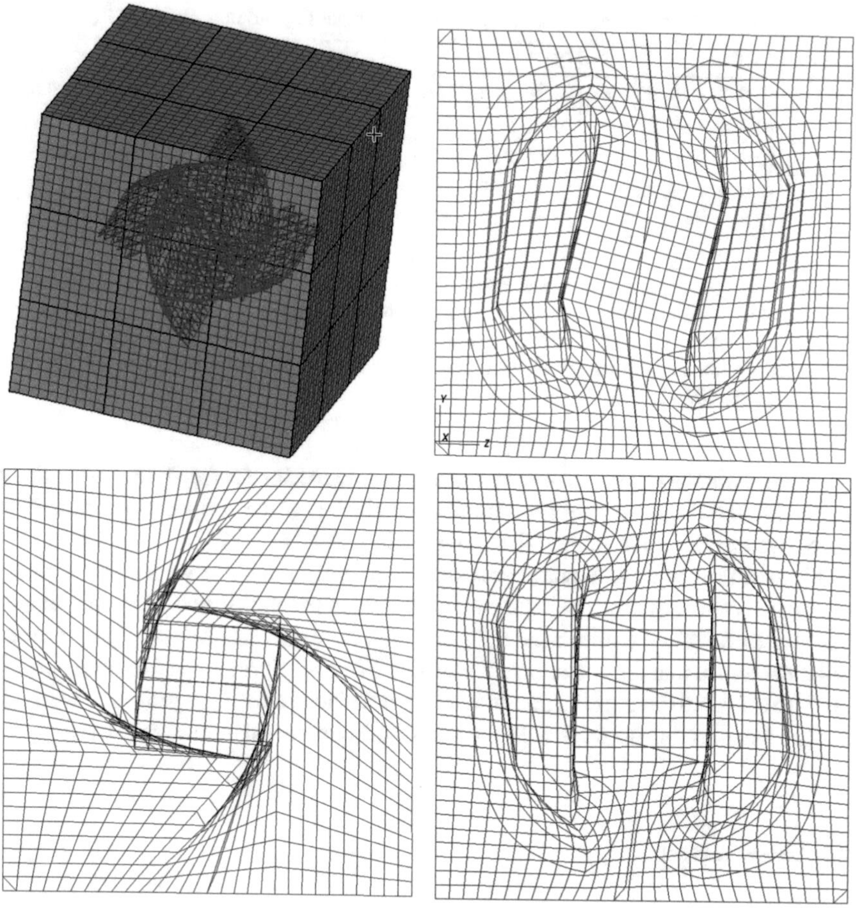

Fig. 11 In the upper left figure, the elements marked by red color are flipped because of the three consecutive $(5\pi/12)$ rotations around x-, y-, and z-axes of the center block of a 3 by 3 by 3 perfect cubic block-mesh. Upper right: sliced by the $x = 0$ plane; lower left: sliced by the $y = 0$ plane; lower right: sliced by the $z = 0$ plane

inverse cosine. The aspect ratio is computed with the longest diagonal divided by the smallest edge size for each element, then taking the maximum among them.

We take a cubic block structure with *3 by 3 by 3 = 27* blocks centered at the origin. Each block is a cube with an edge-length of 2. Initially, the block at the center is rotated by $(5\pi/12)$ around the x-axis, then rotated another $(5\pi/12)$ around the y-axis, finally the center block is again rotated by $(5\pi/12)$ around the z-axis. Each block is then assigned a *10 by 10 by 10* algebraic mesh.

The ideal metrics for a perfectly smoothed cubic mesh would be the smallest relative element size, the smallest angle, and the largest aspect ratio equal to $(1, \pi/2, \sqrt{3})$. The initial mesh metric is (0.0000188721, 3.214596, 1505, 562). Among the 27,000 elements in total, 490 of them are flipped (Fig. 11).

Table 2 Mesh metrics against the number of iterations regarding the initial mesh in Fig. 11

Iters	CND	ORT	ANG	EQP
2^2	(0.001, 1.10, 9088)	(0.24, 24.3, 11.9)	(0.001, 3.52, 8395)	(0.09, 6.02, 36.5)
2^3	(0.045, 7.89, 23.2)	(0.60, 52.7, 3.39)	(0.16, 12.9, 16.9)	(0.37, 27.9, 6.24)
2^4	(0.417, 37.0, 4.77)	(0.76, 69.6, 2.42)	(0.42, 35.4, 5.07)	(0.63, 50.1, 3.33)
2^5	(0.670, 62.3, 2.83)	(0.89, 80.9, 1.99)	(0.65, 58.0, 3.02)	(0.81, 68.0, 2.38)
2^6	(0.805, 76.5, 2.20)	(0.97, 87.2, 1.80)	(0.816, 74.7, 2.24)	(0.924, 80.6, 1.96)
2^7	(0.890, 84.2, 1.92)	(0.997, 89.6, 1.74)	(0.912, 84.5, 1.89)	(0.979, 87.4, 1.78)

'CND' stands for condition number; 'ORT' for the proposed method; 'ANG' for angle-based; and 'EQP' for equi-potential (Winslow-Crowley). The first number in the metrics is the smallest relative side length, the second number is the minimum 2D angle in degrees, the last one is the aspect ratio defined by the longest element-diagonal divided by the shortest side-length. The ideal metrics is (1.0, 90.0, 1.732)

Table 2 shows comparison between mesh metrics for four different smoothing methods including the condition-number method and the proposed method. The other two are the equal-potential and the angle-based methods.

Clearly, the proposed method provides the best mesh metrics with fixed numbers of the smoothing steps. The equi-potential comes second, followed by the angle-based method. The condition-number method converges the slowest. Further computation shows that for the smallest mesh size to get within a relative threshold of 10^{-3} to the ideal case, 180 smoothing steps is sufficient with the proposed method, in contrast, about 1000 steps with the condition-number method is necessary.

5 Conclusion

We propose a mesh smoothing algorithm which is parallel to the condition-number method, with the target function replaced by the sum of an asymptotic expression of cosine squares of angles defined in a simplified regular stencil, plus a position control function similar to the one used in the Laplacian smoothing. Numerical results show that while achieving a similar quality of element squareness, the proposed method converges much more quickly, provide globally more uniform mesh sizes, and is more perturbation resistant, compared to the condition-number method. Overall, the proposed method also converges faster and results in more uniform mesh-sizes than the angle-based method and the Winslow-Crowley method in our numerical tests.

Acknowledgement This work was performed under the auspices of the U.S. Department of Energy by Lawrence Livermore National Laboratory under Contract DE-AC52-07NA27344.

References

1. P. Knupp, S. Steinberg, *Fundamentals of Grid Generation* (CRC, Boca Raton, 1993)
2. J.F. Thompson, B.K. Soni, N.P. Weatherill, *Handbook of Grid Generation* (CRC, Boca Raton, 1998)
3. P.M. Knupp, A method for hexahedral mesh shape optimization. Int. J. Numer. Methods Eng. **58**, 319–332 (2003)
4. S.A. Canann, J.R. Tristano, M.L. Staten, An approach to combined Laplacian and optimization-based smoothing for triangular, quadrilateral, and quad-dominant meshes, in *Proceedings of the 7th International Meshing Roundtable* (1998)
5. T. Zhou, K. Shimad, An angle-based approach to two-dimensional mesh smoothing, in *Proceedings of the 9th International Meshing Round-table, Sandia National Laboratories* (2000), pp. 373–384
6. A.M. Winslow, Numerical solution of the quasi-linear poisson equation in a nonuniform triangular mesh. J. Comput. Phys. **1**, 149–172 (1967)
7. J. Yao, D. Stillman, An equal-space algorithm for block-mesh improvement. Proc. Eng. **163**, 199–211 (2016)
8. J. Yao, *A Mesh Relaxation Study and Other Topics*. Lawrence Livermore National Laboratory, Technical Report, LLNL-TR-637101 (2013)

Dual Surface Based Approach to Block Decomposition of Solid Models

Zhihao Zheng, Rui Wang, Shuming Gao, Yizhou Liao, and Mao Ding

Abstract A high quality block structure of the solid model can support many important applications. However, automated generation of high quality block structure is still a challenging problem. In this paper, a dual surface based approach to automated and valid block decomposition of solid models is proposed. First, an optimized frame field is constructed on background tetrahedral mesh and three kinds of degenerated singularities are corrected. Then, dual loops for block decomposition are generated with the help of the optimized frame field. After that, a required dual surfaces set, whose dual surfaces can suitably separate all boundary elements of solid model and singularities of frame field, is constructed based on dual loops by min cut algorithm. Finally, a valid block structure is obtained by performing dual operations along the dual surfaces on the hex mesh generated by splitting the tetrahedral mesh of the solid model. Experimental results show the effectiveness of the proposed approach.

1 Introduction

A high quality block structure of the solid model can support many important applications. First, high quality hexahedral meshing, which is still a very challenging problem, can be easily achieved based on it. Second, a high quality block structure of a solid model plays an important role in the parameterization required for isogeometric analysis. Finally, this structure can be used to support multi-grid solvers to accelerate computations. It is due to such important applications that block decomposition of solid models has attracted more and more attention. However, automatic and high quality block decomposition for arbitrary shape is a challenging problem.

Z. Zheng · R. Wang · S. Gao (✉) · Y. Liao · M. Ding
State Key Laboratory of CAD&CG, Zhejiang University, Hangzhou, China
e-mail: smgao@cad.zju.edu.cn

X. Roca, A. Loseille (eds.), *27th International Meshing Roundtable*,
Lecture Notes in Computational Science and Engineering 127,
https://doi.org/10.1007/978-3-030-13992-6_9

The dual operation based methods can always guarantee the topological validity of the hexahedral structure, that is, each cell of the decomposed structure is a hexahedron. Kowalski et al. [7] first obtained the initial hexahedral mesh of the model using a tet-to-hex method, then inserted the fundamental sheets into the mesh and finally extracted other sheets to obtain the final block structure. Although this method can generate a block structure for the model, it seems that the method hardly guarantees the geometric validity of block structure, that is, the final block structure cannot fully capture all the boundary elements of the solid model, because only the inserted fundamental sheets cannot sufficiently capture all the geometric information sometimes. To this end, Wang et al. [16] improved this method by adding curve-related sheets besides the fundamental sheets and obtaining the block structure by carefully extracting the sheets. The insertion of curve-related sheets ensures the geometric validity of the block structure while improving the quality of the final block structure. However, this method cannot deal with solid models with free-form surfaces. Gao et al. [5] obtained the base complex in the input hexahedral mesh, iteratively removed the appropriate sheets or chords, and performed geometric optimization after each deletion step, to finally obtain a conformal, non-inverted, coarse hexahedral mesh. This method can simplify the global structure of the complex hexahedral mesh robustly and efficiently, but it seems that the result is sensitive to the quality of the input hexahedral mesh.

Kowalski et al. [8] was the first to propose the frame field based method to decompose the body into blocks. The author set up the frame field by propagating the frames firstly defined along the geometric curves over the domain and smoothing the initial frame field. Then, the singular graph were extracted to obtain the block structure. However, it seems that this method does not guarantee the validity of the block structure for the models with degenerated frame fields.

Lei et al. [10] proposed a meshing algorithm based on the surface foliation theory for high-genus surfaces. This work proved the existence of a structured hexahedral mesh solution for high-genus models, but mainly considered the topological aspect. Other methods, such as skeleton driven [12] and generalized-sweeping-based [4] block decomposition, seem that are not suitable for complex model.

Campen et al. [2] proposed a method for surface quad layout construction based on dual loops. First, they made use of the method proposed by Bommes et al. [1] to create a consistent smooth cross field. Second, they constructed admissible loops to suitably separate the singularities of the cross field. Finally, the quad layout was obtained by layout primalization.

In order to automatically and effectively generate valid block structures of solid models, in this paper we propose a dual surface based approach to block decomposition of solid models. In general, our method has the following contributions:

1. The method automatically guarantees the geometric and topological validity of the final block structure by constructing a required dual surfaces set and performing dual operations.

2. The method ensures the high quality of the every block generated by using the high quality dual surfaces whose construction is based on the dual loops and high quality frame field.

2 Preliminaries and Approach Overview

2.1 Preliminaries

Before describing our block decomposition method, we firstly introduce some related concepts.

2.1.1 Basic Concepts Related to Dual Space

Given a hex mesh H and a mesh edge e in H, the set of mesh edges E_s which consists of all mesh edges topologically parallel to e recursively found is called a **sheet**. E_{bs} is a set consisting of all the mesh boundary edges in E_s. If $E_{bs} \neq \emptyset$, the curve which traverses all the edges in E_{bs} is called a **dual loop**. The manifold surface which traverses all the mid points of the edges in E_s is called a **dual surface**. We call two opposite mesh edges in a mesh face topologically parallel edges. In Fig. 1a, the blue curve is a dual loop while the yellow surface is a dual surface.

The dual operations refer to the operations modifying the topology of the hex mesh. They mainly include the sheet inflation, sheet extraction and column collapse. For the details, the readers can refer to [9, 14].

2.1.2 Frame Field

The frame field will be a promising tool to guide the block decomposition [8]. A 3D **frame F** is a 3-tuple $\{\mathbf{u}, \mathbf{v}, \mathbf{w}\}$, where $\mathbf{u}$, $\mathbf{v}$, and $\mathbf{w}$ are three unit vectors such that $\mathbf{u}.\mathbf{v} = 0$, $\mathbf{w} = \mathbf{u} \wedge \mathbf{v}$. In our algorithm, with a background tet mesh of the solid model, a **frame field** is a frame per cell of the tet mesh, as shown in Fig. 1c.

Between two frames F_s and F_t, there is a **matching matrix** $\Pi_{st} \in \mathcal{G}$ (the chiral cubic symmetry group). There is an orthonormal matrix $type(e, t_0)$ [6] representing the type of an oriented tetrahedral mesh edge e, where t_0 is a cell adjacent to the edge e. If $type(e, t_0) \neq I$, the e is a **singularity edge**.

A **simple singular polyline** is a set of concatenate singular mesh edges, which has the property that the number of singular edges adjacent to each end vertex is not two if it is not a closed polyline. To facilitate the construction of high quality dual surfaces, we divide all simple singular polylines into five categories:

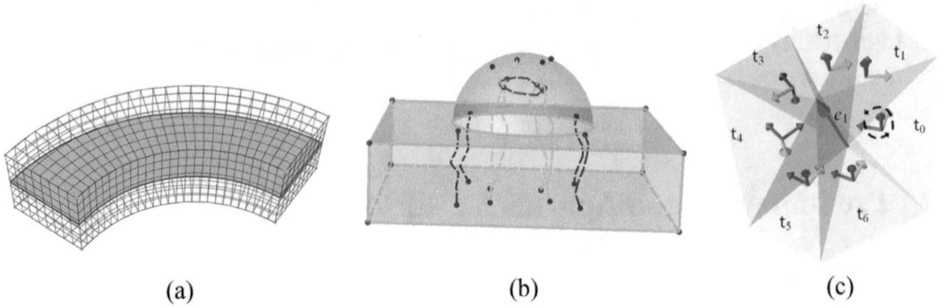

Fig. 1 (**a**) Dual loop and dual surface; (**b**) All simple singular polylines of test case 2 model: the cyan, blue, green and red lines are type 1, 2, 3 and 4 simple singular polylines, respectively; (**c**) A frame is associated with each cell and the edge e_1 is a singular edge with valence 3

Type 1 All the singular edges of the simple singular polyline are inside the body, and only one of the polyline's endpoints is on the boundary surface, as shown by the cyan polylines in Fig. 1b;

Type 2 All the singular edges of the polyline are inside the body, while both of the two endpoints are on the boundary surface, as shown by the blue polylines in Fig. 1b;

Type 3 All the singular edges of the polyline are on the boundary surface, as shown by the green polylines in Fig. 1b;

Type 4 All of the singular edges of the polyline are inside the body, and both endpoints are not on the boundary surface. For a closed simple singular polyline, if all its singular edges are inside the volume, it is also defined as a type four polyline. Type four simple singular polylines are red polylines shown in the Fig. 1b;

Type 5 The simple singular polylines cannot be classified as one of the above four categories.

2.2 *Overview of Approach*

Inspired by the work of Campen et al. [2], we solve the problem of block decomposition in the dual setting, namely, achieving the automatic and valid block decomposition of solid models based on dual surfaces and dual operations.

In order to effectively decompose a solid model into high quality blocks based on dual surfaces, three critical issues need to be addressed:

1. How to construct the dual loops required by the construction of dual surfaces toward valid block decomposition;
2. How to determine the required dual surfaces set based on dual loops;
3. How to generate the high quality dual surfaces that can ensure the high quality of every generated block.

For the first issue, the construction of dual loops is guided by the cross field for block decomposition to meet the needs. For the second issue, we determine the dual surfaces set by enabling all the dual surfaces in the set to separate each boundary elements of solid model and simple singular polylines so that the resulting block structure can be geometrically valid. If two elements are not separated by any dual surface, they will be merged into one in the final block structure. For the third issue, we use the min cut algorithm to construct dual surfaces based on dual loops, so as to ensure the high quality of every dual surface, which leads to blocks of high quality.

The input of our approach is a solid model with boundary representation, and the output is a block structure. The following are the main steps of our method, as shown in Fig. 2:

1. A tet mesh is generated as background mesh;
2. A smooth frame field is built and three types of singularity degeneracy are corrected;
3. Dual loops for block decomposition are constructed;
4. Dual surfaces are constructed based on the dual loops and the singularity structure;
5. A block structure is generated based on the dual surfaces and the dual operations.

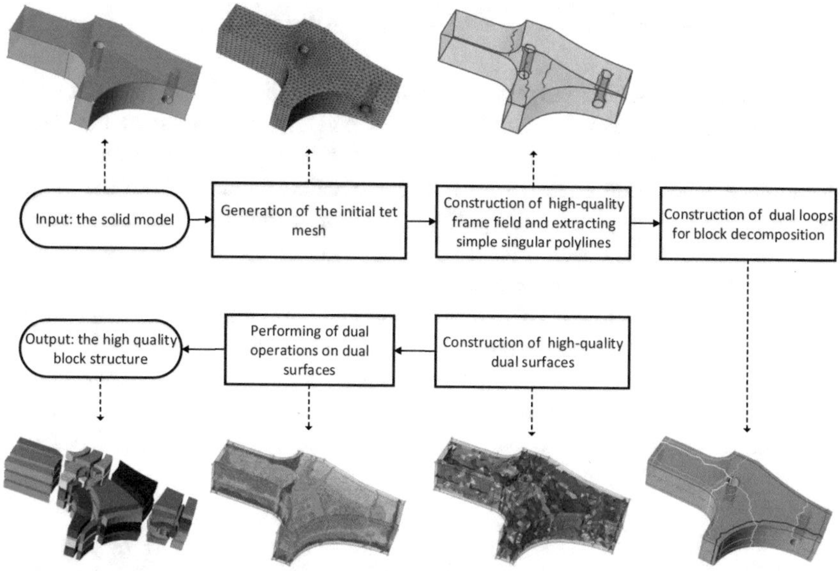

Fig. 2 Pipeline of dual surfaces based block decomposition

3 Frame Field Construction and Singular Polylines Correction

To support the construction of high quality dual loops and dual surfaces, we first construct a high quality frame field of the solid model. First, we initialize the frame field by converting the cross field designed on the surface into frame field and propagating it inside the volume. Then we smooth the frame field by minimizing a non-convex object function using L-BFGS [11]. Due to limited space, we recommend the interested readers to refer to [10]. After the frame field is smoothed, we extract simple singular polylines in the frame field and correct degenerate cases.

In order to ensure that the optimized frame field for block decomposition is as effective as possible, we automatically identify and correct possible degenerate singularities. Specifically, the following three types of degeneracy are identified and corrected: compound singularity, zigzag singularity and 3–5 simple singular polylines. For the first two types of degeneracy, we directly use the fixing strategy proposed in [6] to correct them.

3.1 *Identification of 3–5 Simple Singular Polylines*

The **3–5 simple singular polyline** is the simple singular polyline whose two end edges are 3 and 5 valent singular edges respectively. As shown in Fig. 3a, the highlighted polyline is a 3–5 simple singular polyline with 3 valent end edge (in blue) and 5 valent end edge (in red). The cross field induced from the singular structure with 3–5 simple singular polylines cannot be used to construct all the dual loops for block decomposition (cf. Fig. 4a, c). Therefore, we have to identify and correct each 3–5 simple singular polyline.

According to the definition of 3–5 simple singular polyline, the key to identifying it is to determine the valence of the end edges in each simple singular polyline, and we determine the valence of an edge as follows: for each non-degenerate singular oriented mesh edge e, the **fixed axis** $\tilde{v}$ is one of member vectors of frame in tet t_0 such that $type(e, t_0)\tilde{v} = \tilde{v}$. If its deviation from the e is smaller(respectively, not

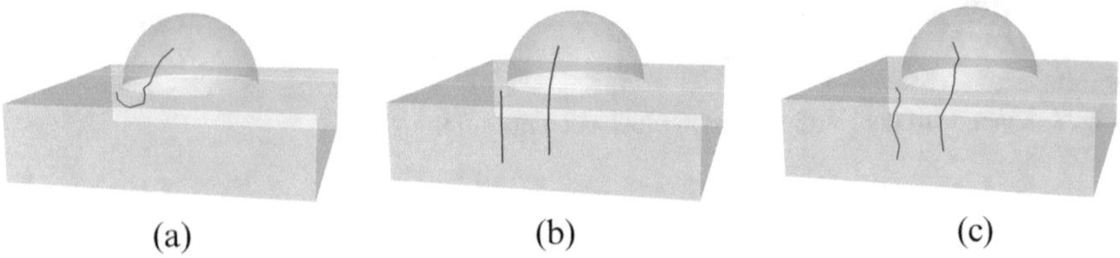

Fig. 3 (**a**) A 3–5 simple singular polyline in test case 2; (**b**) Two related streamlines of the 3–5 simple singular polyline; (**c**) The 3–5 simple singular polyline is replaced with two polylines

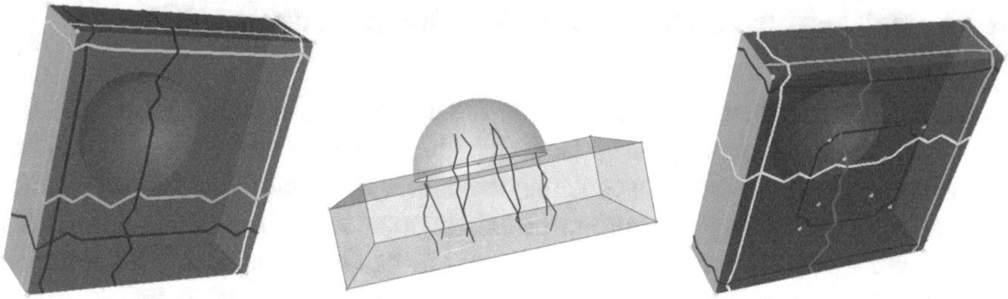

(a) (b) (c)

Fig. 4 (**a**) the dual loops are constructed based on the cross field without adjustment; (**b**) the yellow polylines are the shortest paths connecting the pairs of the intersection points; (**c**) the dual loops are constructed based on the cross field after adjustment

smaller) than $\frac{\pi}{2}$ and $type(e, t_0)$ corresponds to a counterclockwise rotation by $\frac{k\pi}{2}$ along the $\tilde{v}$, then e is a $4 - k$(respectively, $4 + k$) valence singular edge, where $k \in \mathbb{Z}\backslash\{0\}$. In our cases, the valence of singular edges in the optimized frame field after correcting zig-zag and compound singularities is only 3 and 5, and there is theoretical possibility that higher valence of singular edges appears. According to the valence determination, the edge e_1 in Fig. 1c is a 3 valent singular edge.

3.2 Correction of 3–5 Simple Singular Polylines

According to the principle [15] that a simple singular polyline should be consistent with the direction of a frame field, we correct each 3–5 simple singular polyline based on the streamlines of the frame field as follows:

1. Tracing of two related streamlines. For each endpoint of the 3–5 simple singular polyline, one related streamline who starts from this endpoint and takes the direction of frame field closet to the directed end edge from the endpoint as the initial direction is traced to the boundary. The two related streamlines are shown in Fig. 3b.
2. Replacement of the 3–5 polyline. First, for each streamline, we find the corresponding polyline in mesh who are the shortest path connecting two endpoints of the streamline. Second, we replace the original 3–5 simple singular polyline with two polylines corresponding to two related streamlines. Two polylines are shown in Fig. 3c.

Our strategy is heuristic and may not be able to deal with general case.

4 Construction of Dual Loops for Block Decomposition

Since it is very difficult to directly generate a high quality dual surface, we first constructs the boundary of dual surfaces, i.e., dual loops, to support the construction of the dual surfaces. In order to construct the dual loops required by the construction of dual surfaces, we need to improve the existing dual loop construction method [2]. The necessary condition for dual surfaces set to be able to separate all the simple singular polylines and all the boundary elements of solid model is that boundary loops of dual surfaces must be able to separate all the end points of singularity polylines and boundary elements of the surface. The basic idea of improvement is to construct extra dual loops based on surface cross field matching the frame field in volume. Specifically, the dual loops construction process is divided into two steps: first, the surface cross field is built based on the frame field in volume; second, dual loops are constructed to separate all the singular points and the boundary elements of the surface.

4.1 *Construction of Frame Field Based Cross Field*

We first construct a surface cross field that matches the singularity structure in volume, and the specific process consists of the following two steps.

1. Initial construction of the surface cross field based on frame field in volume. The cross direction in each triangle is set as the non-normal direction of frame in adjacent tetrahedral element; the period jump [13] between two adjacent crosses is determined by the matching matrix type between the corresponding frames.
2. Surface cross field adjustment. For each pair of simple polylines induced from a 3–5 simple singular polyline, the period jumps on shortest concatenate mesh edges connecting the new pair of intersection points between adjusted polylines and mesh boundary are adjusted so that a pair of 3–5 valence singular points meeting the requirements appears on the surface. The yellow polylines in Fig. 4b are the paths connecting the pairs of the intersection points while the loops in Fig. 4c are the dual loops constructed based on the cross field after adjustment.

4.2 *Construction of Dual Loops for Dual Surfaces*

We now construct the dual loops based on the cross field with the goal to separate all the target elements. In [2], only the singular points are set as the target elements to be separated. In our work, we set the singular points and the boundary elements of the surface (i.e. the geometric points and geometric edges of the solid model) as target elements. The specific steps are as follows:

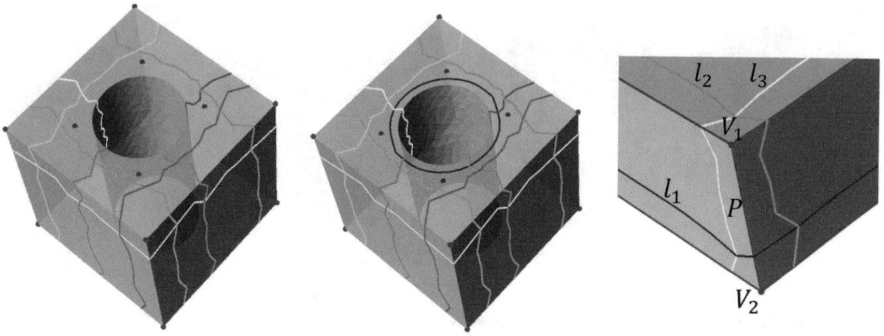

(a) (b) (c)

Fig. 5 (**a**) The result dual loops are constructed based on SIs without boundary elements of the surface taken into consideration. (**b**) The result dual loops are constructed based on SIs with boundary elements of the surface taken into consideration. (**c**) The singular point V_1 and V_2 are all associated with the loop l_1, l_2 and l_3

1. Mesh preprocess before dual loops construction. The corresponding tetrahedron are refined so that there are at least two mesh edges connecting each pair of elements to be separated, so that the elements are separated by dual loops.
2. Construction of separation indicators (SIs) [2]. In order to separate pairs of target elements efficiently, we first construct SIs. For each pair of singularities, SIs are paths representing a homotopy class each, as [2] done. For each pair of boundary elements of the surface with other elements or singular points, SIs are shortest paths between them.
3. Construct the dual loops to cut all SIs by the greedy algorithm in [2].

The constructed dual loops above divide the boundary surface into different regions, each containing at most one singular point. If a region contains a singular point, we associate all the dual loops that surround the region with the singular point. As shown in the Fig. 5c, the dual loops l_1, l_2 and l_3 are associated with the singular points V_1 and V_2 respectively. If all endpoints of a simple singular polyline on boundary surface are associated with a loop and located on same side of the loop, then this loop is associated with this polyline. As show in the Fig. 5c, the loops l_2 and l_3 are associated with the simple singular polyline P, respectively.

5 High Quality Dual Surfaces Construction

In this work, we decompose the solid model into a block structure based on dual surfaces. Therefore, constructing a set of high quality dual surfaces to separate all the simple singular polylines and boundary elements of the solid model is the key to accomplishing the valid block decomposition. In order to facilitate the construction of the dual surfaces, we classify all dual surfaces into two categories, closed and

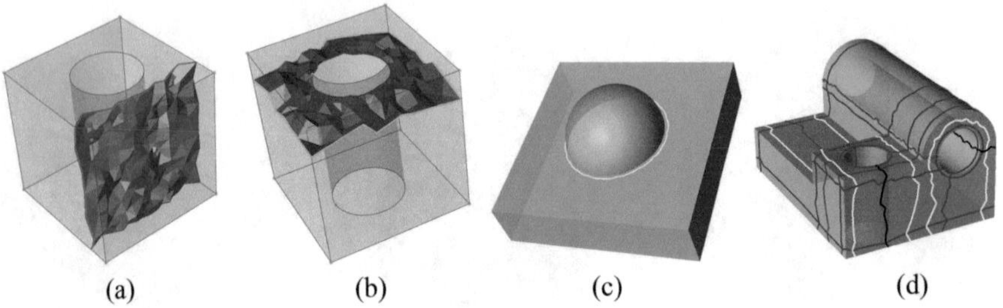

Fig. 6 (a) A simple dual surface. (b) A complex dual surface. (c) The yellow loop is a trivial loop but the loop is an M-loop. (d) The yellow loops are trivial loops and also S-loops. The black loops are loops in the set **B**. The red loops are non-trivial but S-loops. The blue loops are M-loops

open. A dual surface without boundary is called a **closed dual surface** and a dual surface with boundary is called an **open dual surface**. According to the fact that the boundary of an open dual surface must be the dual loops on the boundary surface, we construct each open dual surface by first identifying its corresponding dual loops. As for the closed dual surface, it will be constructed directly according to its separation effect.

5.1 Determination and Construction of Open Dual Surfaces

In the following, we refer to the open dual surface with only one boundary loop as **simple dual surface**, as shown in Fig. 6a, and refer to the open dual surface with multiple boundary loops as **complex dual surface**, as shown in Fig. 6b. In our dual surface construction algorithm, the determination of complex dual surface requires the aid of some simple dual surfaces. Therefore, we first construct the simple dual surfaces, and then construct the complex dual surfaces.

5.1.1 Classification of Dual Loops

In order to facilitate the construction of open dual surfaces, the dual loops are classified into two types:

S-loop An S-loop (short for the boundary loop to construct dual surface) is a loop that can span a required surface in the interior of the solid model by itself;

C-loop An C-loop (short for the multiple loops to construct dual surface) is a loop that can not span a required surface in the interior of the solid model by itself.

A required surface refers to the surface that can separate boundary elements of solid or/and simple singular polylines. Obviously, S-loops and C-loops are the boundary of simple and complex dual surfaces respectively.

There is a special kind of loop on surface called the **trivial loop** which alone can bound a subset of the surface; that is, a trivial loop can divide the boundary into two regions. According to our analysis, the dual loops are classified based on the following heuristic rules:

1. If a trivial loop is associated with all the endpoints of some simple singular polylines, then it is an S-loop, otherwise it is an C-loop. Though every trivial loop can span a surface in the interior of the solid model, only the surface spanned by the trivial loop meeting the above condition can separate simple singular polylines with endpoints associated. Yellow loops in Fig. 6d are trivial loops which can span desired surfaces while the yellow loop in Fig. 6c is a trivial loop which can not span a desired surface.
2. If a nontrivial loop is the shortest loop passing through a hole, then it is an S-loop. Generally, such loop can also span a surface which can separate boundary elements or/and simple singular polylines. And we insert all such S-loops (as black loops shown in Fig. 6d) into a set **B**.
3. If a nontrivial loop which is not in **B** has no intersection with loops in **B**, then it is an S-loop (as red loops shown in Fig. 6d), otherwise it is an C-loop(as blue loops shown in Fig. 6d). Generally, such S-loop can span a surface which separates boundary elements and simple singular polylines.

5.1.2 Construction of Simple Dual Surfaces

After identifying all the S-loops, we can construct simple dual surfaces accordingly. To obtain the block structure by dual operations, the dual surfaces need to be constructed in a hexahedral mesh. So a hex mesh is transformed from the tet mesh by using the tet-to-hex method, i.e., splitting each tetrahedron into four hexahedra. At the same time, the frame field of hex mesh inherits the frame field of tet mesh trivially.

The simple singular polylines are elements to be separated by dual surfaces, so during the dual surfaces construction the simple singular polylines are taken into consideration if the polylines are associated with the input loops. Due to the fact that the determination of the dual surfaces on the mesh is analogous to finding a cut on a graph [3], we use the min cut algorithm for input dual loop(s) to construct dual surface(s) by the following steps:

1. Determination of the source and the target hexahedra sets for min cut algorithm.
   a. Quads division on the boundary surface. The input dual loops divides the quads on the boundary surface into two sets, which are called source and target quads sets respectively.
   b. Source and target hexahedra sets determination. The hexahedra with boundary mesh face in the source(target) quads set are inserted into the source(target) hexahedra set.

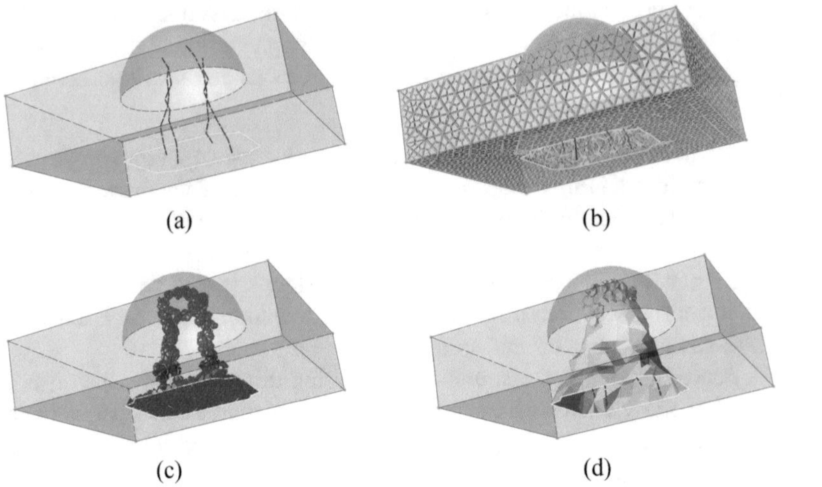

Fig. 7 Illustration of simple dual surface construction: (a) the yellow loop is associated with four type 1 simple singular polylines; (b) the hexahedra in source set; (c) the hexahedra in target set; (d) the yellow surface is constructed by min cut algorithm

   c. Source and target hexahedra sets completion. In order to take the associated simple singular polylines into consideration, the source and target hexahedra sets need to be completed. The hexahedra adjacent to the associated simple singular polylines are inserted into the source (target) hexahedron set if its boundary endpoints belong to the source (target) quadrilateral set. The source and target hexahedra set are shown in Fig. 7b, c, respectively.

2. Construction of desired digraph and acquisition of dual surfaces by min cut algorithm.
   a. Nodes determination. A graph is built by considering all the hexahedra as the nodes, where the nodes corresponding to source and target hexahedra are denoted as s-nodes and t-nodes, respectively.
   b. Arcs determination and weight assignment. Two nodes are considered adjacent if their corresponding cells share at least one mesh face. For the s-node, a directed arc from it to each of its adjacent nodes is added; for the t-node, a directed arc from each of its adjacent nodes to it is added; for the rest pair of adjacent nodes, a pair of opposite directed arcs are added. The weight of each arcs above is set as the area of the common mesh face between corresponding cells.
   c. Dual surface determination. The minimum cut is found between s- and t-nodes using the min cut algorithm, and this cut corresponds to the simple dual surface in the hex mesh.

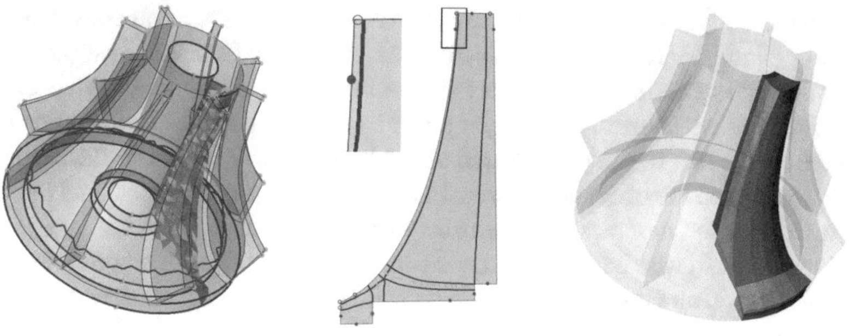

(a) (b) (c)

Fig. 8 Illustration of determining the dual loops group: (**a**) the target dual surface in test case 6; (**b**) the existing dual loops and newly added dual loops represented by solid red circles and hollow red circles in smoothed dual surface; (**c**) the final blocks traversed by the dual surface

5.1.3 Determination and Construction of Complex Dual Surfaces

After all the simple dual surfaces constructed, the complex dual surfaces are constructed with the assist of existing simple dual surfaces. The intersection of two dual surfaces forms the dual curves and the intersections of the boundary loops of these dual surfaces are the endpoints of the corresponding dual curve. Therefore, we use the quad layout of the dual surfaces to determine the C-loops groups. For efficiency, only the dual surfaces spanned by the chosen dual loops and the shortest loops intersecting with the trivial undesired loops are considered. The following are the steps to determine the dual loops groups.

1. Extraction of the dual surfaces' quad layout based on cross field. First, the cross field on the smoothed dual surface is established by projecting the frames associated to cells on one side of surface. Second, the singular points on the surface are determined according to the intersection of the original dual surface and simple singular polylines, as shown in Fig. 8a. Finally, the streamlines are traced from singular points as the separatrices and quad layout is obtained, as shown in Fig. 8b.
2. Dual loops grouping. According to the quad layout, the intersecting loops at opposite side of each chord need to be grouped. Before doing this, some dual loops have to be added to balance the number of intersecting dual loops at both end of the chord. One dual loop per side of the chord is added for the chord who has no loop at neither side, except the trivial chord whose deletion will not influence the singular points configuration in quad layout, as golden dotted lines shown in Fig. 8b. If two dual loops groups intersect, they are merged into one. Figure 8c shows the final blocks traversed by the dual surface.

Since the uneven surface will result in undesired quad layout, we perform Laplacian smoothing to the dual surface before dual loops grouping. If the endpoints of a chord have unequal dual loops intersection numbers, we add dual loops as follows(the endpoint of the chord having larger intersection number is called the start side, and another endpoint is called the end side). First, shortest paths from each intersecting vertices on the start side to the end side are computed. Second, a dual loop for each shortest path is computed by anisotropic front propagation in [2] with the end vertex of the path chosen as the loop's start vertex. The first two steps are performed on the original tet mesh. Finally, the newly constructed dual loops are converted into the hex mesh and replace with original dual loops on the end side. As for each chord who has no intersecting loop at neither sides, the two end vertices of the shortest path connecting the two side are chosen as start vertices to add dual loops.

After determining the dual loops groups, the complex dual surfaces are constructed by the min cut algorithm as simple dual surfaces done in Sect. 5.1.2.

5.2 Closed Dual Surface Construction

Since construction of open dual surfaces only takes boundary dual loops and simple singular polylines into consideration, the type 4 simple singular polylines may not be sufficiently separated. Hence, closed dual surfaces must be constructed to make up for this omissions. Currently, we only consider the general situation that the type 4 polylines are not separated from the boundary surfaces. Therefore, the method first detects whether there is any type 4 polyline unseparated from the boundary surface of model and then constructs corresponding dual surface to separate the polyline and the surface if necessary. The specific algorithm is as follows:

1. Check whether a closed dual surface is needed. First, decompose the hexahedral mesh by the existing dual surfaces. Then, traverse the parts and record the boundary surface if there is a part contains both hexahedra with mesh face on this surface and the singular edges of a type 4 simple singular polyline. There are 12 polylines unseparated from the boundary surface in Fig. 9a.

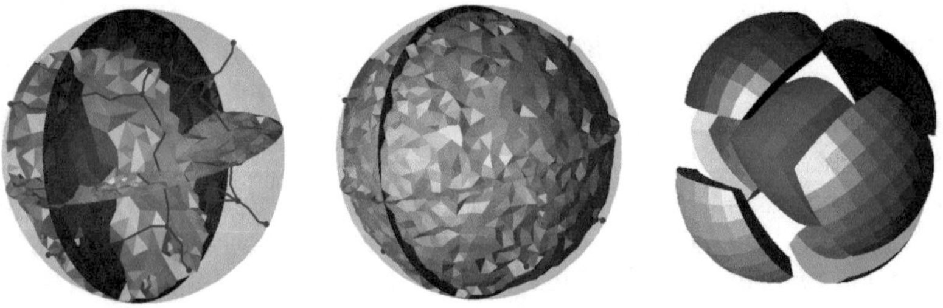

(a) (b) (c)

Fig. 9 Illustration of the closed dual surface construction: (**a**) 12 simple singular polylines with type 4 not separated from boundary surface; (**b**) a closed dual surface constructed; (**c**) the final blocks generated

2. Construct the closed dual surface if necessary. First, the hexahedra whose boundary mesh faces are in the recorded surface of solid are inserted into the source hexahedra set, while the rest hexahedra are inserted into the target set. Then, the closed dual surface is generated by min cut algorithm as described in Sect. 5.1.2. A closed dual surface is inserted in Fig. 9b and the corresponding block structure of the model is shown in Fig. 9c.

6 Block Structure Generation Through Dual Operation

After the required dual surfaces set has been constructed, as shown in the Fig. 10a, the corresponding sheets are generated through sheet inflation operations applied along the dual surfaces, as shown in Fig. 10b. Sheet set extraction [16] is used to remove the remaining sheets in the current hex mesh and final block structure is obtained, as shown in Fig. 10c.

7 Experimental Results and Limitations

The proposed method has been implemented using C++ as programming language and ACIS as the geometric engine. Tet meshes are generated using commercial software Abaqus. According to the max-flow min-cut theorem, we compute the min cut of digraph by the Ford-Fulkeson method. And all the shortest paths are computed by the Dijkstra algorithm. In order to visualize the block structure and measure the scaled Jacobian value, the final block structure generated through dual operations are refined and optimized. Apart from the test case 1 which has been shown in Fig. 2, the other test cases and results are shown in Table 1. And the statistics of the results are given in Table 2.

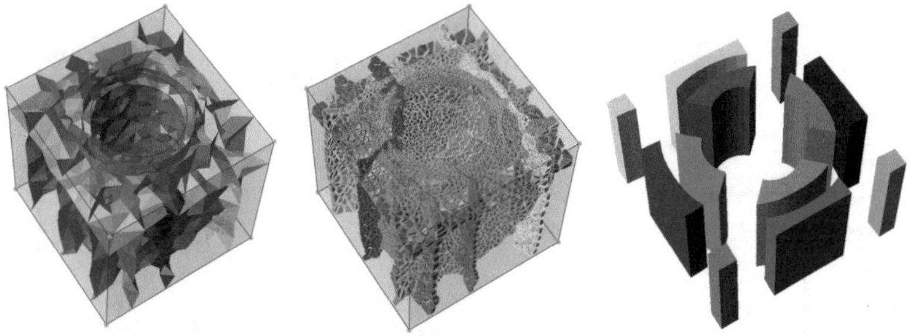

(a) (b) (c)

Fig. 10 Illustration of the block structure generation: (**a**) the dual surface of the model; (**b**) sheets inflated based on the dual surfaces; (**c**) the final block structure

Table 1 The main results of test cases

	Test case 2	Test case 3	Test case 4	Test case 5	Test case 6
Input models	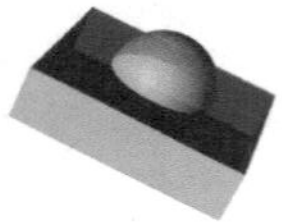				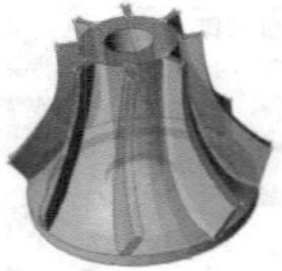
Singular graph	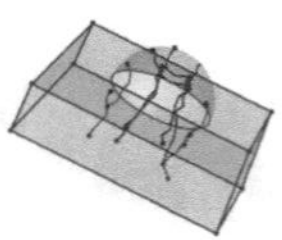	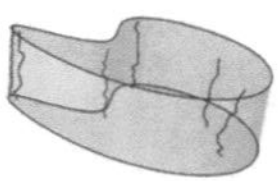	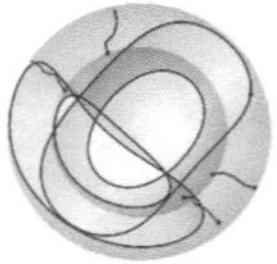	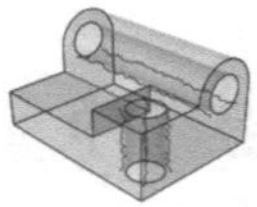	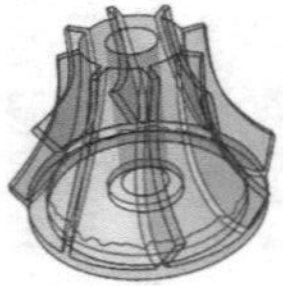
Dual surfaces		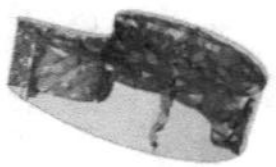	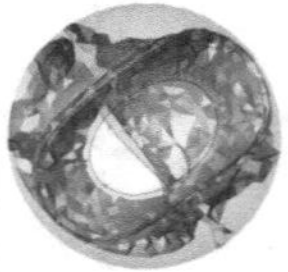		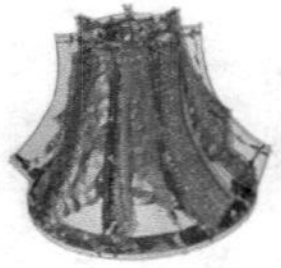
Block structure					

Table 2 The statistics of the results

Model	Dual surfaces	Block	Inner singularity	Scaled Jacobian		Hex Num.
				Min	Mean	
Test case 1	21	75	20	0.434	0.969	8288
Test case 2	10	23	20	0.723	0.971	2432
Test case 3	6	11	6	0.411	0.908	5472
Test case 4	6	7	4	0.784	0.982	4710
Test case 5	17	67	16	0.681	0.988	14252
Test case 6	30	128	16	0.325	0.972	8800

The numbers of dual surfaces constructed, blocks of final structure, inner singular edges are in columns **Dual Surfaces, Block** and **Inner Singularity**, respectively. The fifth and sixth columns show the minimum and average scaled Jacobian values of blocks, and the last column shows the cell numbers of the refined hex mesh used

Comparison with Kowalski et al.'s Method [8] In column 2 of Table 1, the input model, singular graph, all the dual surfaces and final block structure of test case 2 are shown. The test case 2 is a model that can not be handled by the Kowalski's method. After correcting the 3–5 simple singular polylines, a frame field without degenerate singularities can ensure the validity of the final block structure by our method, and the obtained blocks is high quality as shown in Table 2.

Comparison with Wang et al.'s Method [16] The models of test cases 3 and 4 in Table 1 all contain free-form surfaces, which can not be handled by Wang's method. Guided by the optimized frame field, the dual surfaces set are constructed, which not only ensure the geometric validity but also ensure the quality of the blocks generated.

The models of test case 1 and 5 with two through holes are decomposed into high quality blocks. In column 6 of Table 1, the input model, singular graph, some dual surfaces constructed and final block structure of test case 6 are shown. The test case 6 is a relatively complex model and the final block structure of it has 128 blocks. The refined mesh's minimum scaled Jacobian value is 0.325, as shown in Table 2. We think that the reason why these low quality elements is that our geometric optimization algorithm is not good enough for this example.

Limitations There are several limitations existing in the current method. First, there is still the possibility that our frame field optimization and degeneracy correction may fail for complex model. Second, the dual loops constructed may be not enough for block decomposition. For example, there is no dual loop on torus (as shown in the right inset). Third, our dual loops categorization algorithm is currently heuristic based.

8 Conclusion and Future Work

In this paper, we propose a novel approach to block decomposition of solid models. The approach automatically generates a valid block structure of a solid model based on dual surfaces and dual operations. Compared to the previous methods, our approach has the following characteristics:

1. Through constructing the required dual surfaces set based on dual loops and simple singular polylines and enabling all the dual surfaces constructed to separate every simple singular polylines and boundary elements of the solid, the geometric validity of final block structure is guaranteed. In addition, through performing the dual operations on the intermediate hexahedral mesh, the topological validity of final block structure is guaranteed.
2. By constructing dual surfaces based on dual loops for block decomposition and min cut algorithm, the dual surfaces constructed are of high quality, which furthermore ensures the high quality of the final blocks.
3. By the identifying and correcting three types of singularity degeneracy of the frame field, including the 3–5 simple singular polyline introduced in this work, the dual loops generated based on such frame field can support the construction of the necessary dual surfaces as effectively as possible.

The following shortcomings need to be overcome in our future work:

1. Our current approach deals with the three types of degenerate singular polylines that can be observed up to now. However, it is also necessary to test more complex models and theoretically make sure whether there exists any other degeneracy of the singularities in the frame field and, if so, to provide corresponding correction strategy.
2. At present our dual loops classifying algorithm is heuristic based. In the future, a more general algorithm will be studied.
3. Since the construction of optimized frame field and dual loops are all sensitive to the input tet mesh, in the future, the reasonable tet mesh discretization of the input solid model will be studied.

Acknowledgement We thank all anonymous reviewers for their valuable comments. This research is supported by the NSF of China (Nos. 61572432).

References

1. D. Bommes, H. Zimmer, L. Kobbelt, Mixed-integer quadrangulation, in *ACM Transactions on Graphics (TOG)*, vol. 28 (ACM, New York, 2009), p. 77
2. M. Campen, D. Bommes, L. Kobbelt, Dual loops meshing: quality quad layouts on manifolds. ACM Trans. Graph. TOG **31**(4), 110 (2012)
3. J. Chen, S. Gao, R. Wang, H. Wu, An approach to achieving optimized complex sheet inflation under constraints. Comput. Graph. **59**(C), 39–56 (2016)
4. X. Gao, T. Martin, S. Deng, E. Cohen, Z. Deng, G. Chen, Structured volume decomposition via generalized sweeping. IEEE Trans. Vis. Comput. Graph. **22**(7), 1899–1911 (2016)
5. X. Gao, D. Panozzo, W. Wang, Z. Deng, G. Chen, Robust structure simplification for hex re-meshing. ACM Trans. Graph. TOG **36**(6), 185 (2017)
6. T. Jiang, J. Huang, Y. Wang, Y. Tong, H. Bao, Frame field singularity correction for automatic hexahedralization. IEEE Trans. Vis. Comput. Graph. **20**(8), 1189–1199 (2014)
7. N. Kowalski, F. Ledoux, M.L. Staten, S.J. Owen, Fun sheet matching: towards automatic block decomposition for hexahedral meshes. Eng. Comput. **28**(3), 241–253 (2012)
8. N. Kowalski, F. Ledoux, P. Frey, Smoothness driven frame field generation for hexahedral meshing. Comput. Aided Des. **72**, 65–77 (2016)
9. F. Ledoux, J. Shepherd, Topological and geometrical properties of hexahedral meshes. Eng. Comput. **26**(4), 419–432 (2010)
10. Y. Li, Y. Liu, W. Xu, W. Wang, B. Guo, All-hex meshing using singularity-restricted field. ACM Trans. Graph. TOG **31**(6), 177 (2012)
11. D.C. Liu, J. Nocedal, On the limited memory BFGS method for large scale optimization. Math. Program. **45**(1–3), 503–528 (1989)
12. M. Livesu, A. Muntoni, E. Puppo, R. Scateni, Skeleton-driven adaptive hexahedral meshing of tubular shapes, in *Computer Graphics Forum*, vol. 35, pp. 237–246 (Wiley Online Library, Hoboken, 2016)
13. N. Ray, B. Vallet, W.C. Li, B. Lévy, N-symmetry direction field design. ACM Trans. Graph. **27**(2), 10:1–10:13 (2008)
14. M.L. Staten, J.F. Shepherd, F. Ledoux, K. Shimada, Hexahedral mesh matching: converting non-conforming hexahedral-to-hexahedral interfaces into conforming interfaces. Int. J. Numer. Methods Eng. **82**(12), 1475–1509 (2009)
15. R. Viertel, M.L. Staten, F. Ledoux, Analysis of non-meshable automatically generated frame fields, in The 25th International Meshing Roundtable (2016)
16. R. Wang, C. Shen, J. Chen, H. Wu, S. Gao, Sheet operation based block decomposition of solid models for hex meshing. Comput. Aided Des. **85**, 123–137 (2017)

Automatic Blocking of Shapes Using Evolutionary Algorithm

Chi Wan Lim, Xiaofeng Yin, Tianyou Zhang, Yi Su, Chi-Keong Goh, Alejandro Moreno, and Shahrokh Shahpar

Abstract This work focuses on the use of evolutionary algorithm to perform automatic blocking of a 2D manifold. The goal of such a blocking process is to completely partition a 2D region into a set of conforming and non-intersecting quadrilaterals to facilitate the generation of an all-quadrilateral, or more preferably an ideal quadrilateral mesh configuration covering the closed 2D region. However, depending on the input shape, the optimal blocking strategy is often unclear and can be very user-dependent. In this work, a novel approach based on evolutionary algorithm is adapted to search for a potential set of such ideal configurations. Based on a selection within a set of candidate vertices from a pre-computed pool, blocking configurations can be derived and ranked based on the collective quality of its blocks. The quality of a block is computed based on objective functions relating to its interior angles and opposite length ratios. Using multi-dimensional ranking criteria, inferior solutions can be slowly filtered away with each successive generation. Based on observations on a range of turbomachinery test cases, it is possible to derive and improve near-optimal blocking configurations by utilizing a large number of generations. This work has the potential to be extensible to 3D shapes as well.

1 Introduction

Structured mesh generation has traditionally been the preferred form of mesh generation for computational fluid dynamics (CFD) simulation in turbomachinery applications. For aerodynamic analysis, it is has been shown that structured meshes

C. W. Lim (✉) · X. Yin · T. Zhang · Y. Su
A*STAR, Institute of High Performance Computing, Singapore, Singapore
e-mail: limcw@ihpc.a-star.edu.sg

C.-K. Goh
Rolls-Royce, Singapore, Singapore

A. Moreno · S. Shahpar
Rolls-Royce Plc, Coventry, UK

X. Roca, A. Loseille (eds.), *27th International Meshing Roundtable*, Lecture Notes in Computational Science and Engineering 127,
https://doi.org/10.1007/978-3-030-13992-6_10

provide greater accuracy and efficiency compared to unstructured meshes [1]. In addition, structured meshes are able to form elements that flow along boundaries and align to flow features, as well as facilitate implementation of higher-order numerical schemes. However, the application of ideal structured meshing is restricted to relatively simple geometries or usage of templates [2]. Consequently, the automatic generation of structured mesh for an arbitrarily-shaped object is divided into two stages: the first stage involves a blocking process that decomposes the object into well-shaped blocks suitable for structured meshing, and the second stage involves the generation of the mesh for each of these blocks.

The objective of this work focuses on the first stage, by developing an automatic multi-blocking algorithm to decompose a given 2D manifold into high quality quadrilateral blocks that are suitable for simulation. The goal is to search for a set of blocking solutions, wherein each solution uniquely minimizes a set of multi-dimensional objective functions. In this paper, an approach is presented using evolutionary algorithm that evolves candidate blocking solutions over many generations with an aim to improve the quality of the solutions. The contribution of this work lies in the novel generation of candidate points which have the potential to form good quality blocks, thereby greatly reducing the search space. This new algorithm has been applied to a diverse range of test cases, ranging from simple shapes to complex geometries of actual turbomachinery parts, consistently producing high quality blocking solutions.

2 Previous Work

A 2D multi-blocking problem has many similarities with unstructured quadrilateral mesh generation problem, albeit with a significantly coarser resolution and a less stringent demand on the quality of the resultant quadrilaterals, hereinafter termed as "blocks". As such, solutions to both problems share many common strategies and techniques. In general, there are two main categories for unstructured quadrilateral meshing: direct and indirect approaches [3]. In direct approaches, a domain is directly meshed by procedural placement of quadrilateral elements into the domain or its subdivided version. For indirect approaches, the domain is triangulated and conversion techniques are employed to merge triangles into quadrilaterals.

The grid-based, advancing front and decomposition methods are some of the existing blocking techniques that belong to the category of direct approaches. Grid-based blocking approaches are computationally-robust but the quality of the mesh is highly dependent on the orientation of the background grid [4] and tends to yield poor elements at the boundary of the domain. Improvement such as using recursive adaptive quad-tree refinement scheme [5, 6] has also been explored. Advancing front methods such as [7, 8] can produce meshes that adapt well to the flow of the boundary, but unequally-spaced vertices on thin boundaries can result in the formation of elements with poor qualities. Decomposition methods [9–12] employ a divide-and-conquer strategy by reducing a complex polygon into smaller and simpler components. Medial axis [13–15] is a popular variant of the decomposition approach.

For indirect approaches, the general concept is to first create a triangle mesh, typically using the Delaunay triangulation method and then converting it into a quadrilateral mesh. This can be achieved either by the topological subdivision of triangles [16] or quadrangulation (i.e., forming quadrilaterals by merging pairs of triangles). Methods based on quadrangulation [17–21] differentiate themselves by the heuristics used for determining which pairs of triangles are to be prioritized for merging, and more importantly, how isolated triangles are handled.

It has been argued that the key to multi-blocking is the placement of singularities, and their connectivity with each other in order to form blocking configurations [22]. Recent works by Kowalski et al. [23], Fogg et al. [24, 25] are attempts to optimize the placement of such singularities using cross-fields, a four-way rotationally symmetric field made up of four unit vectors that form a regular cross. However, the approach to form a blocking solution from the identified singularities can be a challenging problem, such as reducing the total number of blocks, preventing the formation of narrow blocks and spiraling behavior. In spite of these challenges, in most cases cross-field approaches are able to produce high quality blocking configurations and are a good source of comparison with our proposed method, see Fig. 9.

Depending on the nature and intrinsic features of the input shape, different approaches would yield different levels of success, with no one particular method that works perfectly for all input shapes. In our proposed approach, there is no fixed method in which the blocks are derived, but rather solutions are evolved over time using the evolutionary algorithm approach.

3 Overview of Automatic Multi-Blocking Algorithm

The aim of this work is to make use of evolutionary algorithm (EA) to search for an optimal 2D blocking configuration. To achieve that, this work presents a methodology that formalizes the generation of blocking configuration in terms of a fixed boundary (present in every solution) and a set of candidate points, which the evolutionary algorithm selects from. Figure 1 shows the flowchart depicting the overview of the automatic 2D multi-blocking algorithm. The algorithm begins with the pre-processing stage in which the input is the boundary of a 2D closed polygon, represented as a series of linear or curved segments. There are two main objectives in the pre-processing stage: the first is to reduce the 2D polygonal boundary into a set of *fixed boundary points*. The size of this set of fixed boundary points should be kept as small as possible, but yet large enough to describe the input boundary with adequate fidelity. These fixed boundary points will henceforth be present in all solutions generated by the genetic algorithm. The second objective is to use this set of fixed boundary points to generate a set of *candidate points*. The candidate points

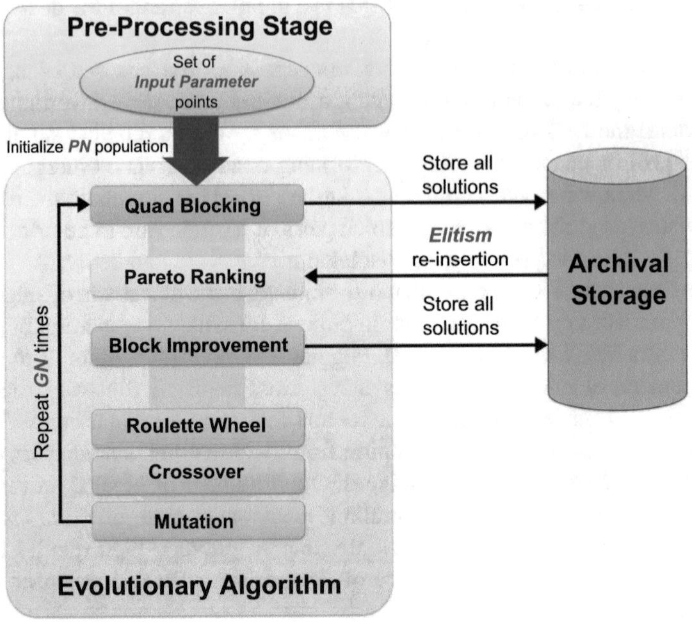

Fig. 1 Multi-blocking algorithm flowchart. The flowchart depicts the process flow of the algorithm. A pre-processing stage downsamples the input boundary, generating fixed boundary and candidate points. Evolutionary algorithm processes are then used to guide the search for an optimal solution. An archival storage is employed to keep track of the best solutions found to date

are generated based on the criteria that they can potentially form good quality blocks with the fixed boundary points.

With the completion of the pre-processing stage, the evolutionary algorithm begins the search for optimal blocking solutions. The two parameters that determine the search space of the EA are the population size (*PS*), and the number of generations (*GN*). *PS* controls the number of solutions that the algorithm holds at each generation, while *GN* determines the number of iterations. The EA initializes the population with *PS* number of elements, where each element is filled up with a randomized subset from the input parameter point set. A quad blocking module is then used to generate a multi-block solution for each population element based on its randomized input set. The fitness of the multi-block solution is quantified by a two-dimensional quality measure that is used by a Pareto fitness ranking scheme to sort and rank solutions in the population based on their blocking optimality.

The ranked population set is then added to the archival storage, and go through another round of ranking, and then trimmed down to a size of *PS* by discarding inferior solutions. Based on an elitism percentage, the top few solutions are added back to the population. Duplicate solutions, if any, are also removed. Similarly, the population set is trimmed down to the size of *PS*. This process of adding all solutions to the archival storage and adding back a number of best-ranked solutions at the start of each generation ensures that the population set always contains the best solutions found so far before applying the evolutionary operators.

There are three types of evolutionary operators used in this work: roulette wheel selection, crossover, and random mutation. Essentially, their goal is to use the higher-ranked solution variants as a starting base to generate newer variants, usually by making minor incremental changes. The whole process is repeated for *GN* number of generations.

4 Pre-processing

The input to the multi-blocking algorithm is a 2D polygonal closed boundary represented as a series of connected and discretized segments. This section describes the process to generate two sets of points, the *fixed boundary points* (minimally set of points for boundary representation) and the *candidate point set* (used for selection by EA).

4.1 Generation of Fixed Boundary Point Set

For simplicity, the 2D polygonal closed boundary is assumed to be represented as one contiguous chain of n number of points, formed by merging all resampled input segments, defined as $P = \{p_1 \dots p_n\}$. The goal is to reduce P to as small as possible using appropriate breakpoints while maintaining an adequate level of fidelity to the original shape. There are three types of breakpoints: *large angle breakpoints* (capturing main feature points), *curvature based breakpoints* (handling curved regions), and *bounding box breakpoints* (reducing deviation from original boundary). For clarity, Fig. 2 illustrates the convention used while referring to angles formed by three consecutive points.

Large angle breakpoints are characterized by any point p with $\alpha_p > 30°$. The point p is then used as a breakpoint to split the chain of points into smaller segments. Assuming that m number of breakpoints exist, represented by the index set $\mathbf{b} =$

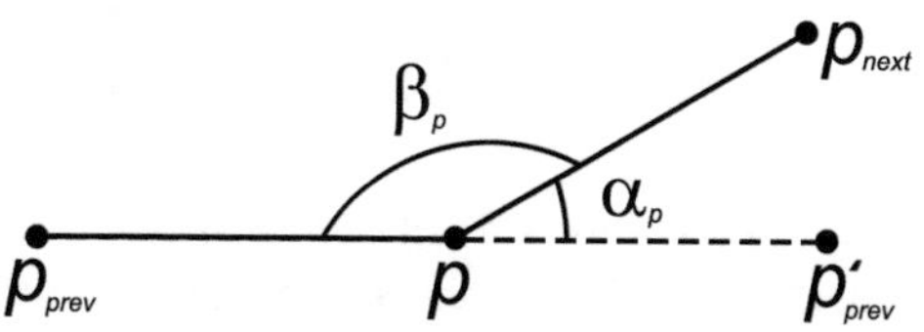

Fig. 2 Angles nomenclature. For any point $p = p_i$, p'_{prev} refers to the reflection of p_{prev} at p. The angle α_p is formed by the triplet points of $\{p_{next}, p, p'_{prev}\}$, while β_p is formed by $\{p_{prev}, p, p_{next}\}$

$\{b_1 \ldots b_m\}$, segments can be defined as $\boldsymbol{S} = \{S_1 \ldots S_m\}$, where

$$S_i = \begin{cases} \{p_{b_i}, p_{b_i+1}, \ldots, p_{b_{(i+1)}}\} \; for \; 1 \le i < m \\ \{p_{b_i}, p_{b_i+1}, \ldots, p_n, p_1, \ldots, p_{b_1}\} \; for \; i = m \end{cases} \tag{1}$$

For a curved segment S_i, it is further split using a curvature based approach to approximate its curved geometry. Assuming that S_i consist of r number of points, $\{p_{i,1} \ldots p_{i,r}\}$, the total accumulated curvature change ($TACC$) of S_i is computed as

$$TACC(S_i) = \sum_{j=2}^{r-1} \alpha_{p_{i,j}} \tag{2}$$

Using $TACC(S_i)$, we define the curvature change interval, $CCI(S_i)$ as

$$CCI(S_i) = \frac{TACC(S_i)}{\left\lceil \frac{TACC(S_i) - 45^\circ}{90^\circ} \right\rceil + 1} \tag{3}$$

The placement strategy is to add one additional breakpoint at the first 45° accumulation, and then place one additional point for every 90° accumulation thereafter. Curvature-based breakpoints can thus be added at every $CCI(S_i)$ interval of accumulated curvature change along the segment S_i. All curvature-based breakpoints are added to the index set $\boldsymbol{b}$, and the segment set $\boldsymbol{S}$ is regenerated again using Eq. (1).

Each segment S_i can be further split using a recursive bounding box approach. Assuming that S_i consist of t number of points, $\{p_1 \ldots p_t\}$, it is recursively split S_i into two subsets at index k, where $k \in 2 \ldots t-1$, if it fulfills two conditions: p_k has the largest perpendicular distance to the line formed by p_1 and p_t, and the ratio of the perpendicular distance to the line exceeds a pre-defined constant of 0.15.

Finally, the bounding box breakpoints are added to the index set $\boldsymbol{b}$ and the updated set $\boldsymbol{b}$ is used as the index set for the fixed boundary points. Figure 3 shows the generation of the three types of breakpoints for a *marine propeller blade* dataset.

4.2 Generation of Candidate Point Set

To form other blocking configuration variants, other than the fixed boundary points, it is necessary to define the candidate point set, which comprises of the additional boundary and interior candidate points, from which the EA can draw from. The performance of the automatic multi-blocking algorithm depends heavily on the characteristic of the candidate point set; a large set size will unnecessarily lengthen its computational time, and the quality of points based on their positions determines

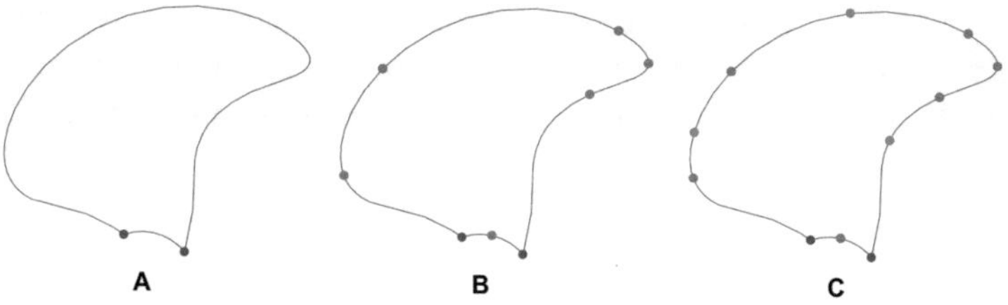

Fig. 3 Generation of fixed boundary points. The breakpoints generated for the *marine blade* dataset (side view of a marine propeller) by using large angle (blue), curvature (green), and bounding box (orange) criteria is shown in (**a**), (**b**), and (**c**), respectively

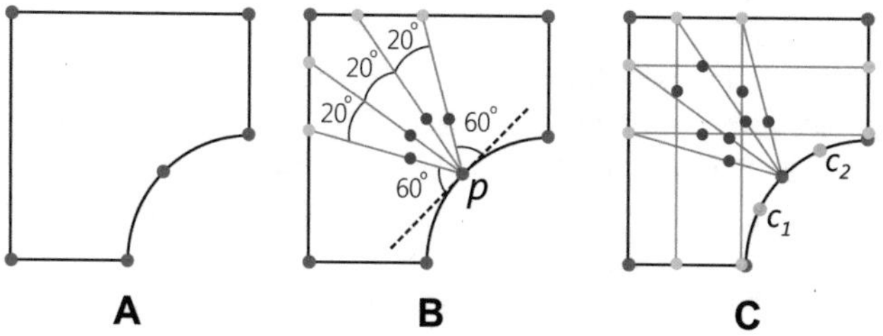

Fig. 4 Generation of additional candidate boundary and interior points. In (**a**), a simple example shape is shown with the fixed boundary points shown in red. In (**b**), candidate points (blue for interior points, green for boundary points) are created using the ray projection approach. In (**c**), a second ray projection pass creates additional candidate points. Orange points (c_1 and c_2) are points that caters to "dead zone"

the quality of the resulting blocking configurations. Hence, each point included in the candidate point set is selected based on their potential to form quadrilateral blocks of good quality.

For the generation of the candidate point set, the following methodology is adopted: For each point $\boldsymbol{p}$ in the fixed boundary set, its interior angle with respect to the polygon is computed. For any point with an interior angle of more than 120°, a ray is projected inwards at every step interval of 20°, with a 60° buffer from either side (see Fig. 4b). The first intersection of the ray with the boundary is then marked as a boundary candidate point (green). At one-third of the length along the projected ray, the location is marked as an interior candidate point (blue). At the marked candidate boundary point, an additional ray is reflected along the normal direction to the boundary, and thereby obtaining another pair of boundary (green) and interior candidate point (blue) (see Fig. 4c). Lastly, if there are no boundary candidate between any pairs of fixed boundary points ("dead zones"), then additional boundary candidate are added at the midway point.

This approach is based on the premise that by forming internal edges with good angle property with each boundary point $\boldsymbol{p}$, it can potentially lead to the formation of good quality quadrilaterals. For each ray emanating from $\boldsymbol{p}$, a pair of boundary and interior candidate point is generated. Assuming that an edge is formed between them and $\boldsymbol{p}$, this edge is likely to form good quality quadrilaterals in the resulting blocking solution, since it would have at least more than 60° on either side of $\boldsymbol{p}$. Lastly, the rationale for the biased placement (one-third distance) of interior candidate points is so that it slightly favors connectivity with $\boldsymbol{p}$, since the ray can potentially be arbitrarily long.

5 Generation of a Multi-Blocking Solution

This section details the process by which a blocking solution is formed, based on a combination of the fixed boundary points and a subset of the candidate points selected during the EA search. The block generation module creates a candidate blocking solution and returns the corresponding two-dimensional fitness score, i.e., TQ and AQ (see next section for definition). The generation of a blocking solution can be summarized into four major steps: constrained Delaunay triangulation of the input points, triangle-based mesh smoothing, quad mesh conversion, and quadrilateral-based mesh smoothing.

A constrained Delaunay triangulation is used to generate a triangle mesh from the input points. The edges to be constrained are the boundary of the shape formed by the fixed boundary points and the boundary candidate points (see Fig. 5a for an example). The triangles generated at this stage will be used to form quadrilaterals by merging pairs of adjacent triangles. A triangle mesh smoothing operation [26] is

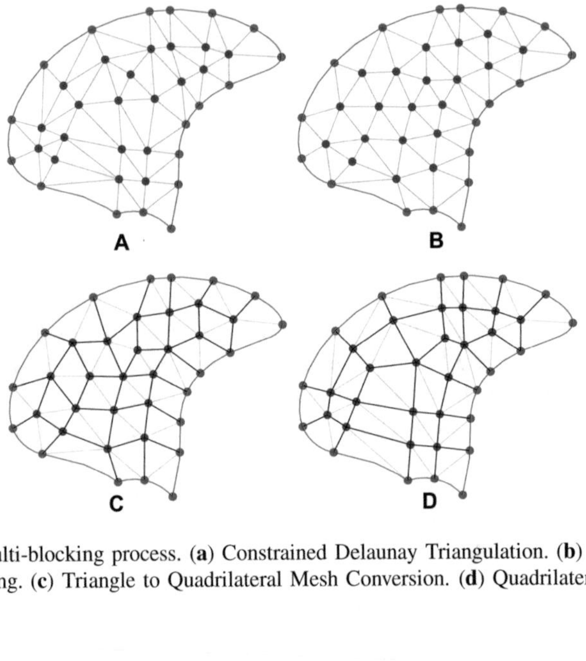

Fig. 5 The multi-blocking process. (**a**) Constrained Delaunay Triangulation. (**b**) Triangle-based Mesh Smoothing. (**c**) Triangle to Quadrilateral Mesh Conversion. (**d**) Quadrilateral-based Mesh Smoothing

then applied to modify the position of all interior candidate points in order to form a mesh with better angle quality, see Fig. 5b.

Subsequently, to convert a triangular mesh into a quadrilateral mesh, pairs of adjacent triangle elements are merged to form one quadrilateral element, necessitating the constraint of having an even number of triangles. By ensuring that there is an even number of points on the boundary, based on Euler's formula, even number of triangles will be formed, independent of the number of interior vertices. The method to convert triangle-pair to quadrilateral is based on the algorithm by Tarini et al. [20], an example of which is shown in Fig. 5c.

After the blocking solution is generated, its quality can be further improved by optimizing the resultant quadrilateral blocks using a smoothing procedure. The aim is to optimize blocks based on their TQs and AQs. For each interior candidate point located at coordinates (x, y), the AQs of its adjacent quadrilaterals are summed up. The point is then shifted in four different small discrete steps, ϵ, in both positive and negative Cartesian directions. The discrete step that best improves the overall AQ of all the adjacent blocks is used to modify the position of the point. This process is repeated for a pre-determined number of iterations, alternating between improving AQ and TQ measure. Assuming that the function $Q(x, y)$ is used to represent the collective TQ or AQ (depending on the current iteration) of its adjacent quadrilaterals, the optimization process can be formulated as

$$\begin{aligned} & Q(x, y) \longrightarrow min(\Delta Q_x, \Delta Q_y) \\ & where \\ & \Delta Q_x = Q(x + \Delta x, y) - Q(x, y) \quad , \quad \Delta x \in (-\epsilon, \epsilon) \\ & \Delta Q_y = Q(x, y + \Delta y) - Q(x, y) \quad , \quad \Delta y \in (-\epsilon, \epsilon) \end{aligned} \tag{4}$$

This process is repeated for a pre-determined number of iterations, alternating between improving AQ and TQ measure (Fig. 5d), and terminates when no more improvements to Q can be achieved. Note that this smoothing process can also be applied to the boundary candidate points, constrained along the boundary.

6 Pareto Ranking of Blocking Solutions

The ranking mechanism determines the characteristics of the resulting blocking solution found by EA, characterized by the collection of quadrilateral blocks formed. Two fitness scores for assessing quadrilateral blocks are defined: one relates to the taper of the block, and the other relates to the internal angles of each block. Referring to Fig. 6, the taper quality (TQ) of the block is defined as the maximum ratio of lengths derived from opposite edges, and the angle quality (AQ) is defined as the sum of all squared deviation from 90° of the four internal angles:

Note that in the computation of AQ, special consideration has to be given for blocks that are adjacent to the boundary. Using Fig. 7 as an example, when

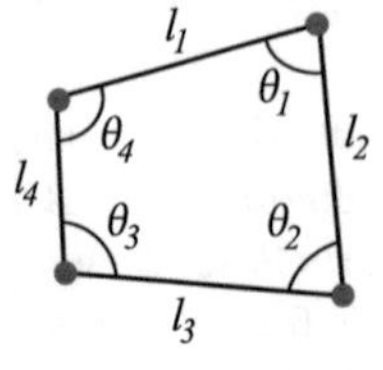

$$Taper\ Quality\ (TQ) = max\left\{\frac{l_1}{l_3}, \frac{l_3}{l_1}, \frac{l_2}{l_4}, \frac{l_4}{l_2}\right\}$$

$$Angle\ Quality\ (AQ) = \sum_{i=1}^{4} (\theta_i - 90^o)^2$$

Fig. 6 Computation of block quality. The two measure qualities of a block is based on its internal angles and ratio of opposite lengths

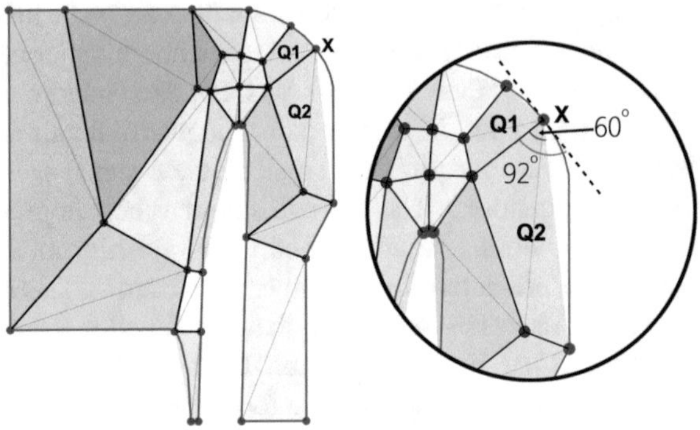

Fig. 7 Computation of angle quality. An illustration of the effect of the curved boundary has on the computation of angle qualities for block ***Q1*** and ***Q2***

computing the interior angle at point x for blocks ***Q1*** and ***Q2***, the boundary tangent at x is used to determine the interior angle. The difference for ***Q1*** might not be noticeable, but for ***Q2*** the disparity is large (60° verses 92°). This approach for computing the interior angle reflects the actual quality because individual mesh elements have to be fitted along the boundary during the meshing stage.

To compute the fitness score of a blocking solution, a simple approach is to use the average TQ and AQ of all the blocks within the solution. However, this approach favors solutions that contain many quadrilateral blocks since a higher number of blocks will lower the fitness score. On the other hand, using the TQ and AQ of the worst block as a representation of fitness will unfairly penalize solutions with a single bad quality block, which in some cases might be unavoidable due to the shape of the boundary. A better balance can be achieved by taking the average quality measure of the worst 10% of blocks in both categories to represent of the fitness of the blocking solution. The superiority of one solution over another is then determined based on the Pareto fitness ranking scheme, introduced by Fonseca and Fleming [27, 28] based on the concept of dominance. A solution is deemed to be dominated when another solution has an equal or better value for all objective functions. The rank of a solution is then determined by the number of solutions that dominates it plus itself. Additional fitness functions, such as alignment to flow direction, can be included to reflect the blocking goals of the user.

7 Evolutionary Algorithm

The general approach of EA is to repeatedly select the best solutions from a large cohort and applying evolutionary operators to generate newer (superior) solutions over many generations. The EA initializes by setting up a population of size *PN* within which each element is filled with a randomized subset from the candidate point set. This population is fed into the multi-blocking module where a blocking solution is generated for every element. At the same time, an archival storage holding a size of *PN* is also set up and filled with the initialized population elements. At the end of each generation, the top few ranked solutions in the population set are then added to the archival storage, where it too undergoes an internal ranking process. Thereafter, the top few ranked solutions (based on an *elitism ratio*) is added back into the population set at the start of the next generation, thus ensuring that the best solutions discovered so far will always be present in the population set.

7.1 Evolutionary Operators

The roulette wheel selection is based on a probability approach to sift out bad solutions from the population. After the population has been sorted from S_1 to S_{PN} based on the Pareto ranking, a solution S_a is assigned a probability value favoring higher ranked solutions based on the following function

$$Probability(S_a) = \frac{PN - a + 1}{\sum_{i=1}^{PN} i} \tag{5}$$

Based on these probability values, a random number generator is invoked PN times to re-populate the population. Superior solutions will tend to be selected more than once, while inferior solution will have a lower chance of being selected again into the population.

The crossover operator takes two solutions (**S1** and **S2**) and exchanges their input parameters to form two new child solutions (**C1** and **C2**) where each child solution resembles a part of both parents. To implement this operation, a random line is used to divide the fixed boundary and candidate point sets in both solutions. Based on this division, the point sets are then interchanged to form child solutions **C1** and **C2**.

The random mutation evolutionary operator is applied separately to both the interior and boundary candidate points. With equal probability, the random mutation operator performs one of the following: does nothing, removes an existing point, or inserts a new point. For boundary candidate points, removal and insertion of points have to be done in pairs, in order to maintain an even number of boundary points.

8 Results

The automatic multi-blocking algorithm is applied on four test datasets, on a population size of 100 and evolved over 15,000 generations. Top ranked solutions are presented in Figs. 8, 9 and 10. As observed, the solution with the best AQ is deemed to be superior. This is to be expected as AQ, rather than TQ, is most often used as a mesh suitability criteria for simulation purposes. Nevertheless, it should be highlighted the evolutionary process is highly dependent on both criteria. By keeping solutions that are good in either angle or taper quality high in the Pareto ranking, opportunities are created for these to produce new child solutions that contain good features from both parents. This increases the variability of solutions and therefore aids the discovery of newer and better solutions, as compared to only using a single objective function.

In general, the auto-blocking algorithm is able to generate blocks with excellent angle qualities. The average AQ of the blocks for the top solutions are 511.32 (Test Geometry), 751.76 (Intake), 521.31 (Symmetrical Shape), and 629.36 (Seal Cavity). The average angle deviation (from 90°), minimum and maximum (internal angles) are as follows: Test Geometry-(11.30°, 54.84°, 127.14°), Intake-(13.7°, 50.67°, 118.63°), Symmetrical Shape-(11.41°, 62.21°, 124.57°) and Seal Cavity-(12.54°, 66.43°, 137.84°). The color of the block provides an additional visual indication of the AQ, with a darker color representing lower AQ blocks. It should be noted that although some blocks can be further improved by shifting the boundary points, these are fixed boundary points which by definition is required to be stationary in order to preserve the shape of the boundary.

With respect to the blocking configurations, it is interesting to observe that the blocks flow along the shape and boundary of the domain (top AQ solutions), much

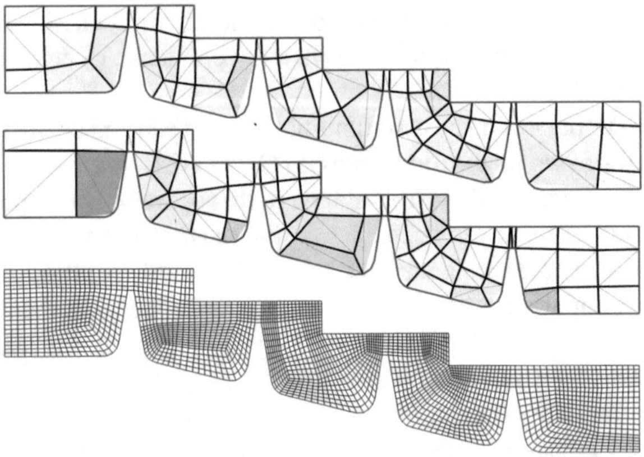

Fig. 8 Multi-blocking results-seal cavity. The blocking configuration on the top depict the best solution with the best AQ, while the middle depict the solution with the best TQ. The bottom image is the meshed output, using the MSC Software, Patran[29]

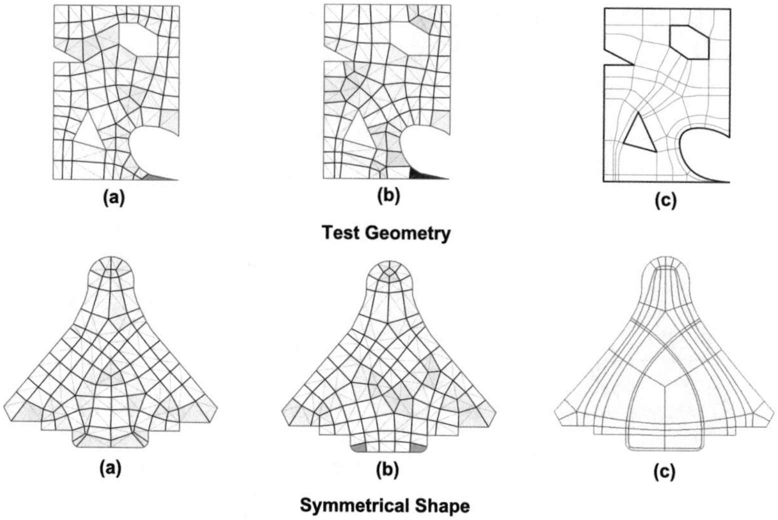

Fig. 9 Multi-blocking results and comparison-test geometry and symmetrical shape. Results from the two test datasets, based on a population size of 100 and 15,000 generations, are shown on (**a**) and (**b**) columns, best angle quality and best taper quality respectively. The third column (**c**) depicts the blocking solution based on cross-field approach, as reported in [23] and [25]

like the formation of a medial-axis. For the *Test Geometry* dataset, the challenging area to resolve is located at the sharp tip at the bottom-right. The best AQ solution creates a block at the sharp tip with reasonable quality, though at the expense of bad TQ, similar to the result presented by Fogg et al. [25] using cross-field, reproduced in Fig. 9 for visual comparison. On the other hand, the best solution based on TQ contains a poorly shaped quadrilateral block at the same location in order to achieve reasonable TQ. Lastly for the *Symmetrical Shape* dataset, a highly symmetrical configuration can be seen for the best AQ solution, which is very similar to the result obtained by using a PDE-based approach as presented in the paper by Kowaslski et al. [23], as illustrated in Fig. 9. In the paper by Ali and Tucker [30], different blocking solutions are presented using various methods such as d-MAT and TopMaker for the same geometries of *Intake* and *Seal Cavity* considered here. Medial-axis based solutions tend to produce blocking configurations that split 90° corners (d-MAT) or unnecessary adaptions at rounded corners (TopMakers), while our solutions do not exhibit these features.

In Fig. 11, a trendline depicting the evolution of the best solution for the *Intake* dataset, in terms of both AQ and TQ, over 15,000 generation is shown. This is a typical trend for the other three datasets, where an approximate solution is discovered very quickly (typically under 1000 generation) and thereafter very little incremental progress is made. This observation suggests that the evolutionary algorithm does not need to run for a large number of generations before a near-optimal solution is discovered. Figure 10 illustrates the sequence of the solution

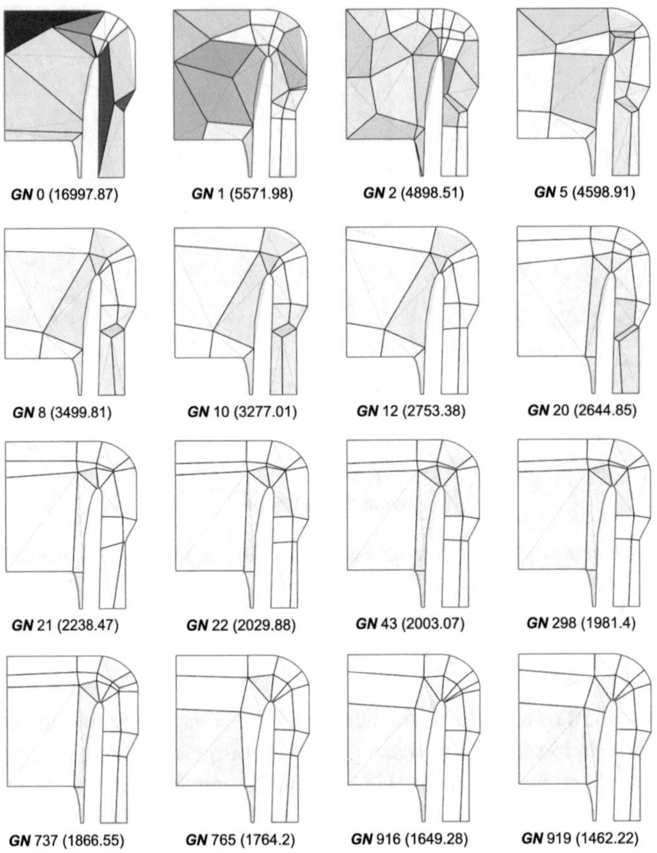

Fig. 10 Evolution of best solutions. This image illustrates the evolution of solutions for the *Intake* dataset using the evolutionary algorithm. The figures below each solution indicate the generation number which it is discovered and the number in brackets are the corresponding angle quality value

evolution, in which the best solution is discovered at generation number 919 (out of a possible 15,000 generation). Although this affirms the convergence efficiency of the algorithm, it also highlights the fact that the evolved solution could very possibly occur at a local minimum which hinders discovery of other superior variants. Fortunately, such a situation can be mitigated by conducting parallel runs of the multi-blocking algorithm. This is because the input is randomly initialized at each instance and thus, it is able to discover multiple solution variants. This observation suggests that the evolutionary algorithm does not need to run for a large number of generations before a near-optimal solution is discovered.

In Fig. 12, solutions generated over five different runs for the *Test Shape* dataset are shown. Even though these variants have very different blocking configurations, they are still useful for the purpose of quadrilateral mesh generation. For quantifica-

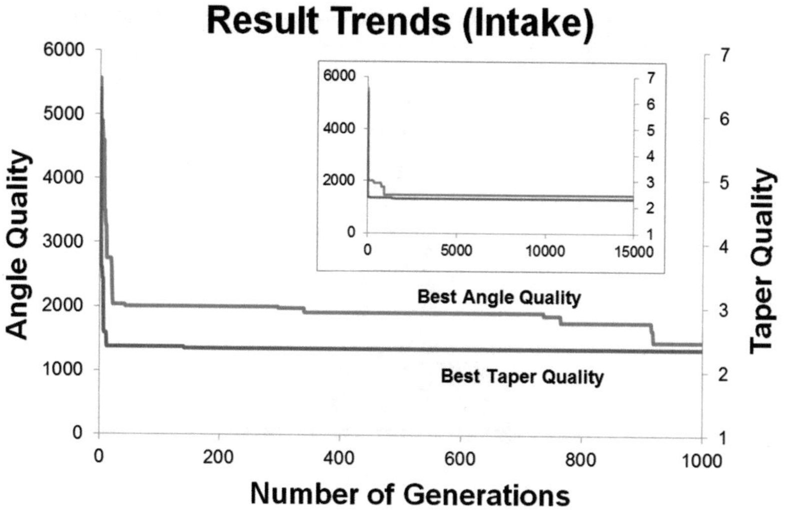

Fig. 11 Solution trendlines-intake. The graph depicts the evolution of the best solutions (in terms of AQ and TQ criterion) over the first 1000 generations for the *Intake* dataset. The full 15,000 generation trend is plotted in the inset image

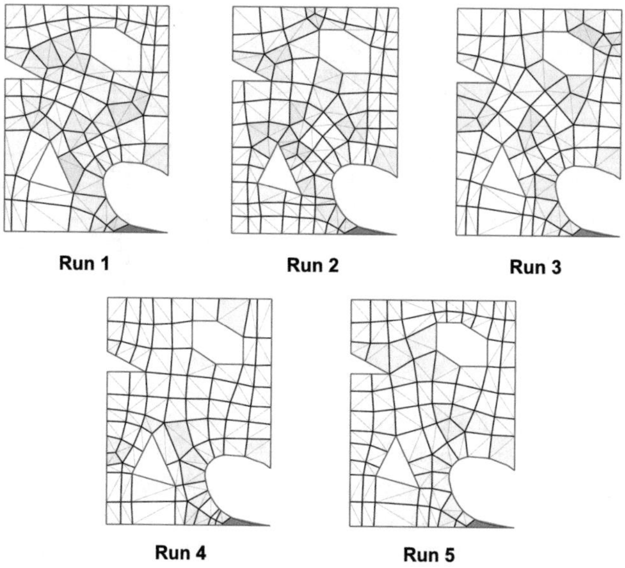

Fig. 12 Solution variability. Different solution variants generated over five different runs

tion, the combined interior angle values of the solutions are tabulated and shown in Fig. 13. Most meshing algorithms stipulate that quadrilateral elements should ideally have interior angles in the range of 45° to 135°. The solutions generated by our method are able to satisfy this angle constraint.

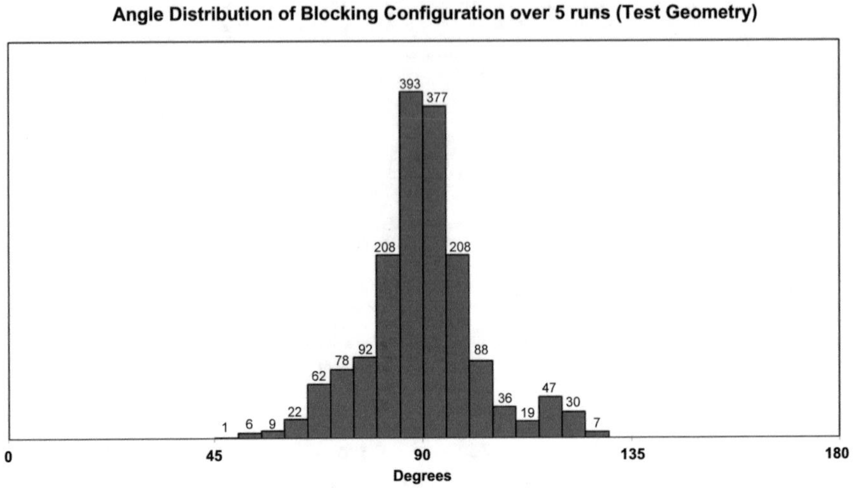

Fig. 13 Angle distribution. The combined interior angles values of the five solutions variant generated in Fig. 12 is tabulated

Table 1 Breakdown of computational time over five runs for a population size of 100

	Symmetrical shape	Intake	Test geometry	Seal cavity
Candidate point size	125	80	258	272
Quad blocking	75.68%	75.39%	78.74%	83.3%
Pareto ranking	4.94%	20.62%	5.12%	7.48%
Block improvement	3.19%	1.77%	4.23%	7.60%
Roulette wheel	0.5%	1.69%	0.56%	0.72%
Crossover	15.68%	0.52%	11.35%	0.87%
Mutation	0.003%	0.01%	0.006%	0.006%
Computational time per generation (s)	3.49	0.77	2.65	2.48

Table 1 shows a breakdown of the computational time of the various algorithm components over five different runs, based on a sequential implementation, with no multi-threading included. The trend is consistent throughout the four different test datasets, where the quad blocking module is found to incur the majority of the computational time. A near-optimal solution can usually be obtained in less than an hour (500–1000 generations) of computational time. This can either mean that an optimum solution has been discovered, or the search has stagnated in a local minimum. Further improvements can be achieved on two fronts, in terms of running time and productivity (reducing stagnation). The running time can be improved through the use of multi-threading capabilities of modern processors. However, a straightforward implementation of multi-threading (regardless of the number of threads) can only provide an average of three to four-fold speedup. This is largely due to the overheads required to manage the large number of population threads, each of which only has a small computational load. Taking this idea further, a multi-instance version can be implemented where multiple instances of the algorithm are launched simultaneously and independently. Once an instance is experiencing stagnation, it can request for an infusion of top solutions from other instances. With a combined larger population pool, a higher quality solution is likely to be discovered earlier in the generation cycle, thereby further reducing computational time.

9 Adaptation Towards Automatic 3D Blocking

A natural extension of this work would be to solve the problem of 3D multi-blocking for hexahedral mesh generation. However, there is no easy equivalent of merging triangles to form quadrilaterals in 3D, i.e., merging tetrahedra to form hexahedra is a non-trivial problem. A promising piece of work in that regard can be found in a recent work [31] where it is stated that the quality and number of potential hexahedra generated is highly dependent on the input tetrahedral mesh. This is dependent on the available vertices for forming the tetrahedral and hence is analogous to the 2D optimization workflow presented in this work. An initial 3D workflow is applied onto a simple 3D geometry (octant-sphere), shown and explained in Fig. 14.

The objective of this test is to illustrate the potential of adapting the 2D automatic blocking workflow onto a 3D geometry blocking problem. As with the 2D case,

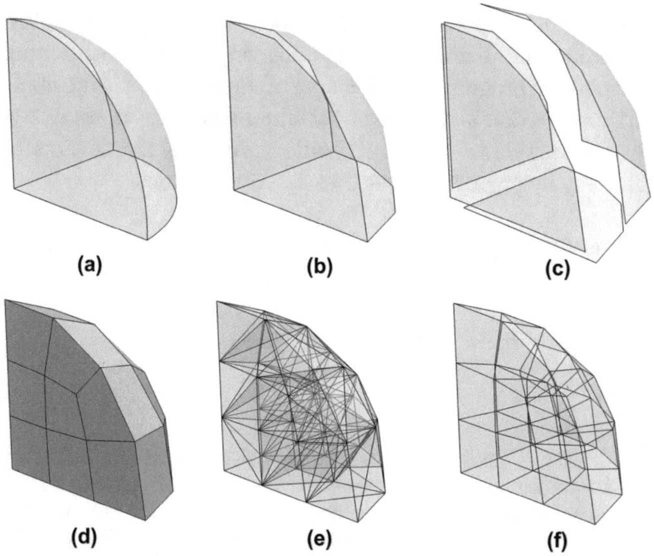

Fig. 14 Tetrahedra mesh conversion to hexahedra mesh. An octant-sphere geometry (**a**) is used for testing the viability of automatic 3D blocking. (**b**) shows the downsampled geometry. (**c**) and (**d**) shows the individual facets and the results of the 2D blocking operation. (**e**) and (**f**) illustrates the tetrahedra mesh and the subsequent conversion to hexahedra mesh

the boundary edges is discretized into fixed boundary points, and then each facet is processed independently through the 2D automatic blocking framework. Curved facets are projected onto planar surfaces before processing. Each facet, through the process of 2D blocking, will have new additional vertices. Together with additional candidate points inserted into the interior of the volume of the octant-sphere, a tetrahedron mesh is generated. Lastly, the software provided by Pellerin et al. [31] is then used to form a hexagonal mesh from it. It is noted that the provided 3D blocking example is based on a relatively simple 3D geometry, and more research work needs to be performed when dealing with more complex and non-2D curved geometries.

10 Conclusion and Future Work

This paper presents a novel method to automatically generate ideal multi-blocking solutions on 2D shapes using an evolutionary algorithm. Utilizing a large population and generation number, the solutions produced well-shaped quadrilaterals. The work has shown that an evolutionary-based approach can be effectively applied to solve the 2D multi-blocking problem, and potentially in 3D. Future work might include other meshing characteristics, such as flow alignment or periodic boundary. The current main drawback of the approach is the large computational time required, and more so for future 3D implementation. However, due to the parallelized nature of the algorithm, the proposed method for both 2D and 3D is highly amenable for multi-threading and distributed computing implementation. A multi-threaded version is currently in progress. Another area of improvement is the implementation of a termination criterion since a near-optimum solution variant can be obtained very quickly in the first few hundred generations. An early termination feature could improve efficiency by preventing unnecessary computations.

Acknowledgement The authors would like to thank Roll-Royce plc for their support and permission to publish the work.

References

1. Z. Ali, P.G. Tucker, S. Shahpar, Optimal mesh topology generation for CFD, in *Computer Methods in Applied Mechanics and Engineering*, vol. 317 (2017), pp. 431–457.
2. A. Milli, S. Shahpar, PADRAM: Parametric design and rapid meshing system for complex turbomachinery configurations, in *ASME Turbo Expo 2012: Turbine Technical Conference and Exposition* (2012), pp. 2135–2148
3. S.J. Owen, A survey of unstructured mesh generation technology, in *International Meshing Roundtable* (1998), pp. 239–267
4. Y. Zhang, C. Bajaj, Adaptive and quality quadrilateral/hexahedral meshing from volumetric data. Comput. Methods Appl. Mech. Eng. **195**(9), 942–960 (2006)

5. X. Liang, M.S. Ebeida, Y. Zhang, Guaranteed-quality all-quadrilateral mesh generation with feature preservation. Comput. Methods Appl. Mech. Eng. **199**(29), 2072–2083 (2010)
6. F.B. Atalay, S. Ramaswami, D. Xu, Quadrilateral meshes with provable angle bounds. Eng. Comput. **28**(1), 31–56 (2012)
7. T.D. Blacker, M.B. Stephenson, Paving: a new approach to automated quadrilateral mesh generation. Int. J. Numer. Methods Eng. **32**(4), 811–847 (1991)
8. D.R. White, P. Kinney, Redesign of the paving algorithm: Robustness enhancements through element by element meshing, in *6th International Meshing Roundtable* (1997), pp. 323–335
9. B. Joe, Quadrilateral mesh generation in polygonal regions. Comput. Aided Des. **27**(3), 209–222 (1995)
10. J.A. Talbert, A.R. Parkinson, Development of an automatic, two-dimensional finite element mesh generator using quadrilateral elements and Bezier curve boundary definition. Int. J. Numer. Methods Eng. **29**(7), 1551–1567 (1990)
11. S.-W. Chae, J.-H. Jeong, Unstructured surface meshing using operators, in *6th International Meshing Roundtable* (1997), pp. 281–291
12. D. Nowottny, Quadrilateral mesh generation via geometrically optimized domain decomposition, in *6th International Meshing Roundtable* (1997), pp. 309–320
13. T.K.H. Tam, C.G. Armstrong, 2D finite element mesh generation by medial axis subdivision. Adv. Eng. Softw. Work. **13**(5), 313–324 (1991)
14. D. Guoy, J. Erickson, Automatic blocking scheme for structured meshing in 2d multiphase flow simulation, in *International Mesh Roundtable* (2004), pp. 121–132
15. H. Xia, P.G. Tucker, Fast equal and biased distance fields for medial axis transform with meshing in mind. Appl. Math. Model. **35**(12), 5804–5819 (2011)
16. E. Catmull, J. Clark, Recursively generated B-spline surfaces on arbitrary topological meshes. Comput. Aided Des. **10**(6), 350–355 (1978)
17. C.K. Lee, S.H. Lo, A new scheme for the generation of a graded quadrilateral mesh. Comput. Struct. **52**(5), 847–857 (1994)
18. S.J. Owen, M.L. Staten, S.A. Canann, S. Saigal, Advancing front quadrilateral meshing using triangle transformations, in *International Mesh Roundtable* (1998), pp. 409–428
19. L. Sun, G.-T. Yeh, F.P. Lin, G. Zhao. Automatic quadrilateral mesh generation and quality improvement techniques for an improved combination method. Comput. Geosci. **19**(2) 1–18 (2015)
20. M. Tarini, N. Pietroni, P. Cignoni, D. Panozzo, E. Puppo, Practical quad mesh simplification. Comput. Graphics Forum **29**(2), 407–418 (2010)
21. T. Itoh, K. Shimada, K. Inoue, A. Yamada, T. Furuhata, Automated conversion of 2D triangular mesh into quadrilateral mesh with directionality control, in *International Mesh Roundtable* (1998), pp. 77–86
22. C.G. Armstrong, H.J. Fogg, C.M. Tierney, T.T. Robinson, Common themes in multi-block structured quad/hex mesh generation. Proc. Eng. **124**, 70–82 (2015)
23. N. Kowalski, F. Ledoux, P. Frey, A PDE based approach to multidomain partitioning and quadrilateral meshing, in *Proceedings of the 21st international meshing roundtable* (2013), pp. 137–154
24. H.J. Fogg, C.G. Armstrong, T.T. Robinson, Multi-block decomposition using cross-fields, in *Proceedings of adaptive modelling and simulation, Lisbon* (2013), pp. 254–267
25. H.J. Fogg, C.G. Armstrong, T.T. Robinson, Automatic generation of multiblock decompositions of surfaces. Int. J. Numer. Methods Eng. **101**(13), 965–991 (2015)
26. T. Zhou, K. Shimada, in *An Angle-Based Approach to Two-Dimensional Mesh Smoothing* (2000), pp. 373–384
27. C.M. Fonseca, P.J. Fleming, Genetic algorithms for multiobjective optimization: formulation discussion and generalization. ICGA **93**, 416–423 (1993)
28. C.M. Fonseca, P.J. Fleming, An overview of evolutionary algorithms in multiobjective optimization. Evol. Comput. **3**(1), 1–16 (1995)
29. MSC Software. *Patran* (2014). http://www.mscsoftware.com/product/patran

30. Z. Ali, P.G. Tucker, Multiblock structured mesh generation for turbomachinery flows, in *Proceedings of the 22nd International Meshing Roundtable* (2014), pp. 165–182
31. J. Pellerin, A. Johnen, J.-F. Remacle, Identifying combinations of tetrahedra into hexahedra: a vertex based strategy. Proc. Eng. **203**, 2–13 (2017)

Multiblock Mesh Refinement by Adding Mesh Singularities

Cecil G. Armstrong, Tak Sing Li, Christopher Tierney, and Trevor T. Robinson

Abstract Several templates for 2D and 3D structured mesh refinement are presented. The templates have the property that the minimum number of irregular points or edges (mesh singularities) are added. For a given set of external division numbers a variety of interior meshes can be generated.

The positions of the internal vertices in the template are calculated explicitly using an extended transfinite mapping scheme, which has previously been shown to be equivalent to iterative iso-parametric smoothing. Since calculating the block vertex positions requires the solution of a small number of linear equations, the optimum mesh in the interior of the template can be evaluated very cheaply before the block structured mesh is generated.

1 Introduction

Despite huge advances in the state of the art of unstructured mesh generation, e.g. [1], there is still a demand for the generation of structured multi-block meshes. A number of authors [2–4] have explored the use of medial axis techniques, since these tend to offer meshes which have close to the minimum number of mesh irregularities or 'singularities'. Other techniques such as frame fields have gained attention, but it was noted in [5] that it is necessary to take into account the global structure of hexahedral meshes. This implies tracking the position and connectivity of singular edges in 3D.

After a multi-block decomposition has been created, adjustment of edge division numbers to achieve an adequate mesh size distribution is required. Frequently mesh

C. G. Armstrong (✉) · C. Tierney · T. T. Robinson
School of Mechanical and Aerospace Engineering, Belfast, UK
e-mail: c.armstrong@qub.ac.uk; christopher.tierney@qub.ac.uk; t.robinson@qub.ac.uk

T. S. Li
School of Science and Technology, The Open University of Hong Kong, Kowloon, Hong Kong
e-mail: tsli@ouhk.edu.hk

X. Roca, A. Loseille (eds.), *27th International Meshing Roundtable*,
Lecture Notes in Computational Science and Engineering 127,
https://doi.org/10.1007/978-3-030-13992-6_11

sizing requirements in one area leads to the propagation of an overly dense mesh elsewhere, requiring the insertion of 'steerbacks' or additional block topology to create transition meshes and local mesh refinement [6]. Solution errors may indicate a need for adaptive refinement of an existing blocking, which is similarly difficult.

One way to avoid this complexity is to implement a non-conformal refinement strategy, using multi-point constraints at incompatible interfaces. This has the benefit of producing higher quality elements in the refinement regions with no propagation through the remaining mesh. However, this can introduce errors at the mismatching interface and is unsuitable for many applications [7].

Current conformal mesh adaptation strategies for quad and hex elements are generally template-based operations focused on 2-refinement or 3-refinements strategies [8]. Schneider [9] used a refinement strategy to subdivide quad elements and hex elements in the refinement region using a quadtree refinement, maintaining associativity by inserting templates in the transition zone. Ebeida [10] introduced a parallel realization of Schneider's 2-refinement strategy for unstructured meshes, whilst Qian [11] extended this approach for non-manifold conformal mesh generation. Other work [12] incorporated conformal refinement and coarsening strategies by combining template-based operations with localized coarsening and quality improvement in a single workflow. While these techniques provide topological mesh adaptation focusing directly on mesh elements they can also be applied to the 'coarser' block decomposition. The issue is that they are generally implemented using a 2-refinement or 3-refinement strategy, meaning arbitrary element numbers require further refinement and the combination of templates that may introduce large numbers of unnecessary singularities (Fig. 1).

The aim of the present work was to identify some generic ways in which an existing block decomposition can be refined to produce meshes which better match mesh density variations. For a given region the minimum necessary number of singular points can be established geometrically, Fogg et al. [13]. Fogg also showed that if additional 2D singularities are to be added, a positive (5-valent) and negative (3-valent) pair must be added simultaneously.

Therefore, in 2D, mesh refinement can be accomplished by incrementally adding pairs of positive-negative singularities to an existing block decomposition. It will be shown that, for a singularity pair, the optimum position of the singular points for a given mesh density can be calculated explicitly. It will also be shown that there are a range of mesh distributions that can satisfy external division number constraints, and it is possible to choose the optimum arrangement at low computational cost.

Other general block arrangements and 3D refinements are also presented.

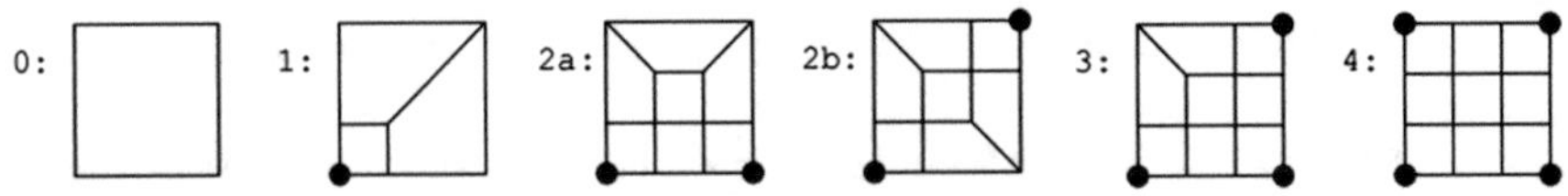

Fig. 1 Two and three refinement templates [9]

2 Laplacian Smoothing and Iso-parametric Mapping

The ‘midpoint subdivision’ technique for decomposing a convex polygon into a series of 2D quadrilateral blocks, or a 3D polyhedron with convex edges and trivalent vertices into hexahedral bricks, has been presented by Li et al. [14]. By joining a face midpoint to all the edge midpoints, one quad is generated at every vertex of the original polygon, Fig. 2b, c. By adjusting the internal mesh division numbers, e.g. n_0, n_1, n_2 in Fig. 2c, a range of external division numbers can be satisfied.

In a subsequent paper [15], efficient mapping methods were developed for determining nodal positions in the resulting mesh based on an extension to transfinite mapping, and it was shown that the result is identical to that obtained by iterative iso-parametric smoothing.

From eqn. (7) of Li et al. [15], the position of a node $\boldsymbol{x}$ at the centre of a k-sided face, Fig. 2, can be identified using

$$\sum_0^{k-1} \frac{\boldsymbol{b}_r + \boldsymbol{b}_{r+1} - \boldsymbol{a}_r}{n_r n_{r+1}} - \sum_0^{k-1} \frac{1}{n_r n_{r+1}} \boldsymbol{x} = \boldsymbol{0},$$

where $\boldsymbol{b_r}$ is the point on edge r that the face midpoint is connected to, and $\boldsymbol{a_r}$ is the opposite corner joining edges r and $r + 1$. n_r is the number of elements (division number) on the radial edge joining the centre point $\boldsymbol{x}$ to edge r.

An alternative form, shown below, allows the position of the face midpoint $\boldsymbol{x}$ to be found from

$$\sum_0^{k-1} \frac{1}{n_r n_{r+1}} \boldsymbol{a}_r - \sum_0^{k-1} \left(\frac{1}{n_r n_{r+1}} + \frac{1}{n_{r-1} n_r} \right) \boldsymbol{b}_r + \sum_0^{k-1} \frac{1}{n_r n_{r+1}} \boldsymbol{x} = \boldsymbol{0}.$$

Here n_{r-1} is the radial division number of the edge preceding edge r in a clockwise traversal of the boundary, whilst n_{r+1} is the radial division number of the following edge.

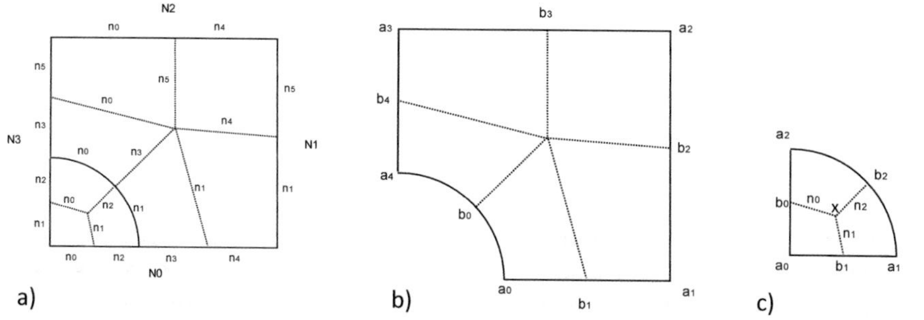

Fig. 2 A pair of singularities in a quad block. (**a**) Singularity pair with internal and external division numbers. (**b**) Positive singularity in a pentagon. (**c**) Negative singularity in a triangle

So, for a 3-sided face, Fig. 2c for example,

$$\left[+\frac{1}{n_0 n_1} + \frac{1}{n_1 n_2} + \frac{1}{n_2 n_0} - \left(\frac{1}{n_0 n_1} + \frac{1}{n_2 n_0}\right)\right.$$

$$\left. - \left(\tfrac{1}{n_1 n_2} + \tfrac{1}{n_0 n_1}\right) - \left(\tfrac{1}{n_2 n_0} + \tfrac{1}{n_1 n_2}\right) + \sum_{r=0}^{2} \tfrac{1}{n_r n_{r+1}}\right] \begin{Bmatrix} \boldsymbol{a_0} \\ \boldsymbol{a_1} \\ \boldsymbol{a_2} \\ \boldsymbol{b_0} \\ \boldsymbol{b_1} \\ \boldsymbol{b_2} \\ \boldsymbol{x} \end{Bmatrix} = \mathbf{0}$$

This representation is potentially useful when some of the positions are unknown, for example in a network of primitives with common edges.

3 Adding a Singularity Pair to an Existing Block

Adding a positive and negative singularity to an existing block can be thought of as partitioning the quad block into a 3-sided and a 5-sided region, Fig. 2a. The difficulty is that, even when the singularities are directly connected, the midpoint on the common edge (b_2 in the triangle and b_0 in the pentagon) affects the position of both face midpoints.

In the equation for the triangular region, Fig. 2c, the positions of $\{\boldsymbol{a_0}\ \boldsymbol{a_1}\ \boldsymbol{a_2}\ \boldsymbol{b_0}\ \boldsymbol{b_1}\}$ are known and can be moved to the right-hand side, so we end up with two unknown positions for the triangle midpoint $\boldsymbol{x^t}$ and the edge midpoint $\boldsymbol{b_2}$ which is shared with the pentagon, as

$$\left[-\left(\tfrac{1}{n_2 n_0} + \tfrac{1}{n_1 n_2}\right) \sum_{r=0}^{2} \tfrac{1}{n_r n_{r+1}}\right] \begin{Bmatrix} \boldsymbol{b_2} \\ \boldsymbol{x^t} \end{Bmatrix} = \boldsymbol{rhs^0}.$$

Similarly, the equation describing the face and edge midpoints in the pentagon reduces to

$$\left[-\left(\tfrac{1}{n_0 n_1} + \tfrac{1}{n_4 n_0}\right) \sum_{r=0}^{4} \tfrac{1}{n_r n_{r+1}}\right] \begin{Bmatrix} \boldsymbol{b_0} \\ \boldsymbol{x^p} \end{Bmatrix} = \boldsymbol{rhs^1}.$$

$\boldsymbol{b_2}$ in the triangle and $\boldsymbol{b_0}$ in the pentagon are of course the same point.

The last step is to calculate the position of triangle point $\boldsymbol{b_2}$ using the same transfinite mapping equations. With reference to Fig. 3, and using the convention that red points are on the triangle and black points are on the pentagon, the

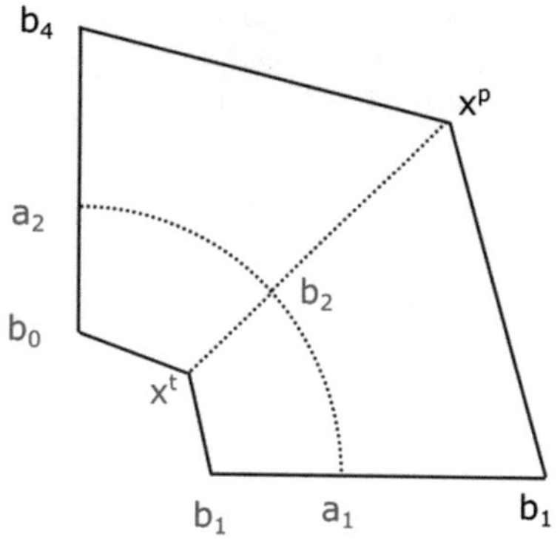

Fig. 3 Quad with common boundary point at the centre. The points are as labelled in Fig. 2b, c

corner points of the 4-sided region surrounding the common edge midpoint b_2 are

$$\left[b_1\ b_1\ b_4\ b_0\right],$$

whilst the edge midpoints are

$$\left[x^t\ a_1\ x^p\ a_2\right],$$

and the triangle and pentagon midpoints are $\boldsymbol{x}^t$ and $\boldsymbol{x}^p$ respectively.

Collecting known terms and moving them to the RHS gives 3 vector equations in 3 unknown positions as

$$\begin{bmatrix} . & . & 0 \\ 0 & 0 & . \\ . & . & . \end{bmatrix} \begin{Bmatrix} b_2 \\ x^t \\ x^p \end{Bmatrix} = \begin{Bmatrix} rhs^0 \\ rhs^1 \\ rhs^3 \end{Bmatrix}$$

Once these equations are solved to find the face midpoints and the common edge midpoint, all the other nodes can be found using Li's existing algorithm.

4 Division Numbers

With the arrangement of the triangle and pentagon shown in Fig. 2 there are six internal division numbers. If target external division numbers on the block are required, this means solving an integer programming problem, but here we will proceed initially by specifying the internal division numbers. On the external edges of the block, the corner and edge midpoint nodes of the triangle and pentagon can be identified. If the division numbers or edge meshes are changed then the position of the face midpoints and the common edge midpoint can be re-calculated as above.

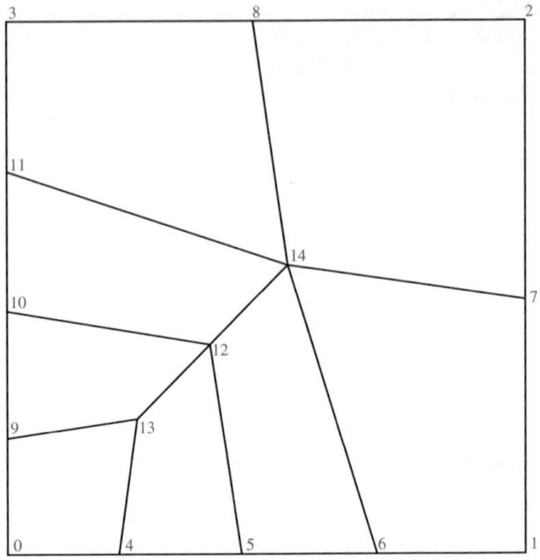

Fig. 4 Labelled corners and midpoints. Internal division numbers n = [1,1,1,1,1,1]

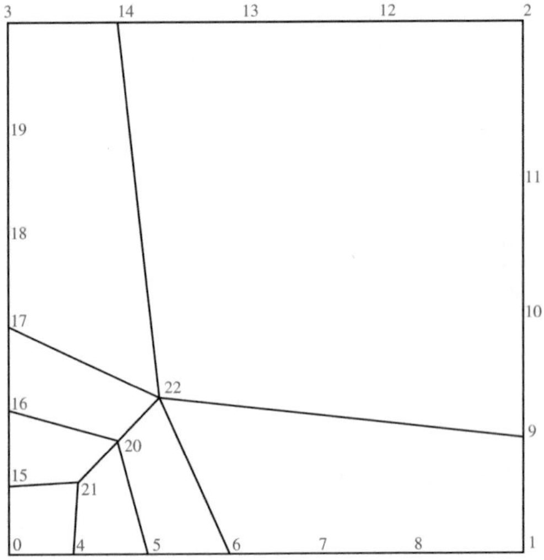

Fig. 5 Internal division numbers = [1,1,1,1,3,3]

Figures 4 and 5 show the block corner points calculated using these equations for two different sets of division numbers. The negative singularity, the positive singularity and the common edge midpoint are at nodes 13, 14 and 12 respectively in Fig. 4 and nodes 21, 22 and 20 in Fig. 5.

5 Common Edge Nodes

After the division numbers have been set, nodes must be generated on the common edge between the triangle and pentagon.

The nodes on the common edge between nodes 9 and 37 in Fig. 6 can be derived using the standard transfinite mapping equations. Based on the position of nodes (0, 4, 5, 6), and nodes (3, 29, 28, 27) representing two opposite logical block edges, the position of nodes 50 and 51 can be calculated since (0, 37, 3) and (6, 44, 27) form the other two sides of a logical 4-sided region. For example, the position of node 50 can be calculated using the equation:

$$\left(\tfrac{1}{7}\ \tfrac{1}{14}\ \tfrac{1}{16}\ \tfrac{1}{8}\ -\left(\tfrac{1}{7}+\tfrac{1}{8}\right)\ -\left(\tfrac{1}{7}+\tfrac{1}{14}\right)\ -\left(\tfrac{1}{14}+\tfrac{1}{16}\right)\ -\left(\tfrac{1}{16}+\tfrac{1}{8}\right)\right)\begin{pmatrix}\boldsymbol{p_0}\\ \boldsymbol{p_6}\\ \boldsymbol{p_{27}}\\ \boldsymbol{p_3}\\ \boldsymbol{p_{37}}\\ \boldsymbol{p_4}\\ \boldsymbol{p_{44}}\\ \boldsymbol{p_{29}}\end{pmatrix}$$

$$+\left(\tfrac{1}{8}+\tfrac{1}{16}+\tfrac{1}{7}+\tfrac{1}{14}\right)\ \left(\boldsymbol{p_{50}}\right)=\boldsymbol{0}$$

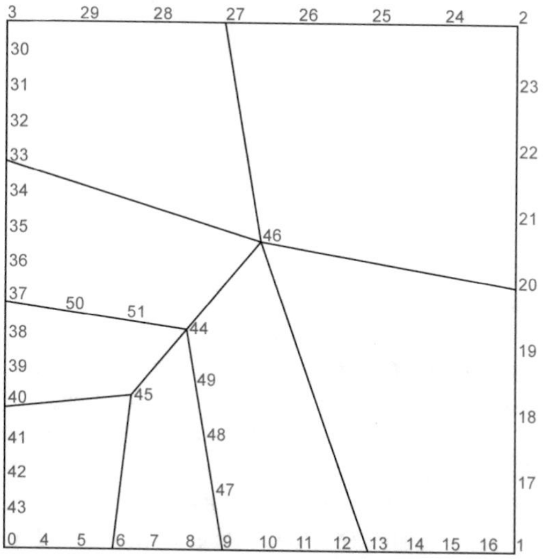

Fig. 6 Nodes on the common edge for the template of Fig. 2 with internal division numbers [3,4,3,4,4,4]

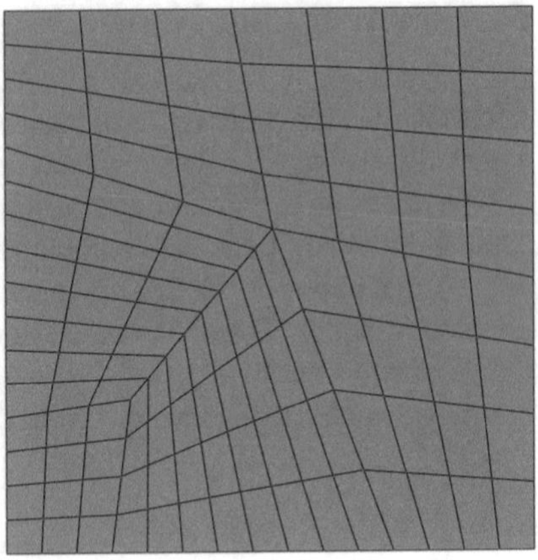

Fig. 7 Final mesh

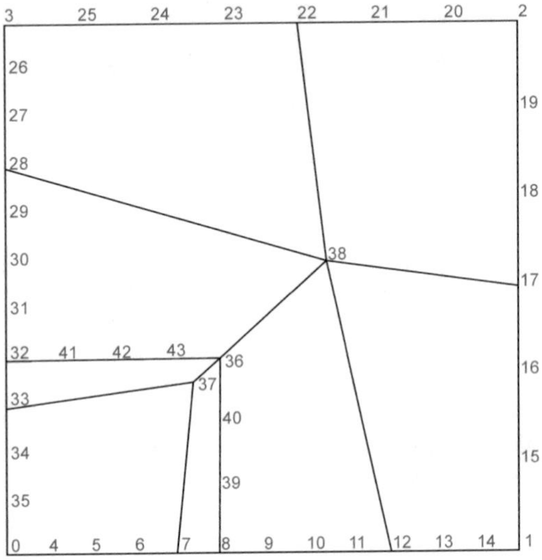

Fig. 8 Blocks for internal divisions n = [4,3,1,4,3,3]. The external numbers are [12, 6, 7, 11]

Similarly, the position of nodes (49, 48, 47) can be found based on the 4-sided region [(0, 9, 1), (1, 17, 18, 19, 20), (20, 44, 40), (40, 41, 42, 43, 0)]. Figure 6 shows the position of these nodes derived for a given set of division numbers.

Once the position of all the boundary nodes for both primitives is established, the generation of the radial edge meshes and the quad mesh for each block can proceed as described in Li's original algorithm. Figure 7 shows the resulting mesh.

For a fixed set of external division numbers, the common edge midpoint is different when n_2 is varied, but if the sum $n_2 + n_3$ is fixed the singular points and the resulting element mesh are the same. Figures 8 and 9 indicate the different block geometries and division numbers for two variations where $n_2 + n_3 = 5$. Thus n_2 and n_3 are not independent variables.

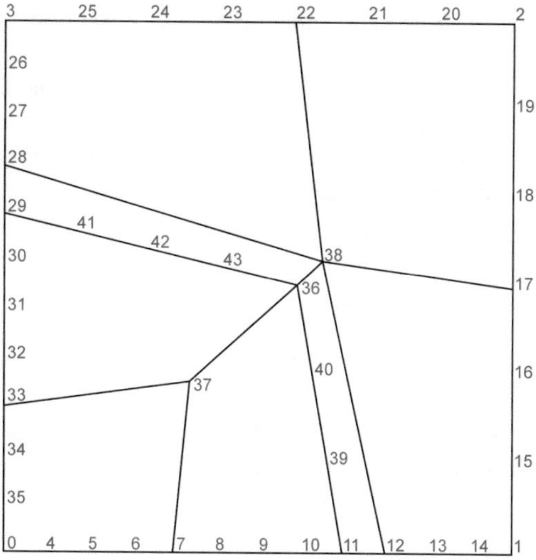

Fig. 9 n = [4,3,4,1,3,3]. Same external division numbers. The singular points 37 and 38 and the final mesh are the same as in Fig. 8

6 Division Number Constraints

From Fig. 2, the internal and external division numbers are related by

$$\begin{bmatrix} 1\,0\,1\,1\,1\,0 \\ 0\,1\,0\,0\,0\,1 \\ 1\,0\,0\,0\,1\,0 \\ 0\,1\,1\,1\,0\,1 \end{bmatrix} \begin{Bmatrix} n_0 \\ n_1 \\ n_2 \\ n_3 \\ n_4 \\ n_5 \end{Bmatrix} = \begin{Bmatrix} N_0 \\ N_1 \\ N_2 \\ N_3 \end{Bmatrix}$$

Summing the rows of the matrix on the left-hand side implies

$$2\,(n_0 + n_1 + n_2 + n_3 + n_4 + n_5) = N_0 + N_1 + N_2 + N_3$$

which is the expected constraint that the sum of the external division numbers is even. The matrix also implies that

$$n_2 + n_3 = N_0 - N_2 = N_3 - N_1$$

This is a measure of the distance between the singularities. Note that it also requires that the difference in division numbers between opposite sides should be the same. As shown above, the same mesh is generated if the sum $(n_2 + n_3)$ is the same.

The division number constraints also imply

$$N_0 + N_1 - N_3 = n_0 + n_4 = N_2$$

This constraint can also be seen by inspecting Fig. 2.

Therefore, the mesh can be parameterized by the position of the edge midpoint on N_2. Varying n_0 whilst keeping the sum $n_0 + n_4$ constant produces the different blocks below for the same external division numbers (Fig. 10).

Similarly

$$n_1 + n_5 = N_1$$

so the mesh can also be parameterized by where the positive singularity is attached to N_1. Figure 11 shows the variation in block geometry for 3 different values of n_1 for $n_1 + n_5 = 6$.

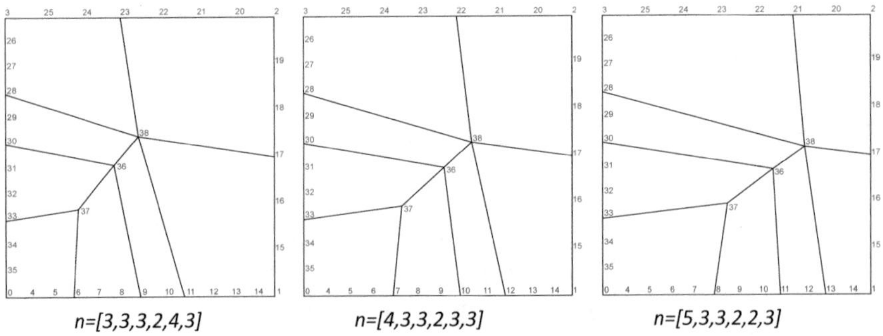

Fig. 10 Range of blocks obtained with the external division numbers [12, 6, 7, 11] whilst varying n_0 for the same $n_0 + n_4$

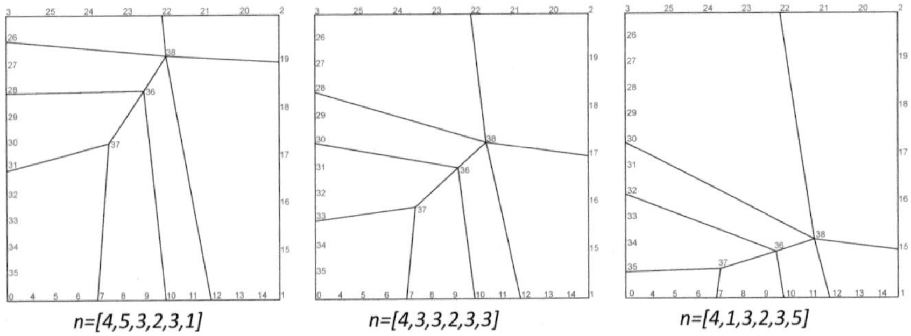

Fig. 11 Range of blocks obtained with the external division numbers [12, 6, 7, 11] whilst varying n_1 for the same $n_1 + n_5$

In summary, for a fixed set of external division numbers, the internal mesh of the template can be varied to create a wide range of meshes with different block shapes, internal element distributions etc.

7 Choosing the Optimum Internal Divisions

The three unknown positions (for the positive singularity, the negative singularity and the common edge midpoint) can be found at low computational cost by solving three linear equations for 2D vectors.

For a given set of external division numbers, there are two parameters which can be used to vary the internal division numbers, whilst maintaining the same external division numbers:

1. The position where the positive singularity is attached to N_1, i.e. varying n_1 for a given $n_1 + n_5$
2. The position where the positive singularity is attached to N_2, i.e. varying n_0 for a given $n_0 + n_4$

Given that it is possible to compute the position of the singular points cheaply, it is feasible to estimate the mesh distortion, either exhaustively or for a substantial sample of the parametric variations. In this study, the minimum corner angle of all the blocks meeting at the singularities was used as the quality measure. Figure 12 show the best and worst quality of internal blocks for a fixed set of external division numbers.

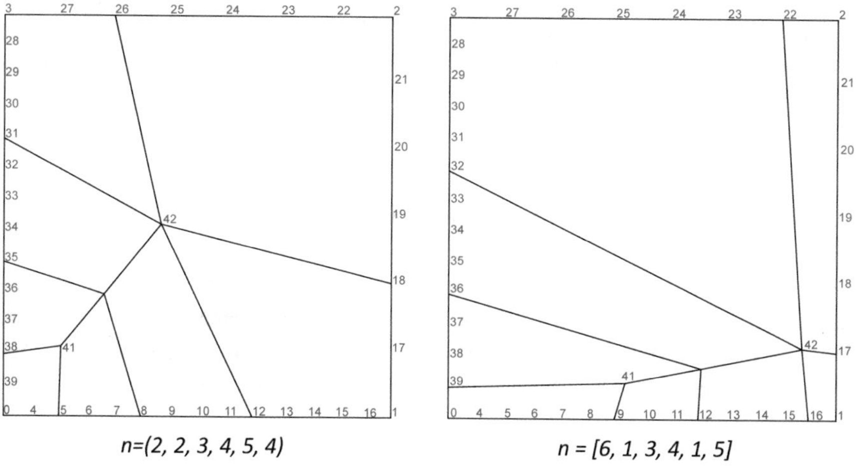

Fig. 12 The best (left) and worst (right) set of blocks obtained with the external division numbers [14,6,7,13] for all parametric variations

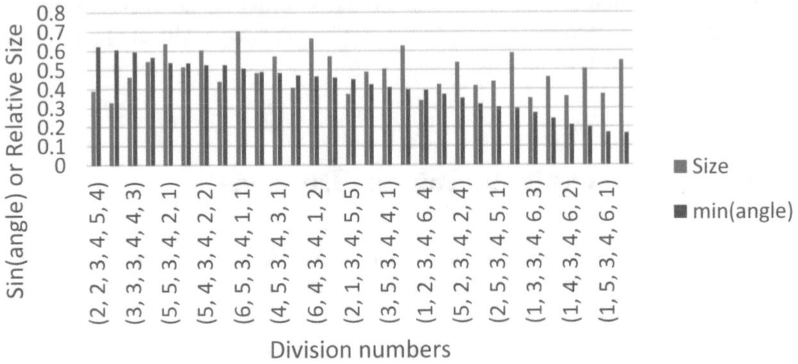

Fig. 13 Variations in mesh quality and size for all parametric variations of external division numbers [14,6,7,13]

Since the division numbers are known before any mesh is generated, the total number of elements can be calculated. For the block arrangement in Fig. 2, the total number of elements is

$$n_{transition} = n_0 * n_1 + n_0 * n_2 + n_1 * n_2 + n_0 * n_3 + n_1 * n_3 + n_1 * n_4 + n_4 * n_5 + n_0 * n_5$$

The number of elements required to achieve the same maximum mesh density in a regular mesh would be

$$n_{regular} = \max(N_0, N_2) * \max(N_1, N_3)$$

Figure 13 shows the minimum corner angle at the singularities for the meshes with this set of external division numbers. The chart also shows the ratio Size$=\frac{n_{transition}}{n_{regular}}$.

8 A Pair of Singularities Which Aren't Directly Connected

Figure 14 shows a potentially more useful transition which also has 6 internal division numbers, but without the redundant constraints between $N_0 - N_2$ and $N_1 - N_3$. By varying the internal division numbers, transitions like the previous one can be implemented. However, it also allows steerback transitions from N_2 to N_0.

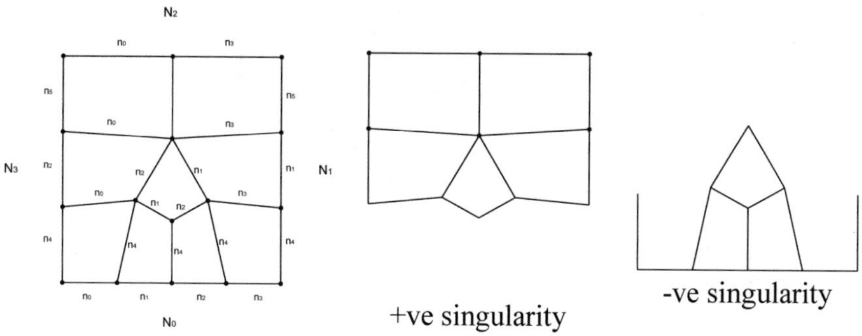

Fig. 14 A more flexible transition

The division number constraints are

$$\begin{bmatrix} 1 & 1 & 1 & 1 & 0 & 0 \\ 0 & 1 & 0 & 0 & 1 & 1 \\ 1 & 0 & 0 & 1 & 0 & 0 \\ 0 & 0 & 1 & 0 & 1 & 1 \end{bmatrix} \begin{Bmatrix} n_0 \\ n_1 \\ n_2 \\ n_3 \\ n_4 \\ n_5 \end{Bmatrix} = \begin{Bmatrix} N_0 \\ N_1 \\ N_2 \\ N_3 \end{Bmatrix}$$

Solving this gives

$$n_2 = \frac{N_o - N_1 - N_2 + N_3}{2}$$

$$n_1 = \frac{N_o + N_1 - N_2 - N_3}{2}$$

$$n_4 + n_5 = \frac{-N_o + N_1 + N_2 + N_3}{2}$$

The position of the pentagon edge midpoint on the top edge can be moved by changing n_0 whilst keeping the total $n_0 + n_3 = N_2$.

The value of $n_4 + n_5$ determines how close the two singularities are—if this is large the singularities are close together. Note that $n_4 \geq 0$ and $n_5 \geq 1$. The resulting constraint that

$$-N_o + N_1 + N_2 + N_3 \geq 1$$

tells us that, if for example we have a boundary layer with a dense mesh that feeds into a far field with a much coarser mesh, once N_0 gets too big we need to add another singularity pair. Figure 15 show how the internal division numbers can be adjusted to obtain mesh transitions in different directions.

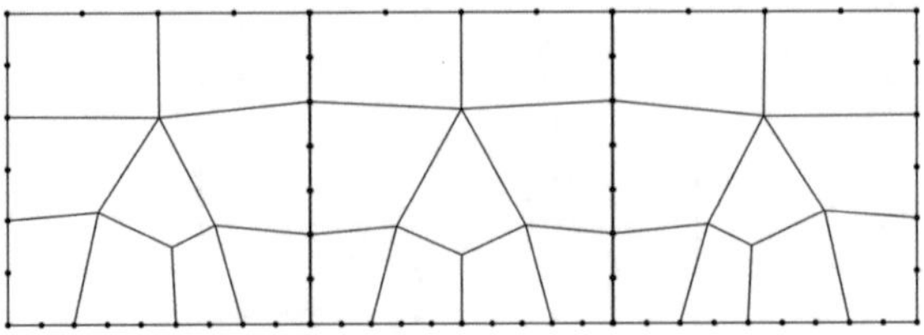

Fig. 15 Three adjacent blocks with singularity pairs adjusted to obtain a mesh transition bottom-right (left), bottom-t4op (middle) and bottom-left (right)

The points which need to be constrained to be the same between the pentagon and the triangle are the two singular points and the two points which act as edge midpoints for both the triangle and the pentagon. This gives four vector equations to determine the positions of the positive and negative singularities and the two triangle edge midpoints. As for the first transition described above, the set of internal division numbers to give the best mesh quality can be computed cheaply for a given block shape.

9 3D Transitions

3D transitions can be constructed as 2D extrusions of mesh singularity arrangements such as are shown in Figs. 2 or 14.

A genuine 3D partitioning is shown in Fig. 16, where a single split is used to partition the block into two of the mesh-able primitives identified in [2]. Each can be meshed by midpoint subdivision [14], where each face midpoint is connected to a body midpoint. The result is a positive/negative singularity pair on three faces of the block, with a regular mesh on the remaining 3 faces. This can be regarded as the 3D equivalent of Fig. 2.

The implementation shown in Fig. 16 was created using a simple Abaqus Python script This employed the following steps:

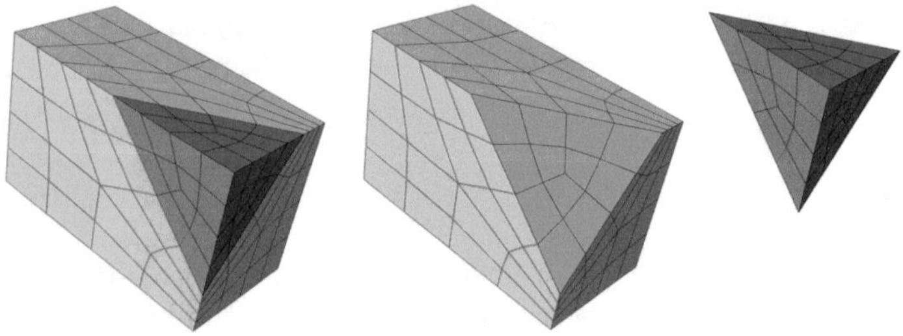

Fig. 16 3D block refinement

1. a hard geometry partition was created from a plane defined by points on 3 edges of the original block. The location of these points is the nodal position which gives the desired edge division numbers and mesh size in each of the primitives. It can therefore be used to produce different mesh distributions.
2. Both the resulting primitives, the tetrahedron and the cube-with-a chip-off, are meshed using the implementation of midpoint subdivision within Abaqus.

The Abaqus implementation enforces any necessary division number constraints on block edges in the model after midpoint subdivision. In 2D, once the radial division numbers from the singular points at the face centers are chosen, the total edge division numbers are defined and the task of the integer programming is to choose the set of radial division numbers which best fits the target external division numbers.

In 3D [15], once the radial division numbers from the singular point at the center of a primitive volume to the face centers are chosen, the radial division numbers from the singular points on the faces to the face edges and the external edge division numbers are defined. The cube-with-a chip-off has seven faces so it contributes seven radial division numbers. The tetrahedron has four faces, but with this refinement template all three division numbers on the common face in Fig. 16 must be the same. It is also expected, as was shown in Sect. 6, that it is the total division number between the singularities which will be significant rather than the two division numbers to the partitioning face. Therefore, there are 7 radial division numbers available to satisfy the 12 edge division numbers of the original cube. Note that if a structured mesh was imposed, only three unique edge division numbers are available.

The Abaqus implementation partitions the block volume using a plane, so the adjustment of nodal positions using transfinite mapping as described in Sects. 3–5 is not available, but this should be a straightforward item of future work.

To prevent propagation of the mesh singularity pairs, the same recipe can be applied to adjacent blocks. A symmetric replica on one adjacent block limits the influence of the mesh singularities to two faces, with the other four faces of the double-sized block having a regular mesh. Patterns of four blocks can be constructed to limit the mesh refinement to one face, whilst eight would limit the effect to the interior of a volume, with all external faces having a regular mesh.

10 Discussion

The illustrations above are for a mesh on a unit cube or block, but the transfinite mapping can be applied to any 2D geometry defined by a sequence of four edge meshes, e.g. Fig. 17. Rotation and reflection transforms can be used to convert the meshes shown to different geometries where the external division numbers have different constraints e.g. that $N_0 < N_2$ or $N_3 < N_1$ in Fig. 2. Graded edge meshes are

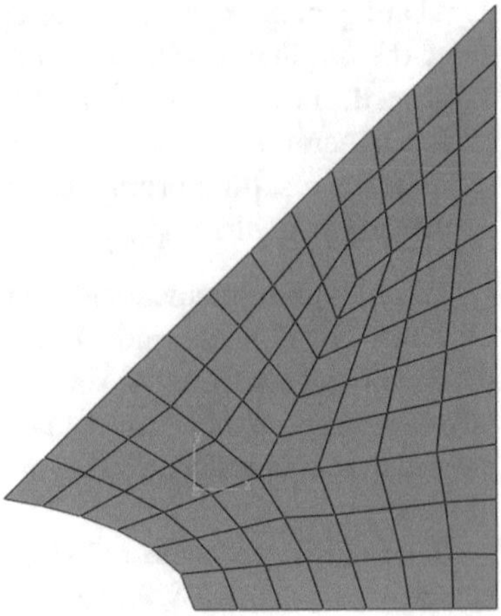

Fig. 17 Mesh in distorted block. Same division numbers as in Fig. 8

handled automatically since only the nodal positions are required, and measures of mesh quality can be computed before the mesh is generated.

Adaptive refinement of block topology and mesh, based on solution error estimates, is an obvious item of future work. There is no fundamental reason why the technique should not be applied to arbitrary block networks—in 2D all that is required is the number of block vertices connected to a given vertex. The same equations as are utilized to compute the face midpoint positions are then applicable.

Multiblock meshes are good at estimating sensitivities to a design change, since a fixed mesh topology is used. The templates here represent the minimal change in mesh topology if a better mesh distribution and solution accuracy is required. The approach could be codified into existing multiblock mesh generators.

The flexible adaptation approach described in this paper enables a wide range of external and internal mesh distributions to be accommodated for given block edge division numbers. This is highlighted in Fig. 18, where it is shown that assigning zero divisions to specific block edges enables the two and three refinement templates from Schneiders to be created. For example, the top left blocking is created by setting n4 as zero.

The 3-refinement template, top middle blocking, can then be created by assigning n5 as zero, and can be used to limit the transition of mesh refinement. In a next step, if n2 is assigned to be zero then the 2-refinement template is created as shown in the top right blocking. This demonstrates the flexibility of the approach described in this paper, e.g. setting both n2 and n3 to have zero division numbers generates the structured quadtree decomposition shown in the blocking of Fig. 19, whose quality can be controlled by altering the division constraints.

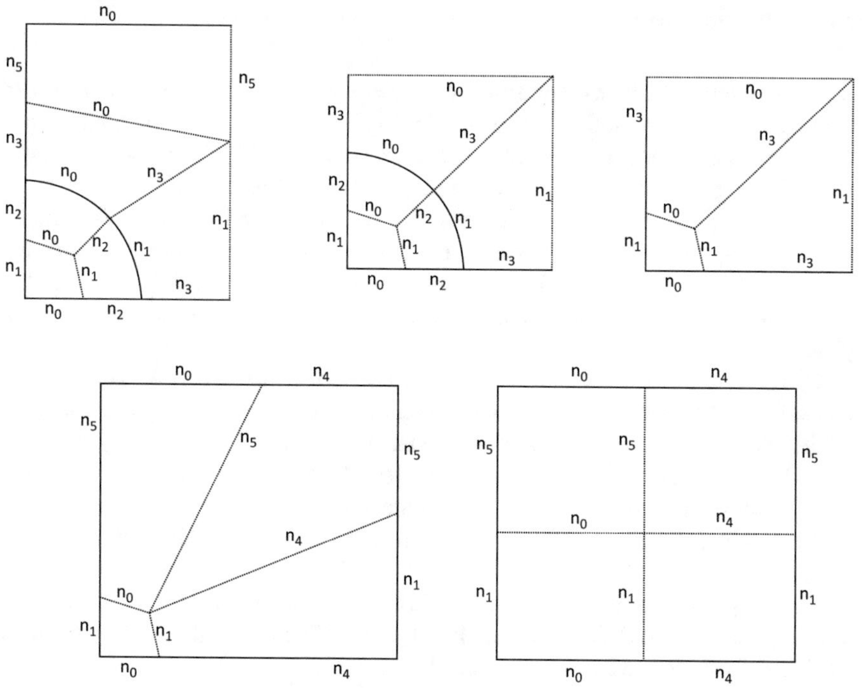

Fig. 18 Reducing block edge to zero division number

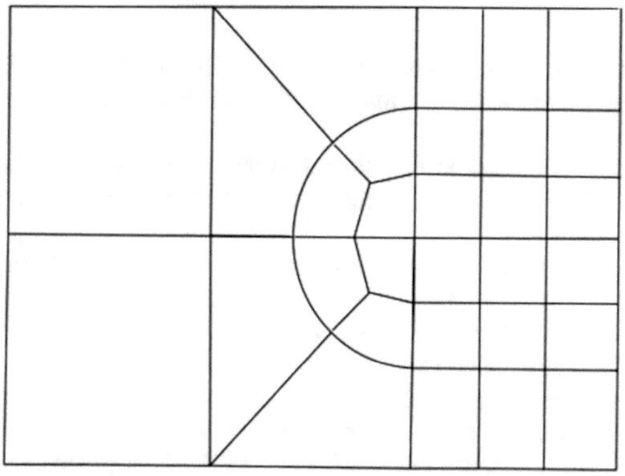

Fig. 19 3-refinement adaptation

11 Conclusions

Structured multiblock decompositions are very expensive to create in terms of engineering time. Adapting the mesh density to improve solution quality in a multiblock decomposition usually requires a change in block division numbers, which has a global effect on the mesh and the solution problem size.

Local templates for block refinement can be constructed which can be implemented cheaply and have only a local effect on mesh density and block structure. For a given set of external division numbers on a block, there is a parametric family of solutions for the internal division numbers in the refinement template.

Using an extended transfinite mapping scheme, a mesh can be generated is equivalent to that which would be generated by iterative iso-parametric smoothing. Whilst the quality may be inferior to that generated using more sophisticated smoothing algorithms, the optimum set of internal division numbers and the validity and quality of the final mesh can be guaranteed before the mesh is generated. It could therefore serve as an initialization step before other smoothing algorithms are applied.

The template refinement procedure adds the minimum number of mesh singularities e.g. a positive and negative singularity pair in 2D. Despite this, a wide range of mesh distributions can be accommodated.

References

1. A. Loseille, Recent Improvements on Cavity-Based Operators for RANS Mesh Adaptation, in *2018 AIAA Aerospace Sciences Meeting*, American Institute of Aeronautics and Astronautics, 2018
2. M.A. Price, C.G. Armstrong, M.A. Sabin, Hexahedral mesh generation by medial surface subdivision: part I. Solids with convex edges. Int. J. Numer. Methods Eng. **38**(19), 3335–3359 (1995)
3. Z. Ali, P.C. Dhanasekaran, P.G. Tucker, R. Watson, S. Shahpar, Optimal multi-block mesh generation for CFD. Int. J. Comput. Fluid Dyn. **31**(4–5), 195–213 (2017)
4. W.R. Quadros, LayTracks3D: a new approach for meshing general solids using medial axis transform. CAD Comput. Aided Des. **72**, 102–117 (2016)
5. N. Kowalski, F. Ledoux, P. Frey, Smoothness driven frame field generation for hexahedral meshing. CAD Comput. Aided Des. **7772**, 65–77 (2016)
6. Gridpro, Multi-Scale Tools, *GridPro website*. [Online]. https://www.gridpro.com/gridpro-advantages. Accessed 27 May 2018
7. A. Keskin et al., On the quantification of errors of a pre-processing effort reducing contact meshing approach, in *53rd AIAA Aerospace Sciences Meeting*, American Institute of Aeronautics and Astronautics, 2015
8. J.S. Sandhu, F.C.M. Menandro, H. Liebowitz, E.T. Moyer, Hierarchical mesh adaptation of 2D quadrilateral elements. Eng. Fract. Mech. **50**(5/6), 727–736 (1995)
9. R. Schneiders, Refining Quadrilateral and Hexahedral Element Meshes, in *5th International Conference on Grid Generation in Computational Field Simulations*, 1996, pp. 679–688

10. M.S. Ebeida, A. Patney, J.D. Owens, E. Mestreau, Isotropic conforming refinement of quadrilateral and hexahedral meshes using two-refinement templates. Int. J. Numer. Methods Eng. **88**(10), 974–985 (2011)
11. J. Qian, Y. Zhang, Automatic unstructured all-hexahedral mesh generation from B-Reps for non-manifold CAD assemblies. Eng. Comput. **28**(4), 345–359 (2012)
12. B.D. Anderson, S.E. Benzley, S.J. Owen, Automatic all quadrilateral mesh adaption through refinement and coarsening, in *Proceedings of the 18th International Meshing Roundtable, IMR 2009*, 2009
13. H.J. Fogg, L. Sun, J.E. Makem, C.G. Armstrong, T.T. Robinson, Singularities in structured meshes and cross-fields. CAD Comp. Aided Des. **105**, 11–25 (2018)
14. T.S. Li, R.M. McKeag, C.G. Armstrong, Hexahedral meshing using midpoint subdivision and integer programming. Comput. Methods Appl. Mech. Eng. **124**(1–2), 171–193 (1995)
15. T.S. Li, C.G. Armstrong, R.M. McKeag, Quad mesh generation for k-sided faces and hex mesh generation for trivalent polyhedra. Finite Elem. Anal. Des. **26**(4), 279–301 (1997)

Part III
Simplicial Meshes

Tuned Terminal Triangles Centroid Delaunay Algorithm for Quality Triangulation

Maria-Cecilia Rivara and Pedro A. Rodriguez-Moreno

Abstract An improved Lepp based, terminal triangles centroid algorithm for constrained Delaunay quality triangulation is discussed and studied. For each bad quality triangle *t*, the algorithm uses the longest edge propagating path (Lepp(t)) to find a couple of Delaunay terminal triangles (with largest angles less than or equal to 120°) sharing a common longest (terminal) edge. Then the centroid of the terminal quadrilateral is Delaunay inserted in the mesh. Bisection of some constrained edges are also performed to assure fast convergence. We prove algorithm termination and that a graded, optimal size, 30° triangulation is obtained, for any planar straight line graph (PSLG) geometry with constrained angles greater than or equal to 30°.

1 Introduction

Lepp bisection algorithm [3, 10] is an efficient reformulation of previous longest edge algorithm for triangulation refinement, that for each target triangle follows the longest edge propagating path (Lepp) to find a couple of terminal triangles sharing a common longest edge (terminal edge), which are then refined by longest edge bisection. Consequently, local refinement operations are used, and conforming triangulations (where adjacent triangles either share a common edge or a common vertex) are maintained throughout the whole refinement process. Due to the properties of the iterative longest edge bisection of triangles, refined triangulations that maintain the triangulation quality (bounded smallest angle) are obtained, while the proportion of quality triangles increases as the refinement proceeds. Based on the

M.-C. Rivara (✉)
Department of Computer Science, Universidad de Chile, Santiago, Chile
e-mail: mcrivara@dcc.uchile.cl

P. A. Rodriguez-Moreno
Department of Information Systems, Universidad del Bio-Bio, Concepción, Chile
e-mail: prodrigu@ubiobio.cl

X. Roca, A. Loseille (eds.), *27th International Meshing Roundtable*,
Lecture Notes in Computational Science and Engineering 127,
https://doi.org/10.1007/978-3-030-13992-6_12

properties of terminal triangles and terminal edges it was also proved that optimal size triangulations are obtained [3].

A Lepp Delaunay algorithm for quality Delaunay triangulation, based on the Delaunay insertion of the midpoint of the terminal edge, was introduced by Rivara [10] and studied by Bedregal and Rivara [2]. An algorithm based on computing the centroid Q of the terminal triangles which is Delaunay inserted, was presented in [11] without proving termination, neither optimal size property. In this paper we study a tuned, order independent algorithm (where the size of the refined triangulation is almost equal independently of the triangle processing order), based on the Lepp centroid algorithm discussed in [11].

Alternative Delaunay refinement algorithms, based on selecting the circumcenter (or a point close to the circumcenter) of each skinny triangle which is Delaunay inserted in the triangulation have been studied by Ruppert [14], Shewchuk [16], and by Erten an Üngor [5]. Lepp Delaunay algorithms and circumcenter based algorithms have analogous practical behavior, as shown in the empirical study of reference [11], where the Triangle software [16] (without later improvement criteria) was compared with Lepp Delaunay algorithms. It is worth noting however that Lepp based algorithms have the advantage of being order independent, in the sense that they construct triangulations of approximately the same size independently of the processing order of the bad quality triangles. Consequently they are simpler methods than circumcenter based algorithms, easy to implement and easy to parallelize. On the other hand, the implementation of circumcenter algorithms is rather cumbersome, and requires processing triangles in bad-quality order. Section 6.3 of reference [4] discusses several recommendations to implement Ruppert's algorithm efficiently, which include maintaining a queue of skinny and oversized triangles throughout the refinement process.

Lepp Algorithms These are longest edge algorithms formulated in terms of the concepts of terminal edges, terminal triangles and longest edge propagating path [2, 3, 10]. An edge E is a *terminal edge* in triangulation τ if E is the longest edge of every triangle that shares E. The triangles sharing E are called *terminal triangles* (edge AB in Fig. 1a). If E is shared by two terminal triangles then E is an interior edge; if E is shared by a single terminal triangle then E is a boundary edge.

For any triangle t_0 in τ, the *longest edge propagating path* of t_0, $\texttt{Lepp}(t_0)$, is the ordered sequence of increasing triangles $\{t_j\}_0^{N+1}$ such that t_j is the neighbor triangle on the longest edge of t_{j-1} and where longestedge $t_j >$ longestedge t_{j-1}, for $j = 1, \ldots, N$. The process ends by finding the terminal edge E and a couple of associated terminal triangles t_N, t_{N+1}. In Fig. 1a, Lepp(t_0) = $\{t_0, t_1, t_2, t_3\}$.

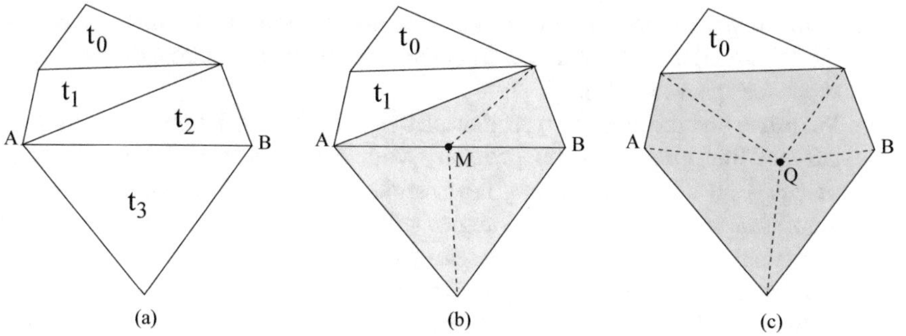

Fig. 1 (**a**) AB is a terminal edge shared by terminal triangles $\{t_2, t_3\}$ and $\texttt{Lepp}(t_0) = \{t_0, t_1, t_2, t_3\}$; (**b**) First step of Lepp-bisection algorithm for refining t_0; (**c**) First step of Lepp Delaunay centroid algorithm

For each target triangle t, the generic Lepp based algorithms find an associated local largest edge shared by a couple of terminal triangles. Then a point is selected inside the terminal triangles (terminal edge midpoint or terminal triangles centroid) and inserted in the mesh. In the Lepp bisection algorithm, the midpoint M of the terminal edge is inserted by longest edge bisection of the terminal triangles as shown in Fig. 1b. In the Lepp centroid Delaunay algorithm, the centroid Q of the terminal triangles is selected and (constrained) Delaunay inserted in the mesh, as shown in Fig. 1c. The process is repeated until the target triangle t is destroyed.

Algorithm Generic Lepp-based algorithm

Input : triangulation τ, set S of triangles to be refined / improved
Output : Refined triangulation τ'

```
for each t in S do
    while t remains in τ do
        Find Lepp(t), terminal triangles t1, t2 and terminal edge E ( t2 can be null)
        Select point P inside terminal triangles, insert P in the mesh and update S
    end while
end for
```

This paper presents improved and new results in the following senses:

(1) In a previous paper, Rivara and Calderon [11] discussed a Lepp Delaunay centroid algorithm where for each selected couple of terminal triangles t_1, t_2 (with non constrained terminal edge) the centroid of the quadrilateral formed by t_1, t_2 is selected and Delaunay inserted in the mesh. In the tuned algorithm of this paper, if t_1 (or t_2) is a bad quality triangle with constrained second longest edge E, the midpoint of E is (constrained) Delaunay inserted in the mesh, (see Sect. 3), which significantly reduces the number of points inserted close to the constrained edges.

(2) In this paper we present new rigorous results on algorithm termination and on the construction of optimal size triangulations, based on the properties of Lepp sequences proved in [3].
(3) We prove that the algorithm produces 30° quality triangulations for any planar straight line graph (PSLG) geometry with constrained angles greater than or equal 30°. This is a strong new result. Note that the proof in Ruppert's algorithm requires constrained angles $\geq$90°, while the modified algorithm of Shewchuk requires constrained angles $\geq$60°.
(4) We prove that the practical behavior of the tuned algorithm is independent of the triangles processing order, which is not the case of circumcircle based algorithms.

More specifically, in this paper we study a tuned Lepp Delaunay centroid algorithm by integrating previous, revisited and new results needed in the algorithm analysis. The following issues are considered:

- The simple insertion of the centroid Q over a couple of Delaunay terminal triangles t_1, t_2, obtained by joining Q with the vertices of t_1, t_2, improves the triangles obtained by longest edge bisection. This is an intermediate operation used in the algorithm analysis. In addition the Delaunay mesh insertion operation of Q improves even more the triangles involved.
- Most of the bad obtuse triangles have largest angle $>$120°, and are eliminated by edge swapping, assuming that an edge swapping Delaunay algorithm is used.
- The average Lepp size is small and tends to be two as the refinement proceeds. This result was proved for triangulations obtained by the Lepp bisection algorithm and extends to the algorithm of this paper.
- The constrained Delaunay triangulation of any PSLG data defines an intuitive edge distribution function which identifies edge details and non edge details in the PSLG geometry. We prove that the algorithm constructs a graded triangulation adapted to the geometry details. The edge details are not refined unless a close smaller detail induces its refinement.
- We use the simple (constrained) Delaunay triangulation associated with the PSLG data, as an intuitive edge distribution function, to prove termination and optimal size property, instead of using the local feature size function introduced by Ruppert [14].
- Our algorithm does not require the edge encroachment test used in Ruppert's algorithm, but a simple test based on triangle constrained edges.
- The mathematical properties of the mesh operations allows us to prove that 30° triangulations are obtained for constrained angles $\geq$30°. Note that Ruppert algorithm requires 90° constrained angles [14], and modified algorithm of Shewchuk requires constrained angles $\geq$60° [16].

2 Previous Results

The iterative longest edge bisection of individual triangles was studied by Rosenberg and Stenger [13] and by Stynes [18, 19]. This process produces a finite number of non-similar triangles with a bounded smallest angle, while the proportion of good triangles (quasiequilateral triangles) increases as the refinement proceeds.

Definition 1 Given a triangle $t(ABC)$ of vertices A, B, C, and edges $AB \geq BC \geq CA$, the longest-edge bisection of t (or simply bisection of t) is performed by joining the midpoint M of AB with the opposite vertex C (see Fig. 2a).

Definition 2 Triangle $t(ABC)$ of edges $AB \geq BC \geq CA$ is *quasiequilateral* if $AC \geq max\{AB/2, CM\}$ and $MC \geq BC/2$ (see Fig. 2b).

Note that for quasiequilateral triangles (see Fig. 2b) after the first median MC is introduced, the next longest edge bisections only produce medians parallel to the edges of the initial triangle ABC, which implies that at most, four similarly distinct triangles are produced. Furthermore the following results hold [13, 18, 19]:

A1. Given any triangle t_0 of smallest angle α_0, the iterative longest edge bisection of t_0 and its descendants produces a finite set $S(t_0)$ of similarly distinct triangles. Furthermore each triangle t in S(t_0) has smallest angle α_t such that $\alpha_t \geq \alpha_0/2$.

A2. For any quasiequilateral triangle t_{qeq}, the triangle set $S(t_{qeq})$ has at most, four similarly distinct triangles, all of which are also quasiequilateral.

A3. For any non quasiequilateral triangle t_0, consider the sequence of triangle sets Q_j defined as follows: $Q_0 = \{t_0\}$, and for j ≥ 1, Q_j is obtained by longest edge bisection of the triangles of Q_{j-1}. Then the triangle sets Q_j improve with j as follows: both the percentage of quasiequilateral triangles and the area of t_0 covered by these triangles, monotonically increase as the iterative refinement proceeds.

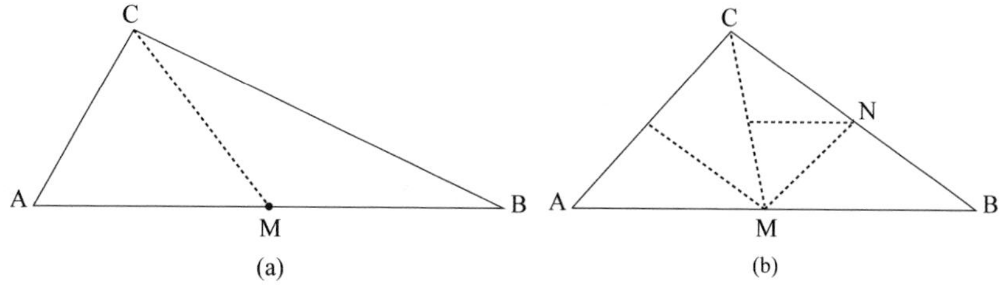

Fig. 2 (**a**) Longest-edge bisection of triangle $t(ABC)$ (**b**) First longest edge bisections that define a quasiequilateral triangle $t(ABC)$

The triangulations obtained by the Lepp bisection algorithm are conforming and inherit properties A1, A2, A3 as follows: the iterative local/global use of the Lepp bisection algorithm (and previous longest edge algorithms) produces sequences of nested, refined and conforming triangulations $\{\tau_j\}$ such that B1, B2 hold [3, 10]:

B1. For any triangle t_0 in τ_0, the refined triangles nested in t_0 belong to a finite set $S(t_0)$ of similarly distinct triangles, all of which have smallest angle $\alpha \geq \alpha_0/2$, where α_0 is the smallest angle of t_0.

B2. The refined triangulations $\{\tau_j\}$ improve with j in the following senses: both the percentage of quasiequilateral triangles, and the area covered by these triangles, increase as the refinement proceeds.

More recently Bedregal and Rivara [3] proved that there exists a close relationship between quasiequilateral triangles and terminal triangles (the proportion of terminal triangles increases as quasiequilateral triangles increases), which imply B3. Furthermore bounds on the number of triangle partitions performed inside a triangle in a Lepp sequence [3] (assertion B4) together with B3, implies B5.

B3. The proportion of terminal triangles increases (approaching 1) as the refinement proceeds and the average length of Lepp(t) tends to be two as the refinement proceeds. Furthermore the Lepp Delaunay algorithms inherit the same properties.

B4. The number of longest edge bisections performed in the interior of a triangle t to make it conforming in a refining Lepp sequence, is constant and less than three in most cases. This constant is bounded by $O(\log^2(1/\alpha))$ for triangles with arbitrary smallest angle α.

B5. Lepp bisection algorithm produces optimal size triangulations.

Finally, the properties of Delaunay terminal triangles [10, 17], play a crucial role in Lepp Delaunay algorithms, and specifically in the algorithm of this paper. Couples of Delaunay terminal triangles ABC, ABD (see Fig. 3) are neighbor triangles that simultaneously satisfy that AB is the common longest edge of the both triangles, and that triangles ABC and ABD are locally Delaunay which implies that vertex D is outside the circumcircle of triangles ABC. Both conditions together imply that vertex D must belong to the shadowed region $\mathscr{R}$ limited by the circumcircle of triangle ABC and the circles of vertices A, B and radius AB. In the case that $\measuredangle ACB = 120°$, $\mathscr{R}$ reduces to one point D' (triangle $AD'B$ is equilateral). Consequently for $\measuredangle ACB > 120°$, $\mathscr{R}$ is empty and the following results hold:

Theorem 1 *For any pair of Delaunay terminal triangles t_1, t_2 sharing a terminal edge AB it holds:*

(a) Largest angle $(t_i) \leq 2\pi/3$ for i=1,2
(b) At most one of the triangles t_1, t_2 is obtuse

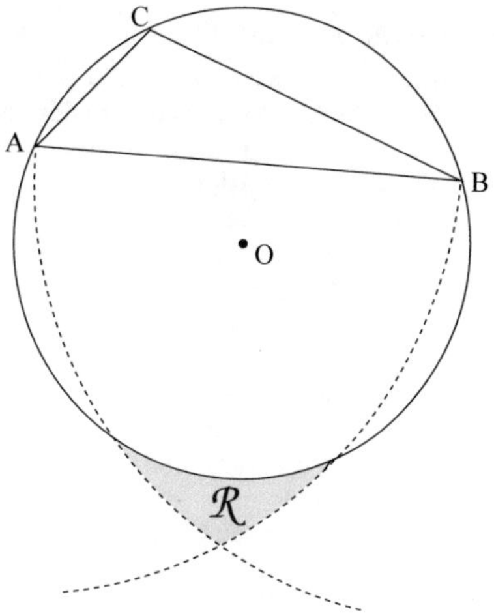

Fig. 3 Delaunay terminal triangles ABC, ABD; vertex D belongs to region R

Since the algorithm of this paper inserts points in the interior of couples of Delaunay terminal triangles, only triangles with largest angle less than or equal to 120° can become a terminal triangle throughout the algorithm processing.

Definition 3 We will say that t is a PD terminal triangle (potentially a Delaunay terminal triangle) if the $largest\ angle(t) \leq 120°$.

3 The Tuned Algorithm

We first introduce the following mesh operation: for bad quality terminal triangle t with constrained second longest edge CB, the constrained Delaunay insertion of the midpoint of CB is performed (see Fig. 4). This operation reduces the number of interior points inserted close to the constrained edges. Note that the previous Lepp Delaunay midpoint algorithm requires this operation to guarantee convergence [10]. The previous Lepp Delaunay centroid algorithm does not use this operation since the centroid selection avoids the introduction of collinear points, but introduces more points than the tuned algorithm of this paper [11].

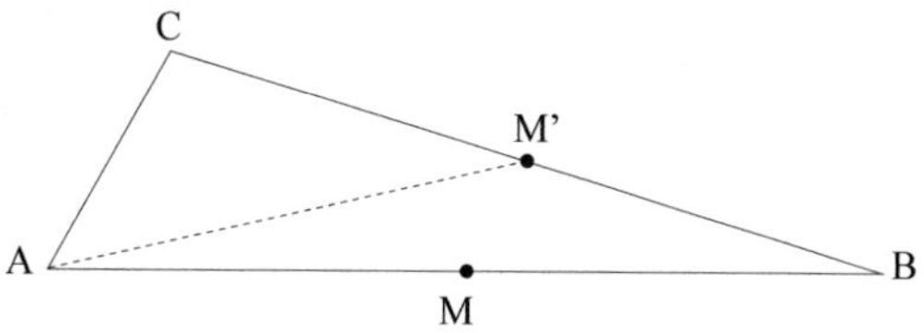

Fig. 4 For constrained second longest edge CB, the midpoint of CB is constrained Delaunay inserted in the mesh

Algorithm Tuned Terminal_Triangles_Centroid_Delaunay_Algorithm

Input: CDT τ associated with PSLG data, angle tolerance θ_{tol}
Output: Refined triangulation τ_f with angles $\geq \theta_{tol}$.

Find S set of bad quality triangles
for each t in S (while $S \neq \varnothing$) **do**
  while t remains unrefined **do**
    Use Lepp(t) to find Delaunay terminal triangles t_1, t_2 and terminal edge E
    if E is constrained (this includes t_2 null) **then**
      Perform Constrained Delaunay insertion of midpoint of E
    else
      if there exists t (t_1 or t_2) such that $\alpha_t < \theta_{tol}$ and second longest edge L is constrained **then**
        Perform constrained Delaunay insertion of midpoint of L
      else
        Compute centroid Q of terminal triangles, and perform constrained Delaunay insertion of Q
      end if
    end if
    Update S
  end while
end for

4 Better Angle Bounds on First Bisections of Triangles

We show that the first longest edge bisection of a triangle produces a better triangle t_B (see Fig. 5) and a bad obtuse triangle t_{OB} [17]. Assume the triangle of Fig. 5 where $AB \geq BC \geq AC$ with the notation shown in this figure.
It is rather easy to see that if t is a right angled triangle then $\alpha_1 = \alpha_0$, $\beta_1 = \beta_0$, $AM = CM$, while if t is an acute triangle then $\alpha_1 < \alpha_0$, $\beta_1 < \beta_0$, $AM < CM$; and if t is an obtuse triangle then $\alpha_1 > \alpha_0$, $\beta_1 > \beta_0$, $AM > CM$. These properties allow proving most of the assertions of Lemma 1 [17]. The bound on α_1 follows from the strong property $A1$.

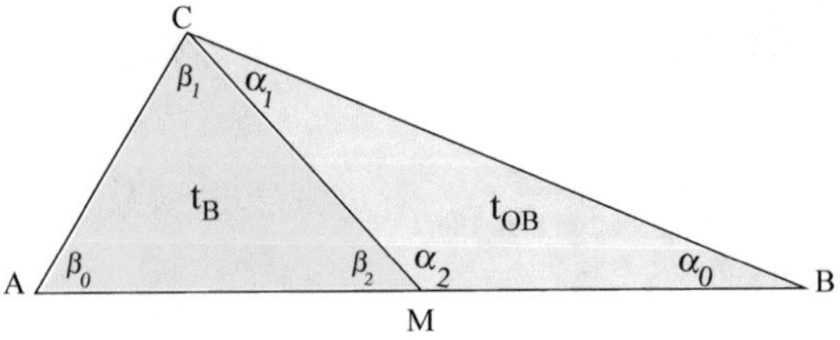

Fig. 5 Notation for longest edge bisection. Angles in longest edge bisection of triangle ABC with $AB \geq BC \geq AC$

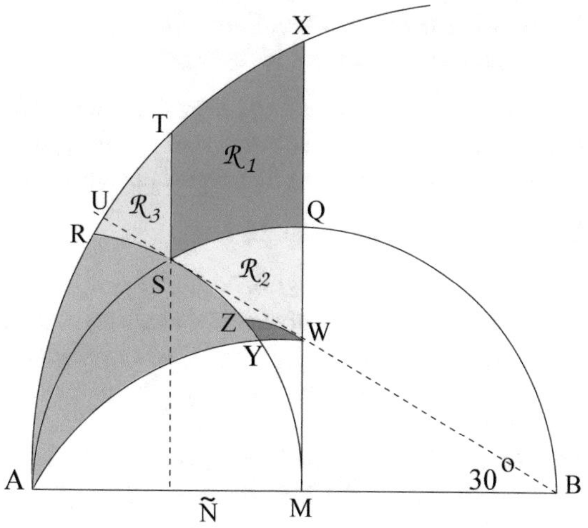

Fig. 6 Taxonomy on longest edge bisection of triangles $t(ABC)$ with $AB \geq BC \geq CA$

Lemma 1 *The following angle bounds hold [17].*

(a) $\alpha_1 \geq \alpha_0/2$, $\alpha_2 \geq 90°$, $\beta_2 \leq 90°$, $\beta_1 \geq \pi/6$, $\beta_1 \geq \alpha_1$
(b) $\beta_2 = \alpha_0 + \alpha_1 \geq 3\alpha_0/2$
(c) if t is obtuse, then $\alpha_1 > \alpha_0$ *and* $\beta_2 \geq 2\alpha_0$
(d) if t is acute, then $\alpha_1 < \alpha_0$ *and t_B is acute*

Next we introduce the taxonomy of Fig. 6, which is a variation of those presented in references [9, 17]. This is obtained by fixing the longest edge AB of triangle ABC considering $AB \geq BC \geq CA$, and studying which is the longest edge of triangle AMC and the longest edge of triangle CMN (see Fig. 2a) according to the position of vertex C. Thus, the half circle of vertex M and radius AM separates obtuse and acute triangles. Arcs AR and MR respectively correspond to isosceles triangles with edges $AM = CM$ and edges $AC = AM$; while arc ZW corresponds to the circle of center $\tilde{N}$ (where $A\tilde{N} = AB/3$) and radious $A\tilde{N}$, corresponding to the triangles for which $CB = 2CM$. The set of quasiequilateral triangles is the union of region $\mathscr{R}_1$ (acute triangles) and region $\mathscr{R}_2$ (obtuse triangles). Finally arc AW corresponds to points C for which the largest angle is equal to 120°, defined by the circle of center W' and radious WW', where points W, W' are symmetric with respect to line AB. By studying the boundaries of regions $\mathscr{R}_1$ and $\mathscr{R}_2$ it is easy to see that $\mathscr{R}_1 \bigcup \mathscr{R}_2$ correspond to quasiequilateral triangles and that most of these triangles (vertex C by above line SW) have the smallest angles $\geq$30°. Only for vertex C in region SZW, smallest angle >27.88° (the worst case corresponds to $C = Z$ where $tg(\alpha_0(Z)) = \sqrt{7}/5$). Note that most of the triangles of $\mathscr{R}_3$ also have the smallest angle $\geq$30°.

Lemma 2 *(a) For quasiequilateral triangles in region $TSWX$, $\alpha_0 \geq 30°$; (b) For quasiequilateral triangles in region SZW, $\alpha_0 > 27.88$.*

Next we extend the notation of Fig. 5. We will call $\alpha_0(t)$, $\alpha_1(t)$ to the angles α_0, α_1 obtained by longest edge bisection of $t(ABC)$; in addition we call $\alpha_0(t_B)$, $\alpha_1(t_B)$ to the α_0, α_1 angles obtained by longest edge bisection of t_B.

Lemma 3 *Given any triangle t, then:*

(a) If $\alpha_0 \leq 30°$, then $\alpha_1 \geq 0.79\alpha_0$ and $\beta_2 \geq 1.79\alpha_0$. Furthermore, the ratio α_0/α_1 increases (α_1 approaching α_0) while α_0 decreases.
(b) If $\alpha_0 \geq 30°$, then $\alpha_0(t_B) \geq 30°$ and t_B is quasiequilateral.
(c) If t is quasiequilateral with $\alpha_0 \geq 30°$, then $\alpha_0(t_B) \geq 30°$ and $\alpha_0(t_{OB}) \geq 27.88°$.
(d) If t is a PD-terminal triangle then t_B is a PD-terminal triangle, and $\alpha_0(t_B) \geq Min\{3/2\alpha_0(t), 30°\}$. Furthermore, if $\alpha_0 \geq 20°$, then smallest $angle(t_B) \geq 30°$.

Proof The proof of item (a) follows by studying the case of acute triangles of region UAS in Fig. 6, where the worst case corresponds to point U for which $\alpha_1 \approx 23.79°$. To prove assertion (d) we consider the case where t is acute. Note that in Fig. 5, $\alpha_0(t_B) = Min\{\beta_0, \beta_1, \beta_2\}$, where $\beta_2 \geq 3/2\alpha_0(t)$. On the other hand, β_0 is the smallest angle of t_B in Fig. 6, when edge AM is the shortest edge of the t_B, which occurs for acute t with C either in region $\mathscr{R}_1$ or in region $\mathscr{R}_2$; and the worst case occurs for the equilateral triangle where $\beta_2 = 30°$ (see Fig. 5). Assertion (b) follows from Lemma 1, while assertion (c) follows from Lemma 2. □

In Lemma 4 we further quantify the notion that the triangle t_B in Fig. 5 is a better triangle than t following the ideas of reference [17]. In Lemma 5 we further characterize PD-terminal triangles. The non PD terminal triangles correspond to very obtuse triangles of region WAM in Fig. 6.

Lemma 4 *If t_B is acute and $\alpha_0 \leq 30°$, then $\alpha_1(t_B) \geq min\{1.4\alpha_0(t), 30°\}$.*

The proof is rather complex and can be found in reference [17].

Lemma 5 *Given any PD-terminal triangle t. Then:*

(a) If t is acute and $\alpha_0 \leq 30°$, then t_{OB} is a non-PD terminal triangle.
(b) If t is obtuse and $\alpha_0 > 22°$, then t_{OB} can be a PD terminal triangle.

Proof Part (a) follows from the fact that for acute triangles $\alpha_1 < \alpha_0$, which in turn implies $\alpha_1 + \alpha_2 < 60°$ and consequently t_{OB} is a non-PD terminal triangle. Part (b) follows from the fact that for obtuse triangles, $\alpha_1 > \alpha_0$. In [12] it was proved that largest angle equal to 120° and $\alpha_1 + \alpha_0 = 60°$ implies that $\alpha_0 > 22°$. Thus, only for some triangles with $\alpha_0 > 22°$ it can hold $\alpha_1 + \alpha_0 > 60°$ and t_{OB} can be a PD terminal triangle. □

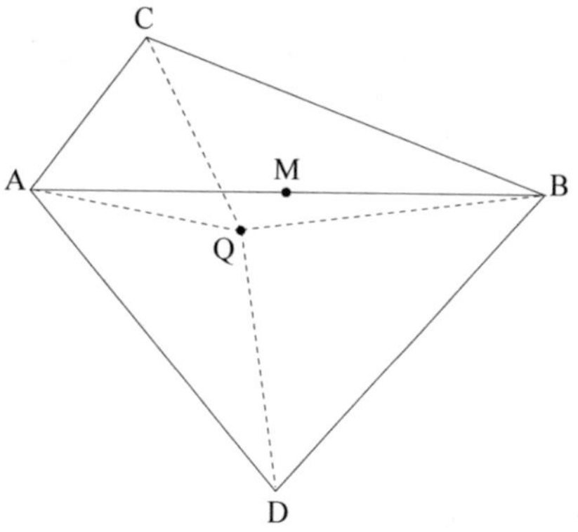

Fig. 7 Centroid refinement of terminal triangles ABC, ADB

5 Improvement Properties of the Centroid Insertion

In what follows we consider that triangle t is good if $\alpha_0 \geq 30°$. The algorithm of this paper in general performs (constrained) Delaunay insertion of the centroid Q of couples of Delaunay terminal triangles (both triangles with largest angle $\leq 120°$). To analyze triangle improvement, the following intermediate simple centroid insertion is needed: consider the centroid Q of a couple of Delaunay terminal triangles as shown in Fig. 7. The simple centroid insertion is then performed by joining Q with the four vertices, instead of performing longest edge bisections. This operation corresponds to a Laplacian smoothing of the terminal edge midpoint M and improves the triangles obtained by longest edge bisection. Note that the Laplacian smoothing works very well for convex geometries [6, 7], and couples of terminal triangles always define a convex quadrilateral.

Lemma 6 *The simple centroid insertion has the following properties: (i) It improves the worst angle obtained by the longest edge bisection; (ii) It avoids the reproduction of a bad quality triangle; (iii) It improves the lightly bad angles obtained by the longest edge bisection of good triangles.*

6 Algorithm Analysis

Consider a general PSLG (planar straight line graph) geometry, defined by a set of points, edges and eventually polygonal objects defining exterior boundaries and interior holes. Any PSLG geometry has edge details and non-edge details. Edge details are small edges in the PSLG data, while non edge details are defined by two close isolated interior points, an isolated point close to an input edge, two edges with close points, constrained angles either over the boundaries or interior to the geometry, and vertices over these angles. For an illustration see Fig. 8a.

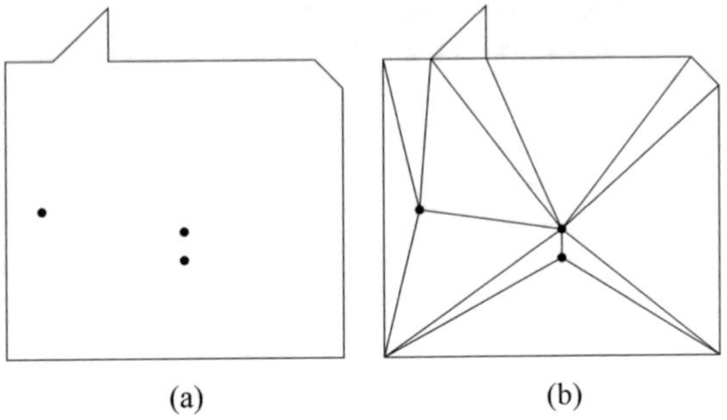

Fig. 8 (**a**) PSLG geometry; (**b**) constrained Delaunay triangulation identifies edge details and non edge details

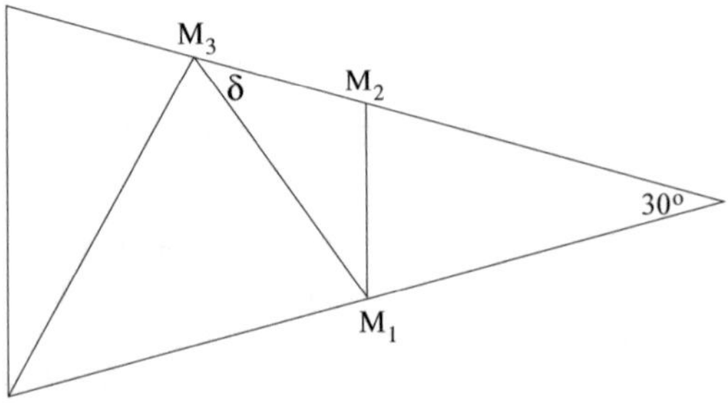

Fig. 9 Worst case of acute isosceles 30° triangle

Note that the constrained Delaunay triangulation of the input PSLG data intuitively defines an edge distribution function to which an optimal size good quality triangulation should be adapted. More specifically this identifies edge details and non-edge details by means of skinny triangles with associated (constrained or non constrained) small edges, very obtuse triangles with largest angled vertex close to an edge data, and triangles with constrained smallest angle. Figure 8b shows the constrained Delaunay triangulation of the example of Fig. 8a. We will prove that the algorithm of Sect. 5 produces a graded quality mesh with smaller good quality triangles around the PSLG geometry details. The following Lemma assures that 30° constrained angles always produce quality triangles:

Lemma 7 *Let t be any triangle with* 30° *constrained angle. Then (a) If t is obtuse, the longest edge bisection of t produces quality triangles (part (c) of Lemma 1); (b) If t is acute the tuned algorithm inserts three points in the constrained edges, as shown in Fig. 9 to produce quality triangles (δ is the worst angle* >34°*).*

Theorem 2 *Consider any PSLG geometry with constrained angles* ≥30° *and the input constrained Delaunay triangulation τ_0 associated with the PSLG data. Then for angle tolerance $\theta_{tol} = 30°$,*

(a) The algorithm finishes with a graded 30° *constrained Delaunay triangulation.*
(b) The final triangulation is size optimal.

Proof Given $\theta_{tol} = 30°$, consider the bad triangles with angles less than 30°. To prove part (a), we will study five cases of triangles processing:

Case 1. Non PD terminal triangles. Each bad quality triangle t (with largest angle $>120°$ either with one or two bad angles) is a non PD terminal triangle and consequently is eliminated by swapping edge AB either by processing t or by processing a Lepp-neighbor bad quality triangle. This operation produces locally more equilateral triangles.

Case 2. Bad PD terminal triangles. Consider a couple of non constrained Delaunay terminal triangles. Let $t(ABC)$ with $AB \geq BC \geq AC$ be the worst triangle in the couple with $\alpha_0 < 30°$. Then:

- According to part(a) of Lemma 3, the longest edge bisection of t introduces midpoint M of AB and a better triangle $t_B(ACM)$ with $\alpha_0(t_B) \geq 1.79\alpha_0$, and a bad obtuse triangle t_{OB}. The simple centroid insertion of Q corresponds to the Laplacian smoothing of point M, which improves the worst angles of t_{OB} (introduced by the longest edge bisection) and avoids the repetition of a triangle similar to triangle ABC. This operation can be seen as a first step of the Delaunay insertion of point Q.
- The centroid Q is Delaunay inserted in the mesh (see Fig. 10). If triangle CQB is a non PD terminal triangle, then triangle CQB is eliminated (and improved) by swapping edge CB, either when Q is Delaunay inserted (if there exists a vertex inside the (big) circumcircle of triangle CMB), or by later processing CBQ, or by processing a neighbor bad quality triangle. If triangle CQB is a PD terminal triangle and still bad, then by processing triangle CQB this can become a terminal triangle and the centroid $\tilde{Q}$ of CQB and its neighbor triangle is inserted, which improves the angles.
- According to part (d) of Lemma 3, for $\alpha_0 < 20°$, CAQ can still be bad. Then for small α_0, a finite sequence of points Q_i can be inserted in the mesh until a good triangle CAQ_n is obtained (see Fig. 11). The process finishes without refining edge AC (AC is a local smallest edge), unless a close smaller edge induces neighbor refinement. See the termination analysis for more details.

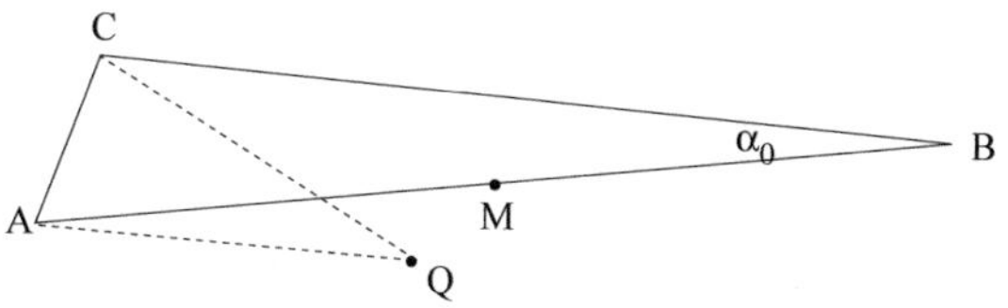

Fig. 10 Triangle ABC with $\alpha_0 < 30°$. Better triangle ACQ and CQB are obtained with respect to the longest edge bisection

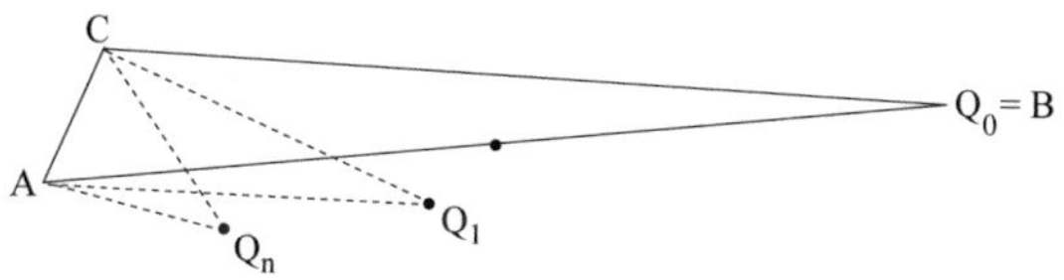

Fig. 11 Points Q_i are introduced until triangle CAQ_n is good

Case 3. Terminal triangles with constrained edges. For Delaunay terminal triangles with constrained terminal edge, the constrained Delaunay insertion of the terminal edge midpoint is performed and the improvement process continues. For bad triangles with constrained second edge E, the simple constrained Delaunay insertion of midpoint of E is performed, which accelerates convergence (Sect. 3).

Case 4. Couples of good Delaunay terminal triangles. For couples of good quality Delaunay terminal triangles with smallest angles $\geq 30°$, the centroid Q of the terminal quadrilateral is inserted, which produces more equilateral triangles than those obtained by longest edge bisection. This is equivalent to a Laplacian smoothing of the terminal edge midpoint introduced by the longest edge bisection of the terminal triangles. This operation improves eventual angles less than 30° that could have been introduced by the longest edge bisection.

Case 5. Triangles with 30° constrained angles. Here, good quality triangles are obtained inside t by inserting one or three points over the constrained edges.

Termination The proof on termination is based on the fact that for skinny triangles, the smallest edge AC is never refined, unless there exists a smaller bad quality triangle t^* such that Lepp(t^*) contains triangle AQ_nC (see Fig. 12). Thus the algorithm stops when every triangle of local smallest edge in τ_0 becomes good (smallest angle $\geq 30°$), and every remaining intermediate bad quality triangle t is processed or eliminated by edge swapping; and every intermediate almost good terminal triangle is improved by centroid insertion. This produces a good quality triangulation graded around the $PSLG$ geometry details.

Optimal Size Property This follows from the termination reasoning together with the fact that the average Lepp size tends to be two as the refinement proceeds.

□

Theorem 3 *The algorithm is order independent, where the mesh size is approximately the same by processing the bad triangles in arbitrary order.*

Proof The set of terminal edges introduces a mesh partition so that every triangle in the partition reaches the same terminal edge. □

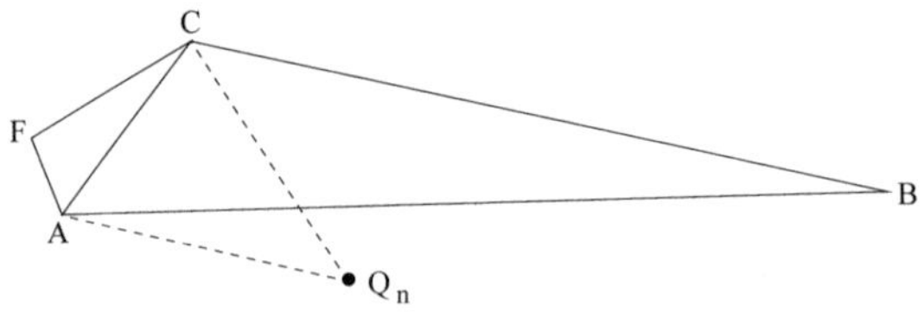

Fig. 12 Neighbor triangle ACF induces refinement of triangle AQ_nC to obtain a graded refined triangulation around edge FA

7 Empirical Study and Concluding Remarks

In Table 1 we compare our algorithm with results reported by Shewchuk [16] on Ruppert's algorithm (without the off-center preprocess of Üngor). Next we present results on the behavior of the Delaunay centroid algorithm for the six geometries of the Fig. 13. Table 2 includes final mesh sizes for $\theta_{tol} = 30°, 33°, 34°, 35°$ obtained with our algorithm. See the final triangulations for $\theta_{tol} = 30°$ for these examples in Fig. 13. Table 3 compares the number of triangles obtained with our software, with respect to those obtained with the current version of Triangle [15] which processes skinny and oversized triangles in order, and includes a boundary preprocess technique due to Üngor [5] to minimize the size of the final triangulation. A negative number means our software introduces less triangles than Triangle, while the $-\infty$ symbol means that Triangle does not converge.

Table 1 Algorithms comparison, key test case, $\theta_{tol} = 33°$

	Del centroid algorithm	Ruppert's algorithm [16]	
Triangle processing	Without order	Without order	Ordering triangles
Final mesh size	229	450	249

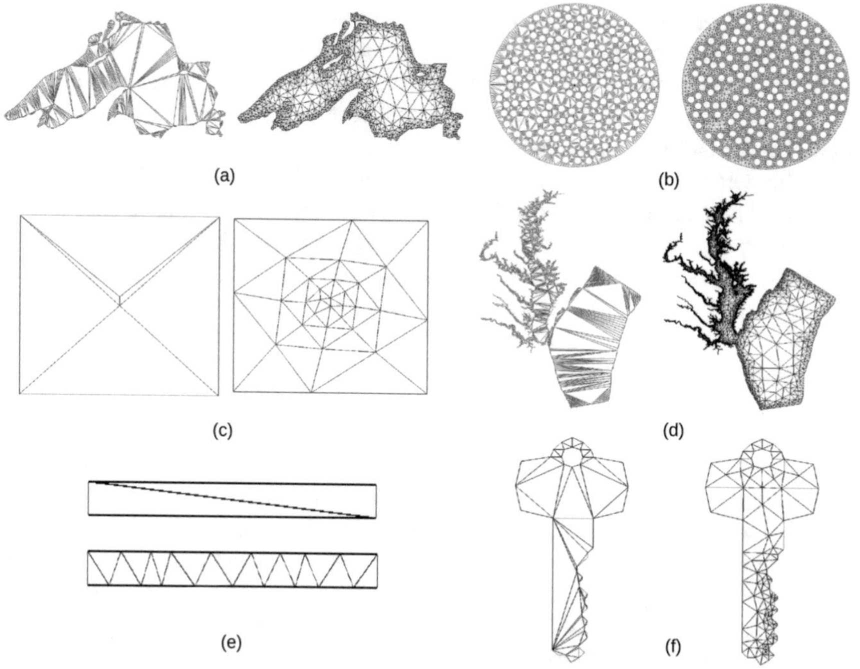

Fig. 13 Quality meshes for $\theta_{tol} = 30°$ (**a**) Superior lake shape; (**b**) Neuss shape; (**c**) Square with skinny triangles; (**d**) Chesapeake bay shape; (**e**) Long rectangle; (**f**) Key shape

Table 2 Mesh sizes for Delaunay centroid algorithm as a function of θ_{tol}

θ_{tol}	Superior lake	Neuss geometry	Square	Chesapeake Bay	Long rectangle	Key geometry
	Size (τ_0)					
	528	3070	9	14,262	2	54
	Size (τ_f)					
30	1835	8338	54	36,803	19	170
33	2273	9939	65	45,883	22	229
34	2512	11,054	70	52,027	25	262
35	3017	12,742	81	63,138	27	349

Table 3 Percentage of triangles added with respect to current version of Triangle[a]

θ_{tol}	Superior lake	Neuss geometry	Square	Chesapeake bay	Long rectangle	Key geometry
30	0.44	13.18	24.07	4.82	−15.79	22.94
33	−5,28	12,01	16.92	2.41	0.00	10.92
34	−5.29	−2.70	20.00	3.61	−68.00	−8.78
35	$-\infty$	$-\infty$	24.69	$-\infty$	−207.41	5.44

[a]Triangle processes skinny and oversized triangles in order and uses a boundary preprocess step

It should be noted that: (1) our results are not far from those obtained by the current optimized version of Triangle; (2) our software works properly until $\theta_{tol} = 35°$ for all the test cases, while Triangle fails for 50% of the test cases ($-\infty$ symbol) for $\theta_{tol} = 35°$; (3) Note that for the key test case and $\theta_{tol} = 33°$, our algorithm produces a final triangulation with 229 triangles against 450 triangles obtained with pure Ruppert algorithm (first-come first split bad quality triangle) and 249 triangles by always processing the worst existing triangle, as reported by Shewchuk [16].

Furthermore, for all the test cases, the average Lepp size is less than three from the beginning and quickly becomes less than 2.5, as the refinement proceeds. The algorithm is an easy to implement, order independent, robust method, suitable for use in adaptive finite element methods where good quality meshes are needed to assure convergence. With an adequate triangle data structure that keeps information on neighbor triangle, the refinement is of cost $O(N)$ where N is the number of points inserted.

In three dimensions, Balboa et al. [1] have introduced a simple and effective mesh improvement algorithm for tetrahedral meshes, which generalizes some of the ideas presented in this paper. Note that for any tetrahedron t, Lepp(t) corresponds to a submesh with several (more than two) terminal edges, and associated terminal stars. We call terminal star to a set of tetrahedra that share a common (terminal) largest edge in the mesh. Two new terminal star operations are alternatively used in the mesh improvement algorithm: the simple centroid insertion of the terminal star (that generalizes the simple centroid insertion of Sect. 5 in two dimensions), and swapping of the terminal edge as described by Freitag and Olliver-Gooch [8], but selectively applied to the terminal stars. The operation that most improves the mesh is performed whenever significant improvement is achieved. For more details see reference [1]

Acknowledgements Work partially supported by Departamento de Ciencias de la Computación, Universidad de Chile, Departamento de Sistemas de Información, Research Group GI150115/EF, and Research Project DIUBB 172115 4/R, Universidad del Bio-Bio.

References

1. F. Balboa, P. Rodriguez-Moreno, M.C. Rivara, Terminal star operations algorithm for tetrahedral mesh improvement, in *Proceedings 27th International Meshing Roundtable, Albuquerque, USA* (2018)
2. C. Bedregal, M.C. Rivara, New results on Lepp-Delaunay algorithm for quality triangulations. Procedia Eng. **124**, 317–329 (2014)
3. C. Bedregal, M.C. Rivara, Longest-edge algorithms for size-optimal refinement of triangulations. Comput. Aided Des. **46**, 246–251 (2014)
4. S.W. Cheng, T.K. Dey, J.R. Shewchuk, *Delaunay Mesh Generation*. CRC Computer and Information Science Series (CRC Press, Boca Raton, 2013)
5. H. Erten, A. Üngor, Quality triangulations with locally optimal Steiner points. SIAM J. Sci. Comput. **31**, 2103–2130 (2009)
6. D.A. Field, Laplatian smoothing and Delaunay triangulations. Commun. Appl. Numer. Methods **4**(6), 709–712 (1988)
7. L.A. Freitag, On combining Laplatian and optimization-based mesh smoothing techniques. ASME Appl. Mech. Div. Publ. AMD **220**, 37–44 (1997)
8. L.A. Freitag, C. Olliver-Gooch, Tetrahedral mesh improvement using swapping and smoothing. Int. J. Numer. Methods Eng. **40**, 3979–4002 (1997)
9. C. Gutierrez, F. Gutierrez, M.C. Rivara, Complexity of the bisection method. Theor. Comput. Sci. **382**(2), 131–138 (2007)
10. M.C. Rivara, New longest-edge algorithms for the refinement and/or improvement of unstructured triangulations. Int. J. Mumer. Methods Eng. **40**(18), 3313–3324 (1997)
11. M.C. Rivara, C. Calderon, Lepp terminal centroid method for quality triangulation. Comput. Aided Des. **42**(1), 58–66 (2010)
12. M.C. Rivara, N. Hitschfeld, R.B. Simpson, Terminal edges Delaunay (smallest angle based) algorithm for the quality triangulation problem. Comput. Aided Des. **33**(3), 263–277 (2001)
13. I.G. Rosenberg, F. Stenger, A lower bound on the angles of triangles constructed by bisecting the longest side. Math. Comput. **29**(130), 390–395 (1975)
14. J. Ruppert, A Delaunay refinement algorithm for quality 2-dimensional mesh generation. J. Algorithms **18**(3), 548–585 (1995)
15. J.R. Shewchuk, Triangle: engineering a 2D quality mesh generator and Delaunay triangulator, in *Applied Computational Geometry*, ed. by M.C. Lin, D. Manocha. Lecture Notes in Computer Science, vol. 1148 (1996), pp. 203–222
16. J.R. Shewchuk, Delaunay refinement algorithms for triangular mesh generation. Comput. Geom. **22**(1–3), 21–74 (2002)

17. R.B. Simpson, M.C. Rivara, Geometrical mesh improvement properties of Delaunay terminal edge refinement, Technical report CS-2006-16, *The David Cheriton School of Computer Science* (University of Waterloo, Waterloo, 2006), pp. 536–544
18. M. Stynes, On faster convergence of the bisection method for certain triangles. Math. Comput. **33**(146), 717–721 (1979)
19. M. Stynes, On faster convergence of the bisection method for all triangles. Math. Comput. **35**, 1195–1201 (1980)

Local Bisection for Conformal Refinement of Unstructured 4D Simplicial Meshes

Guillem Belda-Ferrín, Abel Gargallo-Peiró, and Xevi Roca

Abstract We present a conformal bisection procedure for local refinement of 4D unstructured simplicial meshes with bounded minimum shape quality. Specifically, we propose a recursive refine to conformity procedure in two stages, based on marking bisection edges on different priority levels and defining specific refinement templates. Two successive applications of the first stage ensure that any 4D unstructured mesh can be conformingly refined. In the second stage, the successive refinements lead to a cycle in the number of generated similarity classes and thus, we can ensure a bound over the minimum shape quality. In the examples, we check that after successive refinement the mesh quality does not degenerate. Moreover, we refine a 4D unstructured mesh and a space-time mesh (3D + 1D) representation of a moving object.

1 Introduction

In the last three decades refinement of 2D and 3D unstructured simplicial meshes [1–14], based on red/green refinement [1–7] and bisection [8–14], has been shown to be a key ingredient on efficient adaptive loops. Although one could expect the same in 4D, a case of special interest for space-time adaption, this line of research has not been extensively explored.

For our space-time applications, we are interested in conformal bisection methods since they are really well suited to implement fast geometrical multi-grid conformal solvers. Moreover, bisection methods have ensured either a maximum number of generated similarity classes [11–13] or a minimum lower quality bound over the generated elements after successive refinements [8–10, 14]. Regarding 4D refinement, only a non-conformal local refinement method for pentatopic meshes

G. Belda-Ferrín · A. Gargallo-Peiró (✉) · X. Roca
Computer Applications in Science and Engineering, Barcelona Supercomputing Center, Barcelona, Spain
e-mail: abel.gargallo@bsc.es

X. Roca, A. Loseille (eds.), *27th International Meshing Roundtable*, Lecture Notes in Computational Science and Engineering 127,
https://doi.org/10.1007/978-3-030-13992-6_13

has been proposed [15]. Unfortunately, existent conformal 4D (nD) bisection methods with a bound over the number of generated similarity classes [11, 12] cannot be applied to general unstructured meshes.

The main contribution of this work is to propose a local bisection procedure, with a bound over the number of generated similarity classes, for conformal refinement of 4D unstructured simplicial meshes. Specifically, we propose a recursive refine to conformity procedure, in two stages, based on marking bisection edges on different priority levels (Sect. 3.1). The marking procedure allows classifying the pentatopes in different types (Sect. 3.2) and hence, determining different refinement templates (Sect. 4), in an analogous manner to the 3D bisection method proposed in [13].

The refinement method is composed of two stages (Sect. 4). Two successive applications of the initial stage of the bisection strategy (Sect. 4.1), based on the proposed element classification, ensure that any initial 4D unstructured simplicial mesh can be conformingly refined. After the two initial refinements our recursive refine to conformity strategy switches to the second stage (Sect. 4.2). This final stage is analogous to Maubach's algorithm, when it is successively applied to a single pentatope. Therefore, we can ensure a bound over the number of generated similarity classes. Thus, the minimum quality of the refined mesh is bounded, independently of the number of performed refinements. The main advantage and difference of our method when compared to Maubach's algorithm [11] is the first stage of the method, which allows the application of the method to any 4D unstructured simplicial mesh.

In all the examples (Sect. 5), we show that the proposed methodology leads to a periodic evolution of the minimum element quality (shape quality measure [16]) illustrating the lower bound of the quality through successive refinement. We first illustrate how to check that an implementation of the proposed method is valid by successively refining a pentatope. With our implementation, we show that the proposed bisection technique can be used to refine general unstructured 4D meshes. Finally, we also illustrate our application of interest, the refinement of a 4D mesh corresponding to a space-time representation, with varying resolution, of the temporal evolution of a 3D moving object.

2 Preliminaries

In this section, we state some preliminary notions required for the rest of this work. First, we detail how a pentatope in four dimensions is represented in the 2D plots of this paper. Second, we state the definition of bisection and finally, we introduce the strategy used in this work to refine a given mesh through edge bisection.

The element type considered in this work is the *pentatope* (4D simplex) which is defined as the convex hull of a set of five points $\{x_0, x_1, x_2, x_3, x_4\}$ in $\mathbb{R}^4$. To represent a given pentatope (4D) in the plane (2D), we focus on a perspective where the edges connecting the vertices have a minimal number of edge crossings. Herein, the pentatope $[x_0x_1x_2x_3x_4]$ is displayed plotting the edges of the tetrahedron

$[x_0x_1x_2x_3]$ and the edges that connect the vertices of the tetrahedron $[x_0x_1x_2x_3]$ with the extra vertex x_4 located in the center of the tetrahedron, see Fig. 1a. This representation is used to display the edge marking procedure for bisection proposed in this work. The boundary of a pentatope is formed by five tetrahedra: the outer tetrahedron $[x_0x_1x_2x_3]$ and the four inner tetrahedra $[x_0x_1x_2x_4]$, $[x_0x_1x_4x_3]$, $[x_0x_2x_3x_4]$ and $[x_1x_2x_3x_4]$.

Once detailed the representation of a given pentatope, we particularize the definition of bisection to 4D simplicial elements. In particular, for a given pentatope σ with vertices $[x_0x_1x_2x_3x_4]$, the element vertices are reordered so that the refinement edge is $[x_0x_1]$. Let v be the midpoint of $[x_0x_1]$. The bisection of σ by $[x_0x_1]$ corresponds to removing the element $[x_0x_1x_2x_3x_4]$ and generating two new elements by joining v with the tetrahedral faces $[x_0x_2x_3x_4]$ and $[x_1x_2x_3x_4]$ (Fig. 2).

We highlight that the tetrahedral face $[vx_2x_4x_3]$ is shared between the two children. This shared face has three inherited edges ($[x_2x_3]$, $[x_2x_4]$ and $[x_3x_4]$) and three new edges ($[vx_2]$, $[vx_3]$ and $[vx_4]$). We denote the new edges of the shared face as *potential edges* of the initial element. These potential edges are displayed in Fig. 1b colored in red. This definition is required in Sect. 4.2 to characterize the proposed mesh refinement templates.

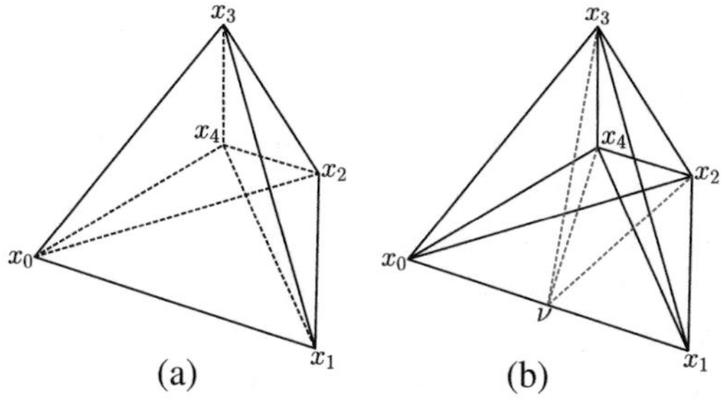

Fig. 1 (**a**) Three dimensional representation of a pentatope $[x_0x_1x_2x_3x_4]$, where the fifth vertex x_4 is plotted in $\mathbb{R}^3$ inside the tetrahedron $[x_0x_1x_2x_3]$. (**b**) Potential edges $[vx_2]$, $[vx_3]$ and $[vx_4]$ of the bisection of the pentatope $[x_0x_1x_2x_3x_4]$

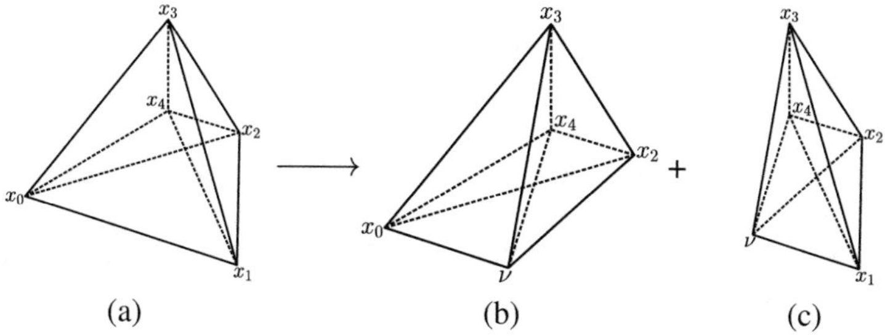

Fig. 2 Bisection of a pentatope (**a**) into two children (**b**) and (**c**)

Procedure 1 Refinement of a mesh ensuring conformity

Input: Marked mesh M
Output: Marked mesh M'
1: **function** RefineToConformity(M,S)
2: **if** $S \neq \emptyset$ **then**
3: $\bar{M} =$ `BisectPentatopes`(M, S)
4: $S = \{\sigma \in \bar{M} \mid \sigma \text{hasahangingnode}\}$
5: $M' =$ `RefineToConformity`$(\bar{M})$
6: **else**
7: $M' = M$
8: **end if**
9: **end function**

Finally, we introduce the algorithm proposed in this work to refine a given mesh by edge bisection. This algorithm uses a refine to conformity strategy similar to the 3D refinement method proposed in [13]. Given a marked mesh M and a set of elements to refine S, the mesh is refined according to Algorithm 1. In this algorithm, while there is not an empty set of elements to refine, Line 2, the mesh is refined as follows. In Line 3, the process `BisectPentatopes` bisects each pentatope in S:

$$\texttt{BisectPentatopes}(M, S) = (M \setminus S) \cup \bigcup_{\sigma \in S} \texttt{Bisect}(\sigma), \tag{1}$$

where `Bisect` performs the element bisection taking into account the element marks (refinement edge) and sets the proper marks to the two generated elements. In Sects. 3 and 4 we will present the marking procedures proposed in this work for pentatopic meshes, and the marks that are assigned to the two children.

Following, in Line 4 the set of elements to refine in the next step is set as the elements with hanging nodes. In Line 5 the Algorithm `RefineToConformity` is called recursively. These recursive calls are continued until there are no more elements with hanging nodes in the mesh. We show that the marking processes presented in Sects. 3 and 4 lead to a conformal mesh.

3 Edge Marking and Element Classification for Compatible Refinement

In this section, we first present in Sect. 3.1 an edge marking process compatible between neighboring elements for conformal mesh refinement. Next, in Sect. 3.2 we present a classification of the elements of the mesh depending on the marks assigned to their edges.

3.1 Edge Marking for Compatible Refinement

In this work, we use a marking procedure organized by levels to determine the priority of the bisection edges used during the element refinement. Following, we present a procedure to mark the edges of the pentatopes of a conformal mesh. These marks are devised to ensure that for a given face shared between two pentatopes, successive bisection of surrounding elements determines the same mesh from both sides of the shared face. Hence, this ensures mesh conformity along the bisection process. We define three levels of marks in a pentatope. The level 0 features one edge, which corresponds to the refinement edge of the current pentatope. The level 1 features two edges, which correspond to the refinement edges of the two children of the first pentatope. Finally, the level 2 features four edges, which correspond to the refinement edge of the four grandchildren of the original pentatope.

Herein, to determine the marks assigned to each of the edges of the element, we prioritize the edges in terms of their length with a well-defined tie-breaking rule. For a given element we define its *consistent bisection edge* as the edge of longest length and lowest global index. The lowest global index is a tie-breaking rule that ensures that if there exist multiples edges with the same length, we select as the longest edge always the same one, independently of the order in which the edges are compared. In particular, with this tie-breaking rule we ensure that the edges of a common face between two adjacent pentatopes are marked in the same manner from the two of them. We remark that the longest edge is considered in the *consistent bisection edge* sorting rule since it is an heuristic to enforce better element quality. Nevertheless, it is not a key ingredient to ensure conformal mesh refinement. For instance, just by sorting the mesh edges using their global index would also lead to a valid consistent sorting rule.

Following, we detail the marking process for a given pentatope $[x_0x_1x_2x_3x_4]$. The process consists of three steps, illustrated in Fig. 3b:

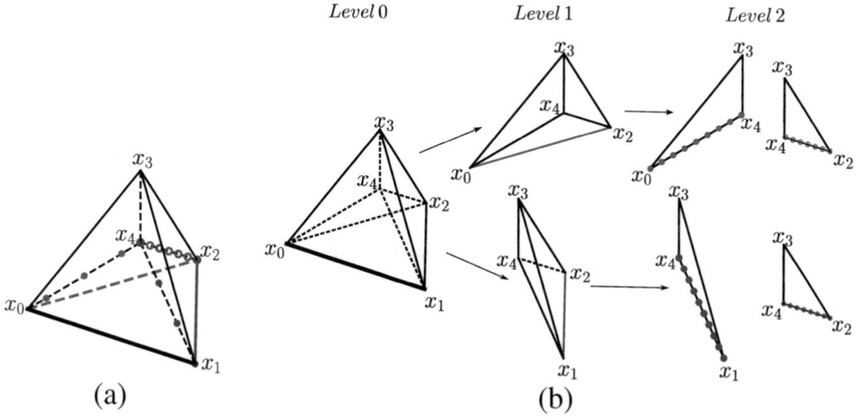

Fig. 3 (**a**) Marked pentatope and (**b**) marking diagram process at different levels

1. Marked edge of level 0: *consistent bisection edge* of the pentatope $[x_0x_1x_2x_3x_4]$. In the bisection process, the marked edge of level 0 corresponds to the bisection edge of the element. In this work, the marked edges of level 0 are plotted with a thick black line, see the first column of Fig. 3b.
2. Marked edges of level 1: the two marked edges of level 1 are determined as the *consistent bisection edge* of the tetrahedra defined by $[x_1x_2x_3x_4]$ and $[x_0x_2x_3x_4]$. These two tetrahedra are indeed the opposite tetrahedral faces of the pentatope with respect to x_0 and x_1, respectively. These two tetrahedral faces are the faces of the original pentatope preserved in each child. The two marked edges of level 1 correspond to the bisection edges of the two children of the current element. A particular configuration of the marked edges of level 1 is illustrated in the second column of Fig. 3b. The marked edges of level 1 associated to the first and second node of the bisection edge are colored in red and blue, respectively.
3. Marking edges of level 2: the four marked edges of level 2 are determined as the *consistent bisection edge* of the opposite faces of the marked edges of level 1 in the tetrahedra $[x_1x_2x_3x_4]$ and $[x_0x_2x_3x_4]$. The four marked edges of level 2 correspond to the bisection edges of the four grandchildren of the current marked element. In the third column of Fig. 3b, we illustrate a particular configuration of the marked edges of level 2, coloring them with the same color of the associated marked edge of level 1. In addition, the edge associated to the first node of the marked edge of level 1 is plotted with fully colored circles, and the other edge is plotted with empty circles.

Figure 3a illustrates the resulting marked element for the test example of the marking procedure of Fig. 3b. We highlight an edge, for instance $[x_2x_4]$ in Fig. 3a, can have two marks once all the marks are displayed on the initial pentatope. These two marks indicate that this edge has been marked from both of the faces that remain after bisection. To differentiate them, we have used blue and red colors. After bisecting a marked pentatope, the marked edges of the two children have to be determined.

Remark 1 (Inheritance of Marks) The marked edges of level 1 and 2 of the parent shift marks in the corresponding children and become the marked edges of level 0 and 1 of the two children, respectively. However, it is not straight-forward to determine the marked edges of level 2 from the parent marks. In Sect. 4 two methods are proposed to determine them.

3.2 Classification of Marked Pentatopes

In this section, we present a classification into different types of a pentatope resulting from the marking process detailed in Sect. 3.1. Several types of pentatopes are obtained depending on the marks assigned to their edges. Before detailing the classification, we introduce four definitions that state how the marked edges of level $l+1$ are located with respect to the associated marked edge of level l for $l = 0, 1$.

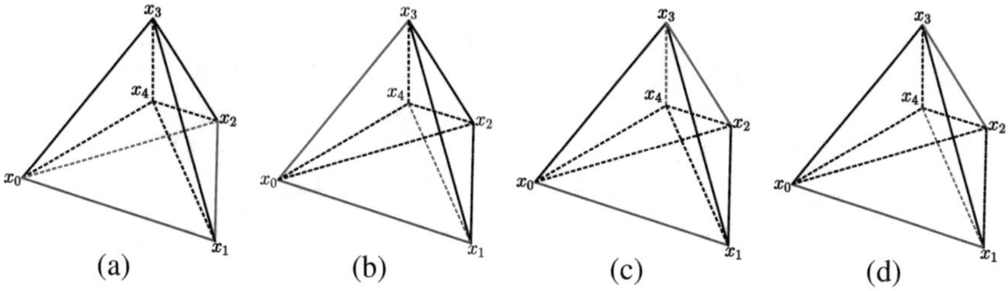

Fig. 4 Type of edge relations between the edges on level $l + 1$ (red) and the edge of level l (blue): (**a**) P, (**b**) A, (**c**) O, and (**d**) M

We propose a classification for different pentatope types, according to the configuration of the marked edges at the different levels. This classification is an extension of the different tetrahedron types proposed in [13], where only two levels of marked edges are required. Figure 4 illustrates the four different configurations between two levels of marked edges, coloring the two marked edges of level $l + 1$ with red color and the marked edge of level l with dark blue color:

- Type P (Planar): the two marked edges of level $l+1$ are coplanar with the marked edge of level l, i.e., the three edges are connected defining a triangle. In Fig. 4a an example of edges of type P is illustrated.
- Type A (Adjacent): each marked edge of level $l+1$ has a common vertex with the marked edge of level l but the two edges of level $l + 1$ do not have any common vertex. In Fig. 4b an example of edges of type A is illustrated.
- Type O (Opposite): the marked edges of level $l + 1$ of the opposite faces of the marked edge of level $l + 1$ do not intersect the marked edge of level l. In Fig. 4c an example of edges of type O is illustrated. We highlight that a possible configuration of edges of type O is that the two edges of level $l+1$ are overlapped. For instance, the edge $[x_2x_3]$ could be the marked edge of level $l + 1$ for the two faces opposite to the edge of level l.
- Type M (Mixed): the marked edges of level $l + 1$ of just one of the opposite faces have a common vertex with one marked edge of level l. In Fig. 4d an example of edges of type M is illustrated. We highlight that it is possible that the marked edges of level $l + 1$ have a common vertex between them. For example, the marked edges of level $l + 1$ could be $[x_1x_4]$ and $[x_4x_3]$.

Herein, in a pentatope we have marked edges of level 0, 1 and 2. We denote by α the edge type determined by how marked edges of level 1 are located with respect to the marked edge of level 0. Additionally, we denote by β and γ the edge relation type between the marked edges of level 2 and the marked edge of level 1. In this manner, a marked pentatope is classified into a type of the form $\alpha_{\beta\gamma}$.

In Fig. 5, we illustrate three different types of marked pentatopes. First, Fig. 5a illustrates a pentatope of type P_{PP}. In particular, the marked edge of level 0 (bisection edge) configures a triangular face together with the marked edges of level 1. Thus, the first index, α, of the element is P. Next, each marked edge of level

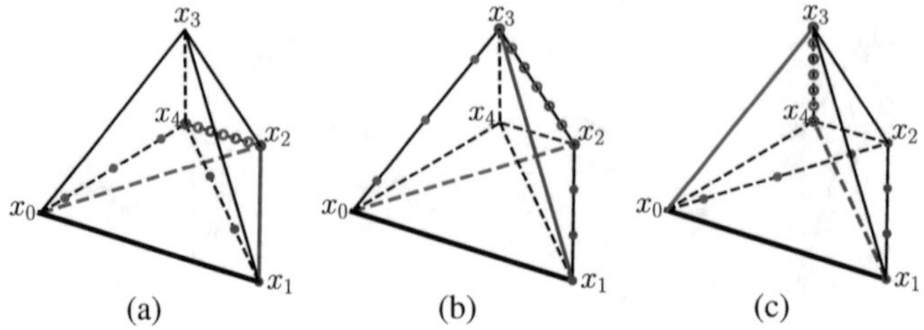

Fig. 5 Three different types of marked pentatopes: (**a**) P_{PP}, (**b**) A_{PP}, and (**c**) A_{AA}

1 defines also a triangular face with the corresponding marked edges of level 2, determining β and γ equal to P. Hence, the element is of type P_{PP}.

Analogously, for the element illustrated in Fig. 5b we detail the same process. For this element, α is A since the red and blue edges share a node with the bisection edge, but they do not share any node between them. In addition, β and γ are equal to P, since each of the blue and red edges determines a triangular face with the corresponding blue and red circled edges. Thus, this element is of type A_{PP}. Similarly, we can conclude that the element illustrated in Fig. 5c is of type A_{AA}.

After bisecting the element, the bisection edge $[x_0x_1]$ is split into two edges, $[x_0v]$ and $[vx_1]$, and thus this edge is not present in any of the children. However, the two adjacent tetrahedral faces to this edge are preserved. Specifically, the face $[x_0x_2x_3x_4]$ is inherited by the child that preserves node x_0 of the bisection edge, and the face $[x_1x_2x_3x_4]$ is inherited by the child that preserves node x_1. Following we detail which marks of the parent pentatope are preserved after its bisection and how these marks are inherited by the two children.

Remark 2 (Inheritance of Element Type) Hence, after bisecting a marked pentatope of type $\alpha_{\beta\gamma}$, where α, β, γ can be $\{P, A, O, M\}$, the obtained children inherit the marked edges of level 1 and 2 of the parent. These marks become the marked edges of level 0 and 1 of the children, see Remark 1. Thus, one child inherits the edge relation type β and the other child the relation type γ. However, the marked edges of level 2 are not determined. Depending on the edges that are selected to be the marked of level 2, the type of element of the children will be $\beta_{\beta_1\beta_2}$ and $\gamma_{\gamma_1\gamma_2}$, where $\beta_1, \beta_2, \gamma_1, \gamma_2 \in \{P, A, O, M\}$.

4 A Refinement Algorithm for 4D Unstructured Simplicial Meshes with Bounded Number of Similarity Classes

In this section, we detail a new procedure composed by two stages for refinement of any 4D unstructured simplicial mesh. Given a mesh, we first mark it using the procedure stated in Sect. 3.1, and following, we classify the elements into the different types stated in Sect. 3.2. Next, given a marked element to be refined, the

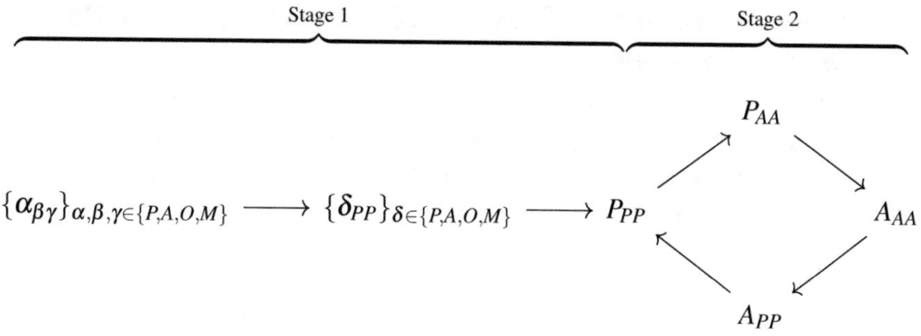

Fig. 6 Refinement process for a pentatope of type $\alpha_{\beta\gamma}$, where $\alpha, \beta, \gamma \in \{P, A, O, M\}$

Procedure 2 Element refinement with global mesh conformity

Input: Pentatope σ, set of marked edges m_σ, descendant level k.
Output: Pentatopes σ_1 and σ_2, set of marked edges m_{σ_1} and m_{σ_2}, descendant level k'

1: **function** BISECT(σ , m_σ, k)
2: $\sigma_1, \sigma_2 =$ `BisectPentatope`(σ, m_σ)
3: $m_{\sigma_1}, m_{\sigma_2} =$ `inheritMarksFromFather`(m_σ)
4: **if** $k < 2$ **then**
5: $m_{\sigma_1}, m_{\sigma_2} \leftarrow$set marked edges of level 2 using Stage 1 from Sect. 4.1
6: **else**
7: $m_{\sigma_1}, m_{\sigma_2} \leftarrow$set marked edges of level 2 using Stage 2 from Sect. 4.2
8: **end if**
9: $k' = k + 1$
10: **end function**

bisection of this element is performed according to Algorithm 2 and the diagram in Fig. 6. Algorithm 2 is used as bisection procedure in the `BisectPentatopes` function, Eq. (1), from the mesh refinement strategy `RefineToConformity` presented in Algorithm 1.

The two stages bisect a given marked element according to the bisection edge, Line 2 from Algorithm 2, and following, according to Remarks 1 and 2, the marked edges of level 0 and 1 of the children are determined from the marked edges of level 1 and 2 of the parent, Line 3. The difference between the two stages is the process to set the marked edges of the children. The marks determine the type of the generated element and at the same time, how the children will be bisected through successive refinement.

The first two times that the element is bisected (Stage 1), Line 4 of Algorithm 2, the marks of level 2 of the children are determined using Sect. 4.1. A child generated with one application of Stage 1 is of type δ_{PP}, being δ any edge relation type. No element enters to Stage 2 before two refinements, and once it is bisected twice in Stage 1, in Sect. 4.1 we show that it is of type P_{PP}. Then, from the second refinement and on, Line 7, Stage 2 is activated, see Sect. 4.2. In Sect. 4.3 the properties of the two-stage method are presented.

4.1 *Stage 1: Refinement from Any Unstructured Marked Mesh to P_{PP} Elements*

Stage 1 determines the marked edges of level 2 following the ideas of the marking strategy presented in Sect. 3.1. From the marking diagram presented in Fig. 3b we observe that two of the edges of the triangular face of the third column remain unmarked in the parent. After the element is bisected, these edges are still present in the tetrahedral faces of the children. These edges can be enforced to be the marked edges of level 2 of the children. This decision is consistent by construction between adjacent elements since it is performed on the face shared between these elements. This approach to determine the marked edges of level 2 of the children leads to a conformal refinement procedure.

Remark 3 (Refinement Towards P_{PP} Elements) Given an element of type $\alpha_{\beta\gamma}$ for $\alpha, \beta, \gamma \in \{P, A, O, M\}$, the application of two refinements of Stage 1 leads to elements of P_{PP}, see Fig. 6.
To show this, we first focus on the initial refinement step. From Remark 2 the children will be $\beta_{\beta_1\beta_2}$ and $\gamma_{\gamma_1\gamma_2}$, where $\beta_1, \beta_2, \gamma_1, \gamma_2 \in \{P, A, O, M\}$ depend on how the marked edges of level 2 are located with respect to the marked edges of level 1. By construction (see Fig. 3b), the marked edges of level 2 have been chosen on the same triangular face of the corresponding marked edge of level 1. Thus, the new marked edges of level 2 are coplanar with the marked edges of level 1 for each child and their edge relation is of type P. Hence, by setting these edges as marked edges of level 2 of the children, we obtain two children of type β_{PP} and γ_{PP}, respectively. Applying this marking strategy again, the grandchildren of the original pentatope are of type P_{PP}.

Although the marking process is consistent between adjacent elements by construction and the marks of level 2 are chosen consistently with the marking process, following we analyze all the possible neighboring configurations between two marked elements to illustrate that the stated bisection procedure is conformal.

Remark 4 (Conformal Refinement) Given two neighbor marked elements, when the shared face is bisected from the two sides, it is bisected by the same edge. That is, the interface between the children of the two elements is still conformal. We analyze three different configurations of the two elements:

- First, let us assume that both elements share a face that contains their *consistent bisection edge*. This edge must be the same for each one of the elements, since in particular, it is the *consistent bisection edge* of the face. Then, it is clear that they are refined by that edge and that the new interface is conformal.
- Second, let us assume that the shared face does not contain the *consistent bisection edge* in any of the two adjacent elements. Following the stated marking procedure, the shared tetrahedral face is marked in the second column of Fig. 3b, containing the marked edges of level 1. Thus, in the first refinement of the elements, the face is not refined and the interface is still conformal. Next, when

Procedure 3 Bisection of a simplex from Maubach [11]

Input: Tagged n-simplex σ .
Output: Tagged n-simplices σ_1 and σ_2.

function BISECTSIMPLEXMAUBACH(σ)
  Set $d' = \begin{cases} d-1, & d > 1 \\ n, & d = 1 \end{cases}$
  Create the new vertex $z = \frac{1}{2}(x_0 + x_d)$.
  Set $\sigma_1 = ((x_0, x_1, \ldots, x_{d-1}, z, x_{d+1}, \ldots, x_n),\ d')$.
  Set $\sigma_2 = ((x_1, x_2, \ldots, x_d, z, x_{d+1}, \ldots, x_n),\ d')$.
end function

we perform the second refinement, the shared face is refined by the same edge from the two elements, ensuring a conformal bisection.

- Finally, the third case to be analyzed is when the face contains the *consistent longest edge* of the pentatope in one element, but does not contain the *consistent longest edge* of the adjacent pentatope. After refining once the elements, the mesh is not conformal, since the face is bisected from one of the elements, but is not bisected from the other one. However, the element that has not bisected the initially shared face, does bisect it after the second refinement, since the *consistent longest edge* of the adjacent pentatope is specifically the *consistent longest edge* of the shared face, and thus it is marked in the level 1 of the second element. Hence, after two iterations the mesh is already conformal.

In addition, in the three different presented configurations, the marks determined on the children are always compatible by construction. Analogously, the same reasoning follows for the case where two pentatopes share a triangular face.

4.2 Stage 2: Conformal Refinement of All-P_{PP} Meshes

In this section, we present a conformal refinement algorithm with a bounded number of generated similarity classes for meshes composed uniquely by elements of type P_{PP}. This algorithm determines the second stage of the refinement method for any unstructured mesh presented in Sect. 4.

The procedure presented in this section is stated in terms of a cycle composed of four steps, presented in Fig. 6. In Fig. 7 the templates for the bisection and setting of the marked edges of the children are presented. Given an element of type P_{PP}, Fig. 7a, this element is split into two P_{AA} elements setting their marks using the templates presented in Fig. 7b, c. After that, the type P_{AA}, Fig. 7d, is bisected into two A_{AA} types applying the templates of Fig. 7e, f. Following, an element of type A_{AA}, Fig. 7g, is bisected into two A_{PP} using the templates presented in Fig. 7h, i. Finally, from the type A_{PP}, Fig. 7j, we obtain again two P_{PP} types applying the templates of Fig. 7k, l.

We highlight that in order to apply the templates of Fig. 7 we need to reorder the vertices of a given P_{PP} element to match the canonical representation of Fig. 7a. Similarly, the two children in Fig. 7b, c have to be reordered to obtain the canonical P_{AA} in Fig. 7d and then apply the corresponding templates. This node reordering has to be performed after each bisection to locate the marks in the canonical representation of the templated fathers. In addition, we highlight that in Figs. 7d, b, and c the marked edges of level 2 are assigned on potential edges (see definition in Sect. 2). Although those edges do not exist on the parent, they exist in the children and grandchildren, where they will be used to determine the bisection edge.

Next, in Remark 5 we detail that this templated refinement procedure is analogous to Maubach's algorithm [11] when applied successively to one element. Maubach's algorithm cannot be applied in general to any given unstructured mesh as detailed in [11, 13]. Thus, finally in Remark 6, we analyze the conformity of the application of our approach for meshes composed of P_{PP} elements.

Remark 5 (Analogy to Maubach's Algorithm) The refinement cycle in Fig. 6 performed using the templates presented in Fig. 7 is analogous to Maubach's algorithm [11] (see Algorithm 3) when applied to a single pentatope. This analogy is interpreted as follows. Given a pentatope to bisect using Maubach's algorithm with a tag d, we consider as marked edge of level 0 the tagged edge. Next, we consider as marked edges of level 1 the tagged edges of the two children in the next application of Maubach's algorithm. Analogously, we consider as marked edges of level 2 the tagged edges of the four grandchildren. Next, we find the permutation of the vertices $[x_0x_1x_2x_3x_4]$ to align the marks on the edges of the element with the canonical representation from Fig. 7. The obtained permutations are presented in Table 1.

Remark 6 (Conformal Refinement for All-P_{PP} Meshes) The refinement using Stage 2 of a marked mesh composed by elements of type P_{PP} leads to a conformal mesh. To illustrate the conformity of the refined mesh, we analyze two different cases:

- First, we analyze the case of the refinement of a single element. Since our method is analogous to Maubach's by Remark 5, it is also conformal when there is a single element successively refined, see details in [11, 13].
- Second, we analyze the conformity between the interface of adjacent elements of type P_{PP} with compatible marks. Extending the reasoning for tetrahedra in [13], it is sufficient to check if the bisection structure determined on a shared face is the same from both sides. Given a P_{PP} element to be refined, if we obtain the same refined mesh on all its tetrahedral faces we can ensure that the refinement of two adjacent P_{PP} is also conformal when using the `RefineToConformity` strategy. In particular, if we refine five times any of the five tetrahedral faces of a given P_{PP} the same tetrahedral mesh is obtained for all of them. This refined face mesh is illustrated in Fig. 8 and is composed by 32 tetrahedra. The same reasoning follows for the case where two pentatopes share a triangular face.

Hence, if a pentatopic mesh can be marked with all elements as P_{PP}, then it can be conformingly refined using our analogy to Maubach's algorithm combined with the

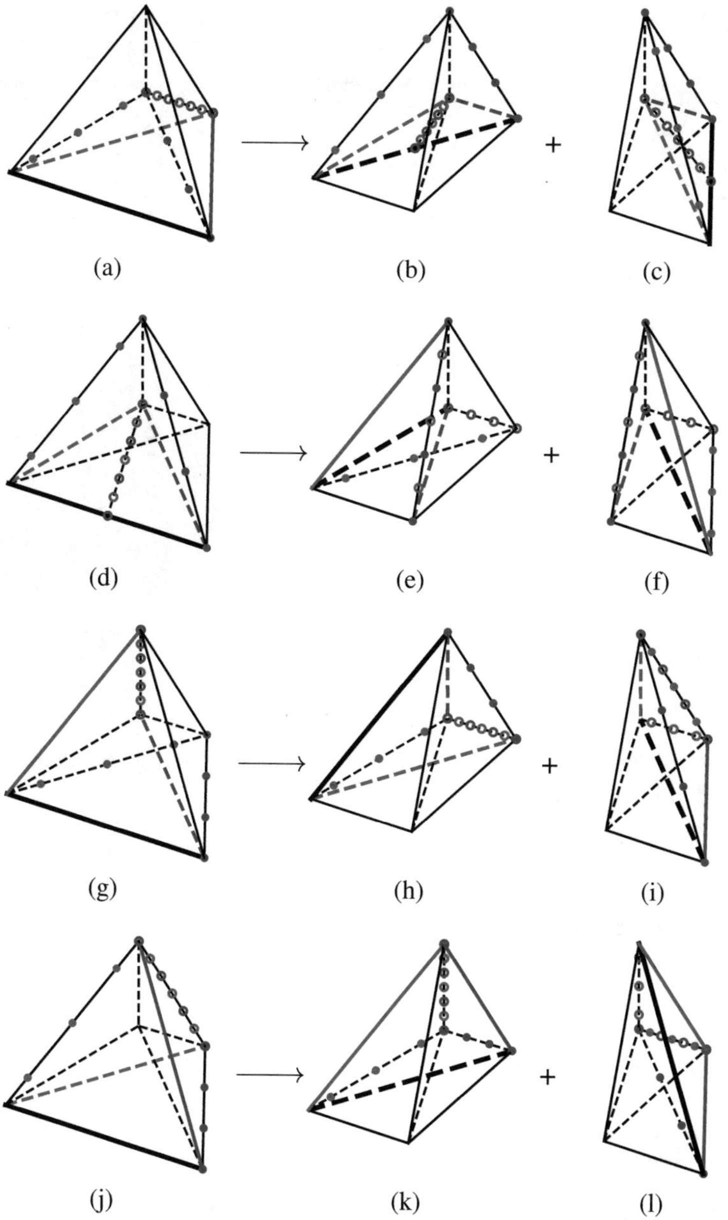

Fig. 7 Templates to perform the refinement cycle presented in Fig. 6. (**a**)–(**c**) An element of type P_{PP} is bisected into two P_{AA}. (**d**)–(**f**) P_{AA} is bisected into two A_{AA}. (**g**)–(**i**) A_{AA} is bisected into two A_{PP}. (**j**)–(**l**) A_{PP} is bisected into two P_{PP}

Table 1 Permutations from the Maubach Algorithm 3 to canonical types in Fig. 7

Canonical type	Tag in Algorithm 3	Permutation to obtain canonical representation
P_{PP}	$d = 2$	(0, 2, 1, 3, 4)
P_{AA}	$d = 1$	(0, 1, 2, 3, 4)
A_{AA}	$d = 4$	(0, 4, 2, 3, 1)
A_{PP}	$d = 3$	(0, 3, 2, 1, 4)

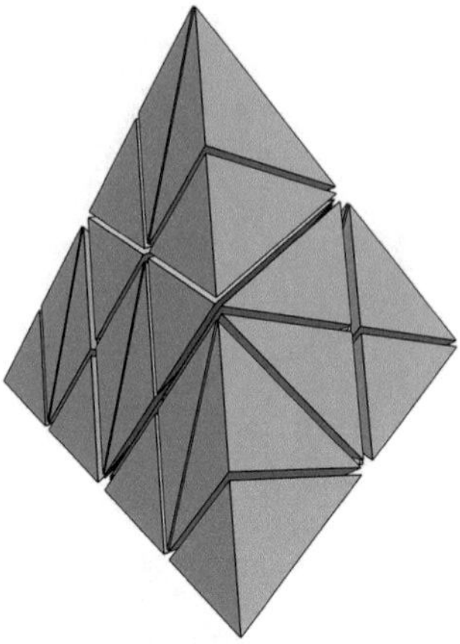

Fig. 8 Tetrahedral face of a pentatope of type P_{PP} after five refinements of the face

`RefineToConformity` strategy. This is the case when any given mesh is refined two times with Stage 1 in Sect. 4.1.

4.3 Properties of the Method

In this section, we analyze the two main properties of the refinement procedure determined by Algorithm 2. Our refinement procedure requires as input a conformal unstructured 4D simplicial mesh. Given a set of elements to refine, the resulting mesh is a locally refined unstructured 4D simplicial mesh that is conformal and has a bounded number of generated similarity classes. These properties are discussed in the following remarks.

Remark 7 (Conformal Refinement) The algorithm presented in Sect. 4 generates a conformal mesh. To show this, we take into account that this algorithm combines two refinement methods. The two first refinement steps in Stage 1 are performed by the algorithm presented in Sect. 4.1. After two refinements the elements are refined in Stage 2 according to the cycle in Fig. 6, see Sect. 4.2. In the worst case scenario, to prove conformal mesh refinement, all the elements of the initial mesh have to be twice refined at Stage 1. At this point, all the elements of the mesh are of type

P_{PP} with compatible marks, as detailed in Remark 4. Then, the conformity of the refinement is ensured by Remark 6.

Remark 8 (Bounded Number of Generated Similarity Classes) The number of similarity classes produced by the repeated application of the cycle presented in Fig. 6 to an element is bounded by 1536. To prove this bound, we take into account that in the refinement scheme of Fig. 6 it is required to perform two bisection steps before entering in the cycle. For each bisection, we generate at most two new similarity classes. Hence, from the given initial element, the bound of the similarity classes after the two first steps is $2 \cdot 2 = 4$. As highlighted in Remark 5 from Sect. 4.2, this second stage is analogous to Maubach's algorithm when applied to a single pentatope. In [13] it is proved that in 4D Maubach's algorithm has a sharp bound of 384 generated similarity classes for an element. Thus, the bound for the procedure of Fig. 6 is $4 \cdot 384$, that is 1536.

5 Results

In this section, we present several results to illustrate the features and the applicability of the presented refinement scheme. In all the examples, we plot the minimum and maximum shape quality [16] in each refinement step of Algorithm 1. To visualize the results we intersect each 4D mesh with a hyperplane to obtain a 3D cut that can be visualized. In Sect. 5.1, we refine an equilateral pentatope with two different initial marking configurations to illustrate that the similarity classes are bounded. In Sect. 5.2, we refine an unstructured 4D mesh to capture a hypersphere and, finally, in Sect. 5.3 we refine a simplicial mesh on a hypercube to capture a moving sphere.We highlight that in all the presented examples it has been explicitly checked that the generated meshes are conformal after the applied `RefineToConformity` strategy by checking that the only boundary faces of the mesh are on the boundary of the domain.

5.1 *Bounded Quality: Iterative Refinement of One Pentatope*

In this example, we check that our implementation of the refinement algorithm does not lead to degenerated elements after successive refinement of a given pentatope type. We enforce an equilateral pentatope to be marked as P_{AA} and a second equilateral pentatope to be marked as A_{PO}. Then, both pentatope types are globally refined 20 times. Figure 9 shows the minimum and maximum element quality at each refinement step. We observe that the minimum quality (vertical axis) decreases on the first refinement steps (horizontal axis) until a minimum value is reached. Then, the minimum and maximum qualities start to cycle every four refinement steps. This is an indicator of the bound of the number of generated similarity classes.

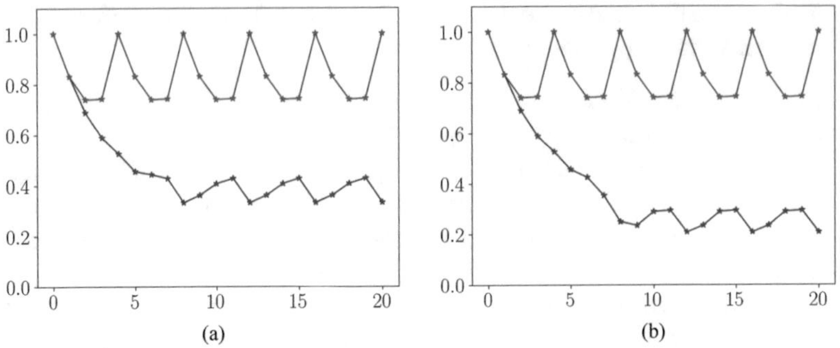

Fig. 9 Quality versus the number of iterations of the `RefineToConformity` algorithm applied to an equilateral pentatope marked as (**a**) P_{AA} and (**b**) A_{PO} type. The blue (red) line corresponds to the minimum (maximum) of the element shape quality at each iteration

5.2 4D Unstructured Mesh: Refining an Extruded Sphere Octant

This example shows that the proposed refinement scheme can be applied to unstructured 4D pentatopic meshes. To this end, we generate an unstructured 4D mesh of a 3D sphere octant, of radius 1 and centered in the origin, extruded one unit along the fourth dimension. Then, we successively refine those elements that intersect a hypersphere of radius 1/4 and centered in the origin. To generate the 4D mesh, we first generate an unstructured 3D mesh of the sphere octant composed by 40 nodes and 95 elements, see Fig. 10a. Then, we embed two copies of the 3D mesh points in the 4D space by setting the fourth coordinate to 0 and 1, respectively. Finally, these 80 points are reconnected, using the implementation of the Delaunay algorithm provided by QHull [17], to obtain the unstructured 4D mesh.

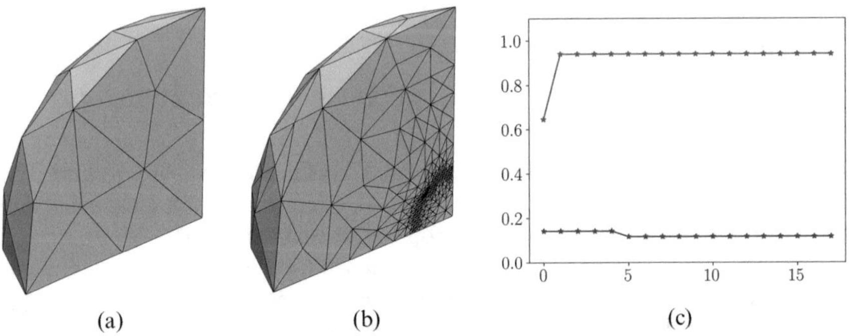

Fig. 10 A slice with the hyperplane $t = 0$ of the 4D simplicial mesh is illustrated for (**a**) the initial configuration and (**b**) after 17 iterations of `RefineToConformity`. (**c**) Minimum (blue) and maximum (red) element quality for each refinement step

After applying 17 times the `RefineToConformity` algorithm, we obtain a 4D mesh composed by 8072909 elements and 433887 nodes. Figure 10b, shows the tetrahedral mesh that corresponds to the boundary of the 4D pentatopic mesh at the base of the extrusion along the fourth dimension. Figure 10c shows the quality at each refinement step, where we can observe a lower quality bound is constant at value 0.11.

5.3 Space-Time Mesh: Refining a Sphere Moving Along the z-Axis

Finally, we illustrate our application of interest, the refinement of a 4D mesh corresponding to a space-time representation, with varying resolution, of the temporal evolution of a 3D moving object. We consider a sphere of radius $1/5$ centered in the origin that moves along the z-axis from 0 to 1 with constant velocity 1. We generate an initial mesh on the hypercube $[0, 1]^4$ composed by 24 pentatopes using Freudenthal-Kuhn algorithm [1–3]. Next, we apply 25 times the algorithm `RefineToConformity` to refine those elements that intersect the 4D sphere extrusion that represents the moving sphere. The final 4D mesh is composed by 5,233,296 pentatopes and 251,457 nodes and it is illustrated in Fig. 11. Figure 11a–c shows three slices of the mesh at $t = 0$, $t = 1/2$ and $t = 1$, respectively. We can observe that each one of the slices on t shows different positions of the moving sphere, from the initial point (0,0,0) at $t = 0$ to the final point $(0, 0, 1)$ at $t = 1$. In contrast with these three slices, in Fig. 11d we show an slice of the mesh at $x = 0$. In the closest quadrilateral face of Fig. 11d we observe the path of the sphere on the surface of dimension 2 defined by the axis z and t at $x = y = 0$. In this quadrilateral face, we can see that the center of the sphere describes a straight line going from the lower left corner $(0, 0, 0, 0)$ up to the top right corner $(0, 0, 1, 1)$. This is so since the sphere goes from $z = 0$ to $z = 1$ with constant velocity starting at $t = 0$ and finalizing at $t = 1$. Specifically, the location on the z-axis of the sphere is $z = t$.

6 Concluding Remarks

In this work, we have presented a new refinement method via edge bisection for 4D pentatopic meshes. This method ensures that the mesh quality does not degenerate after successive refinements of a given element. To develop this method, we require to classify the elements of the mesh into different types in a similar fashion to [13]. Using the pentatope classification we provide four refinement templates to perform a cyclic bisection analogous to Maubach's method [11]. Combining two initializing refinements (Stage 1) with this templated refinement (Stage 2) we obtain a refinement strategy that can be applied to any given pentatopic mesh. Using this

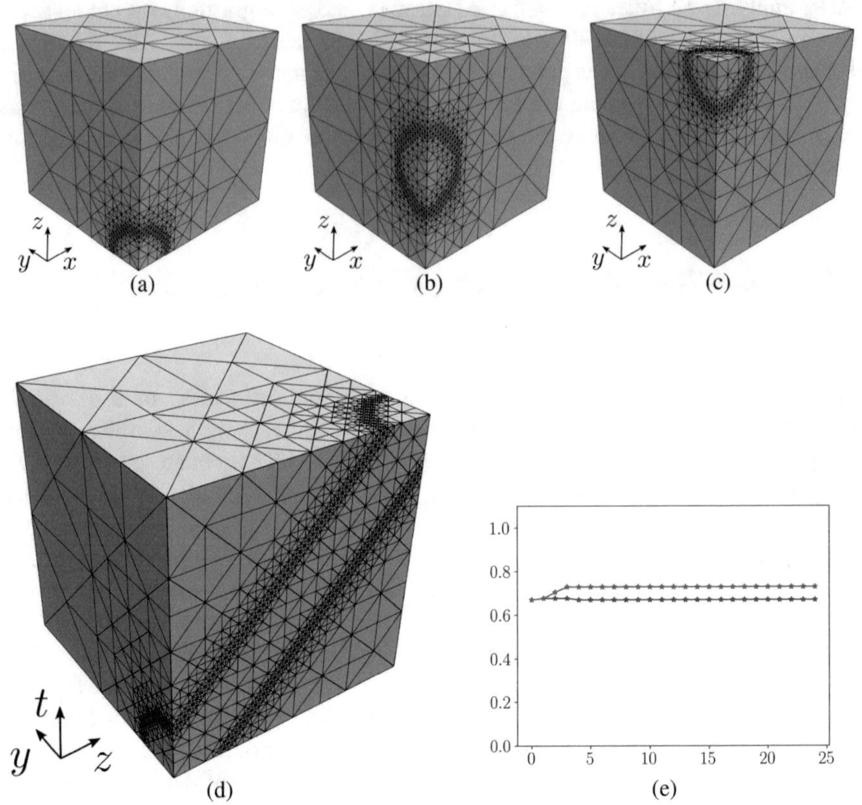

Fig. 11 Slice of the 4D simplicial mesh of the hypercube with the hyperplane: (**a**) $t = 0$, (**b**) $t = 0.5$, (**c**) $t = 1$ and (**d**) $x = 0$. (**e**) Minimum (blue) and maximum (red) element quality for each refinement step

method a finite number of similarity classes are generated when a given element is refined.

We apply the refinement scheme to different meshes to illustrate its features. First, we analyze that the mesh quality of the refinement of different element types does not degenerate. Second, we illustrate the applicability of the technique to refine unstructured 4D simplicial meshes. Finally, we analyze a space-time configuration of a sphere moving along an axis.

Acknowledgements This project has received funding from the European Research Council (ERC) under the European Union's Horizon 2020 research and innovation programme under grant agreement No 715546. This work has also received funding from the Generalitat de Catalunya under grant number 2017 SGR 1731. The work of X. Roca has been partially supported by the Spanish Ministerio de Economía y Competitividad under the personal grant agreement RYC-2015-01633.

References

1. H. Freudenthal, Simplizialzerlegungen von beschrankter flachheit. Ann. Math. **43**(3), 580–582 (1942)
2. H. Kuhn, Some combinatorial lemmas in topology. IBM J. Res. Dev. **4**(5), 518–524 (1960)
3. J. Bey, Simplicial grid refinement: on Freudenthal's algorithm and the optimal number of congruence classes. Numer. Math. **85**(1), 1–29 (2000)
4. R. Bank, A. Sherman, A. Weiser, Some refinement algorithms and data structures for regular local mesh refinement. Sci. Comput. **1**, 3–17 (1983)
5. J. Bey, Tetrahedral grid refinement. Computing **55**(4), 355–378 (1995)
6. A. Liu, B. Joe, Quality local refinement of tetrahedral meshes based on 8-subtetrahedron subdivision. Math. Comput. **65**(215), 1183–1200 (1996)
7. S. Zhang, Successive subdivisions of tetrahedra and multigrid methods on tetrahedral meshes. Houston J. Math. **21**(3), 541–556 (1995)
8. M.C. Rivara, Algorithms for refining triangular grids suitable for adaptive and multigrid techniques. Int. J. Numer. Methods Eng. **20**(4), 745–756 (1984)
9. E. Bänsch, Local mesh refinement in 2 and 3 dimensions. IMPACT Comput. Sci. Eng. **3**(3), 181–191 (1991)
10. A. Liu, B. Joe, Quality local refinement of tetrahedral meshes based on bisection. SIAM J. Sci. Comput. **16**(6), 1269–1291 (1995)
11. J. Maubach, Local bisection refinement for n-simplicial grids generated by reflection. SIAM J. Sci. Comput. **16**(1), 210–227 (1995)
12. C. Traxler, An algorithm for adaptive mesh refinement in *n* dimensions. Computing **59**(2), 115–137 (1997)
13. D. Arnold, A. Mukherjee, L. Pouly, Locally adapted tetrahedral meshes using bisection. SIAM J. Sci. Comput. **22**(2), 431–448 (2000)
14. A. Plaza, M.C. Rivara, Mesh refinement based on the 8-tetrahedra longest-edge partition, in IMR, 2003, pp. 67–78
15. M. Neumüller, O. Steinbach, A flexible space-time discontinuous Galerkin method for parabolic initial boundary value problems. Berichte aus dem Institut für Numerische Mathematik **2**, 1–33 (2011)
16. P.M. Knupp, Algebraic mesh quality metrics. SIAM J. Numer. Anal. **23**(1), 193–218 (2001)
17. C. Barber, D. Dobkin, H. Huhdanpaa, The quickhull algorithm for convex hulls. ACM Trans. Math. Softw. **22**(4), 469–483 (1996)

A Construction of Anisotropic Meshes Based on Quasi-Conformal Mapping

Yuxue Ren, Na Lei, Hang Si, and Xianfeng David Gu

Abstract We propose a novel method which is able to generate planar anisotropic meshes according to a given metric tensor. It is different from the classical metric-based or high dimensional embedding mesh adaptation methods. Our method resolves the anisotropy of a metric tensor field by finding a corresponding Euclidean metric in the plane. This is achieved via quasi-conformal mapping between two Riemannian surfaces. Given a planar source domain together with a metric tensor defined on it, and a target domain with a Euclidean metric, there exists a quasi-conformal mapping between them, such that the mapping is conformal with respect to the metric tensor on the source and the Euclidean metric on the target. A discrete quasi-conformal mapping can be constructed by solving the Beltrami equation on a Riemannian manifold. Our method first computes the Beltrami coefficient which is a complex-valued function from the given metric tensor. It then uses discrete Yamabe flow to construct this quasi-conformal mapping. We then construct an isotropic triangulation on the target domain. The constructed mesh is mapped back to the source domain by the inverse of the quasi-conformal

Y. Ren
School of Mathematics, Jilin University, Changchun, China

Beijing Advanced Innovation Center for Imaging Technology, Capital Normal University, Beijing, China

N. Lei (✉)
DUT-RU ISE, Dalian University of Technology, Dalian, China

Beijing Advanced Innovation Center for Imaging Technology, Capital Normal University, Beijing, China
e-mail: nalei@dlut.edu.cn

H. Si
Weierstrass Institute for Applied Analysis and Stochastics, Berlin, Germany
e-mail: hang.si@wias-berlin.de

X. D. Gu
Department of Computer Science, Stony Brook University, Stony Brook, NY, USA
e-mail: gu@cs.stonybrook.edu

X. Roca, A. Loseille (eds.), *27th International Meshing Roundtable*, Lecture Notes in Computational Science and Engineering 127,
https://doi.org/10.1007/978-3-030-13992-6_14

mapping to obtain an anisotropic mesh of the original domain. This method has solid theoretical foundation. It guarantees the correctness for all symmetric positive definite metric tensors. We show experimental results on function interpolation problems to illustrate both of the features and limitations of this method.

1 Introduction

Many physical problems exhibit anisotropic features, i.e., their solutions change more significantly in one direction than others. Examples include in particular convection-dominated problems whose solutions have, e.g., layers, shocks, or corner and edge singularities. Anisotropic meshes have great importance in numerical methods to solve partial differential equations. They improve the accuracy of the solution and decrease the computational cost.

Anisotropy denotes the way distances and angles are distorted. It is naturally related to approximation theory and is important in function interpolation [5, 11, 21, 22]. For example, it has been shown that for a smooth function the anisotropy is best characterized by the Hessian of that function. In practice, a central question is how to efficiently distinguish the anisotropy of a given problem. Another important question is how to characterize the anisotropy in a such a way that an optimal mesh for a given problem can be defined. These are all difficult questions and are active research subjects.

It is well-understood that anisotropic features can be represented by a metric tenser $\mathcal{M}$ defined on the target space $\Omega \subset \mathbb{R}^d$, where the metric tenser of each vertex is a $d \times d$ symmetric positive definite matrix. $\mathcal{M}$ defines a Riemannian metric on Ω, both lengths and angles can be re-defined according to this metric. This allows the use of classical isotropic mesh adaptation techniques to produce anisotropic meshes. It is one of the major approaches for producing anisotropic meshes, see [1, 3, 9, 12, 16–18, 20, 23]. Although these methods are very successful in practice, there is no theoretical proof that the generated anisotropic meshes are appropriate or good according to the input metric.

Variational mesh adaptation is another useful technique to generate adapted meshes. It is based on the optimisation of a mesh related functional to achieve the best adapted meshes. Such methods are centroidal Voronoi tessellations (CVTs) [10, 15, 24], optimal Delaunay triangulations (ODTs) [4], and monitor functions [13]. Many of these methods are generalised to produce anisotropic meshes by incorporating a metric tensor into the functional. Again, there is no theoretical guarantees on the success of these methods with arbitrary anisotropic metric tensors.

A recent anisotropic meshing technique is through higher dimensional embedding [2, 7, 14, 26]. Instead of using metric tensors, it increases the dimensions to resolves the anisotropy such that it can be treated isotropic in this high dimensional space. The co-dimensions can be flexibly chosen to emphasis the interested quantities. By using the normal component of the surface, this approach can produce curvature-adapted anisotropic surface meshes [6, 14]. By using the gradient of a function, this approach produces well adapted meshes to interpolate anisotropic functions [8]. However, it is not clear how to choose the co-dimensional coordinates for a metric tensor.

In this paper, we propose a novel method to construct anisotropic meshes in the plane. Assume an anisotropic metric tensor M is given on a planar domain Ω, our goal is to construct an anisotropic mesh with respect to M on Ω. Our method is based on the theory of quasi-conformal mapping. The construction of an anisotropic mesh is achieved by building a quasi-conformal mapping φ from Ω to D, where D is a target domain with Euclidean metric. The Beltrami coefficient of the mapping μ_φ is determined by the metric tensor M. Then an isotropic mesh is calculated on D, and pulled back to Ω by φ. This results an anisotropic mesh on Ω. The quasi-conformal mapping φ is achieved by solving the Beltrami equation using the discrete Yamabe flow method. If the metric tensor is symmetric and positive definite, this method guarantees the success of finding a quasi-conformal mapping.

We tested our method using an application of interpolation of anisotropic functions. Our experiments on some published examples showed that this method is able to effectively reduce the interpolation error (measured in L^2 norm) compared with the uniform meshes. Furthermore, the error is consistently reduced with respect to the increase of number of points.

We conducted preliminary comparisons of our results with the results produced by two public codes. The first one is `BAMG`,[1] which is a metric-based anisotropic mesh generator. Another is `Detri2`, developed by the third author.[2] It implements the high dimensional embedding mesh adaptation method [8]. First of all, it is noted that both codes produced high quality anisotropic meshes. Their L^2 interpolation errors are about two orders of magnitude smaller than ours. This shows a strong limitation of our method. Our method has a limitation on the input point set. It does not as flexible as the classical mesh adaptation methods. On the other hand, our method could be seen as an effective anisotropic mesh smoothing step compared with the smoothing algorithms used in `BAMG` and `Detri2`.

The structure of this paper is as follows. Section 2 explains that given a field of metric tensor, how to compute an quasi conformal mapping. Section 3 introduces the theorem of discrete Yamabe flow. All the algorithms can be found in Sect. 4. Section 5 shows the results of experiments.

2 Quasi Conformal Mapping

A quasi conformal mapping is a homeomorphism between plane domains which to first order takes small ellipses of bounded eccentricity to small circles, see Fig. 1 for an example.

[1] Available in `FreeFEM++` (http://www.freefem.org).

[2] http://www.wias-berlin.de/people/si/detri2.html.

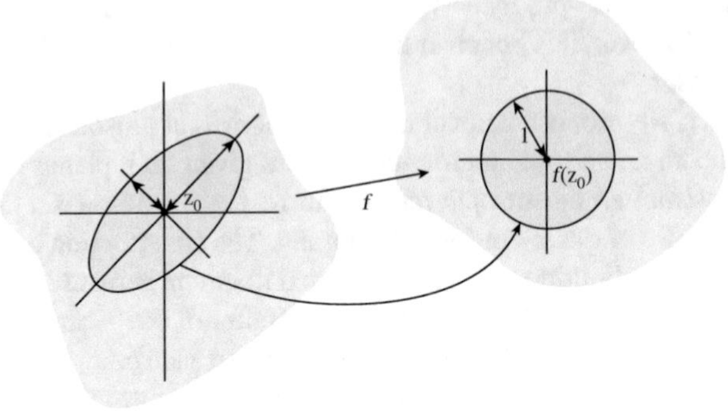

Fig. 1 A quasi-conformal mapping transforms an ellipse into a circle with a bounded eccentricity

Suppose (S_1, g_1) and (S_2, g_2) are Riemannian surfaces, $\{(U_\alpha, z_\alpha)\}, \{(V_\beta, w_\beta)\}$ are their atlas which are compatible with their Riemannian metric g_1, g_2 respectively, then on each local chart (U_α, z_α), $g_1 = e^{2\lambda_1} dz_\alpha d\bar{z}_\alpha$, on each local chart (V_β, w_β), $g_2 = e^{2\lambda_2} dw_\beta d\bar{w}_\beta$. Let $f : (S_1, g_1) \to (S_2, g_2)$ be a diffeomorphism, for each vertex $v \in (S_1, g_1)$, locally f maps a chart $(U, z(v))$ to a chart $(f(U), w(f(v)))$, where U is a neighbourhood of v, then the pull back metric of f near v can be defined as

$$f^* g_2 = e^{2\lambda_2} |w_z dz + w_{\bar{z}} d\bar{z}|^2 = e^{2\lambda_2} |w_z|^2 |dz + \mu d\bar{z}|^2 \tag{1}$$

where $w_z = \frac{\partial w}{\partial z}$, $w_{\bar{z}} = \frac{\partial w}{\partial \bar{z}}$, $\mu = \frac{w_{\bar{z}}}{w_z}$. f is in fact an isometric mapping when it defined on the manifold $(S_1, f^* g_2)$. So for any metric $g_1' = e^{2\lambda_3} |dz + \mu d\bar{z}|^2$ shares the same conformal structure with $f^* g_2$, f is a conformal mapping from (S_1, g_1') to (S_2, g_2), where $\lambda_3 : S_1 \to \mathbb{R}$ is a scalar function defined on S_1.

When μ is bounded, f is called a quasi conformal mapping. Quasi conformal mapping is a generalization of conformal mapping, it is a solution to the Beltrami equation

$$\frac{\partial f}{\partial \bar{z}} = \mu(z) \frac{\partial f}{\partial z}, \ |\mu(z)| \le k. \tag{2}$$

where $\mu(z)$ is called the complex dilatation Beltrami coefficients, describes the distortion of f.

Suppose S_1 be a simply connected domain in $\mathbb{C}$, ∂S_1 has more than one point, $g_1 = dz d\bar{z}$. If $f_0 : (S_1, g_1) \to (S_1, g_1')$ is a quasi conformal mapping, then $f \circ f_0^{-1} : (S_1, g_1') \to (S_2, g_2)$ is a conformal mapping. This provides us a novel way to convert the problem of producing anisotropic mesh to that of producing isotropic mesh via changing the metric. Let $\mathscr{T}$ be a isotropic triangulation of S_1 under metric g_1', then $\mathscr{T}$ will become an anisotropic mesh if we change the metric to be g_1, and it is obvious that the Beltrami coefficients of f_0 convey the anisotropic feature of $\mathscr{T}$ (Fig. 2).

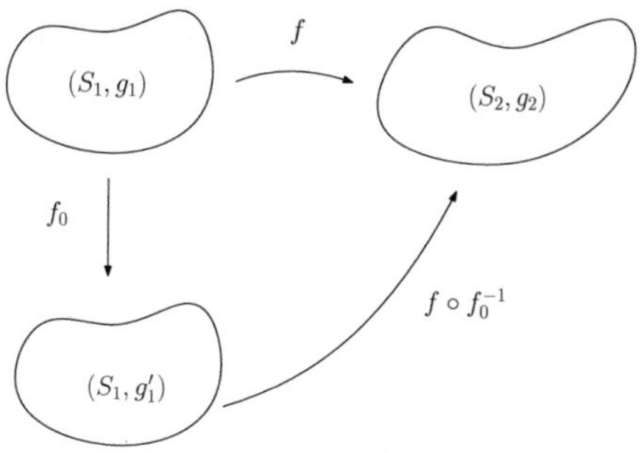

Fig. 2 Convert Quasi-conformal mapping problem to conformal mapping problem

Most of the classical methods focus on the construction of isotropic mesh on (S_1, g_1'), while we provide a new idea, that is, the computation of the quasi conformal mapping f_0, the theorem of discrete Yamabe flow guarantees the existence and uniqueness of f_0. Since $f \circ f_0^{-1}$ is a conformal mapping, f, f_0 shares the same Beltrami coefficients, so we can compute f_0 by solving the Beltrami equation with f's Beltrami coefficients.

2.1 Computation of Jacobian Matrix Based on Metric Tensors

In our problem, S_1 is a simply connected domain in $\mathbb{C}$, the anisotropic feature is represented by a field of metric tensors $\mathcal{M} := \{M(v); v \in S_1\}$, in which $M(v)$s are 2×2 symmetric positive definite matrixes for all $v \in S_1$. Let g_3 be the Riemannian metric defined by $\mathcal{M}$, then for an open curve $c \subset (S_1, g_3)$, the length of c is $\|c\|_{g_3} = \int_{t=0}^{1} \sqrt{v(t)^T M(c(t))v(t)}$, in which $v(t) = \frac{dc(t)}{dt}$. Let $z = x + iy$ be a parameterization of (S_1, g_1), such that $g_1 = dzd\bar{z}$, define $f_0(v) = v, \forall v \in S_1$, then the pull back metric of f_0 can be described as

$$f_0^* g_3 = (dx, dy)M(z)\begin{pmatrix} dx \\ dy \end{pmatrix}.$$

Let $\phi(z) = w$ is the homeomorphism onto (S_1, g_1) such that $f_0 \circ \phi^{-1}$ is conformal, denote $w = u + iv$, then we have

$$\begin{pmatrix} du \\ dv \end{pmatrix} = J(\phi(z))\begin{pmatrix} dx \\ dy \end{pmatrix},$$

in which $J(\phi)$ is the Jacobi matrix of ϕ. Then the pull back metric of $f_0 \circ \phi^{-1}$ with respect to w is

$$(f_0 \circ \phi^{-1})^* g_3 = (du, dv)(J(\phi(z))^{-1})^T M(z) J(\phi(z))^{-1}\begin{pmatrix} du \\ dv \end{pmatrix}.$$

When $f_0 \circ \phi^{-1}$ is conformal, the pull back metric satisfies $(f_0 \circ \phi^{-1})^* g_3 = c dw d\bar{w}$, so $\mathscr{M}, \phi$ satisfies

$$M(z) = cJ(\phi(z))^T J(\phi(z)),$$

in which c is a positive constant.

Since $M(z)$ is a symmetric positive definite matrix for all z, there always exists a matrix N satisfying $M(z) = N(z)^T N(z)$, $N(z)$ can be computed through $M(z)$'s orthogonal decomposition $M(z) = P(z)^T \Lambda(z) P(z)$. $\Lambda(z)$ is a diagonal matrix, its diagonal elements $\lambda_1 > \lambda_2 > 0$ are characteristic values of $M(z)$, P is a 2×2 orthogonal matrix, it can be denoted as

$$\begin{pmatrix} \cos\theta & -\sin\theta \\ \sin\theta & \cos\theta \end{pmatrix} \tag{3}$$

which is a rotation matrix with degree clockwise θ. Its row vectors of $P(z)$ are characteristic vectors of $M(z)$. let $N(z) = \Lambda^{1/2}(z)P(z)$, then $M(z) = N(z)^T N(z)$. So we have $J(\phi(z)) = cN(z)$. $\phi(z)$'s Beltrami coefficients can be computed based on $N(z)$.

2.2 Compute Beltrami Coefficients Based on Jacobian Matrix

Actually, ϕ maps an infinitesimal circle to an infinitesimal ellipse, whose long and short axis' lengthes are characteristic values of $J(\phi(z))$, and the direction of them are $J(\phi(z))$'s characteristic vectors. We know

$$\begin{aligned} J(\phi(z)) &= c\Lambda^{1/2} P \\ &= c\begin{pmatrix} \sqrt{\lambda_1} & 0 \\ 0 & \sqrt{\lambda_2} \end{pmatrix} \begin{pmatrix} \cos\theta & -\sin\theta \\ \sin\theta & \cos\theta \end{pmatrix} \\ &= c\begin{pmatrix} \sqrt{\lambda_1}\cos\theta & -\sqrt{\lambda_1}\sin\theta \\ \sqrt{\lambda_2}\sin\theta & \sqrt{\lambda_2}\cos\theta \end{pmatrix} \end{aligned} \tag{4}$$

so we have $\frac{\partial u}{\partial x} = c\sqrt{\lambda_1}\cos\theta$, $\frac{\partial u}{\partial y} = -c\sqrt{\lambda_1}\sin\theta$, $\frac{\partial v}{\partial x} = c\sqrt{\lambda_2}\sin\theta$, $\frac{\partial v}{\partial y} = c\sqrt{\lambda_2}\cos\theta$.

Denote $\frac{\partial u}{\partial x}, \frac{\partial u}{\partial y}, \frac{\partial v}{\partial x}, \frac{\partial v}{\partial y}$ as u_x, u_y, v_x, v_y. Without loss of generality, assume $\lambda_1 > \lambda_2$ Then

$$dw = w_z dz + w_{\bar{z}} d\bar{z} = (\frac{u_x + v_y}{2} + i\frac{v_x - u_y}{2})dz + (\frac{u_x - v_y}{2} + i\frac{v_x + u_y}{2})d\bar{z}. \tag{5}$$

Therefor,

$$w_z = \frac{u_x + v_y}{2} + i\frac{v_x - u_y}{2}, \; w_{\bar{z}} = \frac{u_x - v_y}{2} + i\frac{v_x + u_y}{2}. \tag{6}$$

Then the Beltrami coefficient is

$$\mu = \frac{w_{\bar{z}}}{w_z} = \frac{\sqrt{\lambda_1} - \sqrt{\lambda_2}}{\sqrt{\lambda_1} + \sqrt{\lambda_2}}(\cos 2\theta - i \sin 2\theta). \tag{7}$$

We can see that the modulus of μ is decided by the characteristic values of $J(\phi)$ everywhere, and the angle of μ is two times of the intersection angle of longer axis with x-axis, while μ have no connection with c. On the other hand, consider ϕ's Jacobian matrix, we have

$$det(J(\phi)) = c^2\sqrt{\lambda_1\lambda_2}cos^2\theta + c^2\sqrt{\lambda_1\lambda_2}sin^2\theta = c^2\sqrt{\lambda_1\lambda_2} > 0,$$

so ϕ is local homeomorphism, furthermore, according to Theorem 3 of [1], ϕ is a global homeomorphism when it maps the boundary ∂S_1 of S_1 homeomorphically onto itself, in Sect. 4, we will discuss that the output of our algorithm satisfies this condition. On the other hand, μ describes the feature of the infinitesimal ellipse, while it cannot describe long and short axis' lengthes. In fact, μ decides the torsion of the mapping, we can control the extent of torsion by controlling the modula of μ, which decides the extent of anisotropic feature. In the following, meshes with different extent of anisotropic are shown.

After computing the Beltrami coefficients, discrete surface Yamabe flow can be used to solve the Beltrami equation to compute the quasi conformal mapping corresponding to Beltrami coefficients.

3 Discrete Surface Yamabe Flow

Yamabe flow is a powerful tool to design Riemannian metric according to prescribed curvature. In this process the conformal structure of the manifold is preserved. In our problem, we can use it to compute an Euclidean metric with the given conformal structure. The Euclidean metric is in fact our target metric, then a new embedding of a mesh can be computed and it forms the image of the quasi conformal mapping we compute.

Let $\mathscr{T}$ be a triangular mesh embedded in $\mathbb{C}$, let e_{ij} be an edge of $\mathscr{T}$, $l_{ij}^{(0)}$ be the initial length. The discrete conformal factor is a function defined on the set of vertices $u : V \to \mathbb{R}$. During the process of Yamabe flow, the length of e_{ij} is defined as

$$l_{i,j} = e^{u(v_i)+u(v_j)} l_{ij}^{(0)}. \tag{8}$$

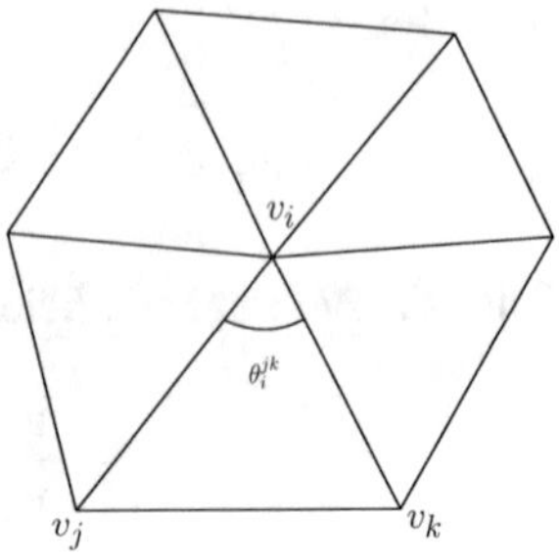

Fig. 3 Geometric interpretation of Gaussian curvature

Let $K_i = K(v_i)$ be the discrete Gaussian curvature of v_i, then $K(v_i) = 2\pi - \sum_{[v_i,v_j,v_k]\in F} \theta_i^{jk}$ if v_i is an interior vertex of $\mathcal{T}$, $K(v_i) = \pi - \sum_{[v_i,v_j,v_k]\in F} \theta_i^{jk}$ if v_i is a boundary vertex, where θ_i^{jk} is a corner angle in the face $[v_i, v_j, v_k]$ at the vertex v_i (Fig. 3).

It is easy to prove the following discrete Gauss-Bonnet Theorem.

Theorem 1 *Suppose $\mathcal{T}$ is a triangular mesh with a discrete metric, then*

$$\sum_{v\in T} K(v) = 2\pi\chi(\mathcal{T}), \tag{9}$$

where $\chi(\mathcal{T})$ is the Euler characteristic of the mesh, $\chi(\mathcal{T}) = |V| - |E| + |F|$, $|V|$, $|E|$, $|F|$ are the numbers of vertices, edges and faces, respectively.

Yamabe flow is the process to update the conformal factor u according to Gaussian curvature K,

$$\frac{du(t)}{dt} = 2(\bar{K} - K(t)), \tag{10}$$

where t is time parameter, $\bar{K}$ is the prescribed curvature which must satisfy Gauss-Bonnet theorem. For our purpose of generating mesh on planar domain, $\bar{K}(v_i) = 0$ for all interior vertices of $\mathcal{T}$. While when v_i is a boundary vertex, its target curvatures are decided by the shape of S_1. For example, if S_1 is a circle domain, the target curvatures on all the boundary vertices are defined as $\frac{2\pi}{m}$, where m is the number of boundary vertices. While when S_1 is an orthogon, there are four vertices on the boundary whose target curvatures are $\pi/2$ and other boundary vertices' target curvatures are zero. It is easy to check that for both cases the total target curvatures are 2π, satisfying Gauss-Bonnet theorem.

The convergence of Yamabe flow has been proven in [19]. In fact, the solution of Yamabe flow can be seen as an extremal point of a convex energy—Yamabe energy. Let $\mathbf{u} = (u_1, u_2, \cdots, u_n)$ be the conformal factor, define a differential 1-form $\omega = \sum_{i=0}^{n}(\bar{K}_i - K_i)du_i$, the differential of ω is

$$d\omega = \sum_{i,j=0}^{n} (\frac{\partial K_i}{\partial u_j} - \frac{\partial K_j}{\partial u_i})du_i \wedge du_j, \tag{11}$$

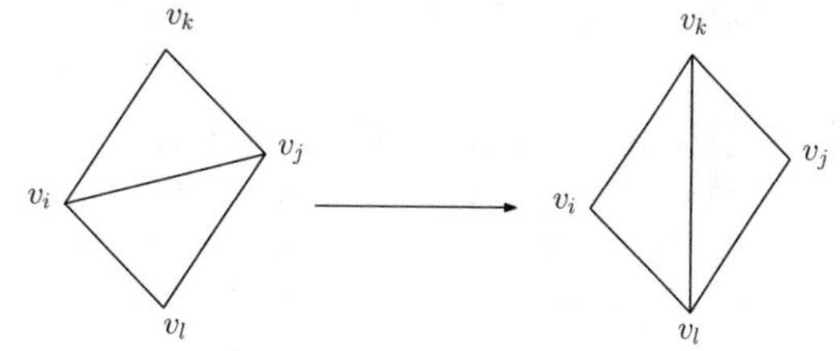

Fig. 4 Edge swap

It is easy to verify that

$$\frac{\partial K_i}{\partial u_j} = \frac{\partial K_j}{\partial u_i}, \tag{12}$$

so $d\omega = 0$, ω is a closed 1-form. The Yamabe energy is defined as

$$E(\mathbf{u}) = \int_{\mathbf{u}_0}^{\mathbf{u}} \sum_{i=1}^{n} (\bar{K}_i - K_i) du_i, \tag{13}$$

$E(\mathbf{u})$ is well defined and convex, so it has an extremal point, we can use Newton's method to solve it. More details of these results can be found in [25].

It is easy to compute the Hessian matrix of $E(\mathbf{u})$. Let e_{ij} be an interior edge of $\mathscr{T}$, $\mathbf{f}_{ijk}$ and $\mathbf{f}_{jil}$ are two faces which are adjacent to the edge, then we can define the weight of e_{ij} as follows

$$\mathbf{w}_{ij} = \cot\theta_k^{ij} + \cot\theta_l^{ij} \tag{14}$$

If e_{ij} is on the boundary of $\mathscr{T}$, there is only one face adjacent to it, the weight is

$$\mathbf{w}_{ij} = \cot\theta_k^{ij}. \tag{15}$$

Then the element on the i-th row and j-th column of the Hessian matrix of $E(\mathbf{u})$ is

$$\frac{\partial^2 E(\mathbf{u})}{\partial u_i \partial u_j} = -\frac{\partial K_i}{\partial u_j} = \begin{cases} \mathbf{w}_{ij} & i \neq j \\ -\sum_k \mathbf{w}_{ik} & i = j. \end{cases} \tag{16}$$

It has been proven by Luo [19] that the Hessian matrix is positive on the linear subspace $\{\mathbf{u} | \sum_{i=1}^{n} u_i = 0\}$. Yin et al. [25] has proved that the admissible metric space for $\mathscr{T}$ with fixed connectivity is not convex, so during the process of Newton's method, the connectivity of $\mathscr{T}$ should be transformed if necessary to ensure that $\mathbf{u}$ is in the admissible metric space during each step. The method that is usually used is edge swap. For an edge e_{ij}, if $\theta_k^{ij} + \theta_l^{ij} > \pi$, we swap it and denote it as e_{kl}, demonstrated in Fig. 4.

4 Algorithm

The problem we consider here is that given a complex domain S_1 and metric tensor $\mathscr{M}$ defined on S_1, to construct an anisotropic mesh with respect to the given metric tensor. The main idea is to convert the construction of an anisotropic mesh to the construction of isotropic mesh through computing a quasi conformal mapping $\phi : (S_1, g_1) \to (S_1, g_3)$. The image of the isotropic mesh for the quasi conformal mapping is an anisotropic mesh with respect to the given metric tensor.

Step one Compute the Beltrami coefficients based on the given metric tensor (Algorithm 1).
Step two Solve Beltrami equation by Yamabe flow method to compute the quasi conformal mapping based on its Beltrami coefficients (Algorithm 2).
Step three Construct an isotropic mesh, compute its image for the quasi conformal mapping (Algorithm 3).

4.1 Compute Beltrami Coefficients Based on Metric Tensor

Given a background triangulation $\mathscr{T}$ on the complex domain (S_1, g_1), discrete metric tensor $\mathscr{M} = \{M(v); v \in V\}$ is a matrix-valued function defined on the vertices set of $\mathscr{T}$. According to the description of Sect. 2, since $M(v)$ is a positive symmetric matrix, there is a matrix $N(v)$ such that $M(v) = N(v)^T N(v)$ which can be computed based on $M(v)$'s orthogonal decomposition. Let ϕ be the quasi conformal mapping which maps infinitesimal ellipse $(x, y)M(v)\begin{pmatrix} x \\ y \end{pmatrix} = \varepsilon$ to infinitesimal circle, as the description of Sect. 2, $N(v)$ is the Jacobian matrix of the quasi conformal mapping ϕ on v, so the Beltrami coefficients of ϕ on v can be computed according to $N(v)$. More details have been provided in Sect. 2.

Algorithm 1 Computation of Beltrami coefficients

Require: A triangular mesh $\mathscr{T}$, metric tensor $\mathscr{M}$.
Ensure: Beltrami coefficients of a quasi conformal mapping with respect to $\mathscr{M}$.

```
1: function BELTRAMI COEFFICIENTS(𝒯, ℳ)
2:     for Each vertex v ∈ V do
3:         compute the orthogonal decomposition of M(v).
4:         Compute the Jacobian matrix of the quasi conformal mapping J(φ(v)).
5:         Compute Beltrami coefficients μ(v) of φ.
6:     end for
7: return μ.
8: end function
```

Algorithm 2 Yamabe flow algorithm

Require: A mesh $\mathcal{T}$ with embedding ϕ_0, Beltrami Coefficients μ, parameter s.
Ensure: $\mathcal{T}$'s new embedding ϕ_1.

function YAMABE FLOW($\mathcal{T}, \mu, s$)
  for $e_{ij} \in E$ **do**
    Compute the initial length of e_{ij}

$$l_{ij}^{(0)} := |(z_j - z_i) + s \cdot \frac{\mu(v_i) + \mu(v_j)}{2} \cdot (\bar{z}_j - \bar{z}_i)|.$$

  end for
  for $v_i \in V$ **do**
    Initialize conformal factor $\gamma_i := 1.0$. Compute the target curvature $\bar{K}_i$, initial curvature K_i.
  end for
  while $max|K_i - \bar{K}_i| > \varepsilon$ **do**
    for Edge $e_{ij} = [v_i, v_j] \in E$ **do**
      Compute edge length $l_{ij} := l_{ij}^{(0)} \cdot \gamma_i \cdot \gamma_j$.
    end for
    Edge swap.
    for $v_i \in V$ **do**
      Compute the Gaussian curvature of v_i.
    end for
    for Edge $e_{i,j} \in E$ **do**
      Compute the edge weight w_{ij} to form the Hessian matrix Δ.
    end for
    $d\gamma = e^{\Delta^{-1}(\bar{K} - K)}$. Normalize $d\gamma$ such that $\Pi d\gamma_i = 1.0$. $\gamma := \gamma \cdot d\gamma$.
  end while
  Compute a new embedding $\phi_1 : V \to \mathbb{C}$ according to $\{l_{ij}; i, j <= |V|\}$.
return ϕ_1.
end function

Algorithm 3 Quasi conformal mapping

Require: A background mesh $\mathcal{T}$ with two embeddings ϕ_0, ϕ_1, a triangular mesh T with an embedding ψ_0.
Ensure: The image of the quasi conformal mapping ψ_1.

function QUASI CONFORMAL MAPPING($\mathcal{T}, \phi_0, \phi_1, T$)
  for Each vertex $v \in V$ **do**
    Find out which domain $\phi_0(\mathbf{f}), \mathbf{f} \in \mathcal{T}$ includes $\psi_0(v)$.
    Compute $\psi_1(v)$ in the domain $\phi_0(\mathbf{f})$.
  end for
return ψ_1.
end function

4.2 *The Algorithm of Computing Quasi Conformal Mapping According to Its Beltrami Coefficients*

Suppose $\phi_0 : \mathcal{T} \to \mathbb{C}$ be an embedding of $\mathcal{T}$. Let $e_{ij} = [v_i, v_j]$ be an edge of $\mathcal{T}$, denote $z_i = \phi_0(v_i)$, $z_j = \phi_0(v_j)$, then define the initial length of e_{ij} in the very beginning of Yamabe Flow as follows

$$l_{ij}^{(0)} := |e_{ij}|_{g_3} = |(z_j - z_i) + s \cdot \frac{\mu(v_i) + \mu(v_j)}{2} \cdot (\bar{z}_j - \bar{z}_i)|. \tag{17}$$

In the formula, $|z|$ is the modulus of a complex number z. This kind of edge length might not correspond to an embedding to the complex space. It changes the conformal structure in the discretion view, we call them the anisotropic lengthes. We add a parameter s to describe the extent of the anisotropic property. Then we use Yamabe Flow to computer a proper value of e^{λ_3} to calculate a new metric, denoted as $|e_{ij}|_{g_4} = e^{\lambda(v_i)+\lambda(v_j)}|e_{ij}|_{g_3}, \forall i, j$, which satisfies the condition that for all interior vertices v of $\mathscr{T}$, $K(v) = 0$, curvatures of vertices on the boundary conform the shape of S_1.

We know that if the lengths of $\mathscr{T}$'s edges are given, all the vertices' discrete gaussian curvature can be calculated according to the cosine formula and discrete Gaussian curvature formula. We denote the Gaussian curvature calculated by $|e|_{g_3}$ as K_0, put K_0 as the initial Gaussian curvature. Put $\bar{K}$ as the target Gaussian curvature, which corresponds to an embedding of $\mathscr{T}$, then $\bar{K}(v) = 0$ for all interior vertices v of $\mathscr{T}$. The shape of S_1 decides $\bar{K}(v)$ for boundary vertices of $\mathscr{T}$.

There is an important essential that should be paid attention to. During the process of Yamabe flow, the connectivity of $\mathscr{T}$ should be transformed if necessary at each step. During the judgement at edge swap, define the length of e_{ij} as $e^{\lambda_t(v_i)+\lambda_t(v_j)}|e|_{g_3}$, in which λ_t is the conformal factor at time t. While when we transform the connectivity of $\mathscr{T}$, all new edges' length in g_3 should be real-time computed. Suppose that there is an new edge e_{kl} connecting v_k, v_l, then

$$|e_{kl}|_{g_3} = |(z_l - z_k) + s \cdot \frac{\mu(v_k) + \mu(v_l)}{2} \cdot (\bar{z}_l - \bar{z}_k)|. \tag{18}$$

so its length at time t is $l_{kl} = e^{\lambda_t(v_k)+\lambda_t(v_l)}|e_{kl}|_{g_3}$.

Our last step is to compute an embedding $\phi_1 : \mathscr{T} \to \mathbb{C}$ of $\mathscr{T}$ according to the metric $|e_{ij}|_{g_4} = e^{\lambda(v_i)+\lambda(v_j)}|e_{ij}|_{g_3}$. In the following we simplify $|e_{ij}|_{g_4}$ to d_{ij}. There are a lot of methods to compute $\mathscr{T}$'s embedding . For example, the problem can be translated to the process of minimizing an energy, or, it can be solved by a branch of computation of intersection points of two circles, see Fig. 5. In our case, we choose the second algorithm. Since our algorithm guarantees that all the triangles of $\mathscr{T}$ are non-degenerate, the lengths of edges satisfy $d_{ij} < d_{ik} + d_{jk}$, so at each step of computation, those two circles must have two intersections.

The choice of solutions decides the direction of f_{ijk}'s normal vector. If we regard f_{ijk} to be a face in $\mathbb{R}^3$, its normal vector is either $(0, 0, 1)$ or $(0, 0, -1)$. During the process of this algorithm, the directions of all the faces' normal vector must be the same.

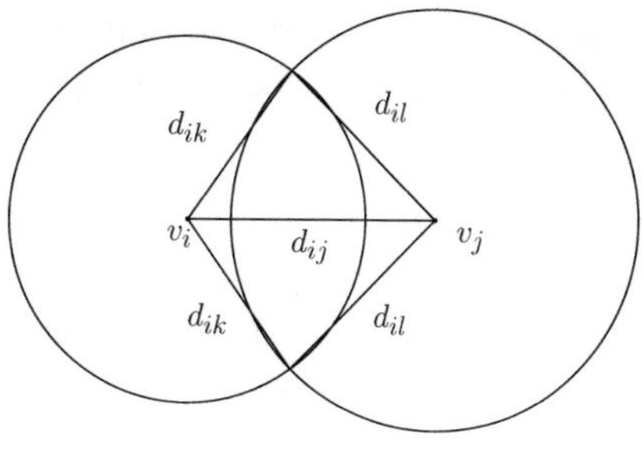

Fig. 5 The intersection points of two circles

4.3 *Construct Anisotropic Mesh Based on Quasi Conformal Mapping*

The discretion of the quasi conformal mapping ϕ we compute is a piecewise linear mapping, it can be represented as two coordinates of $\mathscr{T}$, locally on each cell $\mathbf{f}$ of $\mathscr{T}$, ϕ is a linear mapping which can be decided by the image of $\mathbf{f}$'s vertices. So given a triangulation T, the image of T can be decided by the image of all the vertices of T. Then the problem is converted to the problem of computing the image of a given vertex. Firstly, one should find out which cell $\mathbf{f} \in \mathscr{T}$ includes the given vertex, then compute the image according to the images of vertices of $\mathbf{f}$.

5 Experiments and Comparisons

In this section, we will show experimental results of our method on interpolation of anisotropic functions as well as comparison of our method with other methods implemented in publicly available mesh adaptation codes.

5.1 *A Metric Tensor from the Gradient*

Most of the methods consider to choose the Hessian matrix or recovery Hessian matrix of the input function to be metric tensor. Dassi et al. [8] provides a novel anisotropic mesh adaptation technique based on higher dimensional embedding which contains the information of the function f itself, its gradient and Hessian information. In this section, we describe an anisotropic mesh creation method based on the information with respect to the gradient of the interpolated function.

Let S_1 be a domain in $\mathbb{C}$, define Riemannian metric $g_1 = dzd\bar{z}$ on S_1. Let $f : S_1 \to \mathbb{R}$ be a real-valued C^1 continuous function, then the set $\{(x, y, f(x, y))|(x, y) \in S_1\}$ forms a surface embedded in $\mathbb{R}^3$, denoted as S_2, actually S_2 is a Riemannian surface with Euclidean metric of $\mathbb{R}^3$, denoted as g_2. Define a mapping $\phi : (S_1, g_1) \to (S_2, g_2)$ as

$$\phi(x, y) = (x, y, f(x, y)),$$

obviously ϕ is a diffeomorphism. Let $p := (x_0, y_0, f(x_0, y_0)) \in (S_2, g_2)$, we put an infinitesimal ε-circle on (x_0, y_0), denoted as $\gamma = \{(x_0 + \varepsilon \cos\theta, y_0 + \varepsilon \sin\theta) | 0 \leq \theta < 2\pi\}$, then its image is

$$\phi(\gamma) := \{(x_0 + \varepsilon \cos\theta, y_0 + \varepsilon \sin\theta, f(x_0 + \varepsilon \cos\theta, y_0 + \varepsilon \sin\theta)) | 0 \leq \theta < 2\pi\}. \tag{19}$$

Since f is C^1, we have

$$f(x_0 + \varepsilon \cos\theta, y_0 + \varepsilon \sin\theta) = f(x_0, y_0) + f_x \varepsilon \cos\theta + f_y \varepsilon \sin\theta + O(\varepsilon^2) \tag{20}$$

where $f_x := \frac{\partial f}{\partial x}(x_0, y_0)$, $f_y := \frac{\partial f}{\partial y}(x_0, y_0)$.

Assume that q is a point in $\phi(\gamma)$, $q = (x_0 + \varepsilon \cos\theta_0, y_0 + \varepsilon \sin\theta_0, f(x_0 + \varepsilon \cos\theta_0, y_0 + \varepsilon \sin\theta_0))$, then

$$\begin{aligned} \|q - p\|^2 &= \varepsilon^2 \cos^2\theta_0 + \varepsilon^2 \sin^2\theta_0 + (f_x \varepsilon \cos\theta + f_y \varepsilon \sin\theta)^2 + O(\varepsilon^2) \\ &= (f_x^2 + 1)\varepsilon^2 \cos^2\theta_0 + 2 f_x f_y \varepsilon^2 cos\theta_0 \sin\theta_0 + (f_y^2 + 1)\varepsilon^2 \sin^2\theta_0 + O(\varepsilon^2) \\ &= (\varepsilon \cos\theta_0, \varepsilon \sin\theta_0) \begin{pmatrix} f_x^2 + 1 & f_x f_y \\ f_x f_y & f_y^2 + 1 \end{pmatrix} \begin{pmatrix} \varepsilon \cos\theta_0 \\ \varepsilon \sin\theta_0 \end{pmatrix} + O(\varepsilon^2), \end{aligned} \tag{21}$$

thus $\phi(\gamma)$ is approximately contained by a tiny ellipsoid, its projection to the tangent plane of (S_2, g_2) at p is an infinitesimal ellipse with centre p. It is obvious that, the image of this tiny ellipse

$$\{(x, y) | (x - x_0, y - y_0) \begin{pmatrix} f_x^2 + 1 & f_x f_y \\ f_x f_y & f_y^2 + 1 \end{pmatrix}^{-1} \begin{pmatrix} x - x_0 \\ y - y_0 \end{pmatrix} = \varepsilon^2\}$$

is a circle. So as it described above, if there is a quasi conformal mapping ϕ_0 defined on S_1 which maps the tiny ellipse above to a tiny circle, $\phi \circ \phi_0^{-1}$ is a conformal mapping. So we choose the metric tensor $\mathcal{M} = \{M(z) | z \in S_1\}$ in S_1 to construct the quasi conformal mapping, in which

$$M(z) = \begin{pmatrix} f_x^2 + 1 & f_x f_y \\ f_x f_y & f_y^2 + 1 \end{pmatrix}^{-1}. \tag{22}$$

$\mathcal{M}(z)$ induces an new metric on S_1, denoted as g_3. In fact, the direction of long axis of γ on the surface is the direction in which the function changes fastest and

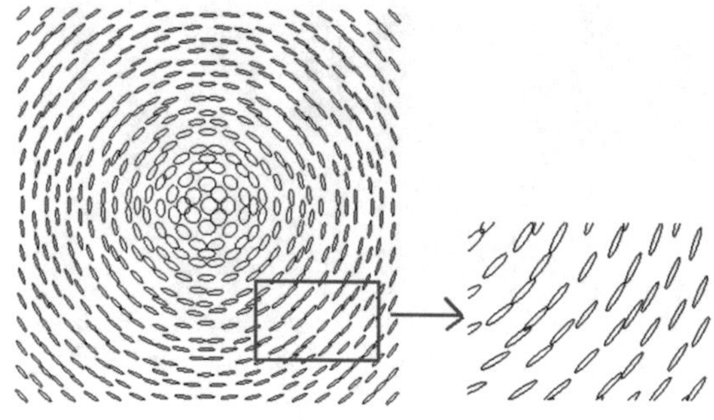

Fig. 6 Ellipse of function $f(x, y) = 4e^{-(x^2+y^2)/2}$

the direction of short axis is the direction in which f changes slowest, it means that they are the principal direction of (S_2, g_2).

Figure 6 shows the ellipses in the domain $[-1, 1]\times[-1, 1]$ with respect to $\mathscr{M}(z)$ of the Gaussian function $f(x, y) = 4e^{-(x^2+y^2)/2}$.

5.2 Experiments on Function Interpolation

In this section, we show experiments on function interpolation using our method. We experimented two functions, the first function is found in the paper [8], which are:

$$f_1(x, y) = \tanh(60x) - \tanh(60(x - y) - 30),$$
$$f_2(x, y) = \tanh(-100(y - 0.4\sin(2\pi x))^2).$$

Figure 7 shows the functions we are going to interpolate.

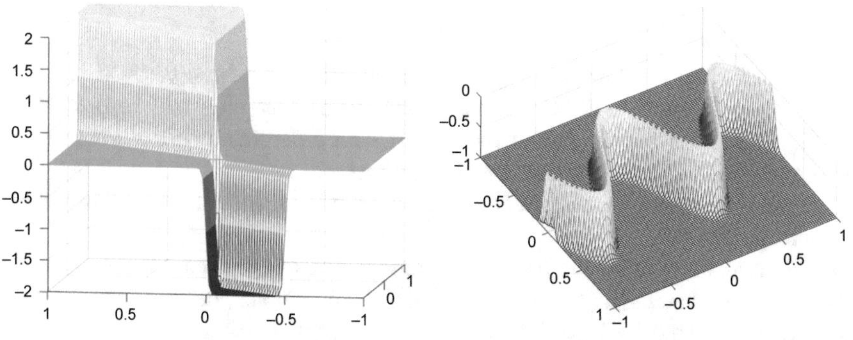

f_1 f_2

Fig. 7 Two functions

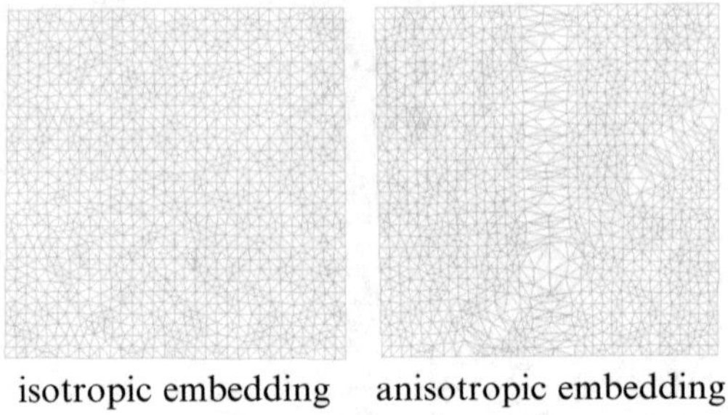

Fig. 8 Two embeddings of background mesh of f_1

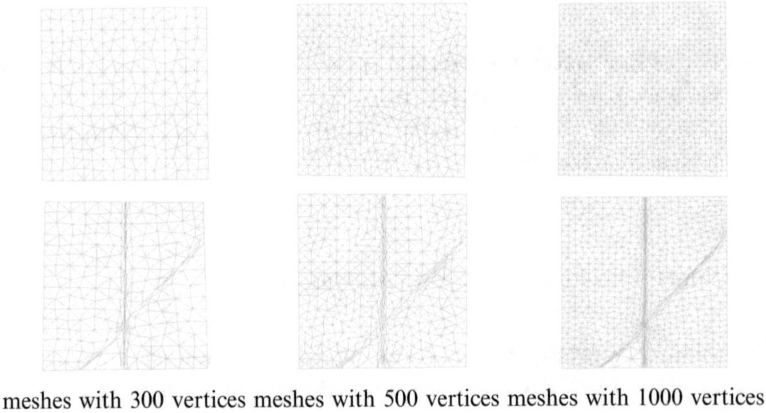

Fig. 9 Anisotropic mesh of f_1

Table 1 The L^2-errors of $f_1(x, y) = \tanh(60x) - \tanh(60(x - y) - 30)$ on different meshes

$\|V\|$	Isotropy	Anisotropy
300	0.434039	0.169213
500	0.3414	0.152131
1000	0.26307	0.0925087

Figure 8 shows two embeddings of a triangulation with 1404 vertices, which is the background mesh to describe the quasi conformal mappings with respect to f_1.

Figure 9 shows the anisotropic mesh of f_1 with different number of vertices and their pre-image of the quasi conformal mapping.

Table 1 reports the L^2-errors of the interpolation functions corresponding to these anisotropic meshes.

We know that during the process of construction of quasi conformal mapping, we add a parameter s to control the extent of "anisotropic", the next figure shows the background mesh of quasi conformal mappings with different parameters with respect to f_2. Figure10 shows the anisotropic mesh of f_2 with different parameters.

Table 2 reports the L^2-errors of the interpolation functions corresponding to these anisotropic meshes.

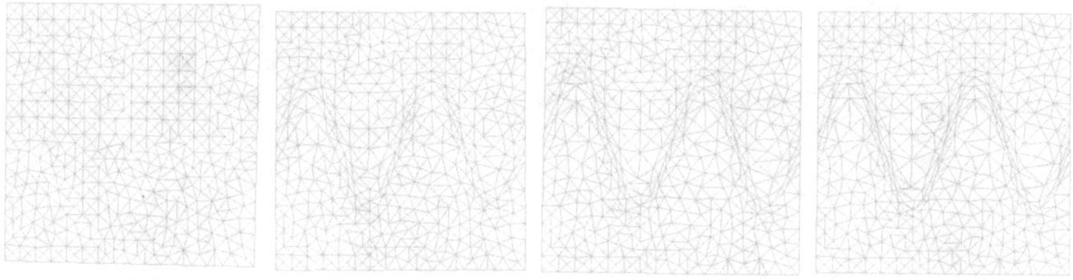

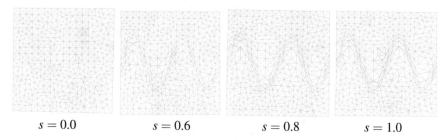

Fig. 10 Anisotropic mesh of f_2 with different parameters

Table 2 The L^2-errors of $f_2(x, y) = \tanh(-100(y - 0.4\sin(2\pi x))^2)$ on different meshes

s	0.0	0.2	0.4	0.6	0.8	1.0
Error	0.250927	0.205359	0.179804	0.160007	0.130494	0.142274

From these experiments, we observe that our method is able to capture the anisotropic features of the interpolated functions. The resulting meshes improved the accuracy of the interpolation compared with uniform meshes (those meshes with $s = 0$). Moreover, the interpolation error consistently decreases according to the increase of the number of points.

5.3 *Comparisons with Other Methods*

In this section, we conducted preliminary comparisons of our results with the results produced by two public codes, one is `BAMG` which implements the classical metric-based methods. Another is `Detri2` which implements the high dimensional embedding method [8]. First of all, it is noted that both codes produced high quality anisotropic meshes, with a much smaller interpolation error which is about two orders of magnitude smaller than those errors of our results, see Fig. 11. This shows the limitation of our method compared to mesh adaptation methods. It is necessary to add/remove vertices, while our method does not change the number of vertices.

On the other hand, our method could be seen as an effective anisotropic mesh smoothing step compared with the heuristic smoothing algorithms used in `BAMG` and `Detri2`. Figure 12 reports the $L2$-error when only using `BAMG`'s smoothing option. We did this experiment by using the `adaptmesh()` function provided in `FreeFEM++`, and we only call this function once, with different iterations of smoothing, `nbsmooth=`xxx, where xxx is the given number of smoothing iterations. We used the parameter `nbvx=`1500 to set a limit of number of points.

From this experiment, we could observe that the mesh smoothing algorithm (which is a heuristic relaxation method) used in *BAMG* has no obvious effect on the resulting meshes. In this case, our method could be used to improve there anisotropic

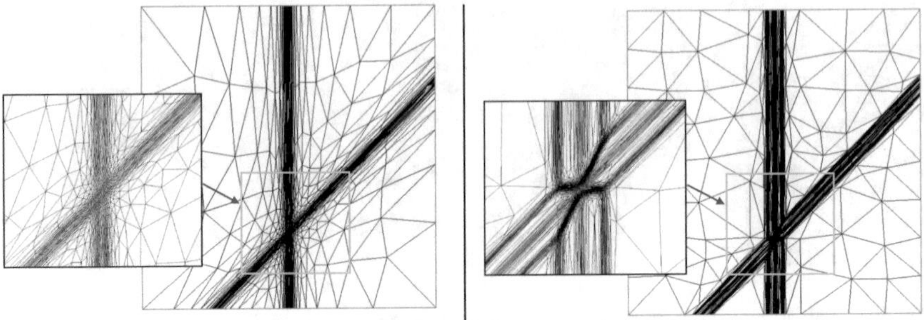

BAMG (884 points, *L*2-error: 0.0040) | *Detri*2 (2346 points, *L*2-error: 0.0044)

Fig. 11 Anisotropic mesh of f_1 generated by BAMG (left) and Detri2 (right)

BAMG (341 points)

nbsmooth	L^2 error
5	0.510075
10	0.51022
20	0.510206
50	0.509879
100	0.509742

Fig. 12 Left: the adapted mesh generated by BAMG with only one iteration. Right: the report of L^2 error on meshes produced by different iterations of smoothing

smoothing algorithm in the mesh adaptation process. This could be an interesting future work.

References

1. F.J. Bossen, P.S. Heckbert, A pliant method for anisotropic mesh generation, in *5th International Meshing Roundtable*, Citeseer, 1996, pp. 63–74
2. G.D. Canas, S.J. Gortler, Surface remeshing in arbitrary codimensions. Vis. Comput. **22**(9), 885–895 (2006)
3. M.J. Castro-Diaz, F. Hecht, B. Mohammadi, O. Pironneau, Anisotropic unstructured mesh adaption for flow simulations. Int. J. Numer. Methods Fluids **25**(4), 475–491 (1997)
4. L. Chen, J.-C. Xu, Optimal Delaunay triangulations. J. Comput. Math. **22**(2), 299–308 (2004)
5. L. Chen, P. Sun, J. Xu, Optimal anisotropic meshes for minimizing interpolation errors in l^p-norm. Math. Comput. **76**, 179–204 (2007)
6. F. Dassi, H. Si, A curvature-adapted anisotropic surface re-meshing method, in *New Challenges in Grid Generation and Adaptivity for Scientific Computing*, vol. 5, ed. by S. Perotto, L. Formaggia (Springer, Cham, 2015), pp. 19–41
7. F. Dassi, A. Mola, H. Si, Curvature-adapted remeshing of CAD surfaces. Procedia Eng. **82**, 253–265 (2014)

8. F. Dassi, H. Si, S. Perotto, T. Streckenbach, Anisotropic finite element mesh adaptation via higher dimensional embedding. Procedia Eng. **124**, 265–277 (2015)
9. C. Dobrzynski, P. Frey, Anisotropic delaunay mesh adaptation for unsteady simulations, in *Proceedings of the 17th International Meshing Roundtable* (Springer, Berlin, 2008), pp. 177–194
10. Q. Du, D. Wang, Anisotropic Centroidal Voronoi tessellations and their applications. SIAM J. Sci. Comput. **26**(3), 737–761 (2005)
11. L. Formaggia, S. Perotto, New anisotropic a priori error estimates. Numer. Math. **89**(4), 641–667 (2001)
12. P.-J. Frey, F. Alauzet, Anisotropic mesh adaptation for CFD computations. Comput. Methods Appl. Mech. Eng. **194**(48–49), 5068–5082 (2005)
13. W. Huang, Mathematical principles of anisotropic mesh adaptation. Commun. Comput. Phys. Citeseer, 2006
14. B. Lévy, N. Bonneel, Variational anisotropic surface meshing with Voronoi parallel linear enumeration, in *Proceedings of the 21st International Meshing Roundtable*, 2012, pp. 349–366
15. Y. Liu, W. Wang, B. Lévy, F. Sun, D.-M. Yan, L. Lu, C. Yang, On Centroidal Voronoi tessellation—energy smoothness and fast computation. ACM Trans. Graph. **28**(4), 1–17 (2009)
16. R. Löhner, J. Cebral, Generation of non-isotropic unstructured grids via directional enrichment. Int. J. Numer. Methods Eng. **49**(1–2), 219–232 (2000)
17. A. Loseille, F. Alauzet, Continuous mesh framework part I: well-posed continuous interpolation error. SIAM J. Numer. Anal. **49**(1), 38–60 (2011)
18. A. Loseille, F. Alauzet, Continuous mesh framework part II: validations and applications. SIAM J. Numer. Anal. **49**(1), 61–86 (2011)
19. F. Luo, Combinatorial Yamabe flow on surfaces. Commun. Contemp. Math. **6**(05), 765–780 (2004)
20. D. Marcum, F. Alauzet, Aligned metric-based anisotropic solution adaptive mesh generation. Procedia Eng. **82**, 428–444 (2014)
21. S. Rippa, Long and thin triangles can be good for linear interpolation. SIAM J. Numer. Anal. **29**(1), 257–270 (1992)
22. J.R. Shewchuk, What is a good linear element? Interpolation, conditioning, and quality measures, in *Proceedings of 11th International Meshing Roundtable*, Ithaca, New York, September 2002 (Sandia National Laboratories, Livermore, 2002), pp. 115–126
23. K. Shimada, A. Yamada, T. Itoh, et al., Anisotropic triangular meshing of parametric surfaces via close packing of ellipsoidal bubbles, in *6th International Meshing Roundtable*, 1997, pp. 375–390
24. D.-M. Yan, K. Wang, B. Levy, L. Alonso, *Computing 2D Periodic Centroidal Voronoi Tessellation*, June 2011 (IEEE, Piscataway, 2011), pp. 177–184
25. X. Yin, M. Jin, F. Luo, X.D. Gu, Discrete curvature flows for surfaces and 3-manifolds, in *Emerging Trends in Visual Computing* (Springer, Cham, 2009), pp. 38–74
26. Z. Zhong, X. Guo, W. Wang, B. Lévy, F. Sun, Y. Liu, W. Mao, Particle-based anisotropic surface meshing. ACM Trans. Graph. **32**(4), 1 (2013)

Terminal Star Operations Algorithm for Tetrahedral Mesh Improvement

Fernando Balboa, Pedro Rodriguez-Moreno, and María-Cecilia Rivara

Abstract We discuss an innovative, simple and effective Lepp terminal-star algorithm for improving tetrahedral meshes. For each bad quality tetrahedron, one branch of the longest edge propagating path (Lepp) is followed to find an associated terminal star, which is a set of tetrahedra that share a common longest edge (terminal edge). Three alternative improvement mesh operations are considered: simple insertion of the centroid Q of the terminal star, or swapping of the terminal edge, or longest edge bisection. The operation that most improves the mesh is performed whenever significant improvement is achieved. Empirical study shows that, using the dihedral angle quality measure, this simple procedure reduces the bad quality tetrahedra by at least a tenth, with low time cost.

1 Introduction

Lepp bisection algorithms and previous longest edge algorithms were designed for local refinement of triangulations in two and three dimensions, for adaptive finite element applications [10–13]. In three dimensions, for each target tetrahedron t to be refined, the Lepp bisection algorithm follows a longest edge propagating path (Lepp) to find a set of largest (terminal) edges, each of them shared by a set of terminal tetrahedra (terminal star). All the tetrahedra are refined by their longest-edge (bisection by the plane defined by the midpoint of the longest edge and the two opposite vertices).

In two dimensions the longest edge bisection guarantees the construction of refined triangulations that maintain the quality of the input mesh, improving the

F. Balboa · M.-C. Rivara (✉)
Universidad de Chile, Department of Computer Science, Santiago, Chile
e-mail: fernando@balboa.cl; mcrivara@dcc.uchile.cl

P. Rodriguez-Moreno
Universidad del Bío-Bío, Department of Information Systems, Concepcion, Chile
e-mail: prodrigu@biobio.cl

X. Roca, A. Loseille (eds.), *27th International Meshing Roundtable*,
Lecture Notes in Computational Science and Engineering 127,
https://doi.org/10.1007/978-3-030-13992-6_15

average triangle quality [10, 12]; and producing optimal size triangulations [1]. Even when the extension of these properties to three dimensions have not been theoretically proved, empirical evidence shows that the three dimensional algorithm behaves analogously to the two dimensional algorithm in practice, as shown by Rivara and Levin [13]. Applications of these algorithms have been discussed by Williams [18], Jones and Plassmann [7], Castaños and Savage [2], Rivara et al. [16].

Lepp Delaunay algorithms for improving two and three dimensional meshes have also been developed [11, 12, 14] which show good practical behavior. Recently, Rivara and Rodriguez-Moreno [15], studied a two dimensional terminal triangles centroid algorithm, proving that the algorithm terminates by producing graded, optimal size, 30° triangulations for planar straight line graph (PSLG) geometries with constrained angles greater than or equal to 30°. The mesh improvement algorithm of this paper generalizes some of the ideas presented in [15] to three dimensions, and is based on some of the ideas presented in [14].

1.1 Previous Algorithms Versus Our Mesh Improvement Algorithm

Previous algorithms for three dimensional mesh improvement, based on smoothing, swapping, optimization techniques and point insertions have been discussed by Freitag and Ollivier-Gooch [6], Klingner and Shewchuk [8], Dassi et al. [4], Misztal et al. [9]. All these algorithms require good distribution of points inside the input mesh. In what follows we will focus the discussion on the algorithms of these papers. Other algorithms have also been developed such as a sliver exudation algorithm discussed by Edelsbrunner and Guoy [5]; and an algorithm based on optimal Delaunay triangulations developed by Chen and Holst [3].

In a known paper in the mesh generation field, Freitag and Ollivier-Gooch [6], presented a complete empirical study on the use of sequences of three dimensional swapping and smoothing techniques, combined with five alternative (rather complex) optimization strategies. This paper also included a set of recommendations to adequately combine (non trivial) sequences of mesh operations to deal with different mesh improvement issues. Computational testing considers application meshes with good distribution of interior points.

Later, Klingner and Shewchuk [8] discussed a mesh improvement method which combines the mesh operations of Freitag and Ollivier-Gooch, with a set of additional mesh operations (insertion of a new point into a bad quality tetrahedron, two operations for improving boundary tetrahedra, multiphase operations and compound operations) and intensive use of optimization. They succeeded in generating very good meshes (with high computational cost), by assuming that "the spacing of the input vertices in the input mesh is already correct". Based on the operations of Klingner and Shewchuk, Dassi et al. [4] discussed an optimization method (without using point insertions) to improve the mesh, for geometries formed by unions/intersections of right parallelepipeds with random points in its interior; while Misztal et al. [9] discussed the inclusion of a new multi-face retriangulation operation that can be performed in the boundary of the mesh.

In this paper we propose a simple and efficient mesh improvement algorithm that uses a Lepp path to find a terminal star (set of tetrahedra that share a common longest edge) over which two alternative improvement operations are performed: a new simple centroid insertion operation, or terminal edge swapping, whenever the mesh is locally improved. Our algorithm is more general than previous methods, since this can be also applied to meshes with bad distribution of interior points.

2 Lepp Bisection and Lepp Centroid Algorithms

In two dimensions, Lepp(t), the longest edge propagating path of a triangle t [11, 12], is a sequence of increasing triangles that allows for finding a unique local largest edge in the mesh (terminal edge) shared by two terminal triangles (one triangle for a boundary terminal edge). For an illustration see Fig. 1a. In 3-dimensions Lepp(t) corresponds to a multidirectional searching process [11, 14] that allows for finding a set of terminal edges.

Definition 1 E is a terminal edge in a tetrahedral mesh τ if E is the longest edge of every tetrahedron that shares E. In addition, we call terminal star *TS(E)* to the set of tetrahedra that shares a terminal edge E.

Definition 2 For any tetrahedron t_0 in τ, Lepp(t_0) is recursively defined as follows: (a) Lepp(t_0) includes every tetrahedron t that shares the longest edge of t_0 with t, and such that longest edge of t is greater than the longest edge of t_0; (b) For any tetrahedron t_i in Lepp(t_0), this Lepp(t_0) also contains every tetrahedron t that shares the longest edge of t_i and where longest edge of t is greater than longest edge of t_i.

Note that Lepp(t_0) is a 3D submesh which has a finite and variable number of associated terminal-edges and terminal stars, as illustrated in Fig. 1b. In the full Lepp bisection algorithm, for each tetrahedron t to be refined, the algorithm computes Lepp(t) and finds an associated set W of terminal edges. Then for each terminal edge E in W, the longest edge bisection of every tetrahedron of the terminal star *TS(E)* is performed, which corresponds to a local refinement operation that maintains a conforming mesh (where the intersection of pair of adjacent tetrahedra is either a common vertex, or a common edge, or a common face). This process is repeated until the target tetrahedron t is refined.

One Path Lepp Bisection Algorithm In this paper we consider algorithms based on partial Lepp computation, following one Lepp branch until one terminal edge is found, according to the following definition:

Definition 3 For any tetrahedron t_0, of longest edge L_0, compute One Branch Lepp(t_0) as follows:

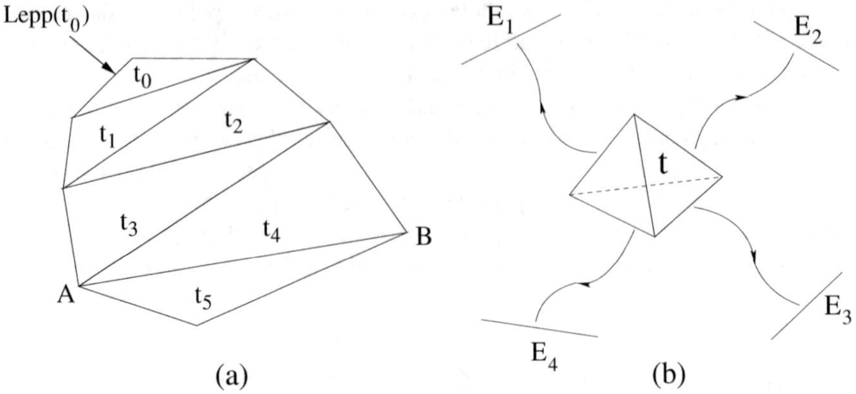

Fig. 1 (**a**) Lepp in 2-dimensions: Lepp(t_0) = $t_0, t_1, t_2, t_3, t_4, t_5$, AB is the terminal edge; (**b**) Lepp(t) in 3-dimensions has several terminal edges E_i

(a) OneBranchLepp(t_0) includes t_0. Then define processing tetrahedron t_{proc} equal to t_0 (of longest edge L_{proc}).
(b) Add to OneBranchLepp(t_0) one tetrahedron t with the greatest longest edge L_t, selected between the set of tetrehedra that share edge L_{proc} and having longest edge L_t greater than L_{proc}.
(c) Repeat for t_{proc} equal to t, while there exists a tetrahedron t in step (b).

Remark Note that by using Definition 3, in most cases we expect to reach the largest terminal edge in the Lepp submesh.

Algorithm 1 One path Lepp bisection algorithm (τ, S)

```
Input: τ mesh of tetrahedra; S set of tetrahedra to be refined
Output: refined mesh τ_f
while S ≠ φ do
   For each tetrahedron t0 ∈ S.
   while t0 remains in the mesh do
      Compute OneBranchLepp(t0), terminal edge E and terminal star TS(E)
      Perform longest edge bisection of each tetrahedron in terminal star TS(E)
   end while
end while
```

One Path Lepp Centroid Algorithm Here we introduce the simple centroid algorithm in three dimensions. Instead of selecting the terminal edge midpoint, the centroid Q of the terminal star $TS(E)$ is computed. Then Q is simply inserted in the mesh by joining Q with the exterior triangular faces of $TS(E)$, whenever this is a valid three dimensional mesh operation. The algorithm is as follows:

Algorithm 2 One path Lepp centroid algorithm (τ, S)

Input: τ mesh of tetrahedra; S set of tetrahedra to be refined
Output: refined mesh τ_f
while $S \neq \phi$ **do**
  For each tetrahedron $t_0 \in S$
  while t_0 remains in the mesh **do**
    Compute OneBrachLepp(t_0), terminal edge E and terminal star $TS(E)$
    Compute centroid Q of terminal star $TS(E)$
    if simple insertion of Q is valid operation **then**
      Perform simple insertion of centroid Q
    else
      Perform longest edge bisection of the tetrahedra in TS(E)
    end if
  end while
end while

As reported in [13, 14], refinement algorithms based on the bisection of tetrahedra tends to improve the meshes in practice. Experimentation with the Lepp centroid algorithm in the context of this research, shows better behavior than pure longest edge bisection algorithms. However, without considering smart operations, the mesh size increases too much to be competitive for mesh improvement. In the new algorithm of this paper, only smart centroid insertions are performed whenever local mesh improvement is attained.

Simple Centroid Insertion It is worth noting that the simple centroid insertion is related with smart laplacian smoothing as follows: assume that the terminal edge midpoint M is inserted in the mesh by longest edge bisection of the tetrahedra of the terminal star; then the laplacian smoothing of vertex M is equivalent to the simple insertion of centroid Q.

3 Swapping of the Terminal Edge

According to the definition, each terminal star is formed by a variable number of tetrahedra sharing a common longest edge. Consequently, each terminal star is a polyhedron very appropriate for performing edge swapping in a tetrahedral mesh. A full discussion on edge swapping operations can be found in reference [6]. Note that for a set of N tetrahedra sharing an edge, the edge swapping operation replaces the N tetrahedra by a set of 2N-4 tetrahedra. Thus, for N = 3, 4, 5, 6, the associated swapping operations to be used correspond to 3-2, 4-4, 5-6, 6-8 tetrahedra; and so on. In Sect. 5.3 we include a statistical discussion on the practical use of these operations in our improvement algorithm.

4 Selective (Terminal Tetrahedra) Centroid/Swapping Algorithm

The algorithm is formulated in terms of three alternative terminal star operations: simple centroid insertion, terminal star swapping, longest edge bisection (applied to constrained terminal edge). Each one of these operations modifies the interior of the terminal star producing a new tetrahedra set. The algorithm chooses the best operation whenever this locally improves the mesh by using a given factor; otherwise this continues by processing another target tetrahedron.

For each tetrahedra set (tetrahedra in the terminal star, tetrahedra obtained by simple centroid insertion, tetrahedra obtained by terminal edge swapping, tetrahedra obtained by longest edge bisection), we use a function Quality(set) that computes a worst dihedral angle measure as follows. For each tetrahedron t in the set, we compute Dangle as

$$\text{Dangle} = \text{Min}\{\alpha, 180^\circ - \beta\}$$

where α, β are the smallest and the largest dihedral angles respectively for tetrahedron t, with the conditions $\alpha \leq \theta_1$ and $\beta \geq \theta_2$. Note that θ_1, θ_2 are user given tolerance parameters for the smallest and largest dihedral angles (see the Algorithm 3). Then the Quality(set) function is equal to the smallest Dangle value for the tetrahedra in the set. The algorithm can be schematically described as follows.

Several remarks are in order:

- We have adjusted the improvement parameter Factor to 1.1. Thus, the local operation is accepted if a 10% of improvement is achieved with respect to the terminal tetrahedra set.
- After one step of the preceding algorithm is performed, certain tetrahedra remain in the mesh, for which the improvement task did not succeed. Then a next step of the algorithm is performed for an actualized set S. The process finishes either if all the tetrahedra in S can not be improved, or after a user fixed number of steps is performed. In practice, after five refinement steps with Factor = 1.1 are performed, no significant mesh improvement is achieved.
- It is worth noting that the algorithm does not include special operations for improving boundary tetrahedra.
- We follow one branch of the Lepp path, according to the Definition 3, to find a terminal star. In practice the results are not different by using alternative Lepp branches.
- We have used a quality criterion that mixes smallest and largest dihedral angles. We use dihedral angles since this is the most used measure in practical mesh generation [7, 8].

Algorithm 3 Selective centroid swap algorithm (τ, θ_2, θ_2)

```
Input: tetrahedral mesh τ, θ2, θ2 are tolerances for smallest and largest dihedral angles
Output: improved mesh τf
Find S set of tetrahedra with dihedral angle ≤ θ2 or dihedral angles ≥ θ2.
for each tetrahedron t in S do
   while t remains in mesh do
      compute OneBranchLepp(t), terminal edge E, terminal star TS
      compute Quality(TS)
      compute centroid Q and Quality(insertion)
      (Quality is 0 if insertion is not a valid operation)
      compute swapping of TS and Quality(Swap)
      if Quality(insertion) > Quality(Swap) and Quality(insertion > Factor * Quality(TS))
      then
         Perform insertion of centroid Q
      else
         if Quality(swap) > Factor * Quality(TS) then
            Perform swapping of the terminal tetrahedra
         else
            set Nothing equal to true
         end if
      end if
      if Nothing is true and Quality(bisection) > Factor * Quality(TS) then
         perform longest edge bisection of TS
      end if
   end while
end for
```

5 Empirical Study

Taking as a basis, three dimensional Lepp-based C++ code, developed by Rodriguez-Moreno [17], our (terminal star operations) improvement algorithm was implemented in a notebook computer with intel core i7 6500U processor without using parallelism of any kind. To study the practical behavior of the algorithm we have considered the complex test problems of Fig. 2, and different sets of random points (inside a box).

5.1 *Comparison of Centroid/Swap Algorithm with Only Centroid and with Only Swap Techniques*

We compared the full algorithm, which chooses the best (smart) operation between insertion and swapping, against only (smart) insertion and only (smart) swapping. For all the test problems of Fig. 2, the full algorithm performs better than the algorithms based on isolated operations. Table 1 illustrates the extreme dihedral angle distribution for the final meshes of the three techniques for the elephant case.

Fig. 2 Test problems

Table 1 Comparison of insert/swap with only insert and only swap

Case	<5°	<10°	<20°	<30°	>150°	>160°	>170°	Mesh size
Initial	0.08	0.54	3.9	10.51	1.6	0.54	0.07	1905
Insert/swap	0	0	0.39	3.63	0.16	0.017	0	2834
Insert	0	0.05	1.38	6.54	0.48	0.09	0	3092
Swap	0	0.07	2.04	7.8	0.67	0.14	0	1819

Elephant mesh

Note that, as expected, when only swapping is performed, the final mesh size is smaller than that obtained for the combined Insert/Swap algorithm.

5.2 *Improvement Performance of the Centroid/Swap Algorithm*

Here we present improvement results for the algorithm of Sect. 4, for the nine test problems of Fig. 2 and for one random points mesh. Table 2 summarizes the mesh sizes (initial and final meshes) for these problems, Table 3 includes the refinement

Table 2 Mesh sizes

Mesh sizes (# tetrahedra)						
Mesh		Retinal	Elephant	*P*	N090	Angel
Size	Initial	1374	1905	926	2623	13,509
	Final	2663	2834	1080	10,014	24,822
Mesh		Helmet	Dragon	Spine	Triceratops	Rand2000
Size	Initial	1268	7209	3089	46,202	13,016
	Final	2105	12,104	7338	85,379	31,030

Table 3 Refinement time

Refinement time (s)					
	Retinal	Elephant	P	N090	Angel
Time	5.7	1.59	0.20	18.52	63.88
	Helmet	Dragon	Spine	Triceratops	Rand2000
Time	1.74	31.5	24.91	116.98	77.13

Table 4 Distribution of extreme dihedral angle

%Dihedral angles								
Mesh		<5°	<10°	<20°	<30°	>150°	>160°	>170°
Retinal	Initial	0.06	1.8	10.46	22.31	3.71	0.92	0.13
	Final	0	0.06	1.87	8.12	0.99	0.12	0
Elephant	Initial	0.008	0.54	3.9	10.51	1.6	0.54	0.07
	Final	0	0	0.39	3.63	0.16	0.017	0
P	Initial	0.25	0.41	1.94	5.45	0.76	0.41	0.14
	Final	0	0	0.062	1.33	0.016	0	0
N090	Initial	6.46	12.73	24.22	31.48	10.08	6.58	2.16
	Final	0.73	2.24	6.93	13.95	2.99	1.33	0.29
Angel	Initial	0.04	0.64	5.01	12.1	1.44	0.43	0.047
	Final	0.005	0.034	0.56	4.41	0.21	0.039	0.007
Helmet	Initial	0.005	0.47	3.4	9.96	0.85	0.25	0.026
	Final	0	0.056	0.46	4.18	0.16	0.055	0
Dragon	Initial	0.053	0.74	5.64	13.19	1.95	0.61	0.067
	Final	0.007	0.075	0.91	5.07	0.43	0.091	0.007
Spine	Initial	1.56	5	13.19	21.36	4.49	2.48	0.77
	Final	0.1	0.36	2.15	7.98	0.84	0.29	0.075
Triceratops	Initial	0.15	0.99	5.52	11.7	1.88	0.72	0.15
	Final	0.015	0.07	0.59	4.27	0.28	0.067	0.015
Rand2000	Initial	1.98	4.34	10.55	18.13	4.41	2.7	1.23
	Final	0.36	0.99	2.49	6.65	1.37	0.84	0.4

time, and Table 4 shows the distribution of the extreme dihedral angles for each one of these problems both for the initial and final meshes. Figure 3 shows the final dihedral angle distributions as compared with the dihedral angle distribution of the initial meshes.

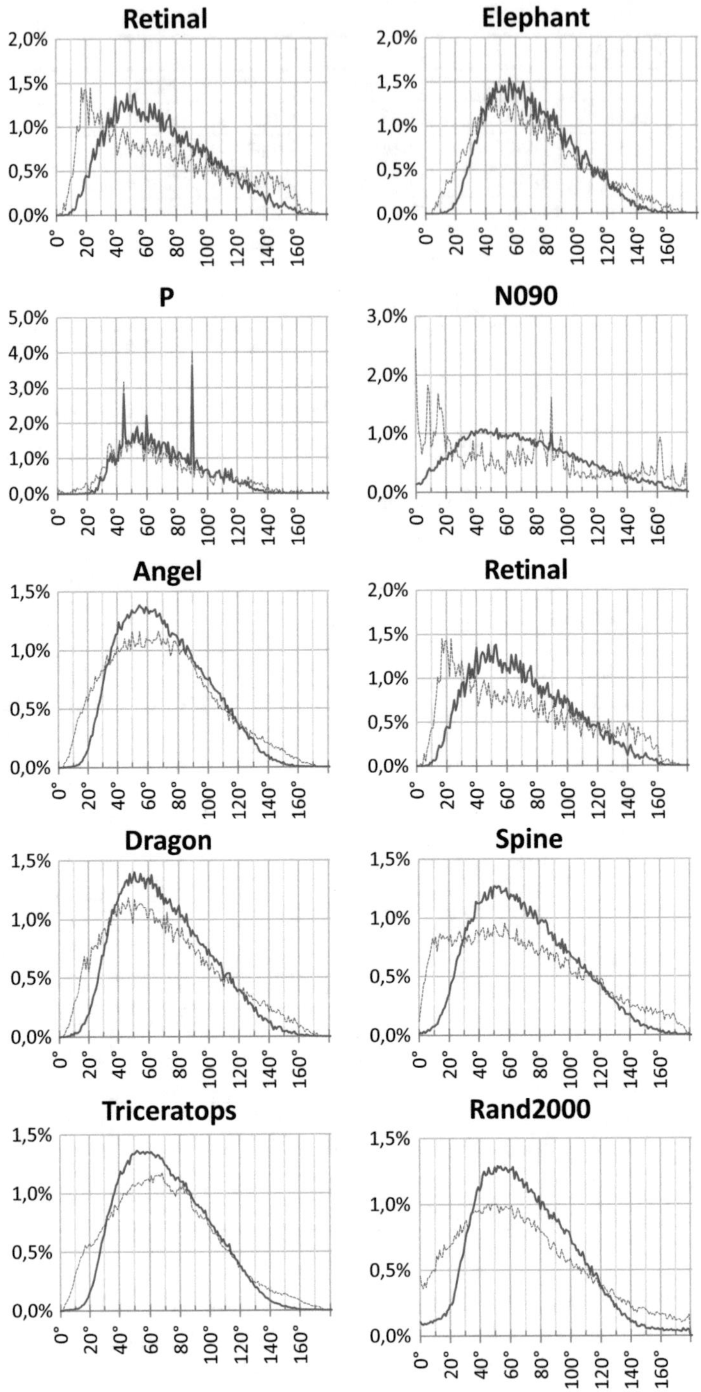

Fig. 3 Histogram of dihedral angles for the input mesh (in blue) and for the output mesh (in red)

5.3 *Performance Behavior of the Terminal Star Operations*

We have performed a complete statistical study on the frequency of the three operations (centroid insertion, star swapping and nothing) over the terminal stars of different sizes. Firstly, we studied the distribution of the different types of terminal stars (according to their sizes), over all the meshes, finding that the distribution is analogous for all of them. Figure 4 summarizes these results (in average) for all the meshes. Then we studied the percentage of the three operations performed over each type of terminal star by our refinement algorithm. Figure 5 summarizes (in average) these results.

Several remarks are in order:

- 70% of the terminal stars have size < 6.
- For most of the stars of size 3, edge swapping is performed.
- The option of doing nothing (to skip the improvement operations) increases with the star size.
- For star sizes ≥ 6, a small amount of improvement operations is performed.

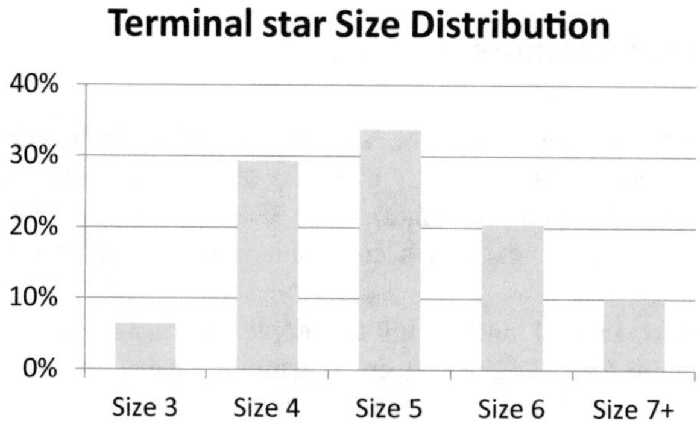

Fig. 4 Size distribution of the terminal stars (in average for all meshes)

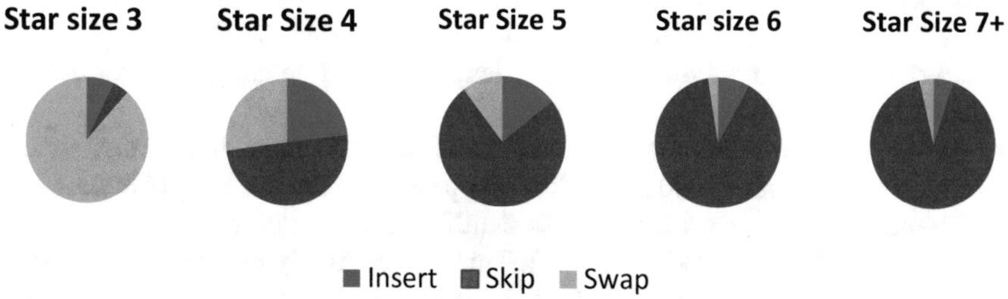

Fig. 5 Local mesh operations (centroid insertion, star swapping, nothing) as a function of the terminal star size

6 Algorithms Comparison

Here we compare our results with those of Klingner and Shewchuk [8], who provided an empirical study (with computing times) for three alternative mesh improvement strategies over a Mac Pro 2.66 GHz Intel Xeon processor, comparable to our hardware. The more recent references [4, 9] did not provide run times. A direct comparison of the computing times shows that our algorithm is 15 to 50 times faster than those of Klingner and Shewchuk. Our algorithm, for a mesh of 12,000 tetrahedra takes 31 s and for a mesh of 85,000 tetrahedra takes 116 s. In exchange the methods of reference [8], for a mesh of 11,000 tetrahedra take 497–940 s, and for a mesh of 50,000 tetrahedra take 950–5823 s. Certainly, better quality meshes are obtained by the methods of reference [8], but these can be only applied to input meshes with good distribution of points.

It is worth noting that our improvement algorithm is more general than previous methods, which can be applied to any valid input mesh independently of the distribution of interior points. Furthermore our algorithm always improves the distribution of dihedral angles independently of the meshing problem.

7 Concluding Remarks

- We have presented an innovative, simple, and effective algorithm for mesh improvement, which works locally over sets of tetrahedra sharing local largest edges in the mesh (terminal stars). This takes advantage of the interesting properties of terminal stars both for point insertion and for terminal edge swapping.
- The method works well both for initial complex meshes having a small number of interior points (elephant, retinal, angel geometries) and for bad meshes having a good distribution of interior points. This is a clear ad-vantage, with respect to previous methods, which require a good distribution of interior points to work [6, 8, 9].
- In our opinion the method has great potential to become the basis of a solid algorithm for three dimensional mesh improvement by adding adequate local operations for improving boundary tetrahedra and intelligent simplification techniques.
- As expected, for the random point meshes, it is more appropriate to use smoothing operations and simplification operations than inserting points.
- The computational cost is low (less than 20 s for most problems and 117 s for the most difficult retinal geometry). This can be highly de-creased at least in the following two senses: (a) Terminal edge swapping operations are critical and expensive operations, whose implementation can be optimized following the guidelines discussed in reference [6]; (b) The computational work can be constrained to terminal stars of size = 6 according to the analysis of Sect. 3.

- The algorithms are very appropriate for parallelization.
- In a next version of this paper we plan to study the algorithm behavior with other tetrahedra quality measures.

Acknowledgements Work partially supported by Departamento de Ciencias de la Computación, Universidad de Chile, and research Project DIUBB 172115 4/R, Universidad del Bío Bío. We are grateful to the referees who contributed to the improvement of this paper.

References

1. C. Bedregal, M.C. Rivara, Longest-edge algorithms for size-optimal refinement of triangulations. Comput. Aided Des. **46**, 246–251 (2014)
2. J. Castaños, J. Savage, PARED: a framework for the adaptive solution of PDEs, in *High Performance Distributed Computing, 1999. Proceedings of the Eighth International Symposium on IEEE*, 1999, pp. 133–140
3. L. Chen, M. Holst, Efficient mesh optimization schemes based on optimal Delaunay triangulations. Comput. Methods Appl. Mech. Eng. **200**, 967–984 (2011)
4. F. Dassi, L. Kamenski, H. Si, Tetrahedral mesh improvement using moving mesh smoothing and lazy searching flips, in *25th International Meshing Roundtable*. Procedia Engineering, vol. 163, 2016, pp. 302–314
5. H. Edelsbrunner, D. Guoy, An experimental study of sliver exudation. Eng. Comput. **18**, 229–240 (2002)
6. L.A. Freitag, C. Ollivier-Gooch, Tetrahedral mesh improvement using swapping and smoothing. Int. J. Numer. Methods Eng. **40**, 3979–4002 (1997)
7. M. Jones, P. Plassmann, Adaptive refinement of unstructured finite-element meshes. Finite Elem. Anal. Des. **25**, 41–60 (1997)
8. B.M. Klingner, J.R. Shewchuk, Aggressive tetrahedral mesh improvement, in *Proceedings 16th International Meshing Roundtable* (Springer, Berlin, 2008), pp. 3–23
9. M.K. Misztal, J.A. Bærentzen, F. Anton, K. Erleben, *Tetrahedral mesh improvement using multi-face retriangulation*, in Proceedings of the 18th International Meshing Roundtable, ed. by B.W. Clark (Springer, Berlin, 2009), pp. 539–V555
10. M.C. Rivara, Algorithms for refining triangular grids suitable for adaptive and multigrid techniques. Int. J. Numer. Methods Eng. **20**, 745–756 (1984)
11. M.C. Rivara, New longest-edge algorithms for the refinement and/or improvement of unstructured triangulations. Int. J. Numer. Methods Eng. **40**, 3313–3324 (1997)
12. M.C. Rivara, Lepp-bisection algorithms, applications and mathematical properties. Appl. Numer. Math. **59**, 2218–2235 (2009)
13. M.C. Rivara, C. Levin, A 3-D refinement algorithm suitable for adaptive and multi-grid techniques. Commun. Appl. Numer. Methods **8**, 281–290 (1992)
14. M.C. Rivara, M. Palma, New LEPP algorithms for quality polygon and volume triangulation: implementation issues and practical behavior, in *Trends in Unstructured Mesh Generation*, AMD-Vol. 220, ed. by S.A. Canann, S. Saigal (ASME, New York, 1997), pp. 1–9
15. M.C. Rivara, P.A. Rodriguez-Moreno, Tuned terminal triangles centroid Delaunay algorithm for quality triangulation, in *Proceedings of the 27th International Meshing Roundtable*, Albuquerque (2018)
16. M.C. Rivara, C. Calderon, A. Fedorov, N. Chrisochoides, Parallel decoupled terminal-edge bisection method for 3D mesh generation. Eng. Comput. 22, 111–119 (2009)

17. P. Rodriguez-Moreno, Parallel Lepp-based algorithms for the generation and refinement of 2D and 3D triangulations, PhD thesis, Department of Computer Science, University of Chile, 160 pages, 2015
18. R. Williams, Adaptive parallel meshes with complex geometry, in *Numerical Grid Generation in Computational Fluid Dynamics and related Fields* (Elsevier, Amsterdam, 1991), pp. 201–213

Part IV
Curved High-Order Meshes

Towards Simulation-Driven Optimization of High-Order Meshes by the Target-Matrix Optimization Paradigm

Veselin Dobrev, Patrick Knupp, Tzanio Kolev, and Vladimir Tomov

Abstract We present a method for simulation-driven optimization of high-order curved meshes. This work builds on the results of Dobrev et al. (The target-matrix optimization paradigm for high-order meshes. ArXiv e-prints, 2018, https://arxiv.org/abs/1807.09807), where we described a framework for controlling and improving the quality of high-order finite element meshes based on extensions of the Target-Matrix Optimization Paradigm (TMOP) of Knupp (Eng Comput 28(4):419–429, 2012). In contrast to Dobrev et al. (2018), where all targets were based strictly on geometric information, in this work we blend physical information into the high-order mesh optimization process. The construction of target-matrices is enhanced by using discrete fields of interest, e.g., proximity to a particular region. As these discrete fields are defined only with respect to the initial mesh, their values on the intermediate meshes (produced during the optimization process) must be computed.

The author "P. Knupp" performed under the auspices of the U.S. Department of Energy under Contract DE-AC52-07NA27344 (LLNL-CONF-752038).

V. Dobrev · T. Kolev · V. Tomov (✉)
Lawrence Livermore National Laboratory, Livermore, CA, USA
e-mail: dobrev1@llnl.gov; kolev1@llnl.gov; tomov2@llnl.gov

P. Knupp
Dihedral LLC, Bozeman, MT, USA

X. Roca, A. Loseille (eds.), *27th International Meshing Roundtable*, Lecture Notes in Computational Science and Engineering 127,
https://doi.org/10.1007/978-3-030-13992-6_16

We present two approaches for obtaining values on the intermediate meshes, namely, interpolation in physical space, and advection remap on the intermediate meshes. Our algorithm allows high-order applications to have precise control over local mesh quality, while still improving the mesh globally. The benefits of the new high-order TMOP methods are illustrated on examples from a high-order arbitrary Lagrangian-Eulerian application (BLAST, High-order curvilinear finite elements for shock hydrodynamics. LLNL code, 2018, http://www.llnl.gov/CASC/blast).

1 Introduction

Discretizations that leverage high-order curved meshes have become increasingly popular in recent years, due to several mathematical and computational advantages. For example, high-order meshes are essential for achieving optimal convergence rates on domains with curved boundaries/interfaces and for symmetry preservation in radial flow [8, 9]. They develop naturally in high-order simulations with moving meshes, including Lagrangian formulations [8, 9, 17]. They play a pivotal role in high-order arbitrary Lagrangian-Eulerian (ALE) methods [3, 7], where maintaining high-order convergence (for smooth flows) and 3D symmetry is generally impossible with linear meshes. High-order meshes can be relatively coarse and still capture curved geometries adequately, leading to equivalent simulation quality for a smaller number of elements. Furthermore, such meshes have increased flexibility and great potential in the context of adaptivity to simulation features.

Our general framework for controlling and improving the quality of *high-order* finite element meshes was presented in [11]. The proposed approach is based on TMOP [19], which is distinguished from similar methods by its emphasis on target-matrix construction methods that permit a greater degree of control over the optimized mesh. Mesh positions are optimized via node movement. Pointwise mesh quality metrics are defined by utilizing sub-zonal information. These metrics are capable of measuring shape, size or alignment of the region around the point of interest. TMOP requires predefined target-matrices as a way for the users to incorporate application-specific physical information into the metric that is being optimized. The combination of targets and quality metrics is used to optimize the node positions, so that they are as close as possible to the shape/size/alignment of their targets. The resulting mathematical problem is posed as global or local optimization of the chosen metric over the mesh.

All optimization methods in [11] are driven explicitly by geometric information. Many high-order applications require an additional capability, namely, adaptivity to certain physical features. The emphasis in this paper is the study of simulation-driven optimization within the context of finite element r-adaptivity on high-order meshes. Copious literature exists for r-adaptivity on first-order finite element meshes (e.g., [18]), but very little for high-order meshes. We extend the methods from [11] by including information about discrete fields in the optimization process. An example of such discrete field is some material's location, which is represented as a finite element function on the starting mesh. Recent advances on this subject were presented in [16, 22]. In contrast, this study addresses the case of general high-order curved triangular/quadrilateral and tetrahedral/hexahedral grids. Furthermore, the concept of adaptivity can be applied towards a general target, e.g., we can adapt not only the size, but also the local shape of the elements.

We enable the use of simulation-specific information by developing target-matrix construction algorithms that include information about discrete solution fields and data of interest. The use of discrete field data is a major challenge, because it is defined only with respect to the initial mesh. The values of the discrete data on the intermediate meshes (produced during the optimization) are obtained either by interpolation in physical space, or by advection/reconstruction on the intermediate meshes [1, 16, 22]. After the discrete values on the current mesh are obtained and the final target-matrices are calculated, we proceed by optimizing a global nonlinear functional via moving the mesh node positions. This nonlinear functional is defined in terms of local quality metrics, which can measure shape/size or alignment in the neighborhood of a given sample point.

Section 2 reviews the representation of the high-order mesh, basic TMOP components, and the objective function. Section 3 describes the construction of adaptivity-based target-matrices. Section 4 considers ways to transition the discrete values of interest from the initial to the current mesh. In Sect. 5 we discuss approaches to control the frequency of mesh optimization steps in time-dependent simulations through leveraging local quality metrics. Section 6 presents numerical results which demonstrate the ability to adapt high-order meshes using basic TMOP target-matrices and quality metrics. Conclusions are presented in Sect. 7.

2 Preliminaries

The goal of this section is to introduce the discrete mesh representation that is used throughout the paper, and briefly recall the purely geometric-based approach for optimization of high-order meshes through TMOP. The complete details can be found in [11]. The changes in the TMOP components related to adaptivity to simulation features are discussed in the following sections.

2.1 Discrete Representation of the High-Order Mesh

Let $d \in \{1, 2, 3\}$ be the space dimension and consider a set of scalar basis functions $\{\bar{w}_i\}_{i=1}^N$ on a reference element $\bar{E}$ of dimension d. Typically, for simplicial elements (triangles, tets), the basis $\{\bar{w}_i\}$ spans P_k, the space of all polynomials of total degree at most k; for tensor product elements (quads, hexes) $\{\bar{w}_i\}$ is chosen to be a basis for Q_k, the space of all polynomials of degree at most k in each variable. The degree k is the mesh order, e.g. $k = 1$ corresponds to linear meshes, $k = 2$ to quadratic meshes, etc. The shape of any element E in the mesh is then fully described by a matrix $\mathbf{x}_E$ of size $d \times N$ whose columns represent the coordinates of the element control points (aka element degrees of freedom). Given $\mathbf{x}_E$, we introduce the map $\Phi_E : \bar{E} \to \mathbb{R}^d$ whose image is the high-order element E:

$$x(\bar{x}) = \Phi_E(\bar{x}) \equiv \sum_{i=1}^{N} \mathbf{x}_{E,i} \bar{w}_i(\bar{x}), \qquad \bar{x} \in \bar{E}, \tag{1}$$

where we used $\mathbf{x}_{E,i}$ to denote the i-th column of $\mathbf{x}_E$. When two elements E and E' share a common mesh entity (vertex, edge, or face) then their control-point matrices $\mathbf{x}_E$ and $\mathbf{x}_{E'}$ are not independent because, to ensure mesh continuity, the descriptions of the common entity (through Φ_E on one hand and $\Phi_{E'}$ on the other) must coincide. This type of interdependence among mesh elements is typically expressed by defining a *global* vector $\mathbf{x}$ of control points (degrees of freedom) and a set of linear operators $\mathbf{x} \to \mathbf{x}_E$, one per mesh element, that define/extract the local coordinates $\mathbf{x}_E$ from the global vector $\mathbf{x}$. Thus, the global continuity of the mesh is ensured for any value of $\mathbf{x}$.

For any element E in the mesh, we can compute the Jacobian of the mapping Φ_E at any reference point $\bar{x} \in \bar{E}$ as

$$A_E(\bar{x}) = \frac{\partial \Phi_E}{\partial \bar{x}} = \sum_{i=1}^{N} \mathbf{x}_{E,i} [\nabla \bar{w}_i(\bar{x})]^T . \tag{2}$$

2.2 *TMOP and Optimization of High-Order Meshes*

At each quadrature point (inside each mesh element), TMOP uses three Jacobian matrices:

- The Jacobian matrix $A_{d \times d}$ of the transformation from reference to physical coordinates, given by (2).
- The *target-matrix*, $W_{d \times d}$, which is the Jacobian of the transformation from the reference to the *target* coordinates. The target-matrices are defined according to a user-specified method prior to the optimization; they define the desired properties in the optimal mesh.
- The *weighted Jacobian* matrix, $T_{d \times d}$, defined by $T = AW^{-1}$, represents the Jacobian of the transformation from the target to the physical coordinates.

The T matrix is used to define the *local quality measure*, $\mu(T)$. The quality measure can evaluate shape, size, or alignment of the quadrature point's neighborhood.

The goal of TMOP is to minimize a global *objective function* that depends on the local quality measure throughout the mesh. In this paper we focus on the following objective function:

$$F(x) := \sum_{E \in \mathcal{M}} \int_{E_t} \mu(T(x_t)) dx_t = \sum_{E \in \mathcal{M}} \sum_{x_q \in E_t} w_q \det(W(\bar{x}_q)) \, \mu(T(x_q)) , \tag{3}$$

where $\mathcal{M}$ is the current mesh, E_t is the target element corresponding to the physical element E, w_q are the quadrature weights, and the point x_q is the image of the reference quadrature point $\bar{x}_q$ in the target element E_t. Note that the right-hand side in (3) depends on the mesh positions x through the Jacobian matrices A used in the definition of T; in addition, we will also allow the target matrices $W(\bar{x}_q)$ to depend on the mesh positions x, see Sect. 3. The integration in (3) is performed over the target elements, enforcing that the integral contribution from a given element E, relative to the contributions from other elements, is only based on the difference with its target E_t (which is measured by $\mu(T)$), and not on its relative size compared to the other elements. In particular, very small elements will not be neglected by the optimization process due to their size. The existence of a minimum for the variational optimization problem (3) has been explored theoretically in [13, 14]. The objective function can be extended by using combinations of quality metrics, space-dependent weights for each metric, and limiting the amount of allowed mesh displacements, see details in [11].

As $F(x)$ is nonlinear, in this paper we minimize it by utilizing Newton's method to solve the critical point equations $\partial F(x)/\partial \mathbf{x} = 0$. This approach involves the computation of the first and second derivatives of $\mu(T)$ with respect to T. Furthermore, boundary nodes are enforced to stay fixed or move only in the boundary's tangential direction. Line search procedure is utilized to guarantee that the Newton updates do not lead to inverted meshes, see details in [11]. Alternative approach for minimizing $F(x)$ is presented in [21], where local optimization problems are solved for each node of the mesh.

In the context of ALE, we will also refer to the initial mesh, which is to be optimized, as the "Lagrangian" mesh. We denote the Lagrangian mesh by $\mathcal{M}_0$, and $\mathcal{M}_n$ is the mesh obtained after n Newton iterations.

2.3 Adaptivity Function

To represent the desired mesh features and to control the adaptivity process, we utilize an adaptivity field $\eta(x)$ (the terms monitor function and metric tensor are also used [6, 18]). In general, the definition of this field depends on some simulation-specific goal of the adaptivity process. Adaptivity fields are constructed by user-defined procedures that combine the simulation's available input. For example, $\eta(x)$ can be a tensor representing multiple (up to 4 in 2D and up to 9 in 3D) point-wise characteristics of the desired mesh element located at the point x.

In this paper, we assume that $\eta(x)$ is derived from the simulation's discrete field data (thus, η is also a discrete field). To illustrate the basic properties of the method in a simplified framework, we define $\eta(x) \in \mathcal{V}$ as a scalar finite element function with values in [0, 1]. We choose the discrete space $\mathcal{V} \subset H^1(\mathcal{M})$ to be the space of continuous finite element functions of degree k defined on the domain mesh $\mathcal{M}$:

$$\mathcal{V} = \begin{cases} \{v \in C^0(\mathcal{M}) \mid v_{|E} \in Q_k(E)\} & \text{forquadrilateral/hexahedral meshes}\,, \\ \{v \in C^0(\mathcal{M}) \mid v_{|E} \in P_k(E)\} & \text{fortriangular/tetrahedral meshes}\,, \end{cases}$$

where E is an arbitrary mesh element, $Q_k(E)$ is the mapped space of polynomials of degree at most k in each variable, and $P_k(E)$ is the mapped space of polynomials of total degree at most k, i.e.

$$Q_k(E) = \{p \circ \Phi_E^{-1} \mid p \in Q_k\} \qquad P_k(E) = \{p \circ \Phi_E^{-1} \mid p \in P_k\}.$$

Unit values of η represent regions of most interest, e.g., smallest local mesh size in the case of size adaptivity, and zero values represent regions of least interest. The choice of the continuous finite element space $\mathcal{V}$ is made to avoid sharp transitions. Furthermore, $\eta \in \mathcal{V}$ uses the same degree k as the mesh, in order to represent adequately the sub-element resolution. One can also choose to discretize η with lower degree polynomials, depending on the level of detail that is expected on sub-element level. Specific formulations and usages of $\eta(x)$ in the context of shape and size adaptivity are presented in Sect. 6.

3 Adaptive Target-Matrices

Once the adaptivity function $\eta(x)$ is known on a given mesh, its information needs to be converted into a set of spatially-varying target-matrices $\{W(q_\ell)\}$ through a Target-Matrix construction algorithm. The set $\{q_\ell\}$ consists of all quadrature points (which depend on the mesh-node locations) in physical space. The specific construction algorithm depends on the type of data contained in the adaptivity field, the corresponding features desired in the adapted mesh, and on properties of the quality metric. In this work we illustrate this idea by utilizing targets of the form

$$W = \begin{pmatrix} W_{11} & 0 \\ 0 & W_{22} \end{pmatrix}, \text{ where } W_{11} > 0 \text{ and } W_{22} > 0 \text{ both depend on } \eta(x).$$

Each of these target elements is orthogonal, has aspect ratio W_{22}/W_{11}, and area $W_{11}W_{22}$. This setup controls shape, if a Shape metric is used, and both shape and size, if a Shape+Size metric is used. Specific values of W_{11} and W_{22} for the above adaptation cases are presented in Sect. 6. Choosing W to be diagonal is not overly restrictive because this form of the target-matrix will be used in conjunction with rotation-invariant shape or shape+size quality metrics.

As W depends on the discrete function $\eta(x)$, a major difference in the resulting optimization problem (3), compared to our previous work on purely geometric mesh quality optimization, is that the gradient of the mesh quality functional, $\partial\mu(T)/\partial\mathbf{x}$, requires the computation of the derivatives of $\{W(q_\ell)\}$ with respect to the mesh-node locations. Namely, its first derivative, at a fixed quadrature point $\bar{x}_q \in \bar{E}$ and a fixed mesh element E such that $q_\ell = \Phi_E(\bar{x}_q)$, has the form (denoting $[\mathbf{x}_E]_{ji} = x_{ij}$,

$i = 1, \ldots, N$, $j = 1, \ldots, d$)

$$\begin{aligned}\frac{\partial \mu(T)}{\partial x_{ij}} &= \sum_{k,l=1}^{d} \frac{\partial \mu(T)}{\partial T_{kl}} \frac{\partial T_{kl}}{\partial x_{ij}} = \frac{\partial T}{\partial x_{ij}} : \frac{\partial \mu}{\partial T} = \frac{\partial \left(AW^{-1}\right)}{\partial x_{ij}} : \frac{\partial \mu}{\partial T} \\ &= \left(\frac{\partial A(\mathbf{x})}{\partial x_{ij}} W^{-1}(\eta(q_\ell)) + A(\mathbf{x}) \frac{\partial W^{-1}(\eta(q_\ell))}{\partial x_{ij}} \right) : \frac{\partial \mu}{\partial T} ,\end{aligned} \tag{4}$$

where $\frac{\partial \mu}{\partial T}$ is a matrix with components $\left[\frac{\partial \mu}{\partial T}\right]_{kl} = \frac{\partial \mu}{\partial T_{kl}}$ and we used the notation $X : Y$ to denote the inner product of two matrices X and Y of the same size, $m \times n$, defined as $X : Y = \sum_{s=1}^{m} \sum_{t=1}^{n} X_{st} Y_{st}$. In the above formula, we can further express

$$\frac{\partial W^{-1}(\eta(q_\ell))}{\partial x_{ij}} = \frac{\partial W^{-1}}{\partial \eta} \frac{\partial \eta}{\partial x_{ij}} = -W^{-1} \frac{\partial W}{\partial \eta} W^{-1} \frac{\partial \eta}{\partial x_{ij}} .$$

The Jacobian matrix $A(\mathbf{x})$ and its derivative with respect to the variable x_{ij} can be expressed as

$$[A(\mathbf{x})]_{kl} = \sum_{s=1}^{N} x_{sk} [\nabla \bar{w}_s(\bar{x}_q)]_l , \qquad \frac{\partial A_{kl}}{\partial x_{ij}} = \delta_{kj} [\nabla \bar{w}_i(\bar{x}_q)]_l .$$

The minimization functional $F(x)$ also depends on $\det(W)$, so we need its derivative as well:

$$\frac{\partial [\det(W)]}{\partial x_{ij}} = \frac{\partial [\det(W)]}{\partial \eta} \frac{\partial \eta}{\partial x_{ij}} = \left(\operatorname{adj}(W)^T : \frac{\partial W}{\partial \eta} \right) \frac{\partial \eta}{\partial x_{ij}} .$$

In the above formulas, the term $\frac{\partial \eta}{\partial x_{ij}}$ can be computed as

$$\frac{\partial \eta}{\partial x_{ij}} \approx \nabla \eta(q_\ell) \cdot \frac{\partial q_\ell}{\partial x_{ij}} , \qquad \text{where} \quad [q_\ell]_k = \sum_{s=1}^{N} x_{sk} \bar{w}_s(\bar{x}_q) .$$

Note that the latter formula is an approximation of the derivative $\frac{\partial \eta}{\partial x_{ij}}$ because it does not take into account the fact that when the mesh nodes move, the discrete values of η on the new mesh will change, as discussed in Sect. 4. Computing the exact numerical derivatives is non-trivial, as the calculation must take into account the transition procedure of the discrete η values. We are planning to address the computations of exact derivatives in our future work. More complicated expressions for η, e.g., tensors, would lead to additional challenges in the computation of the above derivatives. Furthermore, when the Newton's method is used, one also needs to compute second derivatives of the quantities discussed above.

4 Obtaining Discrete Field Values

Adaptive target-matrices must be evaluated on the current mesh configuration $\mathcal{M}_n$. Computing these matrices is not trivial, because the discrete finite element function $\eta(x)$ is typically given on the initial Lagrangian mesh $\mathcal{M}_0$. In this section we propose two alternative methods for evaluating $\eta(x)$ on the mesh $\mathcal{M}_n$.

4.1 *Interpolation*

One way to evaluate $\eta(x)$ is to find the logical image of x on the Lagrangian mesh $\mathcal{M}_0$. More specifically, for a given physical point x, we must find the corresponding element $E_0 \in \mathcal{M}_0$ and reference point $\bar{x}_0 \in \bar{E}_0$, so that

$$\Phi_{E_0}(\bar{x}_0) = x \, .$$

Once the image $\bar{x}_0$ of x is found in some element $E_0 \in \mathcal{M}_0$, we simply perform a standard finite element evaluation of η on the Lagrangian mesh $\mathcal{M}_0$.

The first step in this procedure is to determine a set $\mathcal{S}$ of candidate elements in $\mathcal{M}_0$. Each of these elements would be tested for containing the physical point x. We form this set by including: (1) the element E_0^* whose center point is closest to x, and (2) all vertex-neighbor elements of E_0^*. The above set can also be formed by more advanced algorithms, e.g., Section 7.1.5 in [12].

Testing whether a given element $E_0 \in \mathcal{S}$ contains x, consists of inverting the nonlinear transformation Φ, given by (1). Once solved, if $\bar{x}_0 \in \bar{E}_0$, then the point is found in E_0. Otherwise, the point is outside of E_0 and we proceed with the remaining elements in the candidate set. The inversion problem is solved by a Newton method:

$$\bar{x}_0^{n+1} = \bar{x}_0^n + A^{-1}(\bar{x}_0^n)\left[x - \Phi_{E_0}(\bar{x}_0^n)\right] .$$

In general, there is no guarantee that the above nonlinear iteration will find an element that contains x. Specific implementations of the above interpolation procedure can trade performance for accuracy and robustness, e.g., adjusting convergence tolerances, choosing initial guess within the reference cell, correction strategy when the Newton update overshoots the element boundaries. The specifics of our implementation of the above inversion method can be found in the MFEM finite element library [20], in particular the function *FindPoints* and the class *InverseElementTransformation*.

This interpolation procedure is used to obtain the value of η at a single point. To reconstruct the whole discrete finite element function $\eta(x)$ on $\mathcal{M}_n$, we interpolate the position (with respect to $\mathcal{M}_n$) of every node of the space $\mathcal{V}$.

4.2 Advection Remap

In this section we utilize ideas from [1] to transfer the discrete adaptivity function $\eta_0(x_0)$, defined on the Lagrangian mesh $\mathcal{M}_0$, to the function $\eta(x)$, defined on the current mesh $\mathcal{M}$. The problem is formulated in terms of transporting (or advecting) the adaptivity function between the two meshes over a fictitious (or pseudo) time interval determined by the motion of the mesh between the two configurations.

Assuming that every point $x_0 \in \mathcal{M}_0$ is associated with a unique point $x \in \mathcal{M}$, we can define a continuous transition function $P(x_0, \tau) : \mathcal{M}_0 \times [0, 1] \to \mathbb{R}^d$, such that

$$P(x_0, 0) = x_0 \quad \text{and} \quad P(x_0, 1) = x\,, \quad \forall x_0 \in \mathcal{M}_0\,.$$

We refer to the parameter $\tau \in [0, 1]$ as "pseudo-time". The intermediate meshes, $\mathcal{M}_\tau = \{P(x_0, \tau) : x_0 \in \mathcal{M}_0\}$, are obtained by linear interpolation:

$$P(x_0, \tau) = x_0 + \tau(x - x_0)\,.$$

Next we define the "pseudo-time velocity", u, as

$$u(x_0, \tau) := \frac{\partial P}{\partial \tau} = x - x_0\,.$$

Note that $u(x_0)$ is constant in pseudo-time along each transition trajectory $x_0 \to x$. Having defined such velocity, we can track the values of η in pseudo-time along the trajectories $x_\tau = P(x_0, \tau)$, by utilizing the concept of the advective derivatives (also known as "material" or "Lagrangian" derivatives):

$$\frac{d\eta(x_\tau, \tau)}{d\tau} = \frac{\partial \eta}{\partial \tau} + u \cdot \nabla \eta\,. \tag{5}$$

As the goal of this transfer procedure is to extend η_0 to η without changing its values with respect to pseudo-time, i.e., $\partial\eta/\partial\tau = 0$, we enforce this requirement in (5). That is, we transfer η by solving in pseudo-time the following equation:

$$\frac{d\eta}{d\tau} = u \cdot \nabla \eta\,, \quad \eta(x_0, 0) = \eta_0(x_0)\,. \tag{6}$$

On semi-discrete level, we discretize Eq. (6) by a standard continuous Galerkin formulation (recall that η is considered a continuous finite element function). Our discretization utilizes basis functions that follow the mesh motion, i.e., $dw/dt = 0$. The semi-discrete form of (6) is

$$\boldsymbol{M}\frac{d\boldsymbol{\eta}}{d\tau} = \boldsymbol{K}\boldsymbol{\eta}\,, \quad \text{where} \quad M_{ij} = \int_{\mathcal{M}_\tau} w_i w_j\,, \quad K_{ij} = \int_{\mathcal{M}_\tau} w_i (u \cdot \nabla w_j)\,. \tag{7}$$

This formulation does not need to include any boundary integrals, because every Newton step of our mesh optimization preserves the boundaries of the Lagrangian mesh or only moves nodes in the tangential direction of the boundary, i.e.,

$$(x - x_0) \cdot n = u \cdot n = 0\,.$$

Note that both the mass matrix $\boldsymbol{M}$ and the advection matrix $\boldsymbol{K}$ depend on τ, as the Jacobians of the reference $\to$ physical transformations depend on τ. Thus, these matrices must be reassembled at every pseudo-time step.

The discretization in time of (7) is performed by a standard explicit Runge-Kutta method. Our default choice for all numerical tests in this paper is RK4. The advection time step $\Delta\tau$ is computed by considering the mesh size of the original elements and the magnitude of the mesh advection velocity u as

$$\Delta\tau = 0.5 \min_{\mathcal{M}_0} \frac{\Delta x}{|u|}\,.$$

The above ratio is taken at various quadrature points within each element, and Δx is computed as the determinant (to power $1/d$) of the reference $\to$ physical Jacobian at a given quadrature point.

An important choice that must be considered in this advection approach is the starting point of the advection. When the Newton iteration computes a new mesh $\mathcal{M}_{n+1}$, there are two major options for the calculation of η_{n+1}:

1. Advect from the Lagrangian mesh, i.e., η_0 to η_{n+1}, with $u = x_{n+1} - x_0$.
2. Advect from the latest iteration, i.e., η_n to η_{n+1}, with $u = x_{n+1} - x_n$.

The main advantage of the first approach is that it does not accumulate numerical "noise". Without any nonlinear enhancements, linear transport methods usually produce oscillations [15], especially in the case of high-order finite elements [2]. Our goal in this paper, however, is to derive an advection method that is simple and gives a reasonable representation of η on the new mesh. The first approach will always advect η using the shortest path between x_0 and x_{n+1}, thus avoiding any intermediate stages. Such stages appear during the second approach, when numerous Newton iterations are taken. Each of these intermediate stages would introduce and accumulate oscillations in η, when the second approach is used. For thorough discussion and methods for treating oscillations resulting during transport of high-order finite element fields, see [2]. The advantage of the second approach is that every advection would usually perform less time steps, as x_{n+1} is expected to be closer to x_n than to x_0. In our numerical tests we always utilize the first approach.

After the time evolution of (6) is completed and $\eta(x)$ is computed on $\mathcal{M}_n$, we post-process the solution by adjusting $\eta(x) = \max(1, \eta(x))$ and $\eta(x) = \min(0, \eta(x))$. This is required, because our methods assume the range of η is always $[0, 1]$, and the remap procedure might violate this range, as explained in the previous paragraph.

5 Mesh Optimization Triggers

Moving mesh simulations, such as ALE, typically need to periodically perform mesh optimization steps (referred to as remeshing in ALE), because deterioration of the mesh quality can cause simulation failure. Performing such steps too often, however, usually results in (1) slower computations, and (2) additional numerical errors, because ALE steps are purely numerical and not based on the physics of the problem. In order to trigger remesh steps, applications need procedures to track accurately the mesh quality during the moving mesh phase, in order to produce robust, efficient, and accurate calculations.

The TMOP concepts can be utilized to define such mesh optimization triggers, both in the purely geometric and the simulation-driven case. Similar to specifying target-matrices, the user can specify *admissible* Jacobian matrix, S, which defines the transformation from the reference element to the worst element that can be used during the moving mesh phase. The Jacobian of the transformation between target and admissible coordinates becomes $U = SW^{-1}$. This matrix is then used to calculate $\mu(U)$, the highest admissible mesh quality metric value for the metric of interest μ. Then remesh is triggered whenever $\mu(T) > \mu(U)$ at any mesh quadrature point.

The specific formulation of S is, of course, problem dependent, e.g., it can be used to trigger remesh steps whenever particular mesh configurations occur. Just as the target-matrices W, the admissible matrices can contain information about the local shape, size or alignment, at any quadrature point. They can vary in space/time by depending on a known analytic function or the simulation's discrete fields. As an example, we can define S as

$$S = \begin{pmatrix} 1 & 0 \\ 0 & S_{22} \end{pmatrix}, \text{ where } S_{22} > 0 \text{ is a user} - \text{specified parameter}. \tag{8}$$

When μ is a Shape metric, remesh is triggered based on local aspect ratio. When μ is a Shape+Size metric, remesh is triggered based on local size and aspect ratio. Furthermore, the mesh optimization trigger can be made adaptive by making S_{22} space/time dependent.

6 Numerical Examples

In this section we report numerical results from the algorithms in the previous sections as implemented in the MFEM finite element library [20]. This implementation is freely available at http://mfem.org.

All results below are calculated by utilizing the objective function (3) and Gauss-Legendre quadrature for the resulting integrals. Newton's method is used to solve the nonlinear optimization problems with the modification that the derivatives of the target matrices W are assumed to be zero. The Newton relative tolerance is set to 10^{-12}, and all of the presented results are fully converged to this tolerance. The linear solve inside each Newton step is performed by the standard minimum residual (MINRES) algorithm. Boundary nodes are allowed to move as long as the motion does not perturb the initial domain.

For all tests, values of η on intermediate meshes are obtained by the advection remap method from Sect. 4.2. The interpolation method from Sect. 4.1 was also tested, and it produced similar results. The advection remap method is currently preferred because in our implementation it has better computational performance (it admits easier parallelization). Improvements of our interpolation routines and detailed comparisons between the two methods will be addressed in the future.

6.1 Adaptation to a Static Sinusoidal Region

This example is used to illustrate the utilization of a discrete solution field in TMOP and compare different adaptivity-based target construction methods. We consider three different target constructions. They are oriented towards adaptivity with respect to Size, Shape and coupled Size+Shape. We always start with a Cartesian 16×16 third order mesh. For all of the following tests, our goal is to achieve three times smaller area in regions where $\eta \approx 1$, compared to regions where $\eta \approx 0$.

In all cases, the discrete adaptivity function η is initialized on the initial mesh $\mathscr{M}_0$ by interpolating the following smooth analytic function:

$$\tanh(10(y_0 - 0.5) + \sin(4\pi x_0) + 1) - \tanh(10(y_0 - 0.5) + \sin(4\pi x_0) - 1)\,. \quad (9)$$

Once the above function is interpolated on the initial mesh, the algorithm uses only its finite element version, which is shown in Fig. 1.

Size Adaptation As a first attempt to adapt the local mesh size, we use the following targets and quality metric:

$$W_1(x) = [\eta(x)s + (1 - \eta(x))\alpha s]^{1/2} I\,, \quad \mu_7(T) = |T - T^{-t}|^2\,,$$

where $|T|^2 = \mathrm{tr}(T^t T)$, $\alpha = 7$ is a user-specified size ratio between big and small local size, and s is a local mesh size. Note that μ_7 is a Shape+Size metric. Thus, higher η values should result in smaller local mesh size. The size s is approximated by taking into account the total volume V of the domain, the volume V_η occupied by η, the number N of available elements, and the specified α:

$$\frac{V_\eta}{s} + \frac{V - V_\eta}{\alpha s} = N\,, \quad \text{where} \quad V_\eta = \int_{\mathscr{M}_0} \eta_0(x_0)\,, \quad V = \int_{\mathscr{M}_0} 1\,. \quad (10)$$

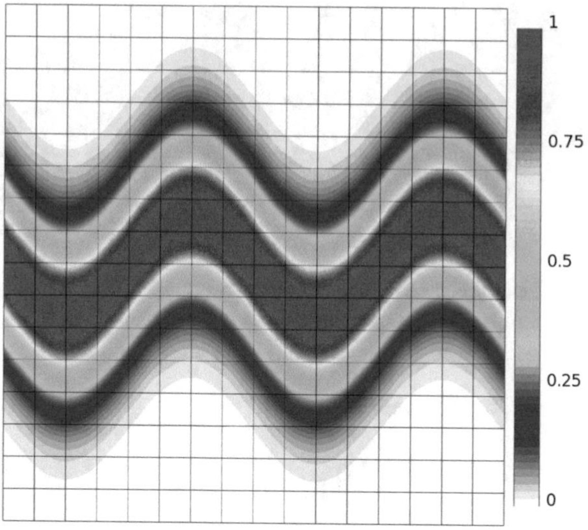

Fig. 1 Third order finite element adaptivity function η on the initial mesh

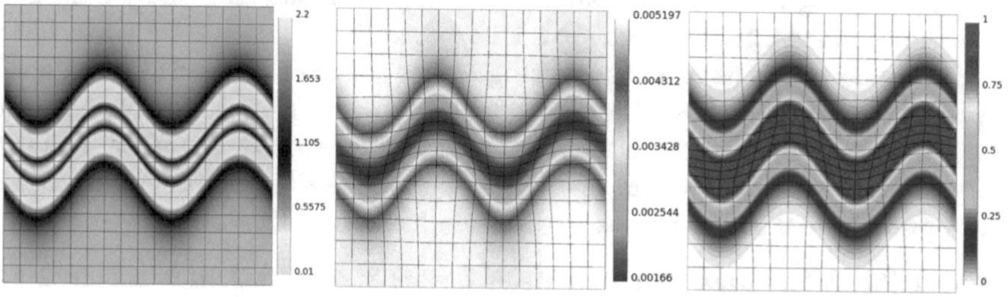

Fig. 2 Size adaptation: initial (μ_9, W_1) metric values (left), $\det(A)$ values on the optimized mesh (middle), and remapped η values on the optimized mesh (right)

Note that the calculation of s is done once using the initial mesh. As usual, the integrals in (10) are approximated by a quadrature rule of order comparable to the order of the mesh and η.

The results of this test are shown in Fig. 2. The objective function (3) is reduced by approximately 47%, from $F(x_0) = 2.165$ to $F(x) = 1.136$. We observe that the mesh size is adapted towards the bigger η values, but to achieve the desired 3-times reduction in size, we need to use a higher ratio in the target construction ($\alpha = 7$). This is caused by the fact that the combination (μ_7, W_1) optimizes not only the local size, but also tries to achieve equal aspect ratio. Nevertheless, combining shape and size optimization is recommended, as pure size optimization usually results in highly distorted meshes, which causes divergence of the Newton iterations.

Shape Adaptation As the above case suggested, to adapt successfully to η, our method must be able to adapt the local aspect ratio. This can be achieved by the

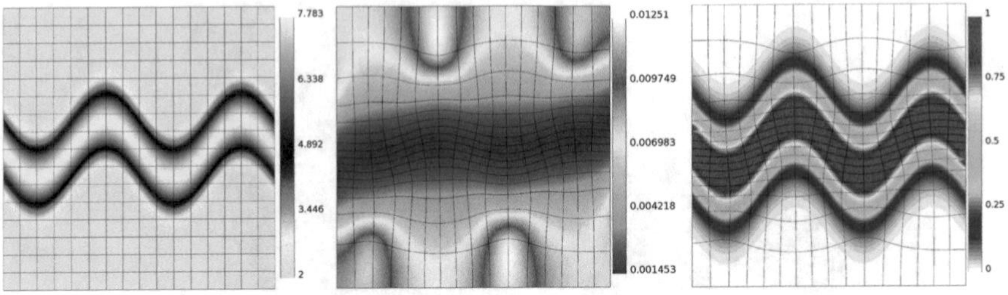

Fig. 3 Shape adaptation: initial (μ_{58}, W_2) metric values (left), $\det(A)$ values on the optimized mesh (middle), and remapped η values on the optimized mesh (right)

following targets and quality metric:

$$W_2(x) = \begin{pmatrix} \eta(x)\beta + (1-\eta(x)) & 0 \\ 0 & 1 \end{pmatrix}, \quad \mu_{58}(T) = \frac{|T^t T|^2}{\det(T)^2} - 2\frac{|T|^2}{\det(T)} + 2,$$

where $\beta = 3$ is a user-specified desired aspect ratio. Note that μ_{58} is a pure Shape metric. Thus, higher η values should result in higher resolution in the y direction, while smaller η values should lead to a 1:1 aspect ratio.

The results of this test are shown in Fig. 3. The objective function (3) is reduced by approximately 22%, from $F(x_0) = 1545.92$ to $F(x) = 1207.33$. We observe that the 3-times reduction in area is achieved in the region specified by η, through the aspect ratio adaptivity. However, optimizing μ_{58} produces bigger local sizes around the top and bottom boundaries, as μ_{58} is not influenced by local size.

Shape+Size Adaptation Our best result for the above problem is achieved by adapting shape and size simultaneously, through the following targets and metric:

$$W_3(x) = \begin{pmatrix} \sqrt{\gamma s} & 0 \\ 0 & \eta(x)\frac{\sqrt{\gamma s}}{\gamma} + (1-\eta(x))\sqrt{\gamma s} \end{pmatrix}, \quad \mu_9(T) = \det(T)|T - T^{-t}|^2,$$

where $\gamma = 3$ is a user-specified desired aspect ratio, and s is computed by (10). Note that μ_9 is a Shape+Size metric. Thus, $\eta = 1$ should result in 3:1 resolution in the y direction, and 3 times smaller local size than $\eta = 0$ values.

The results of this test are shown in Fig. 4. The objective function (3) is reduced by approximately 65%, from $F(x_0) = 0.756$ to $F(x) = 0.258$. We observe that the 3-times reduction in local mesh size and 3:1 aspect ratio is achieved in the region specified by the highest η values, while the 1:1 aspect ratio is preserved in the rest of the domain.

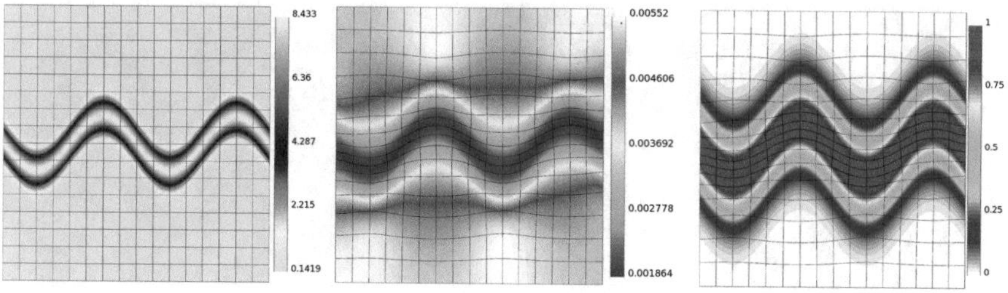

Fig. 4 Shape+Size adaptation: initial (μ_9, W_3) metric values (left), det(A) values on the optimized mesh (middle), and remapped η values on the optimized mesh (right)

6.2 2D Full ALE Simulation of Gas Impact

This problem represents a complete multi-material ALE simulation of the compressible Euler equations. The test is a simplified high velocity impact introduced in [4]. There are three materials that represent an impactor, a wall, and the background. This problem tests the mesh optimization method's ability to couple to full time dependent ALE simulations that use high-order finite elements. The presented simulation is performed by the BLAST code [5], which utilizes our implementation in the MFEM finite element library.

The domain is $[0, 2] \times [0, 2]$ with $v \cdot n = 0$ boundary conditions. The material regions are $0.15 \leq x \leq 0.65$ and $0.9 \leq y \leq 1.1$ for the Impactor, $0.1 \leq x \leq 1.0$ for the Wall, and the rest is Background. The problem is run to a final time of $t = 10$. We use a 60×60 second order mesh. The complete thermodynamic setup of this problem and additional details about our multi-material finite element discretization and overall ALE method can be found in [3, 10].

Remesh steps are performed at different times during the simulation, according to the algorithm presented in Sect. 5. The goal of the mesh optimization is to adapt the size of the mesh towards the locations of the Impactor and the Wall at any given remesh step. To achieve this, we use the following targets and quality metric:

$$W(x) = [\eta(x)s + (1 - \eta(x))\alpha s]^{1/2} I \,, \quad \mu_9(T) = \det(T)|T - T^{-t}|^2 \,,$$

where $\alpha = 10$, and the local size s is computed by (10). The adaptivity field η is constructed from the simulation's discrete volume fractions η_{wall} and η_{imp}. These fields have values in [0, 1], and they represent the positions of the Wall and Impactor at any given spatial point. The adaptivity field η is computed as

$$\eta(x) = \max(\eta_{wall}(x), \eta_{imp}(x)) \,.$$

The time evolution of the material positions and the corresponding mesh can be seen in Fig. 5. We observe that the algorithm adapts well to the moving materials, while preserving good overall shape throughout the domain.

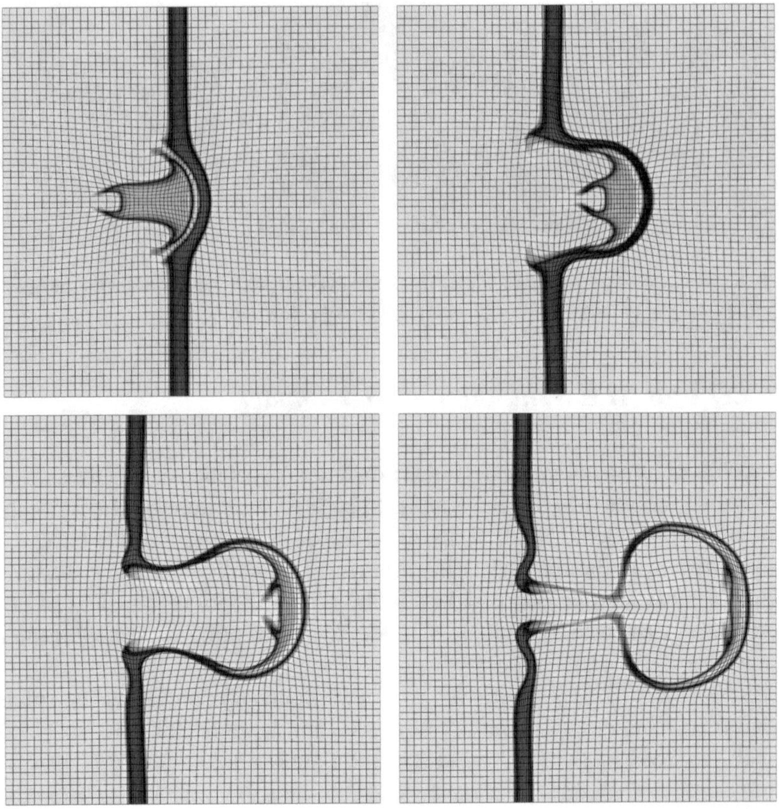

Fig. 5 Time evolution of the materials and mesh positions at times 2.5 (top left), 5 (top right), 7.5 (bottom left) and 10 (bottom right) in the 2D gas impact test case

7 Conclusions

In this paper we outline a framework for optimization of high-order curved meshes that is driven by the state of the overall simulation in which the mesh is being used. This framework is built on the Target-Matrix Optimization Paradigm [19] and [11], and allows high-order applications to have precise control over local mesh quality, while still improving the mesh globally.

While the current paper is intended as an initial extension of the pure geometry-based methods of [11] to the simulation-driven case, we do make the important choice to require only discrete description of the simulation feature to which to adapt to (e.g. provided as a finite element function on the mesh). This is a critical step for the practical applicability of the algorithms we propose and distinguishes us from approaches that require analytical information.

As discrete fields are defined only with respect to the initial mesh, one of the challenges with our approach is that we need to obtain their values on the intermediate meshes produced during the optimization process. We propose two methods to address this challenge. The numerical experiments reported here illustrate results which can be obtained from the second (advection remap) approach. We plan to explore and compare these results to alternative or improved algorithms and perform detailed comparison of accuracy and performance in the future. In addition, we will explore a wider range of applications of our methods beyond ALE. These will require more sophisticated target construction methods.

References

1. R.W. Anderson, V.A. Dobrev, T.V. Kolev, R.N. Rieben, Monotonicity in high-order curvilinear finite element arbitrary Lagrangian–Eulerian remap. Int. J. Numer. Methods Fluids **77**(5), 249–273 (2015)
2. R.W. Anderson, V.A. Dobrev, T.V. Kolev, D. Kuzmin, M.Q. de Luna, R.N. Rieben, V.Z. Tomov, High-order local maximum principle preserving (MPP) discontinuous Galerkin finite element method for the transport equation. J. Comput. Phys. **334**, 102–124 (2017)
3. R.W. Anderson, V.A. Dobrev, T.V. Kolev, R.N. Rieben, V.Z. Tomov, High-order multi-material ALE hydrodynamics. SIAM J. Sci. Comput. **40**(1), B32–B58 (2018)
4. A. Barlow, R. Hill, M.J. Shashkov, Constrained optimization framework for interface-aware sub-scale dynamics closure model for multimaterial cells in Lagrangian and arbitrary Lagrangian-Eulerian hydrodynamics. J. Comput. Phys. **276**, 92–135 (2014)
5. BLAST, High-order curvilinear finite elements for shock hydrodynamics. LLNL code, 2018. http://www.llnl.gov/CASC/blast
6. H. Borouchaki, P.L. George, F. Hecht, P. Laug, E. Saltel, Delaunay mesh generation governed by metric specifications. Part I. Algorithms. Finite Elem. Anal. Des. **25**(1–2), 61–83 (1997). Adaptive Meshing, Part 1
7. W. Boscheri, M. Dumbser, High order accurate direct arbitrary-Lagrangian-Eulerian ADER-WENO finite volume schemes on moving curvilinear unstructured meshes. Comput. Fluids **136**, 48–66 (2016)
8. V. Dobrev, T. Kolev, R. Rieben, High-order curvilinear finite element methods for Lagrangian hydrodynamics. SIAM J. Sci. Comput. **34**(5), 606–641 (2012)
9. V. Dobrev, T. Ellis, T. Kolev, R. Rieben, High-order curvilinear finite elements for axisymmetric Lagrangian hydrodynamics. Comput. Fluids **83**, 58–69 (2013)
10. V.A. Dobrev, T.V. Kolev, R.N. Rieben, V.Z. Tomov, Multi-material closure model for high-order finite element Lagrangian hydrodynamics. Int. J. Numer. Methods Fluids **82**(10), 689–706 (2016)
11. V. Dobrev, P. Knupp, T. Kolev, K. Mittal, V. Tomov, The target-matrix optimization paradigm for high-order meshes. ArXiv e-prints, 2018. https://arxiv.org/abs/1807.09807
12. C. Ericson, *Real-Time Collision Detection* (CRC Press, Boca Raton, 2004)
13. V.A. Garanzha, Polyconvex potentials, invertible deformations, and thermodynamically consistent formulation of the nonlinear elasticity equations. Comput. Math. Math. Phys. **50**(9), 1561–1587 (2010)
14. V. Garanzha, L. Kudryavtseva, S. Utyuzhnikov, Variational method for untangling and optimization of spatial meshes. J. Comput. Appl. Math. **269**, 24–41 (2014)
15. S.K. Godunov, A difference method for numerical calculation of discontinuous solutions of the equations of hydrodynamics. Matematicheskii Sbornik **89**(3), 271–306 (1959)
16. P.T. Greene, S.P. Schofield, R. Nourgaliev, Dynamic mesh adaptation for front evolution using discontinuous Galerkin based weighted condition number relaxation. J. Comput. Phys. **335**, 664–687 (2017)

17. J.-L. Guermond, B. Popov, V. Tomov, Entropy-viscosity method for the single material Euler equations in Lagrangian frame. Comput. Methods Appl. Mech. Eng. **300**, 402–426 (2016)
18. W. Huang, R. Russell, *Adaptive Moving Mesh Methods* (Springer, 2011)
19. P. Knupp, Introducing the target-matrix paradigm for mesh optimization by node movement. Eng. Comput. **28**(4), 419–429 (2012)
20. MFEM: Modular parallel finite element methods library, 2018. http://mfem.org
21. M. Turner, J. Peiró, D. Moxey, Curvilinear mesh generation using a variational framework. Comput. Aided Des. **103**, 73–91 (2018)
22. P. Váchal, P.-H. Maire, Discretizations for weighted condition number smoothing on general unstructured meshes. Comput. Fluids **46**(1), 479–485 (2011)

Curving for Viscous Meshes

Steve L. Karman

Abstract Finite-element flow solvers can utilize high-order meshes to achieve improved accuracy over traditional linear meshes. High order meshes are generally created by elevating linear meshes. For high Reynold's number viscous flows, the linear mesh is tightly clustered to no-slip surfaces. For curved boundaries the high-order mesh must also curve to match the geometry curvature. An optimization-based node perturbation scheme is described that used a two-component cost function to optimize the high order mesh. The first component uses element Weighted Condition Number (WCN) to enforce element shape. The second component uses a normalized Jacobian to enforce element size and validity. The method is applied to several complex linear meshes with highly curved boundaries and tightly clustered normal spacing.

1 Introduction

Significant advances have been made in finite-element (FE) techniques for Computational Fluid Dynamics (CFD) in recent years [1–3]. Finite-element techniques offer increased accuracy over traditional CFD methods, such as finite-volume, with fewer degrees of freedom and increased efficiency. Two popular approaches are Streamline Upwind/Petrov-Galerkin (SU/PG) and Discontinuous Galerkin (DG). Both approaches achieve stable solution by using upwind methods. SU/PG modifies the Galerkin weighting functions to achieve upwinding. DG applies flux jump conditions at element boundaries to approximate a Riemann solution. Linear meshes (polynomial degree 1) result in 2nd order accurate solutions. Higher order accuracy is achieved by introducing additional vertices (new degrees of freedom) to the element edges, faces and interiors that contribute to the integration of the governing

S. L. Karman (✉)
Pointwise Inc., Fort Worth, TX, USA
e-mail: skarman@pointwise.com

X. Roca, A. Loseille (eds.), *27th International Meshing Roundtable*,
Lecture Notes in Computational Science and Engineering 127,
https://doi.org/10.1007/978-3-030-13992-6_17

equations. Arbitrary polynomial order is possible, however typical CFD applications tend to use orders of polynomial degrees 2 (P2), 3 (P3) or 4 (P4).

Elevating a linear mesh to a higher degree is relatively straightforward. The new vertices are inserted into the element at predefined locations, such as equally spaced for Lagrangian-based interpolation schemes. SU/PG methods are continuous, so the nodes shared by adjacent elements are not distinct. DG methods duplicate edge and face nodes, so each element has its own copy. This duplication is typically performed by the flow solver, so the task of mesh generation is to create the continuous version of the high-order mesh.

CFD solutions for viscous computations require clustering of the nodes towards no-slip boundaries in the mesh to properly resolve the viscous boundary layer. This creates high aspect ratio elements near the boundary. If the boundary is curved this complicates the mesh elevation process and requires the elements themselves to be curved. Mesh curving has been attempted in a number of ways to alleviate mesh tangling issues associated with curving meshes, with varying degrees of success. Interpolation methods, such as mean value coordinates [4] and radial basis functions [5] have been used. The more successful approaches tend to use solid mechanics analogies where the mesh is treated as an elastic solid that deforms due to forces acting on the boundaries [6, 7]. Manipulation of the stiffness model or thermo-elastic properties helps maintain valid elements in the mesh interior. More severe cases of high curvature and tight viscous clustering still present challenges. Other efforts focus on the solution to the Winslow equations to perform the interior mesh curving [8]. This approach is a natural application of Winslow smoothing techniques as a copy of the unperturbed, elevated mesh serves as the computational mesh. The solution to the Winslow equations then forces the interior of the physical mesh to take on the same character of the computational mesh, i.e. smooth variation of the spacing and element volume across the physical domain. However, more severely curved cases still present challenges requiring strategies such as freezing the original linear nodes.

A mesh optimization method was presented at IMR25 [9]. The method attempts to optimize a cost related to a distortion measure from an isotropic state, which would be insufficient for viscous mesh curving. An alternate approach for viscous mesh curving was first published in 2016 [10]. This approach relies on a node perturbation/smoothing process first utilized for general isotropic unstructured mesh smoothing [11]. The basic algorithm was generalized for prescribing desired element shapes and extended to the four basic element types [12, 13]. Further enhancements to the original curving technique have been made to integrate with a geometry kernel and enforce positive Jacobians for FE simulations. Surface nodes are projected onto CAD surfaces using a lightweight geometry kernel, Project Geode. The cost function controlling the node perturbations has been expanded to include a normalized Jacobian component that ensures valid positive Jacobians are produced. Details of the mesh curving process are presented. Several examples with a high degree of curvature and tightly clustered viscous meshes are included.

2 Mesh Curving Process

The mesh curving process assumes a valid linear mesh has been created, as opposed to actually curving the mesh during mesh creation. In addition, any curved boundaries require the underlying shape be defined in a Computer Aided Design (CAD) model. As the linear mesh is elevated and as the mesh nodes are smoothed, adherence to the geometry is enforced.

2.1 *Project Geode*

Project Geode is a fourth generation, solid modeling and geometry kernel written in C++. It is an integral part of the commercial mesh generation software, Pointwise [14]. A lightweight version is available for queries such as point projections. Use of Project Geode by the mesh curving process is made robust and efficient by segregating the geometric entities into CAD groups and using a mesh linkage that defines the required geometric entities for each mesh node. Project Geode with the mesh linkage are used by the mesh curving program to perform closest point projections for curved surface edge and interior vertices.

2.1.1 CAD Groups

The geometric entities are sorted into collections of surfaces (NURBS) and curves (B-SPLINES) required by the various mesh surface faces and edge curves. Each group gets stored in a binary split (BSP) tree for fast searches. As a surface mesh interior or edge node is moved it will be projected to the appropriate CAD group in the associated BSP. This provides for efficient projections and robustness. Robustness in the sense that mesh nodes on the top side of a thin geometry, such as a wing trailing edge, are projected to a different CAD group than the mesh nodes on the bottom side of the thin geometry.

2.1.2 Mesh Linkage

The associativity between the mesh entities (points, curves and surface meshes) and the underlying geometry is typically discarded when the final linear mesh is exported to a file. An open source schema is under development that provides information to re-associate mesh entities with the appropriate geometric entity. The schema recognizes and leverages the hierarchical nature of mesh topology, e.g. edges (and their points) are constrained to CAD curves while faces are constrained to CAD surfaces. Furthermore, by allowing a CAD group to contain many CAD entities, the schema provides mesh linkage for cases where mesh topology does not match CAD topology. This schema will be available for any mesh generator to use to write the mesh linkage file. The mesh linkage can then be used by the mesh adaptive or mesh elevation code with the geometry file (IGES, STEP, etc.) to perform the necessary geometric queries of the CAD geometry.

2.2 Mesh Elevation and Initial Perturbation Field

High-order meshes are typically obtained by introducing new nodes on edges, faces and in the interior of elements. The number of new nodes added is dependent on the desired order of accuracy. A hybrid linear mesh is usually comprised of tetrahedra with 4 nodes, pyramids with 5 nodes, prisms with 6 nodes and hexahedra with 8 nodes. All these nodes are located at corners of the elements. These are designated P1 elements as they are used with basis functions defined by first order polynomials, i.e. planar solution variation across the element. Adding mid-edge nodes to each linear edge in the P1 mesh, mid-face nodes to quadrilateral faces in the P1 mesh and a centroidal node for hexahedral P1 elements will create P2 elements. P3 and P4 elements are created in a similar fashion, increasing the number of nodes on each edge, face and element interior. The maximum polynomial order permitted in this endeavor is P4. The CGNS numbering convention is used to identify the ordering of the new nodes in each element [15].

Each new node associated with a boundary edge is projected to its CAD curve group using Project Geode. Each new node associated with a boundary surface is projected to its CAD surface group using Project Geode. For viscous meshes with tight clustering this can invert elements close to convex curved surfaces requiring some form of smoothing to untangle the mesh. To minimize the tangling and provide a better initialization of the interior, a perturbation field is created based on the boundary node perturbations.

All new high-order nodes are initially placed at locations linearly extrapolated from the boundary node perturbations. The process will be described using a two-dimensional section of a Q2 mesh, shown in Fig. 1. The mesh is a collection of quadrilateral elements, but triangles are permitted. In three dimensions all four basic element types are permitted. The high order nodes are depicted in blue. The high order boundary node is red. The boundary geometry is the green curve. The projection of the boundary high order node is shown in Fig. 2.

Each interior node locates the nearest boundary segment or boundary node of the unperturbed mesh, shown in Fig. 3. A surface perturbation vector is computed at these locations by interpolating from the endpoint perturbations. The original linear nodes will have a zero-perturbation vector. This perturbation vector is transferred to the interior node using a linearly decaying function, shown in Eq. (1). The distance

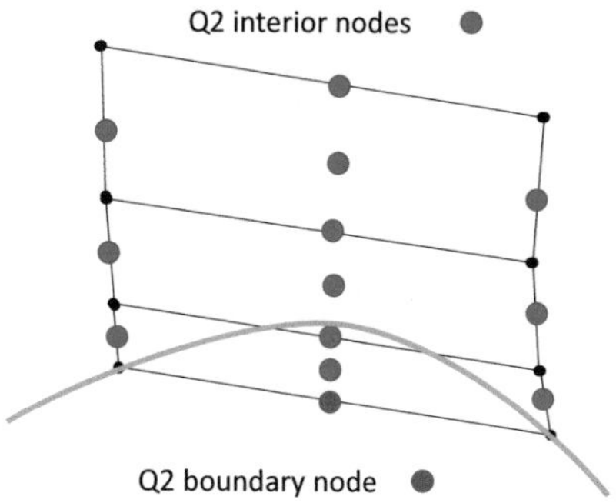

Fig. 1 Section of mesh with Q2 interior and boundary nodes

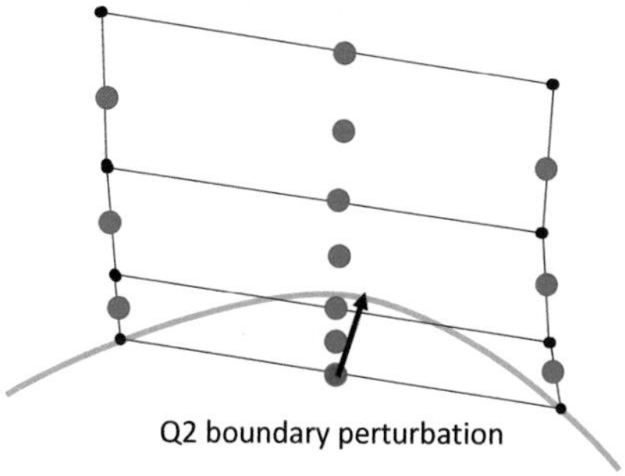

Fig. 2 Perturbation vector for boundary Q2 node

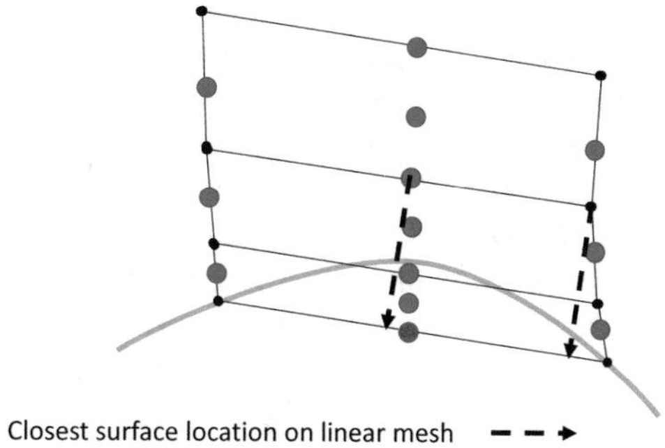

Fig. 3 Closest locations on linear mesh

from the interior node to the boundary location is ds. The factor decays rapidly away from the boundary (Fig. 4).

$$\overrightarrow{P_i} = \overrightarrow{P_b} * MIN\left(1.0, \frac{\left|\overrightarrow{P_b}\right|}{ds}\right) \tag{1}$$

This perturbation field greatly improves the initial curved mesh in the convex and concave regions. Only an extremely small percentage of elements for more complex configurations are invalid prior to smoothing. The resulting initial mesh for the two-dimensional example case is shown in Fig. 5.

Fig. 4 Interior node perturbations

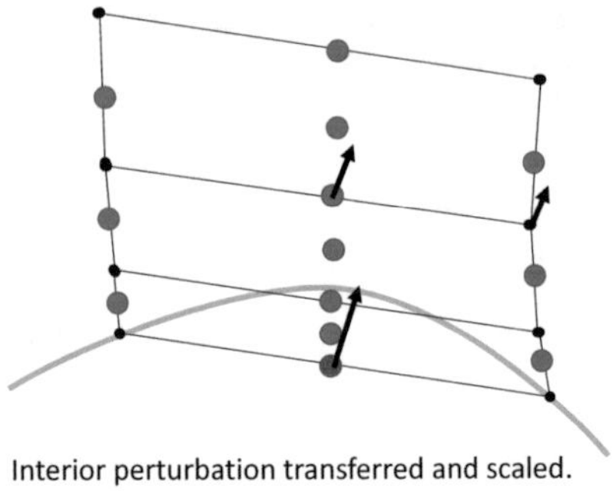

Fig. 5 Final initialized node locations

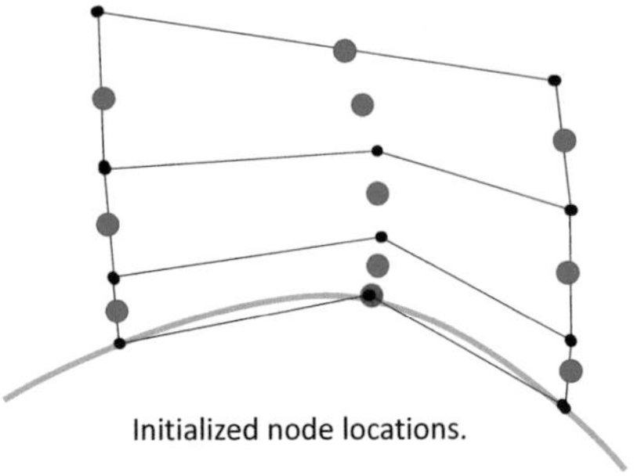

The three-dimensional perturbation field is created in a similar manner using a multi- step process. Each surface high order node has been projected to the appropriate edge curve or geometry surface using Project Geode. This may not be adequate for surfaces that are planar, but the boundary edges are highly curved. So, the first step considers the boundary mesh patches and projects each surface mesh interior point to all other surface patches. If the new perturbation vector is larger than the existing vector, then it is replaced. The volume interior points are then processed by locating the nearest surface node or element. The interpolated perturbation at that location is transferred to the interior location using Eq. (1).

2.3 Optimization-based Smoothing

The smoothing method used on the high-order mesh is a node perturbation scheme that attempts to optimize a two-component cost function. The two components are evaluated on the elements and distributed to the nodes. Both components have optimal values at 1 and indicate element inversion at values of 0 and lower. The first component is based on Weighted Condition Number (WCN) [11] and the second component is based on a normalized Jacobian. The linear combination of components, shown in Eq. (2), is controlled via the α parameter that can vary between 0 and 1. The WCN component is required to maintain element shape but cannot ensure positive Jacobians. The Jacobian component does not maintain the shape but is sometimes required to enforce positive Jacobians. Together they seek to enforce the proper shape and size of the elements.

$$C = \alpha C_{WCN} + (1 - \alpha) \, C_{JAC} \tag{2}$$

The diagrams in Fig. 6 are used to illustrate the locations where each component would be calculated for a two-dimensional triangle. The extension to three dimensions is straightforward. For clarity, the triangular element is straight-sided but that is certainly not required. The upper-left image shows a P2 triangle with the high-order nodes located at the mid-edges. The WCN cost component is computed on the linear sub-elements, shown in the lower left image. This cost component only affects the nodes that define the sub-element. The normalized Jacobian cost component is computed at the survey points shown on the right side of the figure. The Jacobian computation involves all real nodes (linear and high-order) in the element, so it affects all nodes of the element. When operating on pyramids, prisms and hexahedra the WCN cost component is computed on a tetrahedral element constructed at each corner of these elements. The normalized Jacobian cost component is computed over the entire element using similar equally spaced survey points.

In practice the smoothing is applied in one or more stages. In the first stage only the WCN component is used. This is less expensive to compute and provides robustness during mesh untangling at the beginning of the smoothing process. After

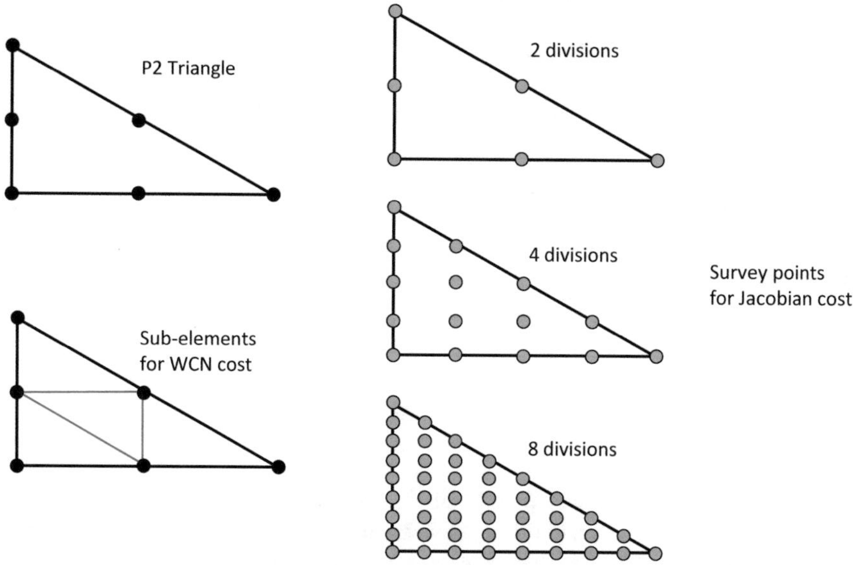

Fig. 6 P2 triangle with the sub-elements and Jacobian survey locations

the first stage the normalized Jacobians are evaluated within each element using a dense survey grid created using number of subdivisions equal to the element polynomial degree times a specified maximum factor, such as 4 for the lower right image. If negative Jacobians are detected, then the second stage is performed with both cost components and the survey grid density corresponds to the polynomial degree, such as 2 for the triangle in the upper left image. If negative Jacobians still exist, then the survey grid density increases for the next stage. Experience has shown the third stage is seldom needed for P2 meshes but needed more for P3 and P4 meshes. The higher polynomial orders permit inflections in the edges and surfaces which can contribute to negative Jacobians. Increasing the survey grid density identifies more regions of the element that might contain small or negative Jacobian values.

WCN smoothing was originally developed for linear mesh smoothing. Each high order element can be subdivided into a collection of linear sub-elements. C_{WCN} is computed on these sub-elements where a desired shape is known from the elevated but unperturbed mesh. Enforcing sub-element shapes will require edge curving near surface curvature but will maintain straight elements away from the surface. Cost values computed on the sub-elements contributes to average and minimum values at the nodes of each sub-element.

The C_{WCN} component takes on two forms, depending on the local Jacobian, Eqs. (3). If the Jacobian is negative the cost is the Jacobian. If the Jacobian is positive the cost is the inverse of the weighted condition number. The delineation at a value of zero is required because the second form does not recognize when the element is inverted. These two forms are C0 continuous at a value of 0.

$$C_{WCN} = J \qquad if \quad J \leq 0.0 \tag{3}$$

$$C_{WCN} = \frac{1}{WCN} \quad if \quad J > \ 0.0$$

The Jacobian, J, defined in Eq. (4) is the magnitude of the determinant of the Jacobian matrix, A, formed by inserting the components of the three edge vectors, u, v and w emanating from the corner of the linear sub-element. A right-hand-rule convention is assumed in the construction of the matrix.

$$J = \begin{vmatrix} u_x & v_x & w_x \\ u_y & v_y & w_y \\ u_z & v_z & w_z \end{vmatrix} \tag{4}$$

The second form of the cost function in Eq. (3) is based on the weighted condition number, WCN, given in Eq. (5). The bracketed quantities are the Frobenius norms of the matrix products. The [A] matrix is the actual Jacobian matrix from Eq. (4). The

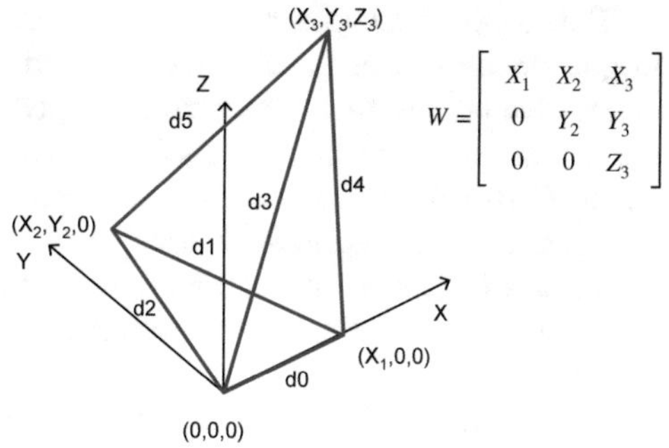

Fig. 7 The weight matrix can be formed from the edge lengths of the corner

[W] matrix is the weight matrix that transforms the reference corner (a right-angled corner with unit length edges) to the desired corner shape.

$$WCN = \frac{\|AW^{-1}\| \, \|WA^{-1}\|}{3} \tag{5}$$

The weight matrix, [W], defines the desired shape. It is constructed from the six edges of a corner tetrahedron, shown in Fig. 7 using the unperturbed (computational) coordinates. If the physical coordinates used in [A] were unperturbed from the computational coordinates the resulting C_{WCN} would be unity. For smoothing of high order meshes the sub-elements of the high-order element are evaluated using Eqs. (3). The elevation and projection processes deform elements near curved boundaries. The C_{WCN} components attempts to recover the original sub-element shapes.

The Jacobian component of the cost function, C_{JAC}, is a normalized Jacobian computed at numerous survey points within each element, Eq. (6). The numerator is the value of the Jacobian, computed using Eq. (7), at a given survey point in the element using the physical coordinates. The denominator is the value of the Jacobian at the same location using the computational (unperturbed) coordinates. The derivative components of the shape factors, N_i, are obtained from basis functions for each element type and polynomial order.

$$C_{JAC} = MIN\left(1, \frac{J_p}{J_c}\right) \tag{6}$$

$$J = \begin{vmatrix} \sum \frac{\partial N_i}{\partial \xi} x_i & \sum \frac{\partial N_i}{\partial \xi} y_i & \sum \frac{\partial N_i}{\partial \xi} z_i \\ \sum \frac{\partial N_i}{\partial \eta} x_i & \sum \frac{\partial N_i}{\partial \eta} y_i & \sum \frac{\partial N_i}{\partial \eta} z_i \\ \sum \frac{\partial N_i}{\partial \zeta} x_i & \sum \frac{\partial N_i}{\partial \zeta} y_i & \sum \frac{\partial N_i}{\partial \zeta} z_i \end{vmatrix} \tag{7}$$

The Jacobian component of the cost function, C_{JAC}, attempts to recover the original Jacobian magnitude of the unperturbed mesh. The possible range of values is negative infinity to 1. The ratio is capped at 1 so larger physical Jacobians are permitted. In essence, this cost component attempts to recover the size of the original element. This form of Jacobian is preferred over other quality measures typically encountered in curved meshes, such as scaled Jacobian. The scaled Jacobian will not respect the viscous clustering. It measures the deviation of the Jacobian within the element. In the optimization process this would drive the mesh towards isotropic element shapes.

C_{JAC} is computed at the survey points within each element. Each value computed contributes to the average and minimum value recorded for each node in the element. The number of survey points within an element can vary. For the smoothing operation it is an equally spaced array of locations derived by specifying the number of divisions on an edge. For instance, if the number of divisions is 2 for a P2 element then the locations will be the corner nodes plus the high-order nodes. Three divisions for P3 elements would also result in calculations at the corners and the high-order nodes. Number of divisions higher than the polynomial order are possible, but generally not required to achieve a valid mesh and would increase the computational cost significantly.

Nodal values of the cost, C, are computed using a blending of the cost for the elements surrounding each node, Eq. (8). The blending combines the worst cost value, C_w, with the arithmetic average of the element cost values, C_{Avg}. This ensures the worst value has significant weight as a surrounding element is inverted or nearly inverted.

$$\beta = 1 - C_w$$

$$C = \beta C_w + (1 - \beta)\, C_{Avg} \tag{8}$$

Smoothing proceeds in an iterative manner where each node is perturbed in the direction of increasing cost, driving towards the ideal value of 1. The direction is determined using operator overloaded math functions which compute values and sensitivities in the same operation. The sensitivity vector is used in a gradient based algorithm to advance the mesh. The step size is limited to a fraction of the local inscribed radius for each element. A user-specified threshold cost value, such as 0.8, is used to activate nodes in the mesh for optimization. Only nodes whose cost is below this threshold and their immediate neighbors are computed and moved during any given iteration. Each element connected to an activated node is also activated during the smoothing iteration. This dramatically improves the performance of the scheme as most of the mesh will have a unit cost value. Only nodes in the vicinity of curved boundaries will have lower cost values. Smoothing stops when all nodal cost values exceed the threshold, or the specified number of iterations are completed.

The cost threshold can greatly influence the curving process run time and resulting quality. Lower values may permit excessive kinks in the edges oriented in the normal direction from curved boundaries, especially very close to the surface. Higher threshold values will increase the run times for curving but will produce higher quality meshes. The default threshold value is 0.8 but higher values approaching 0.95 are commonly used. The α parameter is automatically calculated to be the polynomial degree minus one over the polynomial degree. This would produce values of 1/2, 2/3 and 3/4 for Q2, Q3 and Q4 meshes. In extreme cases this parameter has been lowered to 0.3–0.4 to ensure valid Jacobians are produced. These meshes included some highly curved surface and interior edges as a result.

3 Results

A collection of example cases is shown with multiple curved boundaries and viscous clustering toward these boundaries. The cases increase in size from a small generic shape to a complex air vehicle. All curved boundary shapes are defined using CAD geometry. Surface evaluations are performed using the CAD geometry and Project Geode projection calls. Images were created using the Gmsh application [16] as a mesh viewer for P2, P3 and P4 meshes for the first two cases and ParaView [17] for the final case P2 meshes.

In each case the resulting mesh is valid with no negative Jacobians resulting at the densest survey grid. The minimum combined cost values are greater than 0.5 and approach or exceed the threshold values of 0.9–0.95. For those cases where the combined cost is lower the normalized Jacobian minimum value is in the range from 0.1 to 0.3.

3.1 Weeble

The first case is a small mesh created on the geometry shown in Fig. 8. It is a surface with convex and concave features separated by sharp edges. The outer boundary is

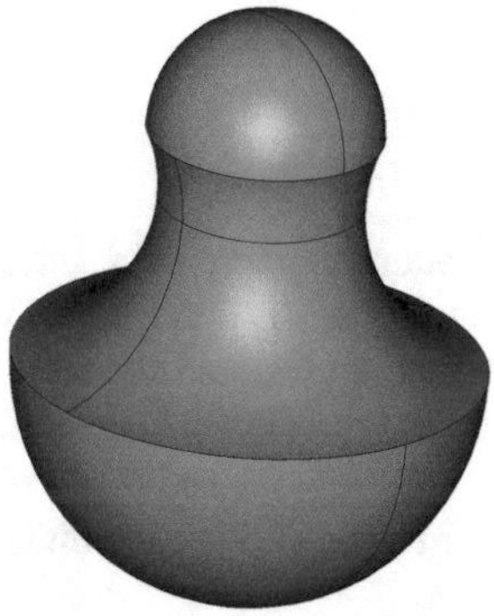

Fig. 8 Surface geometry for weeble

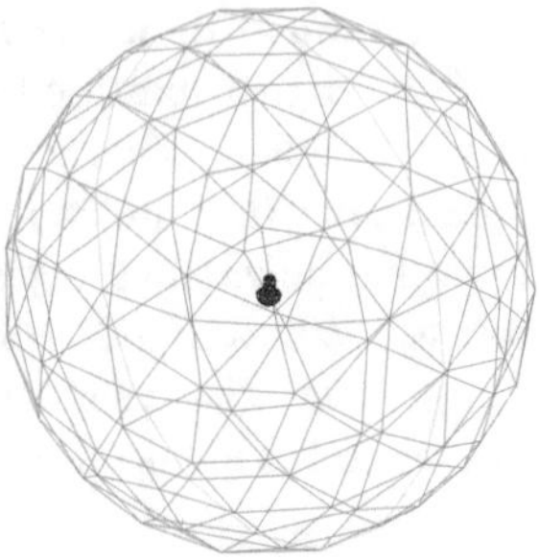
Fig. 9 Outer boundary linear surface mesh for weeble

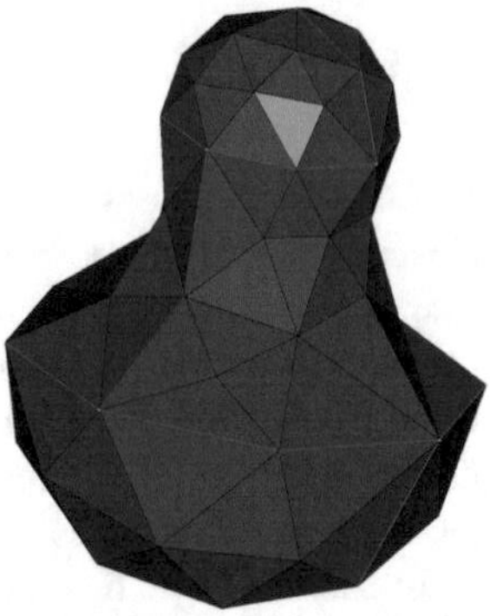
Fig. 10 Weeble surface linear mesh

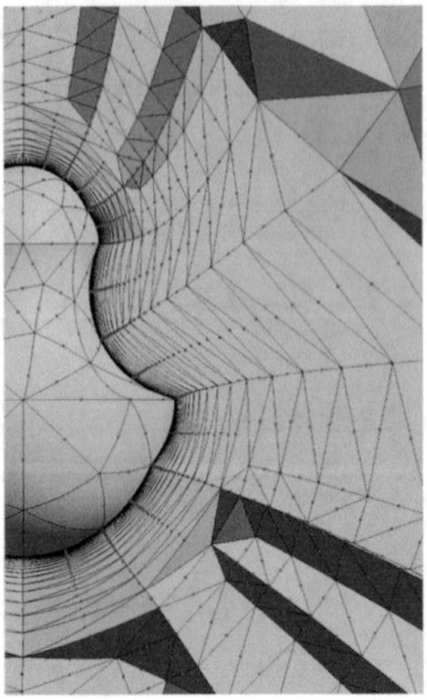
Fig. 11 Weeble P2 mesh with crinkle cut through the centerline

a simple sphere. The linear mesh, shown in Fig. 9, has 4876 points and 28,821 tetrahedra. The mesh on the weeble, shown in Fig. 10, is very coarse. The volume mesh is clustered towards the surface with a normal spacing equal to 1e-04. The maximum element aspect ratio is 5246 for an element on the surface.

The P2, P3 and P4 meshes, are shown in Figs. 11, 12 and 13, respectively. The P2 contains 38,404 points. The P3 mesh has 129,073 points. And the P4 mesh

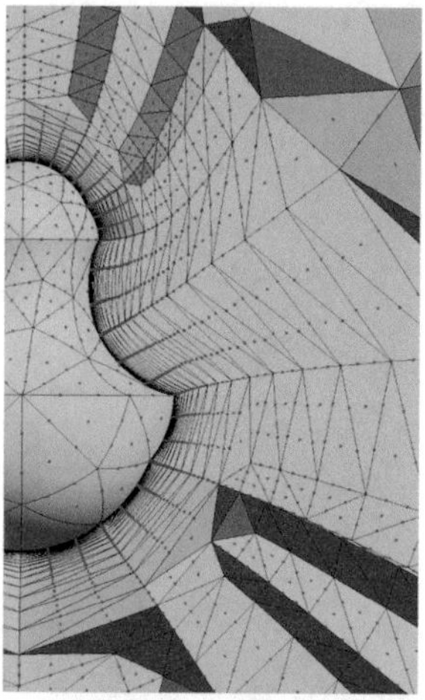

Fig. 12 Weeble P3 mesh with crinkle cut through the centerline

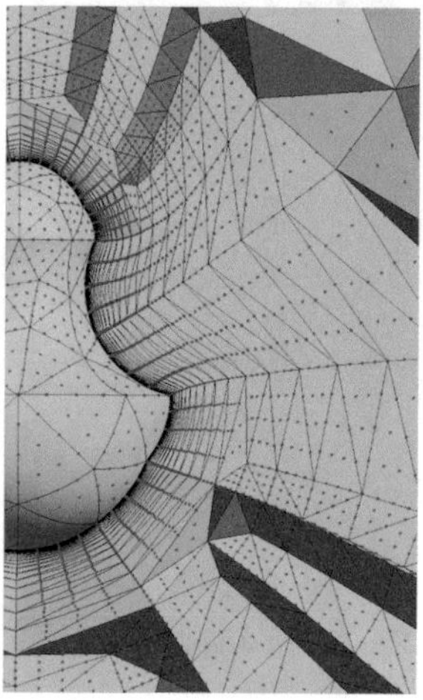

Fig. 13 Weeble P4 mesh with crinkle cut through the centerline

has 305,370 points. The surfaces are smooth, and the sharp edges are retained and curved.

Magnified views of the cut through the volume meshes without the nodes displayed are shown in Figs. 14, 15, and 16. The interior edges show a high degree of curving off the lower sharp edge. The P4 mesh shows an inflection in the edges just above the lower sharp edge on the surface.

Hybrid versions of the meshes containing tetrahedra, pyramid, prisms and hexes were also curved, Figs. 17, 18, and 19. The quadrilateral surface elements are coarser than the triangular elements from the all-tetrahedral mesh.

Fig. 14 P2 volume edges near surface

Fig. 15 P3 volume edges near surface

Fig. 16 P4 volume edges near surface

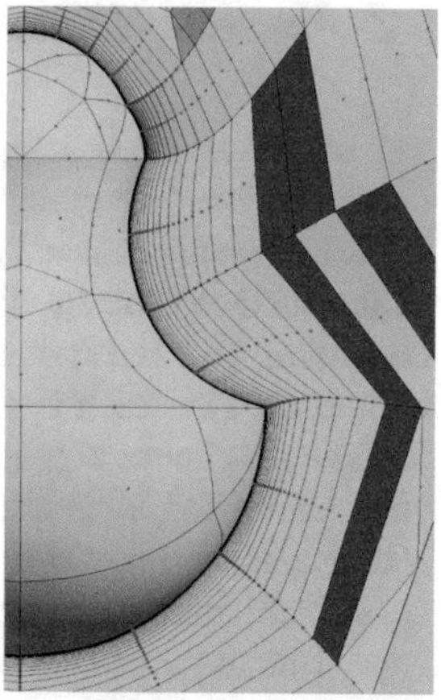
Fig. 17 P2 hybrid volume mesh

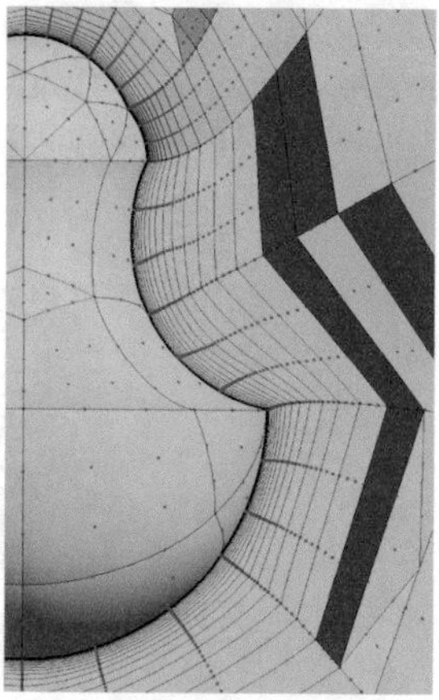
Fig. 18 P3 hybrid volume mesh

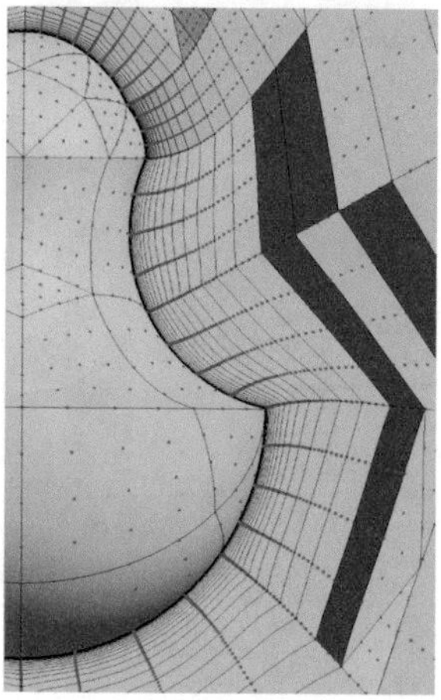
Fig. 19 P4 hybrid volume mesh

3.2 Generic Nozzle

The next example is a slightly bigger mesh with a significant amount of curvature on the surface. The generic nozzle geometry is shown in Fig. 20. The outer nozzle surface contains convex curvature and a sharp break in slope about ¾ of the length of the surface. The interior nozzle surface contains smooth convex and concave curvature. The linear mesh, shown in Fig. 21, has 86,817 points and 506,623 tetrahedra. The mesh is clustered to the nozzle surface with a wall spacing of 1e-03. The maximum aspect ratio is 385.6 located on the inner wall near the convex curvature region. A magnified view of the surface meshes is shown in Fig. 22. The challenging aspect of this case is the nozzle base region, magnified in Fig. 23. It has convex curvature in the plane on the interior edge and concave curvature on the outer edge. This requires more than just projecting the high-order nodes to the surface. It requires smoothing on the surface to create valid curved triangular elements. The smoothing process works on all nodes in the mesh, boundary nodes and interior volume nodes. Any tangling on the surface will be corrected as the volume mesh is smoothed. The boundary nodes are not processed separately from the interior nodes.

Fig. 20 Generic nozzle surface geometry

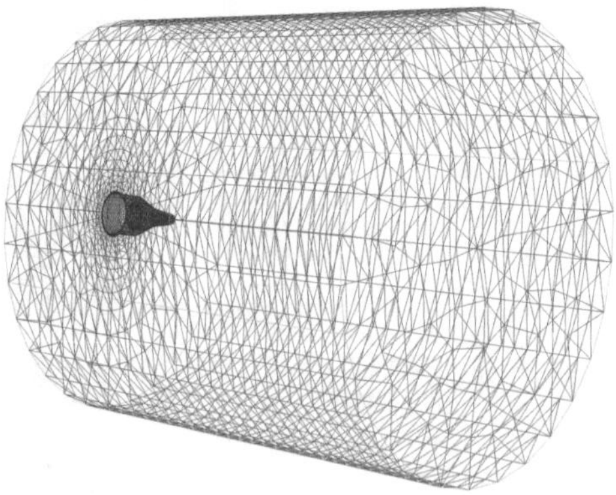

Fig. 21 Generic nozzle linear mesh

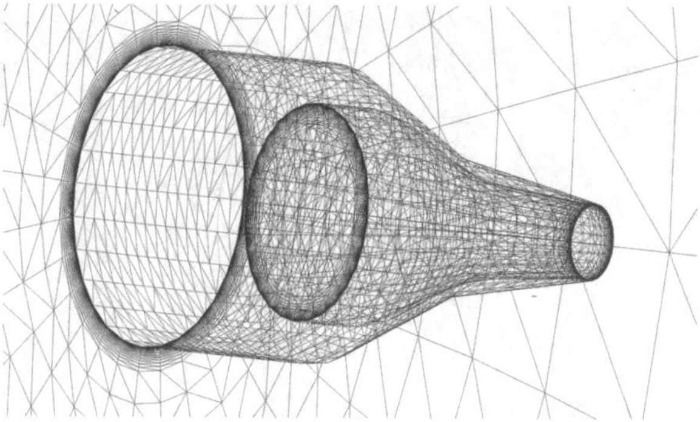

Fig. 22 Magnified view of surface mesh inside and outside of nozzle

Fig. 23 Zoomed in view of nozzle base region

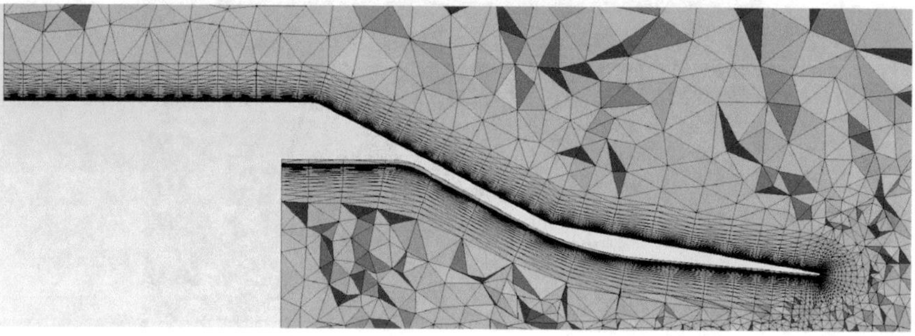

Fig. 24 Crinkle cut through P2 mesh

Crinkle cuts through the centerline of the P2, P3 and P4 meshes are shown in Figs. 24, 25, and 26. The only noticeable difference is at the concave shoulder on the interior side of the nozzle. The P2 surface has a slight bulge that is not seen in the P3 and P4 meshes. This occurred because the length of the elements in the region spanned the transition from the horizontal flat piece to the converging section. The

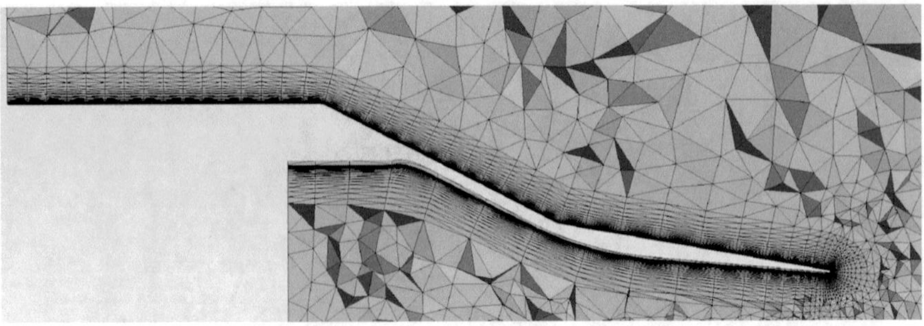

Fig. 25 Crinkle cut through P3 mesh

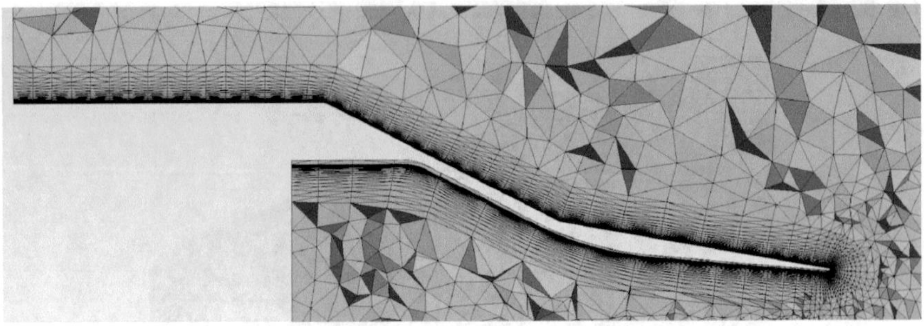

Fig. 26 Crinkle cut through P4 mesh

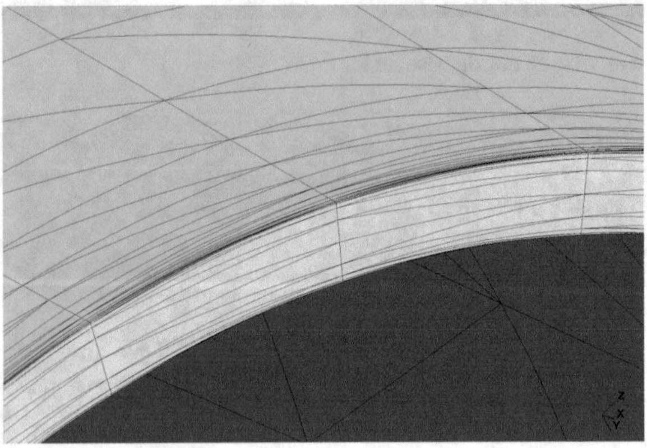

Fig. 27 Base region of P2 mesh

P2 mesh positioned the interior edge node mid-way between the corner nodes while the upstream node was on the flat section. The quadratic surface (P2) creates the bulging section. The P3 and P4 meshes show no bulge.

The nozzle base region for each mesh is shown in Figs. 27, 28, and 29. All three meshes show the same character. The clustering in the plane to the inner and outer

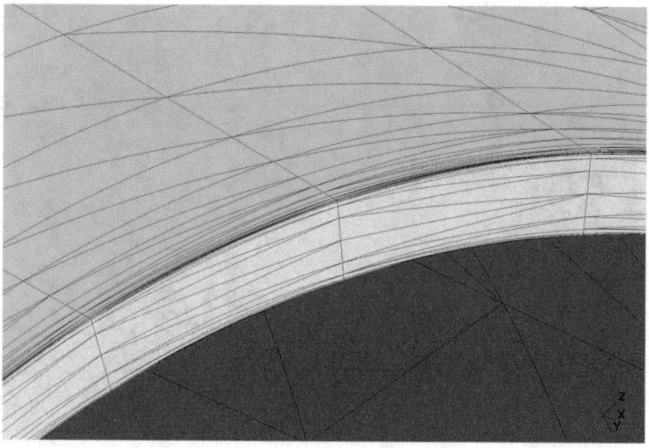

Fig. 28 Base region of P3 mesh

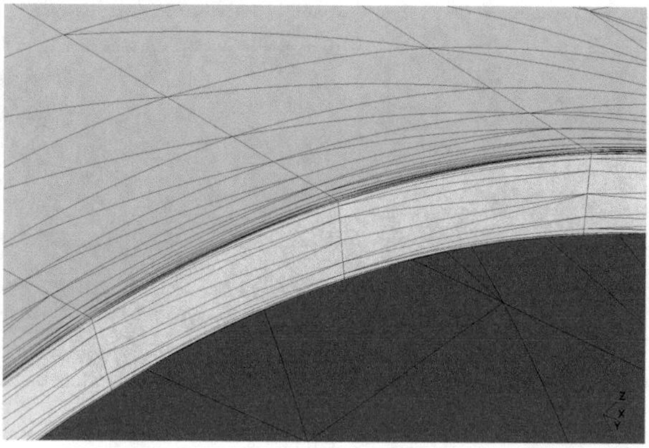

Fig. 29 Base region of P4 mesh

curves is maintained. The triangular elements in the middle of the base tend to maintain the straight-edged shapes.

The volume elements near the base deform to match the surface shape, shown in Fig. 30. The crinkle cut through an X coordinate is just ahead of the base and reveals convex curvature on the outside and concave curvature on the inside. The P2 mesh transitions from curved edges at the boundary to straight edges in a smooth manner.

3.3 *High Lift Common Research Model*

The last case is a complex air vehicle, the NASA/Boeing High-Lift Common Research Model (HL-CRM) [18]. The geometry is shown in Fig. 31, colored by the different components. It contains a fuselage, wing, leading edge slat and two

Fig. 30 Crinkle cut through P2 mesh at nozzle base

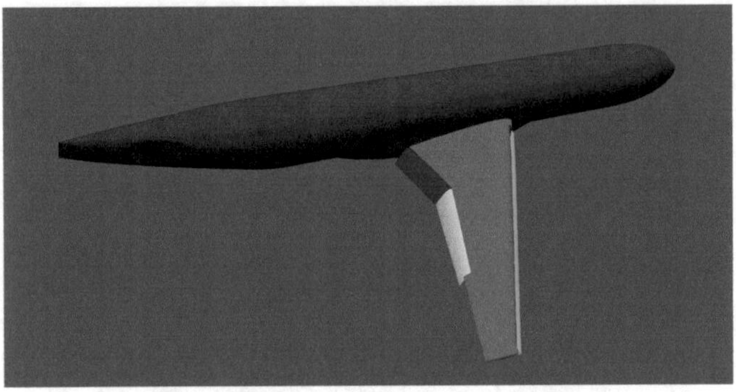

Fig. 31 NASA/Boeing High Lift Common Research Model geometry

trailing edge flaps. The linear mesh contains 5,918,418 nodes and 35,031,781 tetrahedra. This was the "coarse" mesh in a sequence. The elevated P2 mesh contains 47,062,366 nodes.

Surface meshes are shown in Figs. 32, 33, 34, and 35. The viscous clustering, evident in the symmetry plane of Fig. 33, had an initial cell height of 0.00175. The total fuselage length was approximately 2500 units long. The maximum aspect ratio in the symmetry plane mesh was 8941, which is indicative of the viscous clustering over the entire aircraft surface.

The challenging aspect of this case was the viscous clustering and overall size of the mesh. The curving computer code is serial and required approximately 13 GBytes. The surface resolution was comparable to the resolution typically utilized for linear meshes. This results in good resolution of surface curvature which is easily handled by the curving method.

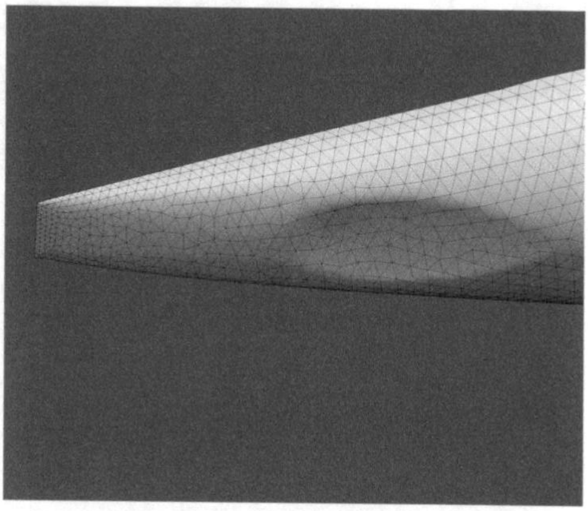

Fig. 32 P2 mesh on HL-CRM aft fuselage

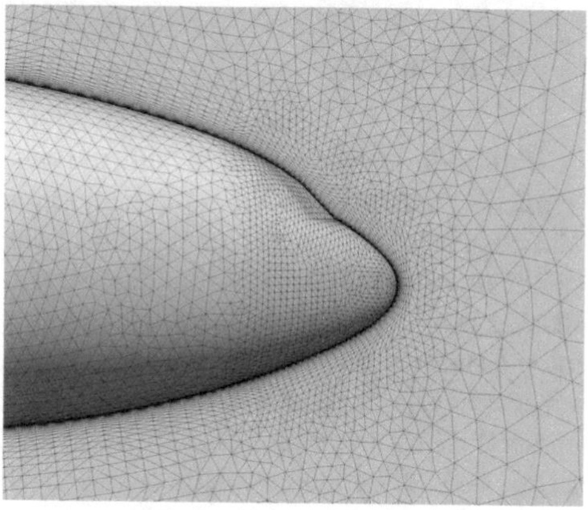

Fig. 33 P2 mesh on HL-CRM wing tip leading edge

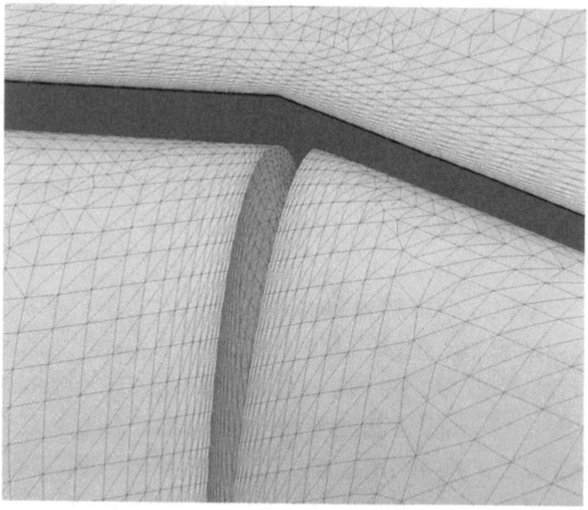

Fig. 34 P2 mesh on HL-CRM flap gap

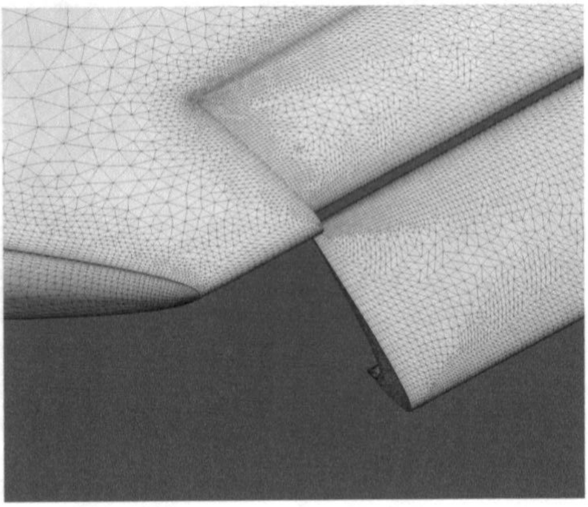

Fig. 35 P2 mesh on HL-CRM forward fuselage and symmetry plane

4 Conclusions

A high-order mesh elevation and curving method has been described that is applicable to high Reynold's number viscous cases for finite element CFD solvers. Viscous clustering introduces significant challenges to mesh curving, especially in regions of high surface curvature. The elevation process in convex regions of viscous meshes can invalidate the high order mesh by inverting volume elements near the curved surface. The curving process uses an optimization-based node perturbation scheme to reposition nodes. The cost function contains two components. The first component uses a Weighted Condition Number function that imposes a desired shape, defined by the elevated but unperturbed mesh. The second cost component uses a normalized Jacobian that imposes sizing restrictions where the high order mesh has been squeezed but the surface curvature. It also guarantees the resulting mesh contains all positive Jacobians; a requirement for the finite element solvers.

Geometry for the curving program is defined by CAD groups (surfaces and edges). Surface node projections are performed using a lightweight geometry kernel, Project Geode. The CAD groups are stored in separate BSP trees for efficient searches. A mesh linkage defines the specific CAD group for each surface mesh node.

Three example cases were curved; all containing viscous clustering to geometries containing high concave and convex surface curvature. High order meshes of polynomial degrees 2, 3 and 4 were generated for the first two cases. A P2 mesh was created for the larger aircraft case. Most of the curved meshes were all-tetrahedral meshes. A hybrid mesh containing tetrahedra, pyramids, prisms and hexes was generated for the first case. All meshes generated were valid with positive Jacobians.

Acknowledgement This work was partially supported by a NASA SBIR contract "High Order Mesh Curving and Geometry Access", Phase I contract NNX17CL83P and Phase II contract 80NSSC18C0109. NASA's support is greatly appreciated.

References

1. R. Glasby, N. Burgess, K. Anderson, L. Wang, S. Allmaras, D. Mavriplis, Comparison of SU/PG and DG finite-element techniques for the compressible Navier-Stokes equations on anisotropic unstructured meshes, 51st AIAA Aerospace Sciences Meeting including the New Horizons Forum and Aerospace Exposition, Aerospace Sciences Meetings, (AIAA 2013-0691) https://doi.org/10.2514/6.2013-691
2. J.T. Erwin, L. Wang, W. Kyle Anderson, S. Kapadia, High-order finite-element method for three-dimensional turbulent Navier-Stokes, 21st AIAA Computational Fluid Dynamics Conference, Fluid Dynamics and Co-located Conferences, (AIAA 2013-2571) https://doi.org/10.2514/6.2013-2571
3. R.S. Glasby, J.T. Erwin, D.L. Stefanski, S.R. Allmaras, M.C. Galbraith, W.K. Anderson, R.H Nichols, *Introduction to COFFE: The Next-Generation HPCMP CREATETM-AV CFD Solver*, AIAA-2016-0567, AIAA SciTech 2016
4. K. Hormann, M.S. Floater, Mean value coordinates for arbitrary polygons, http://www.cs.jhu.edu/misha/ReadingSeminar/Papers/Hormann06.pdf, Submitted to ACM Transactions on Graphics
5. C.H. Chen, A radial basis functions approach to generating high-order curved element meshes for computational fluid dynamics, Master's Thesis, Imperial College, London, 2013
6. P.O. Persson, J. Peraire, Curved mesh generation and mesh refinement using Lagrangian solid mechanics, 47th AIAA Aerospace Sciences Meeting including The New Horizons Forum and Aerospace Exposition, Aerospace Sciences Meetings, (AIAA 2009-0949) https://doi.org/10.2514/6.2009-949
7. D. Moxey, D. Ekelschot, U. Keskin, S.J. Sherwin, J. Peiro, High-order curvilinear meshing using a thermo-elastic analogy. Comput. Aided Des. **2015**. https://doi.org/10.1016/j.cad.2015.09.007
8. M. Fortunato, P.-O. Persson, High-order unstructured curved mesh generation using the Winslow Equations. J. Comput. Phys. **307**, 1–14 (2016)
9. E. Ruiz-Girones, J. Sarrate, X. Roca, Generation of curved high-order meshes with optimal quality and geometric accuracy, 25th International Meshgin Roundtable (IMR25), 2016
10. S.L. Karman, J.T. Erwin, R.S. Glasby, D. Stefanski, High-order mesh curving using WCN mesh optimization, 46th AIAA Fluid Dynamics Conference, AIAA AVIATION Forum, (AIAA 2016-3178) https://doi.org/10.2514/6.2016-3178.
11. L.A. Freitag, P.M. Knupp, Tetrahedral element shape optimization via the Jacobian determinant and condition number, *8th International Meshing Roundtable*, South Lake Tahoe, CA, 1999
12. S.L. Karman, Adaptive optimization-based improvement of Tetrahedral meshes. AIAA J. **54**(5), 1578–1590 (2016). https://doi.org/10.2514/1.J054439
13. S.L. Karman, M.G. Remotigue, Optimization-based smoothing for extruded meshes, 54th AIAA Aerospace Sciences Meeting, AIAA SciTech Forum, (AIAA 2016-1671) https://doi.org/10.2514/6.2016-1671
14. Pointwise, http://www.pointwise.com
15. CFD General Notation System Standard Interface Data Structures, http://cgns.github.io/
16. Gmsh, http://www.gmsh.info
17. ParaView, http://www.paraview.org
18. D.S. Lacy, A.J. Sclafani, Development of the High Lift Common Research Model (HL-CRM): A Representative High Lift Configuration for Transonic Transports, 54th AIAA Aerospace Sciences Meeting, AIAA SciTech Forum, (AIAA 2016-0308) https://doi.org/10.2514/6.2016-0308

An Angular Approach to Untangling High-Order Curvilinear Triangular Meshes

Mike Stees and Suzanne M. Shontz

Abstract To achieve the full potential of high-order numerical methods for solving partial differential equations, the generation of a high-order mesh is required. One particular challenge in the generation of high-order meshes is avoiding invalid (tangled) elements that can occur as a result of moving the nodes from the low-order mesh that lie along the boundary to conform to the true curved boundary. In this paper, we propose a heuristic for correcting tangled second- and third-order meshes. For each interior edge, our method minimizes an objective function based on the unsigned angles of the pair of triangles that share the edge. We present several numerical examples in two dimensions with second- and third-order elements that demonstrate the capabilities of our method for untangling invalid meshes.

1 Introduction

The appeal of high-order methods for solving partial differential equations lies in their ability to achieve higher accuracy at a lower cost than low-order methods. One challenge in the adoption of these high-order methods for problems with curved geometries is the lack of robust high-order mesh generation software [19]. More specifically, to fully leverage the accuracy of high-order methods in the presence of curved geometries, such methods need to be paired with a high-order mesh that correctly reflects the curvature of the geometry, as demonstrated in [1, 8].

M. Stees (✉)
Department of Electrical Engineering and Computer Science, Information and Telecommunication Technology Center, University of Kansas, Lawrence, KS, USA
e-mail: mstees@ku.edu

S. M. Shontz
Department of Electrical Engineering and Computer Science, Bioengineering Graduate Program, Information and Telecommunication Technology Center, University of Kansas, Lawrence, KS, USA
e-mail: shontz@ku.edu

X. Roca, A. Loseille (eds.), *27th International Meshing Roundtable*, Lecture Notes in Computational Science and Engineering 127, https://doi.org/10.1007/978-3-030-13992-6_18

The most common approach for high-order mesh generation methods is to transform a coarse linear mesh [2–6, 9, 10, 12–14, 16–18, 20]. The main challenge of the transformation is obtaining a valid high-order mesh. In general, these methods involve three steps: (1) adding additional nodes to the linear mesh; (2) moving the newly added boundary nodes to conform with the curved geometry, and (3) moving the interior nodes. There are two categories of methods which are especially popular for transforming the initial mesh. The first category involves transforming the mesh based on optimization of an objective function [2, 4, 5, 13–17]. Several of the objective functions proposed in this category include a measure of element validity, which allows them to untangle invalid elements [2, 4, 5, 13, 14, 17]. While they do not guarantee successful untangling, many of them are robust. The second category of methods transform the mesh based on the solution to a partial differential equation [3, 10, 12, 20].

In this paper, we describe an optimization-based approach for untangling invalid second- and third-order meshes. The primary goal of this work is to untangle invalid meshes that result from deforming the newly added boundary nodes to conform with the true boundary. Toward that end, we demonstrate our method on several meshes composed of second- and third-order elements that became invalid following the projection of the boundary nodes onto the true boundary. We also explore the untangling of meshes that became invalid as a result of small deformations. The remainder of this paper is organized as follows. In Sect. 2, we present our new method for high-order mesh untangling. In Sect. 3, we illustrate the performance of our method on several examples. Finally, in Sect. 4, we offer concluding remarks and discuss some possibilities for future work.

2 Untangling High-Order Curvilinear Meshes

In this section, we propose a local edge-based optimization method for untangling high-order curvilinear meshes based on the unsigned angles of curvilinear triangles. For each interior mesh edge, we identify the two triangles that share the edge and compute the distortion of each of the two triangles. For each pair of triangles with a minimum distortion measure less than 0, we solve the following unconstrained optimization problem:

$$x^* = \underset{\mathbf{x}}{\operatorname{argmin}} \sum_{i=1}^{4} \boldsymbol{\alpha}_{\mathbf{i}}(\mathbf{x}), \tag{1}$$

where

$$\boldsymbol{\alpha}_{\mathbf{i}} = \text{the } i\text{th entry of the vector of the four unsigned angles,}$$

$$\mathbf{x} = \text{the nodal positions of the high} - \text{order nodes that lie on the edge}$$

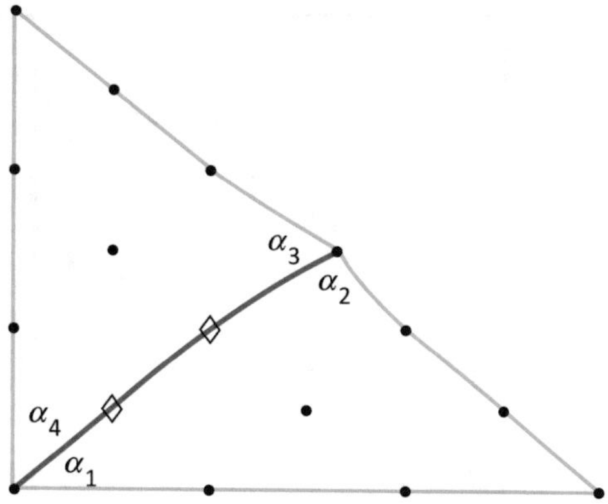

Fig. 1 A pair of triangles showing the interior edge (red), the free nodes (black diamonds), the fixed nodes (black dots), and the four angles α_i

In Fig. 1 we give an example which shows an interior edge in red and the pair of triangles that share that edge in green. We also label the four unsigned angles that are calculated, the nodes that are allowed to move during the optimization (black diamonds), and the nodes that are fixed (black dots).

To better understand the behavior of the objective function, consider the five examples shown in Fig. 2. In Fig. 2a for $\alpha_1 + \alpha_4$ and $\alpha_2 + \alpha_3$, moving the free node (green diamond) will shift the proportion of each term, while leaving the overall sum fixed. In other words, increasing α_1 will cause a corresponding decrease in α_4 while the quantity $\alpha_1 + \alpha_4$ remains the same. Similarly, increasing α_2 will cause a corresponding decrease in α_3 while the quantity $\alpha_2 + \alpha_3$ remains the same. This behavior means that the sum of all four angles cannot be further decreased by moving the free node. Furthermore, this behavior is desirable because patches with no distortion will not be modified since the optimization will not move the free node (as there is no step that will lead to a decrease in the objective function). Fortunately, this behavior holds true as we add minor distortion as well. In Fig. 2b, we moved the bottom node (denoted by a blue square) to increase the distortion of the bottom element. In Fig. 2c, we moved the node slightly further to increase distortion. In both cases, we can see that the overall sum cannot be further decreased by moving the free node. Finally, in Fig. 2d, we move the node to the point that it causes tangling. Now that α_1 is an angle between tangled edges, this angle can be decreased by moving the free node. By decreasing the value of α_1, we decrease the value of $\alpha_1 + \alpha_4$, and thus decrease the overall sum of the four angles. In other words, minimizing our objective function attempts to decrease the value of angles that occur between tangled edges by moving the free node away. In Fig. 2e, we show results of moving the free node to minimize our objective function.

To measure distortion, we use the scaled Jacobian [2]. To solve our unconstrained optimization problem, we use the Broyden-Fletcher-Goldfarb-Shanno (BFGS) quasi-Newton method described in Chapter 6 of [11]. In place of the analytical gradient, we use a 6th order centered finite difference with a step size of 10^{-6}. As our initial Hessian approximation, we use a scaled version of the identity matrix. In Algorithms 1 and 2, we give pseudocode descriptions of our untangling method

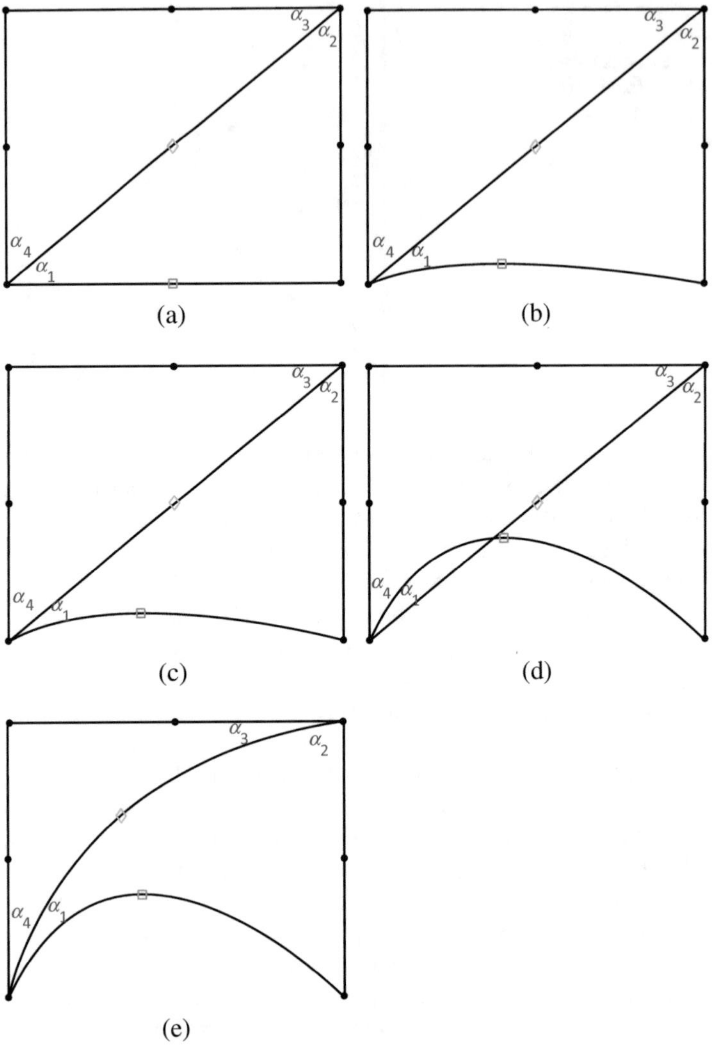

Fig. 2 A simple patch showing the angles (α_i), the free node (green diamond), and the node that is moved to increase distortion (blue square). In (**a**), the patch with no distortion is shown. In (**b**)–(**d**), the amount of distortion is gradually increased. In (**e**), the mesh after applying our method to minimize the sum of the angles is shown

and optimization method, respectively. Our implementation of the BFGS quasi-Newton method uses a backtracking line search. This backtracking approach based on the Wolfe conditions ensures that the step results in a sufficient decrease in our objective function. In the next section, we discuss how the angles $\alpha_i(x)$ of the curved elements are calculated.

Algorithm 1 Pseudocode for our edge-based mesh untangling method

while there are tangled elements or passes < count **do**
  for each interior edge e **do**
    Find the two triangles t_1 and t_2 with e as a common edge
    Compute the element distortions ed_1 and ed_2 for t_1 and t_2, respectively
    if $\min(ed_1, ed_2) < 0$ **then**
      Solve Eq. (1) for x^* using Algorithm 2
      Update nodal positions of the free nodes on e to x^*
    end if
  end for
  passes = passes + 1
end while

Algorithm 2 Pseudocode for our BFGS quasi-Newton method

Given an initial value x_0, an initial value for the Hessian B_0, and a tolerance tol;
while $\|\nabla f(x_k)\| > tol$ **do**
  Compute Cholesky factorization $B_k = LL^T$
  Compute the direction vector d_k by solving $LL^T d_k = -\nabla f(x_k)$.
  $\rho_k = 1.0$
  while $f(x_k + \rho_k d_k) > f(x_k) + 10^{-4}\rho_k \nabla f(x_k)^T d_k$ **do**
    $\rho_k = 0.5\rho_k$
  end while
  $x_{k+1} = x_k + \rho_k d_k$
  $s_k = x_{k+1} - x_k$
  $y_k = \nabla f(x_{k+1}) - \nabla f(x_k)$
  $B_{k+1} = B_k - \dfrac{B_k s_k s_k^T B_k}{s_k^T B_k s_k} + \dfrac{y_k y_k^T}{y_k^T s_k}$
  $k = k + 1$
end while

2.1 Measuring the Angles of Curvilinear Triangles

In order to compute the angle between two curves at a given point, we compute the derivatives of the curves, evaluate the derivatives at the given point, and then compute the angle between the resulting tangent vectors. Following this approach, we will compute the angles between each pair of edges of curvilinear triangles. For our derivation, we use the third-order Lagrange elements. Derivation for the second-order Lagrange elements is similar.

Consider the third-order Lagrange triangle shown in Fig. 3 with shape functions defined as follows:

$$s_1 = \frac{9}{2}(1-\xi-\eta)\left(\frac{1}{3}-\xi-\eta\right)\left(\frac{2}{3}-\xi-\eta\right)$$

$$s_2 = \frac{9}{2}\xi\left(\xi-\frac{1}{3}\right)\left(\xi-\frac{2}{3}\right)$$

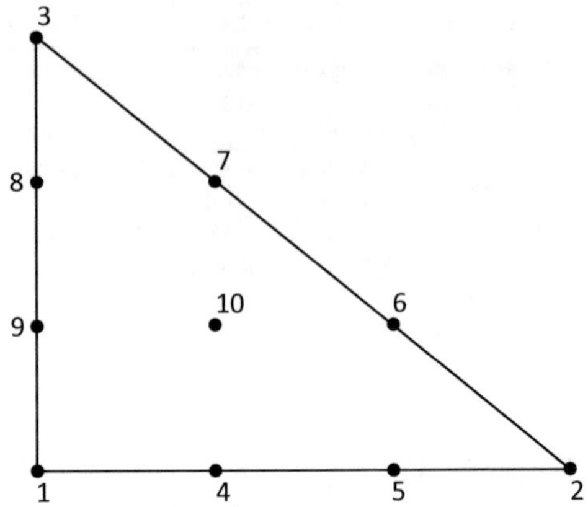

Fig. 3 Third-order Lagrange reference unit triangle

$$
\begin{aligned}
s_3 &= \frac{9}{2}\eta\left(\eta - \frac{1}{3}\right)\left(\eta - \frac{2}{3}\right) \\
s_4 &= \frac{27}{2}(1-\xi-\eta)\,\xi\left(\frac{2}{3} - \xi - \eta\right) \\
s_5 &= \frac{27}{2}(1-\xi-\eta)\,\xi\left(\xi - \frac{1}{3}\right) \\
s_6 &= \frac{27}{2}\xi\eta\left(\xi - \frac{1}{3}\right) \\
s_7 &= \frac{27}{2}\xi\eta\left(\eta - \frac{1}{3}\right) \\
s_8 &= \frac{27}{2}(1-\xi-\eta)\,\eta\left(\eta - \frac{1}{3}\right) \\
s_9 &= \frac{27}{2}(1-\xi-\eta)\,\eta\left(\frac{2}{3} - \xi - \eta\right) \\
s_{10} &= 27\xi\eta\,(1-\xi-\eta)\,.
\end{aligned}
$$

The mapping $\boldsymbol{\phi}(\boldsymbol{\xi}, \boldsymbol{\eta})$ from the reference unit element in Fig. 3 onto the physical element is then given by:

$$
\boldsymbol{\phi}(\boldsymbol{\xi}, \boldsymbol{\eta}) = \sum_{i=1}^{10} \mathbf{x_i}\, s_i(\xi, \eta), \tag{2}
$$

where $\mathbf{x_i}$ are the nodal positions, and (ξ, η) is a point in the reference element. Since we are concerned with the angles between each pair of edges, we need to define mappings from each point on the edges of the reference element to the corresponding point on the edges of the physical element. The edges correspond to third-order Lagrange elements in 1D. The shape functions associated with these elements are defined as:

$$n_1(t) = \frac{9}{2}(1-t)\left(\frac{2}{3}-t\right)\left(\frac{1}{3}-t\right)$$

$$n_2(t) = \frac{27}{2}(1-t)\left(\frac{2}{3}-t\right)(t)$$

$$n_3(t) = \frac{27}{2}(1-t)\left(\frac{1}{3}-t\right)(-t)$$

$$n_4(t) = \frac{9}{2}\left(\frac{2}{3}-t\right)\left(\frac{1}{3}-t\right)(t).$$

The derivatives of these shape functions with respect to t are given by:

$$n_1{}'(t) = \frac{1}{2}\left(-11+36t-27t^2\right)$$

$$n_2{}'(t) = \frac{1}{2}\left(18-90t+81t^2\right)$$

$$n_3{}'(t) = \frac{1}{2}\left(-9+72t-81e^2\right)$$

$$n_4{}'(t) = \frac{1}{2}\left(2-18t+27t^2\right).$$

Using these shape functions, we define the mappings from each edge in the reference element to each edge in the physical element as:

$$\mathbf{f_{12}(t)} = \mathbf{x_1}n_1(t) + \mathbf{x_4}n_2(t) + \mathbf{x_5}n_3(t) + \mathbf{x_2}n_4(t)$$

$$\mathbf{f_{23}(t)} = \mathbf{x_2}n_1(t) + \mathbf{x_6}n_2(t) + \mathbf{x_7}n_3(t) + \mathbf{x_3}n_4(t)$$

$$\mathbf{f_{31}(t)} = \mathbf{x_3}n_1(t) + \mathbf{x_8}n_2(t) + \mathbf{x_9}n_3(t) + \mathbf{x_1}n_4(t).$$

The notation f_{ij} denotes the edge between nodes i and j in Fig. 3. In their expanded forms, each $\mathbf{f_{ij}(t)}$ is a cubic polynomial in the variable t. Next, we need to compute the derivatives of our functions. Straightforward differentiation with respect to t results in the following:

$$\mathbf{f_{12}}'\mathbf{(t)} = \mathbf{x_1}n_1{}'(t) + \mathbf{x_4}n_2{}'(t) + \mathbf{x_5}n_3{}'(t) + \mathbf{x_2}n_4{}'(t)$$

$$\mathbf{f_{23}}'\mathbf{(t)} = \mathbf{x_2}n_1{}'(t) + \mathbf{x_6}n_2{}'(t) + \mathbf{x_7}n_3{}'(t) + \mathbf{x_3}n_4{}'(t)$$

$$\mathbf{f_{31}}'\mathbf{(t)} = \mathbf{x_3}n_1{}'(t) + \mathbf{x_8}n_2{}'(t) + \mathbf{x_9}n_3{}'(t) + \mathbf{x_1}n_4{}'(t).$$

Given these derivatives, we can return to the problem of calculating the angles between edges. As an example, suppose that we want to calculate the angle between edge e_{12} and edge e_{31} in Fig. 3. To calculate the angle in radians, we use the following formula:

$$\theta = \pi - \arccos\left(\frac{f_{12}'(0)\, f_{31}'(1)}{||f_{12}'(0)||\; ||f_{31}'(0)||}\right) = \frac{\pi}{2}.$$

Returning to the calculation of $\boldsymbol{\alpha}_{\mathbf{i}}(\mathbf{x})$ in Eq. (1), we loop over each triangle in the patch and calculate the two angles of each triangle formed by the edges incident to the shared edge between the triangles as described above.

3 Numerical Experiments

In this section, we show the results from performing several numerical experiments to untangle invalid second- and third-order meshes. For each example, we show the initial meshes; the meshes which result after untangling them with our method; the minimum distortion, maximum distortion, average distortion computed over all elements (referred to as Avg1 in figures), and average distortion computed over curved elements (referred to as Avg2 in figures), and the run time needed for our method to untangle the mesh. For each mesh, we show the nodes associated with the element of the given order. We do not show the location of the quadrature points. The code was run using Matlab R2017b, and the execution times were measured on a machine with 8GB of RAM and an Intel Xeon(R) W3520 CPU. All mesh visualizations and distortion calculations were done using Gmsh [7].

Our first example is a third-order annulus composed of 30 elements. During the process of curving the boundary, tangled elements were created near the top and bottom of the inner ring. Figure 4a, c show the initial invalid mesh and the final mesh resulting from our method, respectively. Figure 4b, d, show detailed views of the inner ring from Fig. 4a, c, respectively. In Fig. 4e we give the mesh element distortion.

Our second example is the leading edge of a third-order NACA0012 airfoil. In Fig. 5a, we can see that curving the inner boundary resulted in two tangled elements near the leading edge of the airfoil. In Fig. 5b, we show the final mesh resulting from our method. Finally, Fig. 5c gives the mesh element distortion.

Our third example is a second-order mesh of a mechanical part with several holes. Figure 6a–c shows the initial invalid mesh, the final mesh resulting from our method, and the mesh quality as measured by the distortion metric. In Fig. 6a, we can see that curving the boundaries resulted in tangled elements near the top and bottom holes.

Our fourth and fifth examples are valid meshes of a square plate with a circular hole. To induce mesh tangling in the fourth example, we applied a rotation of 10 degrees counterclockwise to the inner ring followed by a horizontal shear with a shear factor of 0.5. In Fig. 7a, b, d, we show the initial valid mesh, the tangled mesh

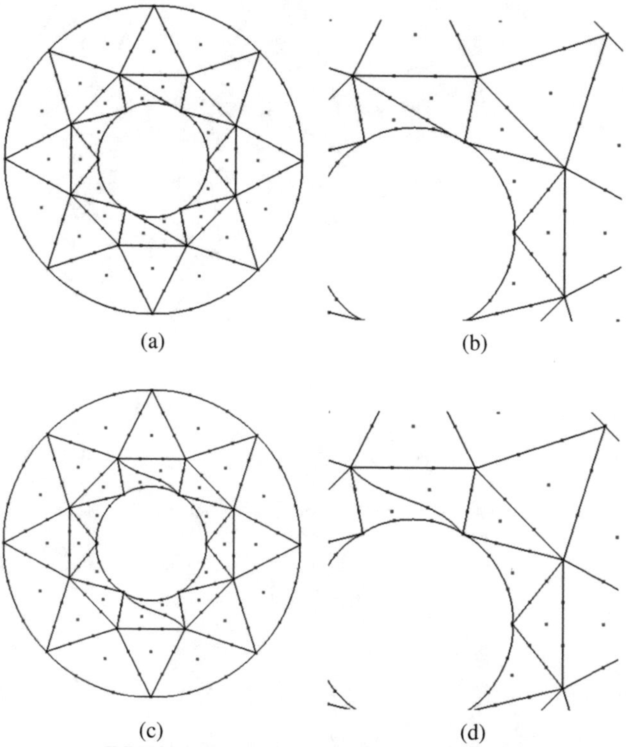

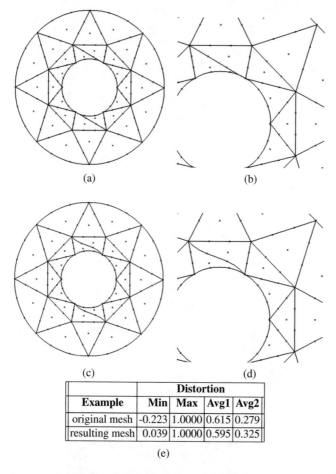

	Distortion			
Example	**Min**	**Max**	**Avg1**	**Avg2**
original mesh	-0.223	1.0000	0.615	0.279
resulting mesh	0.039	1.0000	0.595	0.325

(e)

Fig. 4 Annulus example: (**a**) the tangled third-order mesh; (**b**) a detailed view of one tangled element along the top of the inner boundary; (**c**) the mesh resulting from our method; (**d**) a detailed view of the untangled element from (**c**), and (**e**) the mesh quality as measured by the element distortion metric

resulting from rotation and shearing, and the final untangled mesh resulting from our method. In Fig. 7c, e, we show detailed views of the inner ring. In Fig. 7f, we give the element distortion for the initial, tangled, and final meshes, respectively. In the fifth example, we applied a rotation of 10 degrees counterclockwise to the inner ring followed by a stretching of the bottom half of the plate. In Fig. 8a, b, c, we show the initial valid mesh, the tangled mesh resulting from rotation and stretching, and the final untangled mesh resulting from our method. In Fig. 8d, we give the element distortion for each mesh. In our final example, we show a valid mesh of a two-dimensional beam. To create mesh tangling, we treated the beam as a simply supported beam and applied a center load. After applying the load, we translated

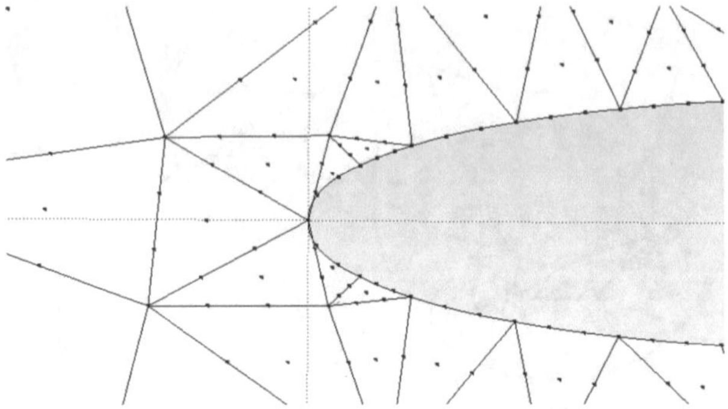

(a)

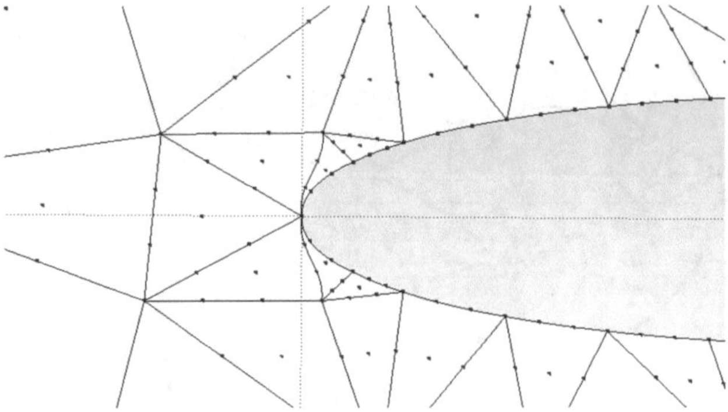

(b)

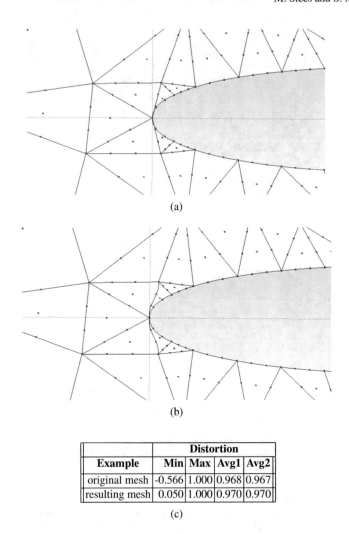

Example	Distortion			
	Min	Max	Avg1	Avg2
original mesh	-0.566	1.000	0.968	0.967
resulting mesh	0.050	1.000	0.970	0.970

(c)

Fig. 5 Airfoil example: (**a**) the tangled third-order mesh; (**b**) the mesh resulting from our method, and (**c**) the mesh quality as measured by the element distortion metric

the left and right sides of the beam. In Fig. 9a, b, c, we show the initial valid mesh, the tangled mesh resulting from our transformations, and the final untangled mesh resulting from our method. In Fig. 9d, e, we show detailed views of the left side of the beam from Fig. 9b, c. Finally in Fig. 9f, we give the mesh element distortion.

While the test cases are relatively straightforward, our goal was to explore the types of tangling that occur as a result of moving the new boundary nodes onto the curved boundary during the typical high-order mesh generation process. We were also interested in tangling that might result from small deformations to a valid mesh. With these points in mind, the examples demonstrate that our method is able

Fig. 6 Mechanical part example: (**a**) the tangled second-order mesh; (**b**) the mesh resulting from our method, and (**c**) the mesh quality as measured by the element distortion metric

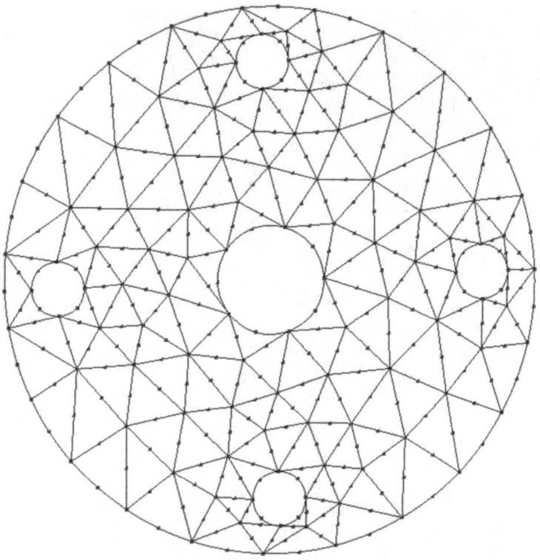

(a)

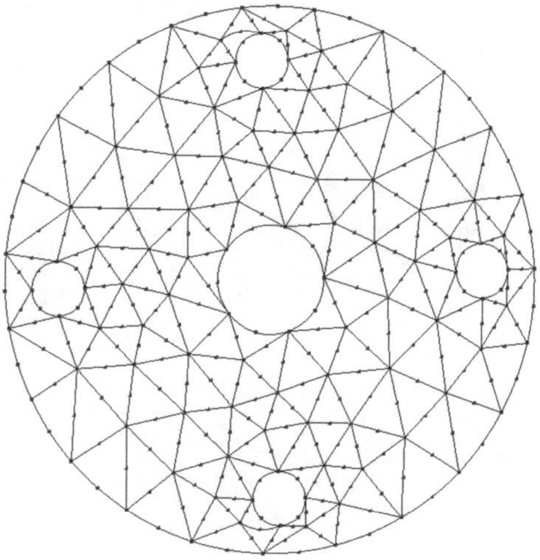

(b)

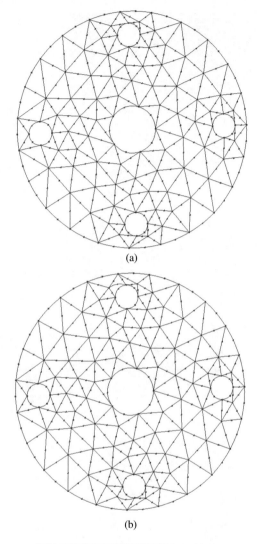

	Distortion			
Example	**Min**	**Max**	**Avg1**	**Avg2**
original mesh	-0.049	1.000	0.904	0.601
resulting mesh	0.008	1.000	0.905	0.625

(c)

to handle the small deformations that might result in tangling for second- and third-order meshes. Additionally, our method only required a single pass for each of the test cases. We demonstrate the runtime performance of our method in Table 1. We list the number of elements and wall clock time for each of our numerical examples

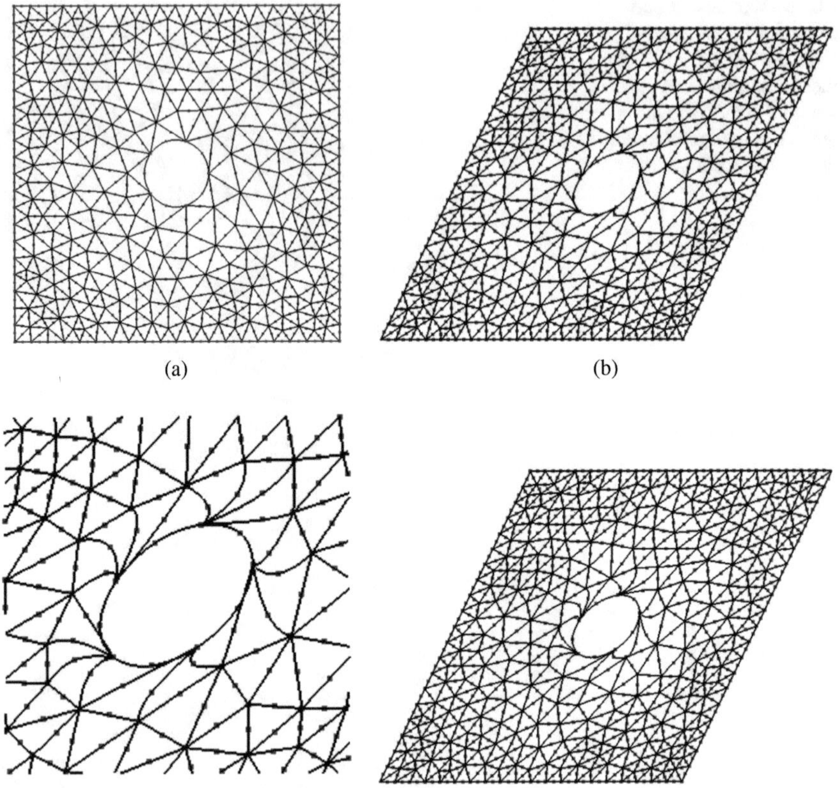

(a) (b)

(c) (d)

(e)

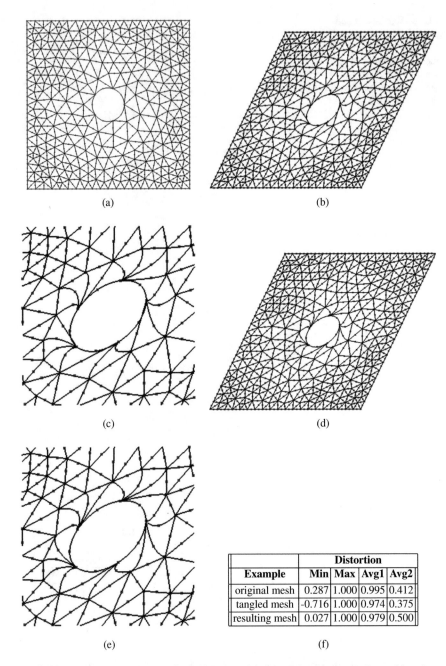

Example	Distortion Min	Max	Avg1	Avg2
original mesh	0.287	1.000	0.995	0.412
tangled mesh	-0.716	1.000	0.974	0.375
resulting mesh	0.027	1.000	0.979	0.500

(f)

Fig. 7 Square plate example: (**a**) the initial second-order mesh; (**b**) the mesh resulting from rotating the inner ring 10 degrees counterclockwise and applying a horizontal shear with a shear factor of 0.5; (**c**) a detailed view of the elements along the inner ring; (**d**) the mesh resulting from applying our method, (**e**) a detailed view of the elements along the inner ring; and (**f**) the mesh quality as measured by the element distortion metric

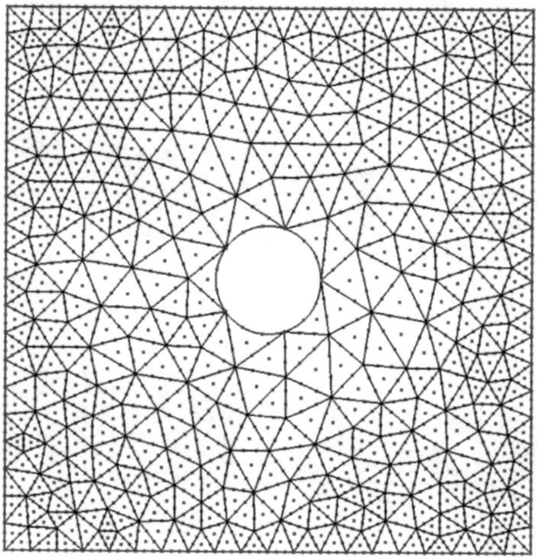

(a)

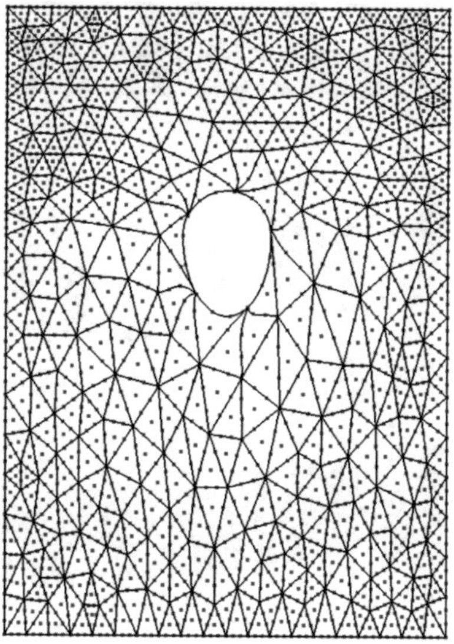

(b)

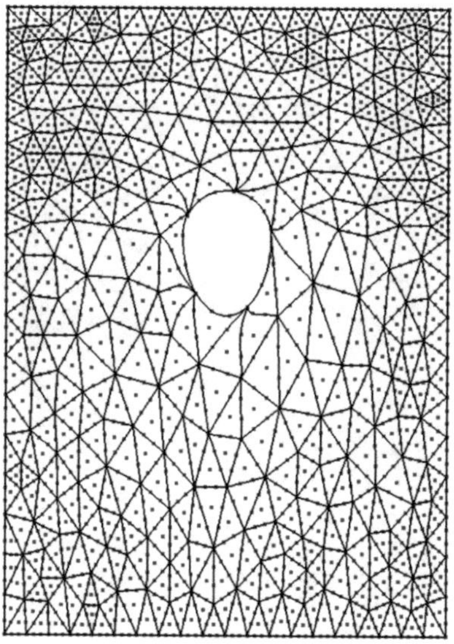

(c)

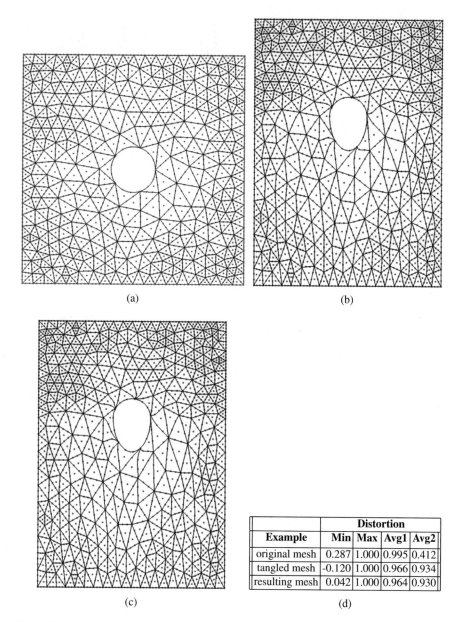

	Distortion			
Example	**Min**	**Max**	**Avg1**	**Avg2**
original mesh	0.287	1.000	0.995	0.412
tangled mesh	-0.120	1.000	0.966	0.934
resulting mesh	0.042	1.000	0.964	0.930

(d)

Fig. 8 Square plate example: (**a**) the initial third-order mesh; (**b**) the mesh resulting from rotating the inner ring 10 degrees counterclockwise and stretching the bottom half of the plate; (**c**) the mesh resulting from applying our method, (**d**) the mesh quality as measured by the element distortion metric

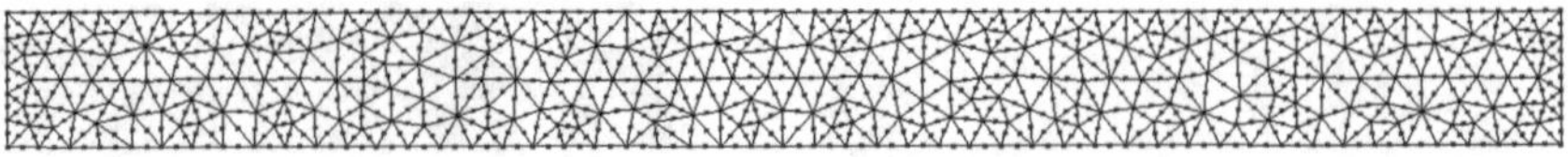

(a)

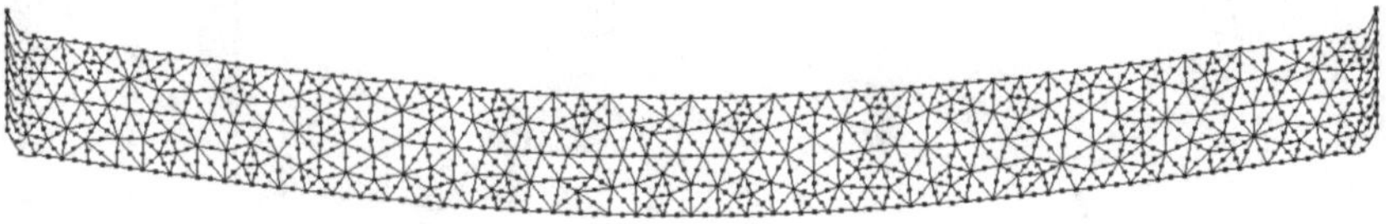

(b)

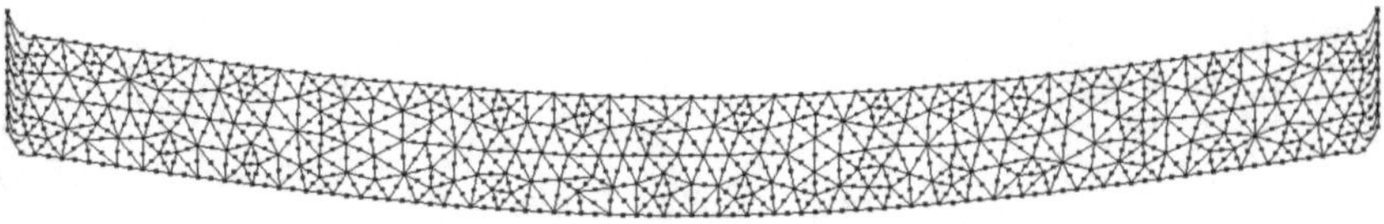

(c)

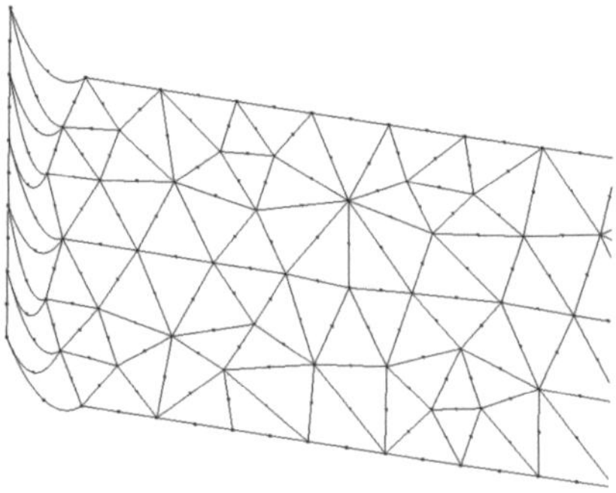

(d)

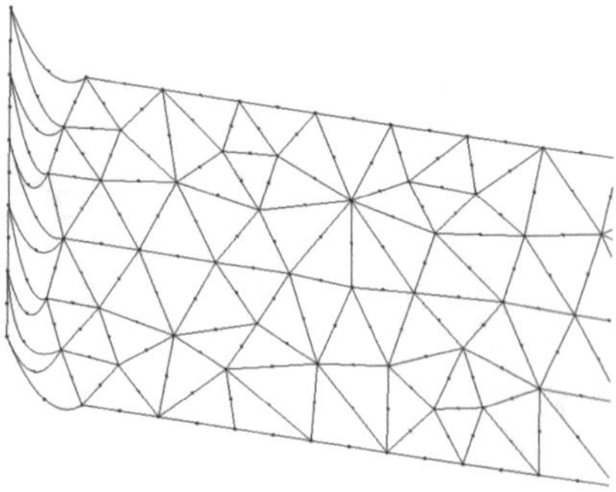

(e)

	Distortion			
Example	**Min**	**Max**	**Avg1**	**Avg2**
original mesh	1.000	1.000	1.000	1.000
tangled mesh	-0.044	1.000	0.982	0.975
resulting mesh	0.032	1.000	0.980	0.973

(f)

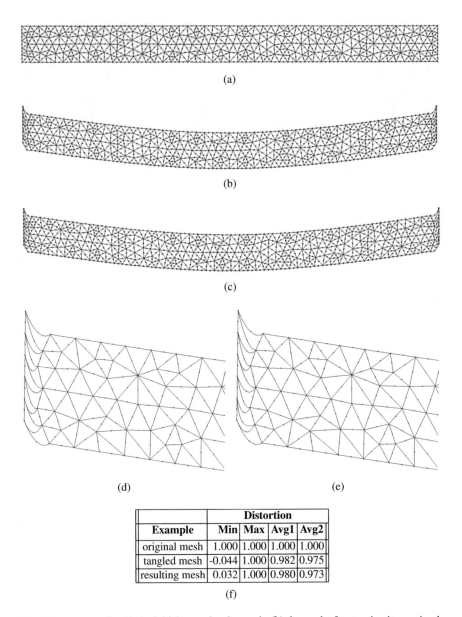

Fig. 9 Beam example: (**a**) the initial second-order mesh; (**b**) the mesh after treating it as a simply supported beam with a center load and translating the left and right ends; (**c**) the mesh resulting from applying our method; (**d**) a detailed view of the left edge of (**b**); (**e**) a detailed view of the left edge of (**c**), and (**f**) the mesh quality as measured by the element distortion metric

Table 1 The number of elements and the wall clock time for each mesh

Example	Number of elements	Wall clock time (s)
Annulus	30	2.56
Airfoil	282	16.63
Mechanical part	182	2.44
Plate 1	597	7.40
Plate 2	597	7.48
Beam	542	6.73

in Table 1. While these times are reasonable, for large meshes, faster run times will be required. Fortunately, there is high potential for improved performance using parallel computing, as our method can be applied to non-adjacent patches simultaneously.

4 Conclusions

We have presented a new optimization-based method for untangling the edges of second- and third-order meshes. The two-dimensional examples have shown that our proposed method based on the unsigned angles of curvilinear triangles is able to successfully untangle several invalid second- and third-order meshes.

We note that presently our method has a few limitations. The first limitation is that it only allows movement of the high-order nodes that lie on the interior edge (e.g. the free nodes show in Fig. 1). That is, it does not allow movement of the endpoints. The second limitation is that non-edge nodes (e.g. like node 10 in Fig. 3) are not moved at all. The final limitation is that our objective function does not measure element validity. Due to these limitations, our method does not guarantee that it will successfully untangle a given tangled patch. With these limitations in mind, our future work will include extending the capabilities of our method to include moving non-edge nodes, as well as allowing the endpoints of edges to move. We will also explore the use of signed angles, where a negative angle indicates that tangling is present. This would allow us to directly check element validity, but would likely require modification of the objective function to achieve the desired untangling behavior. Other future improvements include extending our approach to three dimensions by using the dihedral angles between curved faces of high-order tetrahedral elements, and extending our implementation to allow for elements with $p > 3$.

Acknowledgements The work of the first author was funded in part by the Madison and Lila Self Graduate Fellowship and NSF CCF grant 1717894. The work of the second author was funded in part by NSF CCF grant 1717894.

References

1. F. Bassi, S. Rebay, High-order accurate discontinuous finite element solution of the 2D euler equations. J. Comput. Phys. **138**(2), 251–285 (1997)
2. S. Dey, M.S. Shephard, Curvilinear mesh generation in 3D, in *Proceedings of the 8th International Meshing Roundtable*, 1999
3. M. Fortunato, P.-O. Persson, High-order unstructured curved mesh generation using the Winslow equations. J. Comput. Phys. **307**(2016), 1–14 (2016)
4. A. Gargallo-Peiró, X. Roca, J. Peraire, J. Sarrate, Distortion and quality measures for validating and generating high-order tetrahedral meshes. Eng. Comput. **31**(3), 423–437 (2015)
5. A. Gargallo-Peiró, X. Roca, J. Peraire, J. Sarrate, Optimization of a regularized distortion measure to generate curved high-order unstructured tetrahedral meshes. Int. J. Numer. Methods Eng. **103**(5), 342–363 (2015)
6. P.L. George, H. Borouchaki, Construction of tetrahedral meshes of degree two. Int. J. Numer. Methods Eng. **90**(9), 1156–1182 (2012)
7. C. Geuzaine, J.-F. Remacle, Gmsh: A 3-D finite element mesh generator with built-in pre-and post-processing facilities. Int. J. Numer. Methods Eng. **79**(11), 1309–1331 (2009)
8. X. Luo, M.S. Shephard, J.-F. Remacle, The influence of geometric approximation on the accuracy of high order methods. Rensselaer SCOREC report, 1, 2001
9. D. Moxey, M. Green, S. Sherwin, J. Peiró, An isoparametric approach to high-order curvilinear boundary-layer meshing. Comput. Methods Appl. Mech. Eng. **283**, 636–650 (2015)
10. D. Moxey, D. Ekelschot, Ü. Keskin, S.J. Sherwin, J. Peiró, High-order curvilinear meshing using a thermo-elastic analogy. Comput. Aided Des. **72**, 130–139 (2016)
11. J. Nocedal, S. Wright, *Numerical Optimization*. Springer Series in Operations Research and Financial Engineering, 2nd edn. (Springer, Berlin, 2006)
12. P.-O. Persson, J. Peraire, Curved mesh generation and mesh refinement using Lagrangian solid mechanics, in *Proceedings of the 47th AIAA Aerospace Sciences Meeting Including the New Horizons Forum and Aerospace Exposition* (2009), p. 949
13. X. Roca, A. Gargallo-Peiró, J. Sarrate, Defining quality measures for high-order planar triangles and curved mesh generation, in *Proceedings of the 20th International Meshing Roundtable* (Springer, Berlin, 2012), pp. 365–383
14. E. Ruiz-Gironés, J. Sarrate, X. Roca, Generation of curved high-order meshes with optimal quality and geometric accuracy, in *Proceedings of the 25th International Meshing Roundtable*. Procedia Engineering, vol. 163, 2016, pp. 315–327
15. S.J. Sherwin, J. Peiró, Mesh generation in curvilinear domains using high-order elements. Int. J. Numer. Methods Eng. **53**(1), 207–223 (2001)
16. M. Stees, S.M. Shontz, A high-order log barrier-based mesh generation and warping method, in *Proceedings of the 26th International Meshing Roundtable*. Procedia Engineering, vol. 203, 2017, pp. 180–192
17. T. Toulorge, C. Geuzaine, J.-F. Remacle, J. Lambrechts, Robust untangling of curvilinear meshes. J. Comput. Phys. **254**, 8–26 (2013)
18. M. Turner, D. Moxey, J. Peir, M. Gammon, C.R. Pollard, H. Bucklow, A framework for the generation of high-order curvilinear hybrid meshes for CFD simulations, in *Proceedings of the 26th International Meshing Roundtable*. Procedia Engineering, vol. 203, 2017, pp. 206–218
19. Z.J. Wang, K. Fidkowski, R. Abgrall, F. Bassi, D. Caraeni, A. Cary, H. Deconinck, R. Hartmann, K. Hillewaert, H.T. Huynh, et al., High-order CFD methods: current status and perspective. Int. J. Numer. Methods Fluids **72**(8), 811–845 (2013)
20. Z.Q. Xie, R. Sevilla, O. Hassan, K. Morgan, The generation of arbitrary order curved meshes for 3D finite element analysis. Comput. Mech. **51**(3), 361–374 (2013)

Imposing Boundary Conditions to Match a CAD Virtual Geometry for the Mesh Curving Problem

Eloi Ruiz-Gironés and Xevi Roca

Abstract We present a high-order mesh curving method where the mesh boundary is enforced to match a target virtual geometry. Our method has the unique capability to allow curved elements to span and slide on top of several CAD entities during the mesh curving process. The main advantage is that small angles or small patches of the CAD model do not compromise the topology, quality and size of the boundary elements. We associate each high-order boundary node to a unique group of either curves (virtual wires) or surfaces (virtual shell). Then, we deform the volume elements to accommodate the boundary curvature, while the boundary condition is enforced with a penalty method. At each iteration of the penalty method, the boundary condition is updated by projecting the boundary interpolative nodes of the previous iteration on top of the corresponding virtual entities. The method is suitable to curve meshes featuring non-uniform isotropic and highly stretched elements while matching a given virtual geometry.

1 Introduction

Curved high-order meshes are required for unstructured high-order methods in order to keep their advantages [1–5]. These advantages come in the form of geometrical flexibility, high accuracy, and low numerical dissipation and dispersion. High-order methods feature exponential converge rates and therefore, they have been proved to be faster than low-order methods in several applications [6–14], especially in those problems where an implicit solver is required [15].

E. Ruiz-Gironés (✉) · X. Roca
Computer Applications in Science and Engineering, Barcelona Supercomputing Center, Barcelona, Spain
e-mail: eloi.ruizgirones@bsc.es

X. Roca, A. Loseille (eds.), *27th International Meshing Roundtable*,
Lecture Notes in Computational Science and Engineering 127,
https://doi.org/10.1007/978-3-030-13992-6_19

Usually, to generate a curved high-order mesh an a posteriori approach is used [16–25]. First, a linear mesh with elements of the desired shape and size is generated and then, the mesh boundary is curved to match the target geometry. This step may introduce low-quality and inverted elements that have to be repaired using a high-order mesh curving technique. There are several manners to formulate the mesh curving problem: PDE-based methods like solid mechanics analogies [21, 24, 26–29] or the Winslow equation [25], and optimization-based methods [23, 30–32].

One of the key points of all these methods is the imposition of the boundary displacement, because it drives the insertion of invalid elements that might hamper the convergence and robustness of the mesh curving algorithm. In the most simple scenario, the boundary nodes are directly projected onto the target geometry to interpolate it. In other approaches, after the nodal projection, the nodes can slide along the single geometric entity they belong to [30, 33–36]. Thus, the boundary condition is introduced into the problem formulation by means of the parametric coordinates of the nodes. Although inverted and low-quality elements may still appear in the first stages of the optimization process, the additional freedom of the boundary nodes allows obtaining a *better* mesh that interpolates the target geometry.

Instead of directly moving the boundary nodes to the target position, it is possible to introduce the Dirichlet condition in an incremental manner [21, 25, 27, 28]. In this way, it is possible to mitigate the insertion of inverted elements during the optimization process and therefore, we increase the practical robustness of the mesh curving method. The key ingredient is to define the boundary trajectory and the number of sub-steps to obtain a valid boundary mesh during the curving process.

Alternatively, it is possible to pose the mesh curving as a constrained optimization problem. The boundary condition is introduced into the target function by means of a penalty method or an augmented Lagrangian formulation. The constraint of the problem can be introduced by imposing an interpolation condition [37], or by approximating the target geometry in a weak sense [38, 39].

It is important to point out that standard curved mesh generators do not consider to span the curved elements on top of several CAD entities. Singularly, this option has been explored in the context of the NURBS-enhanced finite element method, where the curved boundary elements are fixed to span on top of several patches of the boundary representation [29]. Nevertheless, the option of sliding the curved elements on top several CAD entities has not been explored yet.

The main contribution of this work is to propose a mesh curving method that allows curved elements to span and slide on top of several CAD entities during the optimization process. Therefore, the curved mesh is not constrained by the topology of the geometric model. This is especially important when the model contains tangent curves with small inner angles that hinder the element quality, or small geometric entities that limit the element size. To accomplish this, we need to group the curves and surfaces of the model into virtual wires and shells using a virtual geometry kernel. Although we manually group the different geometric entities into virtual wires and shells, we can perform this process in an automatic manner by checking the angles between adjacent entities.

To set up the virtual curves and wires, the element sizes, and the initial linear mesh we have used Pointwise [40]. When the distribution of high-order nodes is inserted in the mesh, each boundary node is associated to a virtual wire or virtual shell. Then, we project the boundary nodes onto the corresponding virtual entity, and the boundary condition for the mesh curving problem is defined as the interpolation of the target geometry through the projected points.

Once the boundary condition is obtained, we pose the mesh curving problem as a constrained optimization problem in which we minimize the mesh distortion [23] subject to the boundary condition. To solve this problem, we use a penalty method to transform the mesh curving problem into an unconstrained optimization problem. Note that the boundary condition depends non-linearly on the curved high-order mesh, since we project the boundary nodes. To solve this non-linear problem, we perform a fix-point iteration in which the boundary condition is defined in terms of the curved high-order mesh of the previous iteration. Thus, the boundary condition is not known a priori, and it is updated during the mesh curving process.

In the proposed method, we explicitly minimize the mesh distortion, and we penalize inverted elements in order to preserve the mesh validity during the whole optimization process. Thus, even if the current boundary condition defines an invalid boundary mesh, we obtain a valid volume mesh. Nevertheless, since we update the boundary condition at each iteration of the penalty method, in the presented applications we obtain a valid boundary condition at some point of the curving process.

Our previous curving formulations do not allow to deal with virtual geometry [36, 39]. We first proposed to slide the nodes along the geometric entities by expressing the position of the boundary nodes in terms of the corresponding parametric coordinates [36]. However, since the parametric coordinates are not continuous between different adjacent entities the nodes are not able to jump between adjacent entities. Then, we proposed to morph meshes by solving a constrained minimization problem where the boundary deformation is known and fixed during the whole process [39]. Accordingly, the latter approach does not allow updating the boundary condition in terms of the current iteration of the mesh curving process and therefore, it cannot deal with virtual geometry. On the contrary, in this work we weakly approximate an interpolative condition that depends non-linearly on the high-order mesh. Thus, it is the first time that we show a mesh curving formulation that can deal with virtual geometry.

The rest of the paper is structured as follows. Section 2 introduces several definitions related to the presented work. Section 3 presents the formulation of the proposed high order mesh curving methodology. Section 4 presents several examples to show the capabilities of the proposed formulation. Finally, Sect. 5 details the conclusions and the future work.

2 Preliminaries

2.1 Geometric Model and Mesh Discretization

In this work, we use a geometric model, Ω, composed of several surfaces and curves, $\{\Omega_i^f\}_{i=1,\dots,n_f}$ and $\{\Omega_i^e\}_{i=1,\dots,n_e}$, respectively. Nevertheless, the given decomposition of the geometric model into these surfaces and curves may lead to low-quality meshes. This is the case when the interior angle between two adjacent curves is small, or when small entities limit the maximum element size. For this reason, we group the curves and surfaces of the geometric model into virtual wires and shells, respectively. That is,

$$\Omega_i^w = \bigcup_{j=1}^{n_i} \Omega_j^e \qquad \Omega_i^s = \bigcup_{j=1}^{n_i} \Omega_j^f,$$

where Ω_i^w and Ω_i^s are the wires and shells, respectively. Note that after grouping the surfaces into shells, the shell interior curves are not used to define the wires. That is, not all the curves of the model are used to define the new wires. This *gluing* operation is common when dealing with virtual geometry engines [41, 42]. Currently, we use Pointwise [40] to generate the initial meshes since it can deal with such virtual geometry operation.

We consider that a geometric model, Ω, is defined as the union of its wires and shells in the following manner:

$$\Omega = \bigcup_{i=1}^{N} \Omega_i.$$

Figure 1a shows a CAD model of a sphere with surfaces that limit the maximum element quality. Note that the maximum element quality is bounded by the

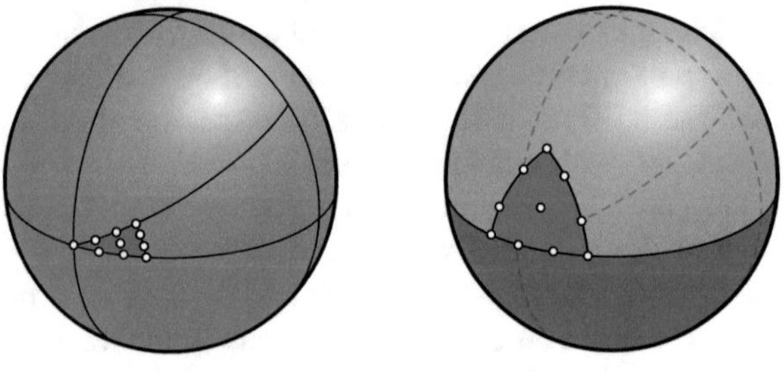

(a) (b)

Fig. 1 High order element approximating a geometric model of a sphere: (**a**) with surfaces limiting the element quality; and (**b**) with virtual shells and wires minimum angle between the curves, and for general CAD models, this angle can be arbitrarily small. To address this issue, we group the top and the bottom surfaces into two different shells, and the four equatorial curves that form the boundary of the shells into one wire, see Fig. 1b. By joining the upper surfaces into a single shell, we are able to generate a high-quality surface element that spans several surfaces of the original model.

The geometric model, Ω, is discretized using an initial linear mesh, $\mathcal{M}^I$, composed of elements of the desired shape and size. The discretization is performed in such a way that

$$\mathcal{M}^I = \bigcup_{i=1}^{N} \mathcal{M}_i^I,$$

where $\mathcal{M}_i^I$ is a discretization of the geometric entity Ω_i. Figure 1a shows a high-order triangle that approximates the surface of a sphere. Moreover, two edges of the triangle are approximating two different curves of the model. In this case, the triangles of the mesh are restricted to belong to a single geometric entity. In contrast, Fig. 1b shows a high-order triangle that approximates a shell of the model. Note that there are high-order nodes of the triangle located in different surfaces of the geometric model. The main advantage of this procedure is that the mesh is no longer constrained to the topology of the geometric model. That is, the nodes can freely slide through different surfaces of the geometry.

2.2 Point Projection Algorithm

The main ingredient to perform the proposed high-order mesh curving technique is the point-projection onto the geometric model. Given a point $\mathbf{x}$ and an entity of the geometric model, Ω_i, we want to compute the orthogonal projection of $\mathbf{x}$ onto Ω_i that satisfies

$$\Pi_{\Omega_i}(\mathbf{x}) = \underset{\mathbf{y}\in\Omega_i}{\arg\min} \|\mathbf{x}-\mathbf{y}\|^2 . \qquad (1)$$

Since Ω_i is defined as a union of sub-entities of the geometric model, to compute the projection $\Pi_{\Omega_i}(\mathbf{x})$ we need to compute the projection in all the contained sub-entities and take the minimum. That is,

$$\Pi_{\Omega_i}(\mathbf{x}) = \underset{\Omega_i^j\in\Omega_i}{\arg\min} \left\{ \min_{\mathbf{y}\in\Omega_i^j} \|\mathbf{x}-\mathbf{y}\|^2 \right\} = \underset{\Omega_i^j\in\Omega_i}{\arg\min} \left\{ \Pi_{\Omega_i^j}(\mathbf{x}) \right\} \qquad (2)$$

3 Formulation of the Mesh Curving Problem

In this section, we pose the mesh curving problem as a constrained optimization problem. Then, we define the boundary condition using the geometric model, and show the proposed high-order mesh curving technique.

3.1 Constrained Minimization of the Distortion Measure

Given an initial linear mesh, $\mathscr{M}_I$, we want to characterize a curved high-order one, $\mathscr{M}_P$, in terms of a diffeomorphism $\boldsymbol{\phi}^*$ [36, 43]. This diffeomorphism should present optimal point-wise distortion, and has to satisfy a prescribed boundary condition. That is,

$$\begin{aligned} M\boldsymbol{\phi}^* &= 1, & \forall \mathbf{y} \in \mathscr{M}_I, \\ \boldsymbol{T}\boldsymbol{\phi}^* &= \mathbf{g}_D\left(\boldsymbol{T}\boldsymbol{\phi}^*\right), & \forall \mathbf{y} \in \partial\mathscr{M}_I, \end{aligned} \tag{3}$$

where $\boldsymbol{T}$ is the trace operator, $\mathbf{g}_D\left(\boldsymbol{T}\boldsymbol{\phi}^*\right)$ is a non-linear Dirichlet boundary condition on $\partial\mathscr{M}_I$ that depends on the values of $\boldsymbol{\phi}^*$, and

$$M\boldsymbol{\phi}^*(\mathbf{y}) = \eta\left(\mathbf{D}\boldsymbol{\phi}^*(\mathbf{y})\right) = \frac{\left\|\mathbf{D}\boldsymbol{\phi}^*(\mathbf{y})\right\|^2}{n\sigma\left(\mathbf{D}\boldsymbol{\phi}^*(\mathbf{y})\right)^{2/n}}$$

is a point-wise distortion measure defined in terms of the shape distortion measure [44] for linear simplices, where $\|\cdot\|$ and $\sigma(\cdot)$ are the Frobenius norm and the determinant, respectively.

Nevertheless, the shape distortion measure presents finite values when the determinant is negative, and this can potentially lead to meshes with inverted elements. To solve this issue, we propose to regularize the shape distortion measure as

$$\eta\left(\mathbf{D}\boldsymbol{\phi}^*\right) = \frac{|\mathbf{D}\boldsymbol{\phi}|^2}{n\sigma_0\left(\mathbf{D}\boldsymbol{\phi}\right)^{2/n}}, \quad \text{where} \quad \sigma_0 = \frac{1}{2}\left(\sigma + |\sigma|\right). \tag{4}$$

In this manner, when the determinant is non-positive, the point-wise distortion takes a value of infinity, and when the determinant is positive, the point-wise distortion takes a finite value.

In order to solve the problem in Eq. (3), we first rewrite it as a constrained optimization one in the following manner

$$\begin{aligned} &\min_{\boldsymbol{\phi}\in\mathscr{V}} E(\boldsymbol{\phi}) = \|M\boldsymbol{\phi}\|^2 \\ &\text{subjectto}: \\ &\boldsymbol{T}\boldsymbol{\phi} = \mathbf{g}_D(\boldsymbol{T}\boldsymbol{\phi}), \end{aligned} \tag{5}$$

where

$$\mathscr{V} = \left\{ \mathbf{u} \in \left[\mathscr{C}^0(\mathscr{M}_I) \right]^n \text{suchthat} \mathbf{u}_{|e_I} \in \left[\mathscr{P}^p(e_I) \right]^n \ \forall e_I \in \mathscr{M}_I \right\}.$$

Being $\mathscr{P}^p(e_I)$ the space of polynomials of degree at most p over the element e_I.Thus,

$$\boldsymbol{\phi} = \sum_{i=1}^{N} \mathbf{x}_i N_i,$$

being $\{N_i\}_{i=1,\ldots,N}$ a Lagrangian basis of element-wise polynomials continuous at the element interfaces. Note that $\boldsymbol{\phi}$ depends on the nodal positions and therefore, so does the functional in Eq. (5).

We have introduced a merit function to measure the distortion of the mapping that transforms the linear mesh into a curved high-order one, and the boundary condition, $\mathbf{g}_D$ is a constraint of the optimization problem. To solve the constrained optimization problem in (5), we use a penalty approach, see [45], in which we introduce the boundary constraint into the objective function in a weak sense as follows

$$\min_{\boldsymbol{\phi} \in \mathscr{V}} E_\mu (\boldsymbol{\phi}) = \frac{E(\boldsymbol{\phi})}{\|1\|^2_{\mathscr{M}_I}} + \mu \frac{\|\boldsymbol{T\phi} - \mathbf{g}_D(\boldsymbol{T\phi})\|^2_{\partial \mathscr{M}_I}}{\|1\|^2_{\partial \mathscr{M}_I}}, \tag{6}$$

where μ is a penalty parameter that enforces the validity of the constraint when it tends to infinity. We have introduced the measures of the initial mesh and its boundary in order to balance the two contributions of the new functional.

3.2 *Definition of the Boundary Condition*

In order to solve the problem in Eq. (6), we need to define the non-linear Dirichlet boundary condition $\mathbf{g}_D(\boldsymbol{T\phi})$. The boundary condition takes into account the geometric model Ω, and we define it by means of the point projection algorithm presented in Sect. 2.2, and the current value of $\boldsymbol{\phi}$. Thus, we define the boundary condition as

$$\mathbf{g}_D(\boldsymbol{T\phi}) = \sum_{i=1}^{N_b} \Pi_{\Omega_{\mathbf{x}_i}}(\mathbf{x}_i) N_i^b = \Pi_\Omega(\boldsymbol{T\phi}), \tag{7}$$

where $\Omega_{\mathbf{x}_i}$ denotes the entity where the boundary node $\mathbf{x}_i$ belongs to, and $\{N\}_{i=1,\ldots,N_b}$ is a Lagrangian basis of shape functions continuous between adjacent boundary faces. Thus, we generate a function $\mathbf{g}_D \in \mathscr{V}_b$, where

$$\mathscr{V}_b = \left\{ \mathbf{u} \in \mathscr{C}^0 \text{such that} \mathbf{u}_{|f_I} \in \left[\mathscr{P}^p(f_I)\right]^n \ \forall f_I \in \partial \mathscr{M}_I \right\},$$

being $P^p(f_I)$ the space of polynomials of degree at most p over the boundary face f_I. Note that the function $\mathbf{g}_D$ is generated as a linear combination of a Lagrangian basis of shape functions. The coefficients of the linear combination are the projection of the boundary high-order nodes onto the geometric entities they belong to. Thus, the boundary condition is defined as the interpolation of the geometric model into the function space $\mathscr{V}_b$. The interpolation points are the projection of the high-order nodes of the curved mesh.

Since we have grouped the curves and surfaces into wires and shells, the interpolation points are free to jump between curves and surfaces of the wire and shell they belong to. The only constraint for the interpolation points is that they need to be associated to the same wire or shell during the whole optimization process.

3.3 High-Order Mesh Curving

To obtain a curved high-order mesh, we optimize the functional in Eq. (6) with an increasing penalty parameter. Nevertheless, as shown in Sect. 3.2, the boundary condition depends on the actual solution of the problem. Thus, we propose to apply a fix-point iteration in which

$$\mathbf{g}_D^k = \Pi_\Omega\left(\boldsymbol{T}\boldsymbol{\phi}^k\right), \qquad \boldsymbol{\phi}^{k+1} = \underset{\boldsymbol{\phi}\in\mathscr{V}}{\arg\min}\, E_{\mu^k}\left(\boldsymbol{\phi}; \mathbf{g}_D^k\right)$$

In this manner, we are able to deal with the non-linearity of the boundary constraint. Algorithm 1 describes the proposed penalty method for high-order mesh curving. The inputs of the algorithm are an initial linear mesh, $\mathscr{M}_I$, a geometric model, Ω, and the tolerances for the non-linear problem and the constraint norm, ω^* and ε^*, respectively. The algorithm stops when a solution is found that satisfies

$$\left\|\nabla E_{\mu^k}\left(\boldsymbol{\phi}^k; \mathbf{g}_D^k\right)\right\| < \omega^* \qquad \text{and} \qquad \left\|\boldsymbol{T}\left(\boldsymbol{\phi}^k\right) - \mathbf{g}_D^k\right\|_{\partial\Omega_I}^2 < \varepsilon^*.$$

The output of the algorithm is a valid curved high-order mesh, $\mathscr{M}_P$, that approximates the target geometric domain. In Line 2 we initialize $\boldsymbol{\phi}^0$ to the identity mapping, **Id**. Note that the identity mapping is optimal with respect to the distortion measure. However, it does not satisfy the boundary constraint. In Line 3, we initialize the boundary condition using the projection of the boundary high-order nodes, according to Eq. (7). The initial penalty parameter is initialized to 10, and the initial tolerance for the non-linear solver is initialized to the norm of the objective function gradient over 10. Lines 6–15 define the main loop of the proposed penalty

Algorithm 1 Penalty method for high-order mesh curving

Input: Mesh $\mathscr{M}_I$, GeometricModel Ω, Real ω^*, Real ε^*
Output: CurvedHighOrderMesh $\mathscr{M}_P$
1: **function** HIGHORDERMESHCURVING
2: $\quad \boldsymbol{\phi}^0 \leftarrow \mathbf{Id}$
3: $\quad \mathbf{g}_D^0 \leftarrow \Pi_\Omega\left(\boldsymbol{\phi}^0\right)$
4: $\quad \mu^0 \leftarrow 10$
5: $\quad \omega^0 \leftarrow \left\|\nabla E_{\mu^0}\left(\boldsymbol{\phi}^0; \mathbf{g}_D^0\right)\right\| / 10$
6: $\quad$ **while** $\left\|\boldsymbol{T}\boldsymbol{\phi}^k - \mathbf{g}_D^k\right\|_{\partial \mathscr{M}_I} > \varepsilon^*$ **and** $\left\|\nabla E_{\mu^k}\left(\boldsymbol{\phi}^k, \mathbf{g}_D^k\right)\right\| > \omega^*$ **do**
7: $\qquad \boldsymbol{\phi}^{k+1} \leftarrow$ OPTIMIZEFUNCTION$\left(E_{\mu^k}\left(\boldsymbol{\phi}^k; \mathbf{g}_D^k\right), \omega^k\right)$
8: $\qquad$ **if** $\left\|\boldsymbol{T}\left(\boldsymbol{\phi}^k\right) - \mathbf{g}_D^k\right\|_{\partial \mathscr{M}_I} > \varepsilon^*$ **then**
9: $\qquad\quad \mu^{k+1} \leftarrow 10\mu^k$
10: $\qquad\quad \omega^{k+1} \leftarrow \omega^k / 10$
11: $\qquad$ **else**
12: $\qquad\quad \omega^{k+1} \leftarrow \omega^*$
13: $\qquad$ **end if**
14: $\qquad \mathbf{g}_D^{k+1} \leftarrow \Pi_\Omega\left(\boldsymbol{\phi}^{k+1}\right)$
15: $\quad$ **end while**
16: $\quad \mathscr{M}_P \leftarrow \boldsymbol{\phi}\left(\mathscr{M}_I\right)$
17: **end function**

method. In Line 7, we optimize the proposed functional. Then, if the norm of the constraint is too large, we increase the penalty parameter and tighten the tolerance of the non-linear solver, Lines 9 and 10. On the contrary, if the norm of the constraint is low enough, we keep the current value of the penalty parameter and we set the tolerance of the non-linear problem to the prescribed tolerance of the mesh curving algorithm. Finally, we update the boundary condition for the next non-linear problem, and iterate the main loop until convergence is achieved.

To optimize each non-linear problem of the proposed penalty method, we use a backtracking line-search method in which the advancing direction is computed using Newton's method and the step-length is set using the Wolfe conditions, see [45] for more details. To solve the linear systems that arise during Newton's method, we use a generalized minimum residual method (GMRES) with a relative tolerance of 10^{-9}, preconditioned with a successive over-relaxation method. The stopping criteria for this optimization process is $\left\|\nabla E_{\mu^k}\left(\boldsymbol{\phi}^k; \mathbf{g}_D^k\right)\right\| < \omega^k$.

One of the advantages of the proposed high-order mesh curving algorithm is that it maintains a valid mesh during the whole process. The main reason is that the proposed functional detects invalid meshes by taking infinite values. Thus, when an invalid configuration is detected, the backtracking line-search reduces the step length until a valid mesh is obtained for the next iteration. Moreover, we do not need to ensure that the boundary condition defines a valid boundary mesh at each iteration of the penalty method. Since we ensure the volume mesh validity and we re-compute the boundary condition at each step, in practical situations we obtain a valid boundary condition at some point of the mesh curving process.

To increase the practical robustness of the mesh curving method, we apply a p-continuation technique in order. In this manner, the optimal configuration of a given polynomial degree is used as the initial condition on to optimize the mesh for the next polynomial degree.This p-continuation technique has allowed us to obtain curved high-order meshes with stretched elements for complex geometries.

4 Examples

This section presents several examples that show the capabilities of the presented high-order mesh curving method. Specifically, we show four three-dimensional examples in which we present isotropic and stretched meshes for two different geometric models.

To generate the initial linear meshes, we have used Pointwise [40]. The mesh curving framework has been implemented in Python [46] using the FEniCS [47] and the petsc4py [48] libraries. To project the boundary high-order nodes we have used both the geode [49] and the Open CASCADE [50] libraries interfaced with a python wrapper using swig [51].

The optimization process has been performed in the MareNostrum4 supercomputer located at the Barcelona Supercomputing Center. It is composed of 3456 nodes, connected using an Intel Omni-Path network. Each node contains two Intel Xeon Platinum 8160 CPU with 24 cores, each at 2.10 GHz, and 96 GB of RAM memory.

In all the examples, we color the elements of the mesh according to its element quality relative to the initial meshes [23]

$$q_{e_I} = \frac{1}{\eta_{e_I}}, \quad \text{where} \quad \eta_{e_I} = \left(\frac{\displaystyle\int_{e_I} (M\boldsymbol{\phi})^2 \, \mathrm{d}\Omega}{\displaystyle\int_{e_I} 1 \, \mathrm{d}\Omega} \right)^{1/2}.$$

The element quality takes values between zero and one, being zero for inverted elements, and one for ideally deformed elements.

4.1 Isotropic Mesh Around a Sphere

In this example we show the generation of an isotropic curved high-order mesh generated for the exterior domain of a sphere. The inner and outer spheres radius are one and 21 units, respectively. The initial linear mesh, see Fig. 2a, contains isotropic elements of size 1.0. Figure 2b shows a curved high-order mesh of polynomial degree five, obtained after applying the proposed mesh curving technique. The mesh

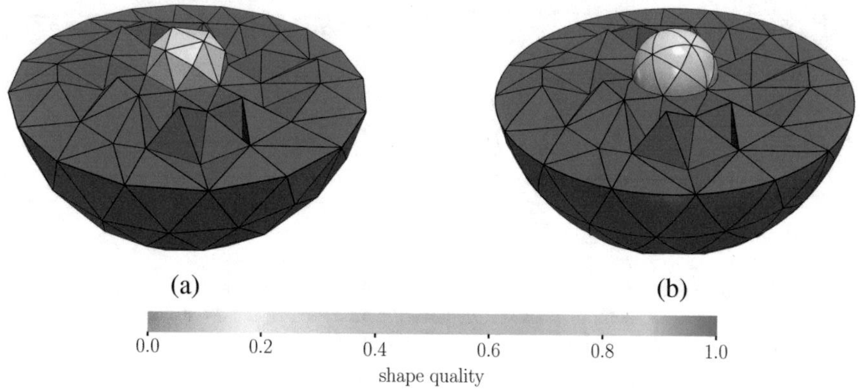

Fig. 2 Isotropic meshes generated around a sphere: (**a**) initial linear mesh; and (**b**) curved high-order mesh of polynomial degree five

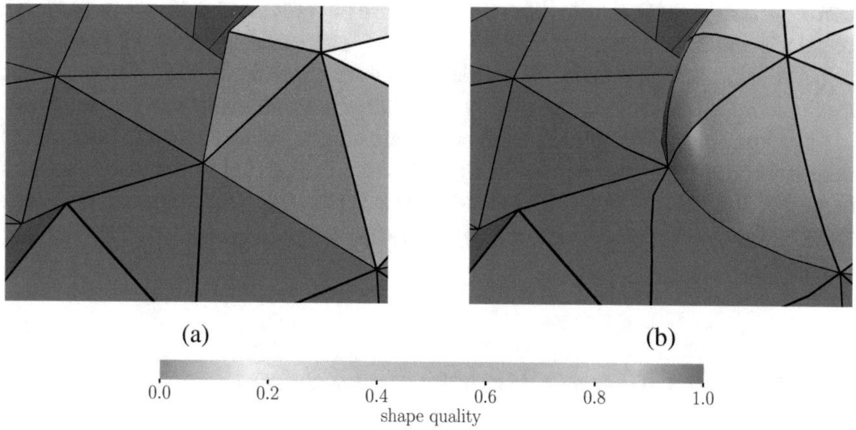

Fig. 3 Detail of the isotropic meshes generated around a sphere: (**a**) initial linear mesh; and (**b**) curved high-order mesh of polynomial degree five

contains 2048 elements and 38,167 nodes. The whole optimization process has been performed using 24 cores and took 1261 s distributed in seventeen iterations of the penalty method. The norm of the constraint at the last iteration is of the order of 10^{-9}, and the minimum element quality of the whole mesh is 0.972. Figure 3a, b shows a detailed view of the initial linear mesh and the curved high-order one near the inner sphere. Note that the curved high-order mesh represents the spherical boundary with a high geometric precision without hampering the element quality.

Figure 4 shows the evolution of the norm of the boundary condition and the minimum element quality of the mesh, against the iterations of the proposed penalty method. In Fig. 2, we depict with gray dots the initial iteration of each polynomial degree. The first gray dot denotes the starting iteration of the mesh curving for the

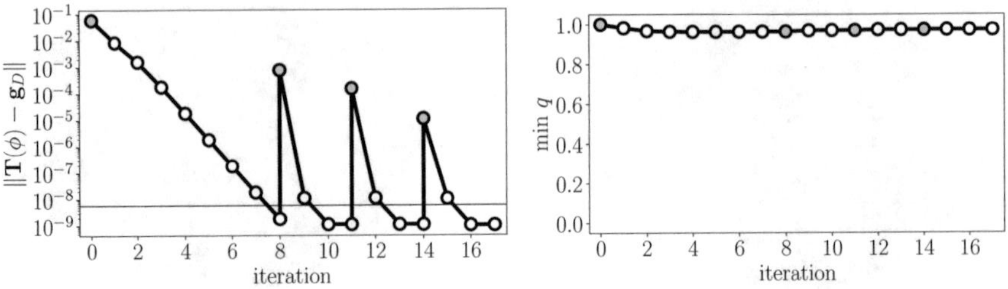

Fig. 4 Evolution of the constraint norm (left) and minimum element quality (right) over the iterations of the penalty method. Thin line denotes the tolerance of the constraint norm, and gray dots denote the first iteration of a new polynomial degree

mesh of polynomial degree two. Note that the norm of the constraint decreases linearly in logarithmic scale with the iterations of the penalty method. At each iteration, the boundary condition norm is reduced one order of magnitude, and we converge it in eight iterations. Then, we increase the polynomial degree of the mesh to three. Since the boundary condition is also expressed in terms of piece-wise polynomials of degree three, it is able to capture more features of the geometric model. For this reason, the norm of the boundary condition increases. Nevertheless, in three iterations of the penalty method, we are able to obtain a mesh of polynomial degree three. In the next iterations, this behavior is repeated to obtain the meshes of polynomial degree four and five. Note that during the whole process, the mesh remains valid, since we are enforcing in the proposed penalty method the validity of the mesh.

4.2 Stretched Elements Mesh Around a Sphere

In this example we present the mesh generated for the exterior domain of a sphere with high stretched elements. The geometric model and the element sizes are the same as the ones in Example (Sect. 4.1). Nevertheless, we introduce a boundary layer around the inner sphere, see Fig. 5a. The boundary layer is defined by a wall distance of 10^{-3}, a growing factor of 1.3, and 24 layers of elements. The maximum stretching of this boundary layer is 10^3. In Fig. 5b we show the optimized curved high-order mesh of polynomial degree five, composed of 3616 elements and 71,072 nodes. The optimization process is performed using 24 cores. The whole process is performed in seventeen iterations and takes 2121 s. At the last iteration, the norm of the constraint is of the order of 10^{-9}, and the minimum element quality is 0.94 (Fig. 6).

The evolution of the constraint norm through the iterations of the penalty method, see Fig. 7, presents a similar behavior than the evolution of the inviscid mesh. Nevertheless, although the presence of highly stretched elements, the minimum element quality is only slightly reduced compared with the isotropic case.

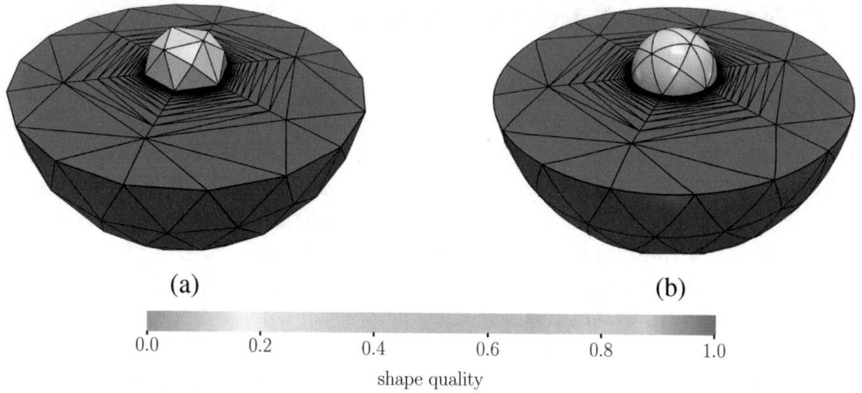

Fig. 5 Stretched meshes generated around a sphere: (**a**) initial linear mesh; and (**b**) curved high-order mesh

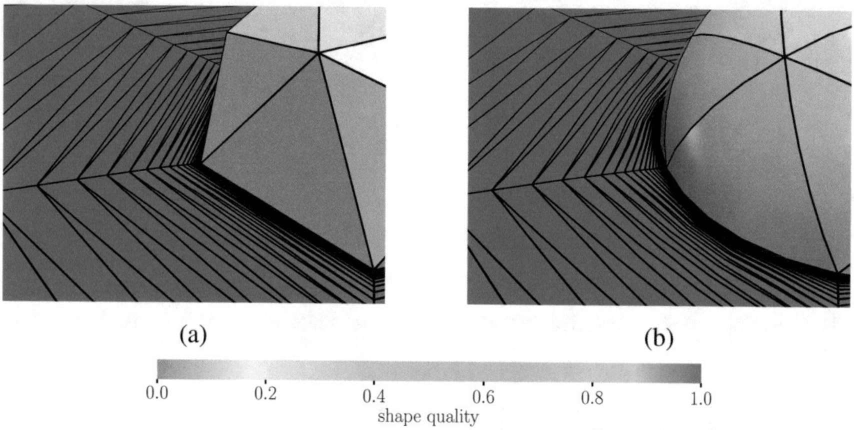

Fig. 6 Detail of the stretched meshes generated around a sphere: (**a**) initial linear mesh; and (**b**) curved high-order mesh

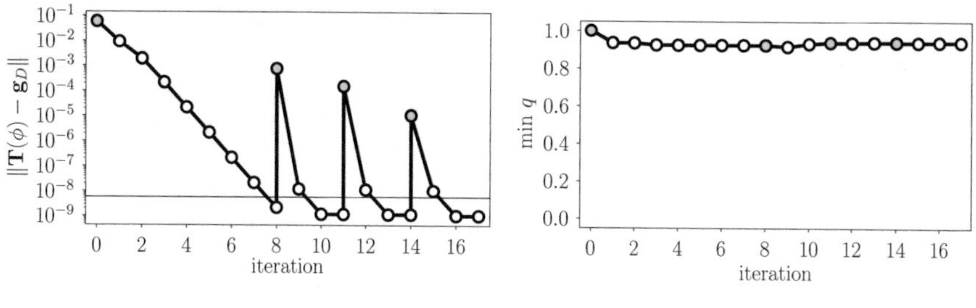

Fig. 7 Evolution of the constraint norm (left) and minimum element quality (right) over the iterations of the penalty method. Thin line denotes the tolerance of the constraint norm, and gray dots denote the first iteration of a new polynomial degree

4.3 Isotropic Mesh Around a Propeller

This example shows the generation of a curved high-order mesh for a complex geometry. In this example, we generate an initial linear mesh around a propeller, see Fig. 8a. We apply the proposed technique to obtain the high-order mesh of polynomial degree three shown in Fig. 8b. It is composed of 227,214 elements and 980,992 nodes. The optimization process is performed using 768 cores, and converges in 8 iterations of the penalty method, and takes 431 s. At the last iteration, the norm of the boundary condition is of the order of 10^{-7}, and the minimum element quality is 0.839 (Fig. 9).

In Fig. 10, we show the evolution of the boundary condition norm and the minimum element quality along the iterations of the penalty method. In the first

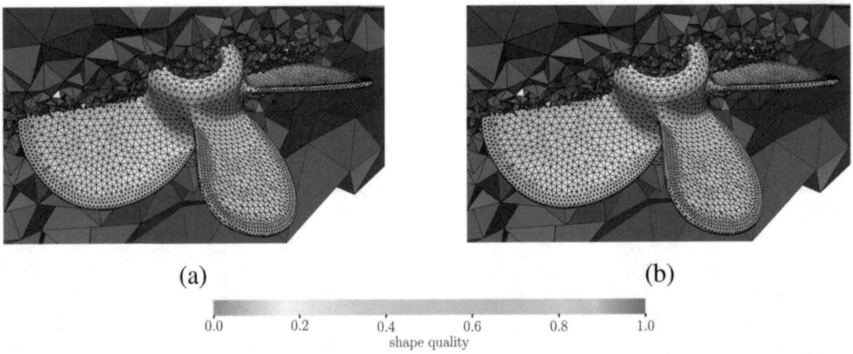

Fig. 8 Isotropic meshes generated around a propeller: (**a**) initial linear mesh; and (**b**) curved high-order mesh

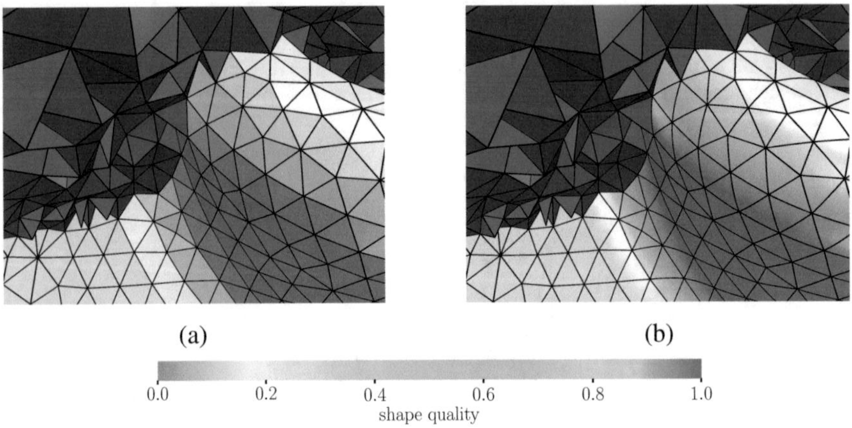

Fig. 9 Detail of the isotropic meshes generated around a propeller: (**a**) initial linear mesh; and (**b**) curved high-order mesh

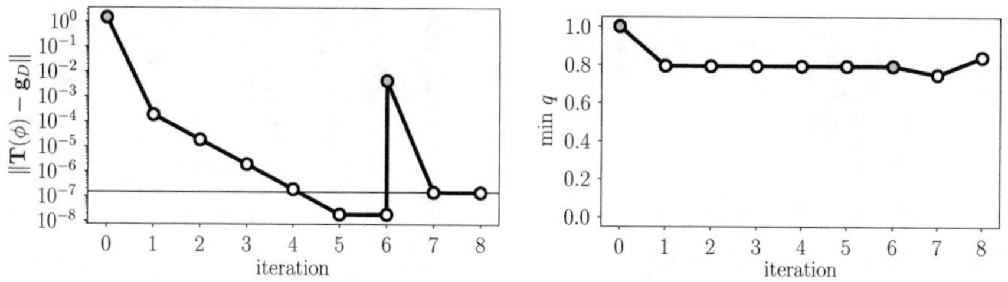

Fig. 10 Evolution of the constraint norm (left) and minimum element quality (right) over the iterations of the penalty method. Thin line denotes the tolerance of the constraint norm, and gray dots denote the first iteration of a new polynomial degree

iterations, the boundary condition norm is reduced linearly in logarithmic scale. In six iterations we obtain the converged quadratic mesh, and in two more iterations we obtain the final curved high-order mesh of polynomial degree three.

4.4 *Stretched Elements Mesh Around a Propeller*

In this example we deal with the generation of a mesh with stretched elements for the exterior domain of a propeller. The boundary layer around the propeller contains 25 layers of elements, a wall distance of 0.002, and a growing ratio of 1.3. This leads to maximum stretching factor of 750. The objective of this example is to show that we are able to generate a high-order mesh with high stretched elements around a complex geometry. Figure 11a shows the initial linear mesh, and Fig. 11b shows the final curved high-order mesh of polynomial degree three. This mesh is obtained in eight iterations of the penalty method, and takes 5452 s (1.5 h) to optimize it using

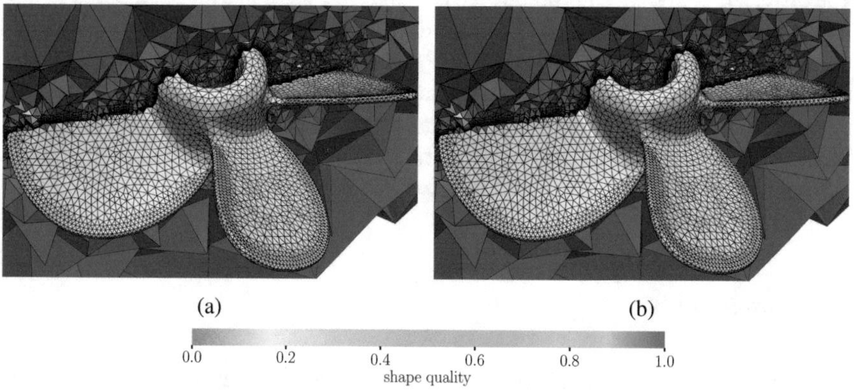

Fig. 11 Stretched meshes generated around a propeller: (**a**) initial linear mesh; and (**b**) curved high-order mesh

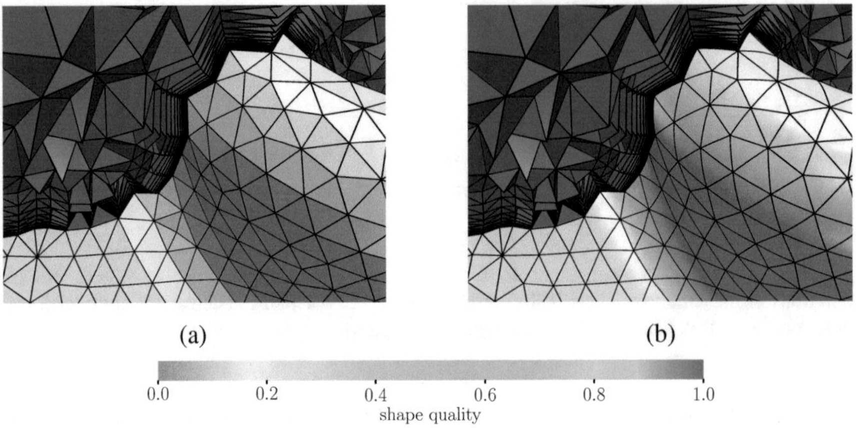

Fig. 12 Detail of the stretched meshes generated around a propeller: (**a**) initial linear mesh; and (**b**) curved high-order mesh

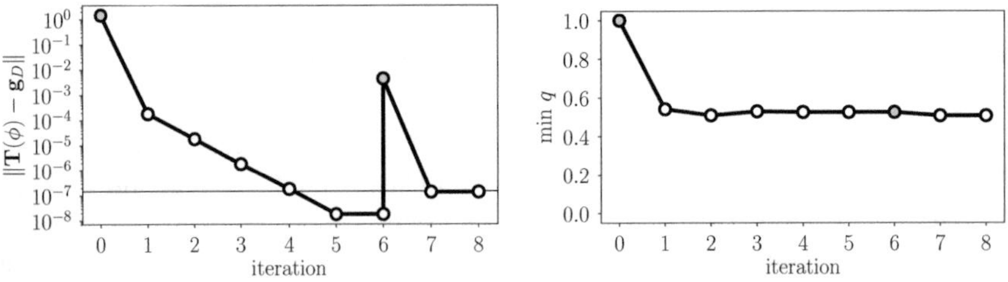

Fig. 13 Evolution of the constraint norm (left) and minimum element quality (right) over the iterations of the penalty method. Thin line denotes the tolerance of the constraint norm, and gray dots denote the first iteration of a new polynomial degree optimization

768 processors. The mesh is composed of 1,648,596 elements and 6,863,825 nodes. At the last iteration, the norm of the boundary constraint is of the order of 10^{-7}, and the minimum element quality is 0.524 (Fig. 12).

The evolution of the norm of the constraint, see Fig. 13, shows a similar behavior than in the inviscid case, even when a boundary layer around the propeller is present. In addition, the complexity of the geometric model and he highly stretched elements induce a lower minimum quality with respect to the isotropic case.

5 Conclusions

We have presented a novel high-order mesh curving technique in which the boundary elements of the high-order mesh are able to span and slide between different geometric entities of the target model by using a virtual geometry kernel.

To accomplish this, we have deduced a novel methodology to introduce the boundary condition of a high-order mesh curving problem based on the projection of the boundary nodes onto the geometric model.

Currently, the proposed algorithm to solve the constrained minimization problem is a penalty method. Nevertheless, we could solve the constrained minimization problem using the augmented Lagrangian method, as we have done in other works. Using the augmented Lagrangian formulation, the value of penalty parameter is bounded and therefore, the condition number of the system matrix does not increase.

Acknowledgements This project has received funding from the European Research Council (ERC) under the European Union's Horizon 2020 research and innovation programme under grant agreement No 715546. This work has also received funding from the Generalitat de Catalunya under grant number 2017 SGR 1731. The work of Xevi Roca has been partially supported by the Spanish Ministerio de Economía y Competitividad under the personal grant agreement RYC-2015–01633.

References

1. B. Szabó, I. Babuška, *Finite Element Analysis* (Wiley, New York, 1991)
2. C. Schwab, *p-and hp-Finite Element Methods: Theory and Applications in Solid and Fluid Mechanics* (Clarendon Press, Oxford, 1998)
3. M.O. Deville, P.F. Fischer, E.H. Mund, *High-Order Methods for Incompressible Fluid Flow*, vol. 9 (Cambridge University Press, Cambridge, 2002)
4. J.S. Hesthaven, T. Warburton, *Nodal Discontinuous Galerkin Methods: Algorithms, Analysis, and Applications*. Texts in Applied Mathematics (Springer, New York, 2007)
5. G. Karniadakis, S. Sherwin, *Spectral/hp Element Methods for Computational Fluid Dynamics* (Oxford University Press, Oxford, 2013)
6. P.E. Vos, S. Sherwin, R. Kirby, From h to p efficiently: implementing finite and spectral/hp element methods to achieve optimal performance for low- and high-order discretisations. J. Comput. Phys. **229**(13), 5161–5181 (2010)
7. C. Cantwell, S. Sherwin, R. Kirby, P. Kelly, From h to p efficiently: strategy selection for operator evaluation on hexahedral and tetrahedral elements. Comput. Fluids **43**(1), 23–28 (2011)
8. C. Cantwell, S. Sherwin, R. Kirby, P. Kelly, From h to p efficiently: selecting the optimal spectral/hp discretisation in three dimensions. Math. Model. Nat. Phenom. **6**(3), 84–96 (2011)
9. R. Löhner, Error and work estimates for high-order elements. Int. J. Numer. Methods Fluids **67**(12), 2184–2188 (2011)
10. M. Yano et al., An optimization framework for adaptive higher-order discretizations of partial differential equations on anisotropic simplex meshes. PhD thesis, Massachusetts Institute of Technology (2012)
11. R. Kirby, S. Sherwin, B. Cockburn, To CG or to HDG: a comparative study. J. Sci. Comput. **51**(1), 183–212 (2012)
12. A. Huerta, X. Roca, A. Angeloski, J. Peraire, Are high-order and hybridizable discontinuous Galerkin methods competitive? Oberwolfach Rep. **9**(1), 485–487 (2012)
13. R. Löhner, Improved error and work estimates for high-order elements. Int. J. Numer. Methods Fluids **72**, 1207–1218 (2013)
14. Z.J. Wang, K. Fidkowski, R. Abgrall, F. Bassi, D. Caraeni, A. Cary, H. Deconinck, R. Hartmann, K. Hillewaert, H.T. Huynh et al. High-order CFD methods: current status and perspective. Int. J. Numer. Methods Fluids **72**(8), 811–845 (2013)

15. A. Huerta, A. Angeloski, X. Roca, J. Peraire, Efficiency of high-order elements for continuous and discontinuous Galerkin methods. Int. J. Numer. Methods Eng. **96**, 529–560 (2013)
16. S. Dey, M. Shephard, J.E. Flaherty, Geometry representation issues associated with p-version finite element computations. Comput. Methods Appl. Mech. Eng. **150**(1–4), 39–55 (1997)
17. S. Dey, R. O'Bara, M.S. Shephard, Curvilinear mesh generation in 3D. Comput. Aided Des. **33**, 199–209 (2001)
18. X. Luo, M.S. Shephard, J.-F. Remacle, R. O'Bara, M. Beall, B. Szabó, R. Actis, p-Version mesh generation issues, in *Proceedings of the 11th International Meshing Roundtable* (Springer, Berlin, 2002), pp. 343–354
19. X. Luo, M.S. Shephard, R. O'Bara, R. Nastasia, M. Beall, Automatic p-version mesh generation for curved domains. Eng. Comput. **20**(3), 273–285 (2004)
20. M.S. Shephard, J.E. Flaherty, K. Jansen, X. Li, X. Luo, N. Chevaugeon, J.-F. Remacle, M. Beall, R. O'Bara, Adaptive mesh generation for curved domains. Appl. Numer. Math. **52**(2–3), 251–271 (2005)
21. P.-O. Persson, J. Peraire, Curved mesh generation and mesh refinement using lagrangian solid mechanics, in *Proceedings of the 47th AIAA*, 2009
22. D. Moxey, M.D. Green, S.J. Sherwin, J. Peiró, An isoparametric approach to high-order curvilinear boundary-layer meshing. Comput. Methods Appl. Mech. Eng. **283**, 636–650 (2015)
23. A. Gargallo-Peiró, X. Roca, J. Peraire, J. Sarrate, Optimization of a regularized distortion measure to generate curved high-order unstructured tetrahedral meshes. Int. J. Numer. Methods Eng. **103**(5), 342–363 (2015)
24. D. Moxey, D. Ekelschot, Ü. Keskin, S.J. Sherwin, J. Peiró, High-order curvilinear meshing using a thermo-elastic analogy. Comput. Aided Des. **72**, 130–139 (2016)
25. M. Fortunato, P.E. Persson, High-order unstructured curved mesh generation using the winslow equations. J. Comput. Phys. **307**, 1–14 (2016)
26. S. Sherwin, J. Peiró, Mesh generation in curvilinear domains using high-order elements. Int. J. Numer. Methods Eng. **53**(1), 207–223 (2002)
27. Z. Xie, R. Sevilla, O. Hassan, K. Morgan, The generation of arbitrary order curved meshes for 3D finite element analysis. Comput. Mech. **51**, 361–374 (2012)
28. R. Poya, R. Sevilla, A.J. Gil, A unified approach for a posteriori high-order curved mesh generation using solid mechanics. Comput. Mech. **58**(3), 457–490 (2016)
29. R. Sevilla, L. Rees, O. Hassan, The generation of triangular meshes for NURBS-enhanced FEM. Int. J. Numer. Methods Eng. **108**(8), 941–968 (2016)
30. T. Toulorge, C. Geuzaine, J.-F. Remacle, J. Lambrechts, Robust untangling of curvilinear meshes. J. Comput. Phys. **254**, 8–26 (2013)
31. S.L. Karman, J.T. Erwin, R.S. Glasby, D. Stefanski, High-order mesh curving using WCN mesh optimization, in *46th AIAA Fluid Dynamics Conference*, 2016, p. 3178
32. M. Stees, S.M. Shontz, A high-order log barrier-based mesh generation and warping method. Procedia Eng. **203**, 180–192 (2017)
33. L. Liu, Y. Zhang, T.J.R. Hughes, M.A. Scott, T.W. Sederberg, Volumetric t-spline construction using boolean operations. Eng. Comput. **30**(4), 425–439 (2013)
34. A. Gargallo-Peiró, X. Roca, J. Peraire, J. Sarrate, A distortion measure to validate and generate curved high-order meshes on cad surfaces with independence of parameterization. Int. J. Numer. Methods Eng. (2015)
35. T. Toulorge, J. Lambrechts, J.F. Remacle, Optimizing the geometrical accuracy of curvilinear meshes. J. Comput. Phys. (2016)
36. E. Ruiz-Gironés, X. Roca, J. Sarrate, High-order mesh curving by distortion minimization with boundary nodes free to slide on a 3D CAD representation. Comput. Aided Des. **72**, 52–64 (2016)
37. A. Kelly, L. Kaczmarczyk, C.J. Pearce, Mesh improvement methodology for 3D volumes with non-planar surfaces, in *Proceedings of the 21st International Meshing Roundtable*, 2011
38. E. Ruiz-Gironés, J. Sarrate, X. Roca, Generation of curved high-order meshes with optimal quality and geometric accuracy. Procedia Eng. **163**, 315–327 (2016)

39. E. Ruiz-Gironés, A. Gargallo-Peiró, J. Sarrate, X. Roca, An augmented lagrangian formulation to impose boundary conditions for distortion based mesh moving and curving. Procedia Eng. **203**, 362–374 (2017)
40. Pointwise Inc., Mesh Generation Software for CFD | Pointwise, Inc. http://www.pointwise.com, 2018
41. T.J. Tautges, CGM: a geometry interface for mesh generation, analysis and other applications. Eng. Comput. **17**(3), 299–314 (2001)
42. R. Haimes, M. Drela, On the construction of aircraft conceptual geometry for high-fidelity analysis and design, in *50th AIAA Aerospace sciences meeting including the new horizons forum and aerospace exposition*, 2012, p. 683
43. A. Gargallo-Peiró, X. Roca, J. Peraire, J. Sarrate, Distortion and quality measures for validating and generating high-order tetrahedral meshes. Eng. Comput. **31**(3), 423–437 (2015)
44. P.M. Knupp, Algebraic mesh quality metrics. SIAM J. Numer. Anal. **23**(1), 193–218 (2001)
45. J. Nocedal, S. Wright, *Numerical Optimization* (Springer, New York, 1999)
46. Python Software Foundation, Python. http://www.python.org, 2018
47. M.S. Alnæs, J. Blechta, J. Hake, A. Johansson, B. Kehlet, A. Logg, C. Richardson, J. Ring, M.E. Rognes, G.N. Wells, The FEniCS Project Version 1.5. Arch. Numer. Softw. **3**(100) (2015)
48. petsc4py, PETSc for Python. https://bitbucket.org/petsc/petsc4py/src/master, 2018
49. Geode, Project Geode: Geometry for Simulation. http://www.pointwise.com/geode/, 2018
50. Open CASCADE, Open CASCADE Technology, 3D modeling and numerical simulation. www.opencascade.org, 2012
51. swig, Simplified Wrapper and Interface Generator. http://www.swig.org/, 2018

Part V
Parallel and Fast Meshing Methods

Exact Fast Parallel Intersection of Large 3-D Triangular Meshes

Salles Viana Gomes de Magalhães, W. Randolph Franklin, and Marcus Vinícius Alvim Andrade

Abstract We present 3D-EPUG-OVERLAY, a fast, exact, parallel, memory-efficient, algorithm for computing the intersection between two large 3-D triangular meshes with geometric degeneracies. Applications include CAD/CAM, CFD, GIS, and additive manufacturing. 3D-EPUG-OVERLAY combines five separate techniques: multiple precision rational numbers to eliminate roundoff errors during the computations; Simulation of Simplicity to properly handle geometric degeneracies; simple data representations and only local topological information to simplify the correct processing of the data and make the algorithm more parallelizable; a uniform grid to efficiently index the data, and accelerate testing pairs of triangles for intersection or locating points in the mesh; and parallel programming to exploit current hardware. 3D-EPUG-OVERLAY is up to 101 times faster than LibiGL, and comparable to QuickCSG, a parallel inexact algorithm. 3D-EPUG-OVERLAY is also more memory efficient. In all test cases 3D-EPUG-OVERLAY's result matched the reference solution. It is freely available for nonprofit research and education at https://github.com/sallesviana/MeshIntersection.

1 Introduction

The classic problem of intersecting two 3-D meshes has been a foundational component of CAD systems for some decades. However, as data sizes grow, and parallel execution becomes desirable, the classic algorithms and implementions now exhibit some problems.

S. V. G. de Magalhães · M. V. A. Andrade
Universidade Federal de Viçosa (MG), Viçosa, Brasil

W. Randolph Franklin (✉)
ECSE Department, Rensselaer Polytechnic Institute, Troy, NY, USA
e-mail: mail@wrfranklin.org

X. Roca, A. Loseille (eds.), *27th International Meshing Roundtable*, Lecture Notes in Computational Science and Engineering 127,
https://doi.org/10.1007/978-3-030-13992-6_20

1. *Roundoff errors*. Floating point numbers violate most of the axioms of an algebraic field, e.g., $(a + b) + c \neq a + (b + c)$. These arithmetic errors cause topological errors, such as causing a point to be seen to fall on the wrong side of a line. Those inconsistencies propagate, causing, e.g., nonwatertight models. Heuristics exist to ameliorate the problem, and they work, but only up to a point. Larger datasets mean a larger probability of the heuristics failing.
2. *Special cases (geometric degeneracies)*. These include a vertex of one object incident on the face of another object. In principle, simple cases could be enumerated and handled. However, some widely available software fails. There are a few reasons.

    a. The number of special cases grows exponentially with the dimension. In 2-D, when intersecting a infinite line l with a polygon, (at least) the following cases occur with respect to the line's intersection with a finite edge e of the polygon: l crosses e's interior, l is coincident with e, and l is incident on a vertex v of e, and the other edge e' incident on v is either coincident with l, on the same side of l as e, or on the opposite side of l as e. In 3-D, the problem is much worse, so that a complete enumeration may be infeasible.
    b. One technique is to reduce the number of cases by combining them. E.g., when comparing point p against line l, the three cases of *above*, *on*, and *below* may be compressed into two: *above or on* and *below*. The problem is to do this in a way that results in higher level functions that call this as a component executing correctly. E.g., does intersecting two polylines where a vertex of one is coincident with a vertex of the other still work?

3. Another problem is that current data structures are too complex for easy parallelization. Efficient parallelization prefers simple regular data structures, such as structures of arrays of plain old datatypes. If the platform is an Nvidia GPU, then warps of 32 threads are required to execute the same instruction (or be idle). Ideally, the data used by adjacent threads is adjacent in memory. That disparages pointers, linked lists, and trees.

Some components of 3D-EPUG-OVERLAY have been presented earlier. PINMESH preprocesses a 3D mesh so that point locations can be performed quickly [31]. EPUG-OVERLAY overlays 2D meshes [30].

Background Kettner et al. [27] studied failures caused by roundoff errors in geometric problems. They also showed situations where epsilon-tweaking failed. (That uses an ϵ tolerance to consider two values x and y to be equal if $|x - y| \leq \epsilon$.) Snap rounding arbitrary precision segments into fixed-precision numbers, Hobby [24], can also generate inconsistencies and deform the original topology. Variations attempting to get around these issues include de Berg et al. [6], Hersberger [23], and Belussi et al. [2]. Controlled Perturbation (CP), Mehlhorn et al. [34], slightly perturbs the input to remove degeneracies such that the geometric predicates are correctly evaluated even using floating-point arithmetic. Adaptive Precision Floating-Point, Shewchuk [39], exactly evaluates predicates (e.g. orientation tests) using the minimum necessary precision.

Exact Geometric Computation (EGC), Li [29], represents mathematical objects using algebraic numbers to perform computations without errors. E.g., $\sqrt{2}$ can be represented exactly as the pair $(x^2 - 2, [1, 2])$, interpreted as the root of the polynomial $x^2 - 2$ that lies in the interval [1, 2]. However this is slow. Even determining the sign of an expression such as $\sqrt{\sqrt{\sqrt{5}+1}+\sqrt{\sqrt{5}-1}-\sqrt[4]{2\sqrt{5}+4}}$ is nontrivial. (Answer: 0.)

One technique to accelerate algorithms based on exact arithmetic is to employ arithmetic filters and interval arithmetic, Pion and Fabri [38], such as embodied in CGAL [4]. Arithmetic operations are applied to the intervals. If the sign of the exact result can be inferred based on the sign of the bounds of the interval, its value is returned. Otherwise, the predicate is re-evaluated using exact arithmetic.

Current Freely Available Implementations One technique for overlaying 3-D polyhedra is to convert the data to a volumetric representation (voxelization), perhaps stored as an octree, Meagher [33], and then perform the overlay using the converted data. This approach has some advantages: first, the volumetric model can be created using any precision, and so, if the application does not demand a high precision, this algorithm can be used to compute a fast approximation of the overlay. Furthermore, it is trivial to perform a robust overlay of volumetric representations. However, the volumetric representation is usually not exact, and the overlay results are usually approximate. Also, oblique surfaces cannot be represented exactly, which impacts fluid flow and visualization. Also, the data structure size tends to grow at least quadratically with the desired resolution. Pavić et al. [37] present an efficient algorithm for performing this kind of overlay.

For exactly computing overlays, a common strategy is to use indexing to accelerate operations such as computing the triangle-triangle intersection. For example, Franklin [12] uses a uniform grid to intersect two polyhedra, Feito et al. [11] and Mei and Tipper [35] use octrees, and Yongbin et al. [41] use Oriented Bounding Boxes trees (OBBs) to intersect triangulations. Although those algorithms do not use approximations, robustness cannot be guaranteed because of floating point errors. For example, Feito et al. [11] use a tolerance to process floating-point numbers, but this is error-prone.

Another algorithm that does not guarantee robustness is QuickCSG, Douze et al. [9], which is designed to be extremely efficient. QuickCSG employs parallel programming and a *k*-*d*-tree index to accelerate the computation. However, it does not handle special cases (it assumes vertices are in general position), and does not handle the numerical non-robustness from floating-point arithmetic, Zhou et al. [42]. To reduce errors caused by special cases, QuickCSG allows the user to apply random numerical perturbations to the input, but this has no guarantees.

Even if an algorithm using floating-point arithmetic can intersect two specific meshes consistently (i.e., without creating topological impossibilities or crashing), some output coordinates may not be exactly representable as floating-point numbers.

Although small errors may sometimes be acceptable, they accumulate if several inexact operations are performed in sequence. This gets even worse in CAD and

GIS where it is common to compose operations. For use when exactness is required, Hachenberger et al. [21] presented an algorithm for computing the exact intersection of Nef polyhedra. A Nef polyhedron is a finite sequence of complement and intersection operations on half-spaces. Although dating from the 1970s, only in the 2000s were concrete algorithms developed, and then embodied into CGAL [4]. One application is the SFCGAL [36] backend to the PostGIS DBMS. SFCGAL wraps the CGAL exact representation for 2-D and 3-D data, allowing PostGIS to perform exact geometric computations. Although these algorithms are exact, they are slow, Leconte et al. [28]. Also, in most cases, the data must be converted into the Nef format.

Bernstein et al. [3] presented an algorithm that tries to achieve robustness in mesh intersection by representing the polyhedra using binary space partitioning (*BSP*) trees with fixed-precision coordinates. It can intersect two such polyhedra by only evaluating fixed-precision predicates. However, in 3D, the BSP representation often has superlinear size, because the partitioning planes intersect so many objects. Also, converting BSPs back to more widely used representations (such as triangular meshes) is slow and inexact.

Recently, Zhou et al. [42] presented an exact and parallel algorithm for performing booleans on meshes. The key is to use the concept of winding numbers to disambiguate self-intersections on the mesh. Their algorithm first constructs an arrangement with the two (or more) input meshes, and then resolves the self-intersections in the combined mesh by retesselating the triangles such that intersections happen only on common vertices or edges. The self-intersection resolution eliminates not only the triangle-triangle intersections between triangles of the different input meshes, but also between triangles of the same mesh. As a result, their algorithm can also eliminate self-intersections in the input meshes, repairing them. Finally, a classification step is applied to compute the resulting boolean operations.

That algorithm is freely available and distributed in the LibiGL package, Jacobson et al. [25]. Its implementation employs CGAL's exact predicates. The triangle-triangle intersection computation is also accelerated using CGAL's bounding-box-based spatial index. LibiGL is not only exact, but also much faster than Nef Polyhedra. However, it is still slower than fast inexact algorithms such as QuickCSG.

2 Our Techniques

Our solution to the above problems combines the following five techniques.

Big Rational Numbers Representing a number as the quotient of two integers, each represented as an array of groups of digits, is a classic technique. The fundamental limitation is that the number of digits grows exponentially with the depth of the computation tree. Our relevant computation comprises comparing the intersection of two lines defined by their endpoints against a plane defined by three vertices. So, this growth in precision is quite tolerable.

The implementation challenges are harder. Many C++ implementations of new data structures automatically construct new objects on a global heap, and assume the construction cost to be negligible. That is false for parallel programs processing large datasets. Constructing and destroying heap objects has a superlinear cost in the number of objects on the heap. Parallel modifications to the heap must be serialized.

Therefore we carefully construct our code to minimize the number of times that a rational variable needs to be constructed or enlarged. This includes minimizing the number of temporary variables needed to evaluate an expression.

Furthermore, we use interval arithmetic as a filter to determine when evaluation with rationals is necessary.

Simulation of Simplicity Simulation of Simplicity (SoS), Edelsbrunner et al. [10], addresses the problem that, "sometimes, even careful attempts at capturing all degenerate cases leave hard-to-detect gaps", Yap [40]. Figure 1 is a challenging case. It consists of two pyramids with central vertices incident at a common vertex v. v is non-manifold and is on 8 faces, 4 from each pyramid. It is not easy to determine which of the 8 faces should intersect the ray that would be run up from v in order to locate v. In the subproblem of point location, RCT gets this point location case wrong; PINMESH is correct because of SoS, Magalhães et al. [31].

SoS symbolically perturbs coordinates by adding infinitesimals of different orders. The result is that there are no longer any coincidences, e.g., three points are never collinear. A positive infinitesimal, ϵ, is smaller than any positive real number (but greater than zero). That violates the Archimedean property for the real field, but we don't need this independent axiom. A second order infinitesimal, ϵ^2, is smaller than any first order infinitesimal. Etc. Linear combinations of reals and infinitesimals work. In 1-D, SoS can be realized by indexing all the input variables of both input objects, and then modifying them thus:

$$x_i \rightarrow x_i + \epsilon^{(2^i)} \tag{1}$$

A coordinate's perturbation depends on its index. The efficient implementation of SoS is to examine its effects on each predicate and then to recode the predicate

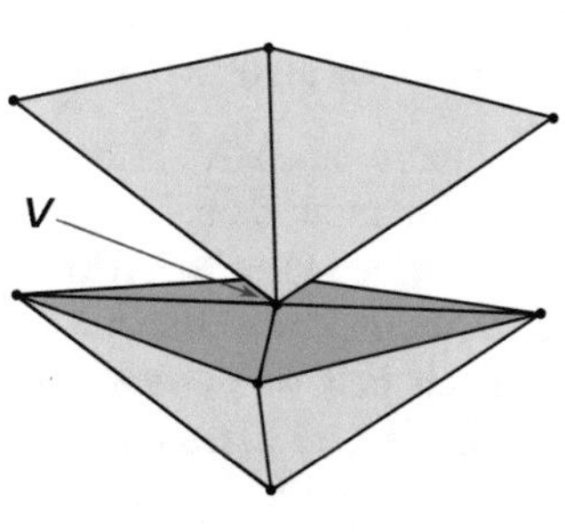

Fig. 1 Difficult test case for 3-D point location

to have the same effect, but without the need to use infinitesimals. E.g., $x_i \le x_j$ becomes $(x_i < x_j) \vee ((x_i = x_i) \wedge (i > j))$.

Because each type of predicate needs to be analyzed and recoded, we construct our algorithm to use only one type of predicate: the sign of a 3×3 determinant, or equivalently the order of four vertices in 3-D.

Minimal Topology The minimal explicit topology required for computing some property of an object depends on the desired property. E.g., testing for point location in a polygon requires only the set of unordered edges. That is still true for multiple and nested components. For computing the area and other mass properties, the set of ordered edges (where we know each edge's inside side) suffices. Alternatively, the set of vertices and their neighborhoods is sufficient, Kankanhalli and Franklin [26], Franklin [14]. That means for each vertex, knowing its location, the directions of the two adjacent edges, and which adjacent sector is inside. A sufficient representation of a 3-D mesh comprises the following:

1. the array of vertices, (v_i), where each $v_i = (x_i, y_i, z_i)$.
2. the array of tetrahedra or other polyhedra, t_i, used solely to store properties such as density, and
3. the array of augmented oriented triangular faces (f_i), where $f_i = (v_{i1}, v_{i2}, v_{i3}, t_{i1}, t_{i2})$. The tetrahedron or polyhedron t_{i1} is on the positive side of the face $f_i = (v_{i1}, v_{i2}, v_{i3})$; t_{i2} on the negative.

It is unnecessary to store any further relations, such as from face to adjacent face, from vertex to adjacent face, edge loops, or face shells.

Note that there are no pointers or lists; we need only several structures of arrays. If the tetrahedra have no properties, then the tetrahedron array does not need to exist, so long as the tetrahedra, which we are not storing explicitly, are consistently sequentially numbered. The point is to minimize what types of topology need to be stored.

Uniform Grid The uniform grid, Akman et al. [1], Franklin et al. [13, 16, 17] is used as an initial cull so that, when two objects are tested for possible intersection, then the probability of intersecting is bounded below by a positive number. Therefore, the number of pairs of objects tested for intersection that do not actually intersect is linear in the number that do intersect. Thus the expected execution time is linear in the output size. Frey and George [18] contains a comprehensive description of spatial data structures.

Our basic algorithm goes as follows.

1. Choose a positive integer g for the grid resolution as a function of the statistics of the input data. Typically, $10 \le g \le 1000$. The goal is for each grid cell, as described later, to have a constant number of intersections with input objects.
2. Superimpose a 3-D grid of $g \times g \times g$ cells on the input data.
3. Each cell will contain an abstract data structure of the set of input objects intersecting it. Call it the *cell intersection set.*

4. For each input object, determine which cells it intersects, and insert it into each of those cells' sets.

A careful concrete implementation of this abstraction is critical. We tested several choices; details are in Magalhães [7]. We also tested an octree, but our uniform grid implementation is much faster.

We also used a second level grid for some cells. This allowed us to use an approximation to determine which faces intersected each cell: enclosing oblique faces with a box and then marking all the cells intersecting that box, which is more cells than necessary.

OpenMP Because the data structures are simple and the algorithms are regular, they are easily parallelizable with OpenMP to run on a multicore Intel Xeon. This should also parallelize well on an NVIDIA GPU, as we have done for other algorithms, Hedin and Franklin [22] and Franklin and Magalhães [15].

3 3-D Mesh Intersection

3D-EPUG-OVERLAY exactly intersects 3-D meshes. Its input is two triangular meshes M_0 and M_1. Each mesh contains a set of 3-D triangles representing a set of polyhedra. The output is another mesh where each represented polyhedron is the intersection of a polyhedron from M_0 with another one from M_1. The key is the combination of five techniques described later. Extra details are in de Magalhães et al. [7, 8, 30–32].

Data Representation The input is a pair of triangular meshes in 3-D (E^3). Both meshes must be watertight and free from self-intersections. The polyhedra may have complex and nonmanifold topologies, with holes and disjoint components. The two meshes may be identical, which is an excellent stress test, because of all the degeneracies.

There are two types of output vertices: input vertices, and intersection vertices resulting from intersections between an edge of one mesh and a triangle of the other. Similarly, there are two types of output triangles: input triangles and triangles from retesselation. The first contains only input vertices while the second may contain vertices generated from intersections created during the retesselation of input triangles. An intersection vertex is represented by an edge and the intersecting triangle. For speed, its coordinates are cached when first computed.

Implementing Intersection with Simple Geometric Predicates To simplify the implementation of the symbolic perturbation, we developed two versions of each geometric function. The first one focused on efficiency, and was implemented based on efficient algorithms available in the literature. The second one focused on simplicity, and was implemented using as only a few orientation predicates.

The idea is that, during the computation, the first version of each function is called. If a special case is detected, then the second version is called. In order to make sure the special cases are properly handled we only need to implement the perturbation scheme on these predicates.

The Mesh Intersection Algorithm This computation uses only local information. The algorithm has three basic steps and a uniform grid is employed to accelerate the computation:

1. Intersections between triangles of one mesh and triangles of the other mesh are detected and the new edges generated by the intersection of each pair of triangles are computed.
2. A new mesh containing the triangles from the two original meshes is created and the original triangles are split (retesselated) at the intersection edges. I.e., if a pair of triangles in this resulting mesh intersect, then this intersection will happen necessarily on a common edge or vertex.
3. A classification is performed: triangles that shouldn't be in the output are removed and the adjacency information stored in each triangle is updated to ensure that the new mesh will consistently represent the intersection of the two original ones.

A two-level 3-D uniform was employed in 3D-EPUG-OVERLAY to accelerate two important steps of the algorithm: the detection of intersections between pairs of input triangles, and the point location algorithm employed in the triangle classification.

After computing the intersections between each pair of triangles, the next step is to split the triangles where they intersect, so that after this process all the intersections will happen only on common vertices or edges. After the intersections between the triangles are computed, the triangles from one mesh that intersect triangles from the other one are split into several triangles, creating meshes M_0' and M_1'.

Retesselation was implemented with orientation predicates, de Magalhães [7], which reduced to implementing 164 functions. A Wolfram Mathematica script was developed to create the code for all the predicates.

Experiments 3D-EPUG-OVERLAY was implemented in C++ and compiled using g++ 5.4.1. For better parallel scalability, the gperftools Tcmalloc memory allocator [20], was employed. Parallel programming was provided by OpenMP 4.0, multiple precision rational numbers were provided by GNU GMPXX and arithmetic filters were implemented using the Interval_nt number type provided by CGAL for interval arithmetic. The experiments were performed on a workstation with 128 GiB of RAM and dual Intel Xeon E5-2687 processors, each with 8 physical cores and 16 hyper-threads, running Ubuntu Linux 16.04.

We evaluated 3D-EPUG-OVERLAY, by comparing it against three state-of-the-art algorithms:

1. *LibiGL* [42], which is exact and parallel,
2. *Nef Polyhedra* [4], which is exact, and

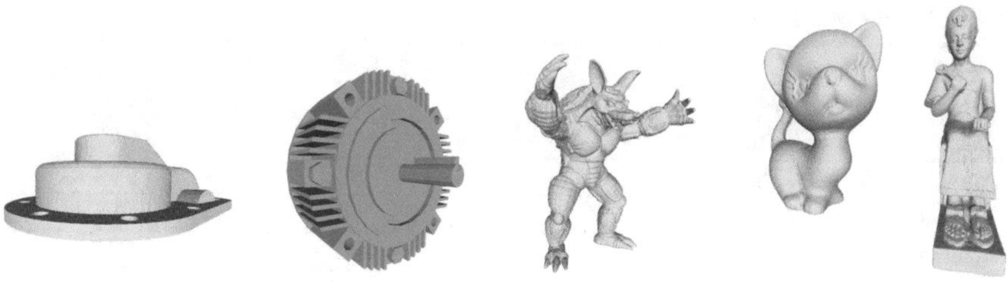

Fig. 2 Some test meshes

Table 1 Test datasets

Mesh	Verts ($\times 10^3$)	Tris ($\times 10^3$)	Polys ($\times 10^3$)	Mesh	Verts ($\times 10^3$)	Tris ($\times 10^3$)	Polys ($\times 10^3$)
Clutch2kf	1	2	–	Casting10kf	5	10	–
Horse40kf	20	40	–	Dinausor40kf	20	40	–
Armadillo52kf	26	52	–	Camel	35	69	–
Camel69kf	35	69	–	Cow76kf	38	76	–
Bimba	75	150	–	Kitten	137	274	–
Armadillo	173	346	–	461112	403	805	–
461115	411	822	–	RedCircBox[a]	701	1403	–
Ramesses	826	1653	–	Ramesses Rot.	826	1653	–
Ramesses Tran.	826	1653	–	Vase	896	1793	–
226633	1226	2452	–	Neptune	2004	4008	–
Neptune Tran.	2004	4008	–	914686Tetra	66	605	281
68380Tetra	107	1067	506	Armad.Tetra[b]	340	3377	1602
Arm.Tet.[b] Tran.	340	3377	1602	518092Tetra	603	5938	2814
461112Tetra	842	8495	4046				

Meshes with the suffix Tetra have been tetrahedralized
Rot. and Tran. mean, respectively, that the mesh has been rotated or translated
[a]Red circular box
[b]Tetrahedralized version of the Armadillo mesh

3. *QuickCSG* [9], which is fast and parallel, but not exact, and does not handle special cases.

Our experiments showed that 3D-EPUG-OVERLAY is fast, parallel, exact, economical of memory, and handles special cases.

Datasets Experiments were performed with a variety of non self-intersecting and watertight meshes; see Fig. 2 and Table 1. The ones with suffix *tetra* were tetrahedralized with GMSH [19]. The sources of the data are as follows: Barki (Clutch2kf, Casting10kf, Horse40kf, Dinausor40kf, Armadillo52kf, Camel69kf, Cow76kf), AIM@SHAPE (Camel, Bimba, Kitten, RedCircBox, Ramesses, Vase, Neptune), Stanford (Armadillo), Thingi10K (461112, 461115, 226633), Thingi10k+GMSH (914686Tetra, 68380Tetra, 518092Tetra, 461112Tetra), and Stanford+GMSH (Armad.Tetra).

Table 2 Pairs of meshes intersected

M_0	M_1	# triangles ($\times 10^3$)			Grid size
		M_0	M_1	Out	G_1 G_2^a
Casting10kf	Clutch2kf	10	2	6	64 2
Armadillo52kf	Dinausor40kf	52	40	25	64 4
Horse40kf	Cow76kf	40	76	24	64 4
Camel69kf	Armadillo52kf	69	52	16	64 4
Camel	Camel	69	69	81	64 4
Camel	Armadillo	69	331	43	64 4
Armadillo	Armadillo	331	331	441	64 8
461112	461115	805	822	808	64 8
Kitten	RedCircBox	274	1402	246	64 8
Bimba	Vase	150	1792	724	64 8
226633	461112	2452	805	1437	64 8
Ramesses	RamessesTrans	1653	1653	1571	64 16
Ramesses	RamessesRotated	1653	1653	1691	64 16
Neptune	Ramesses	4008	1653	1112	64 16
Neptune	NeptuneTrans	4008	4008	3303	64 16
68380Tetra	914686Tetra	1067	605	9393	64 2
ArmadilloTetra	ArmadilloTetraTran.	3377	3377	61325	64 4
518092Tetra	461112Tetra	5938	8495	23181	64 4

[a]Resolution of the first level grid, second level grid

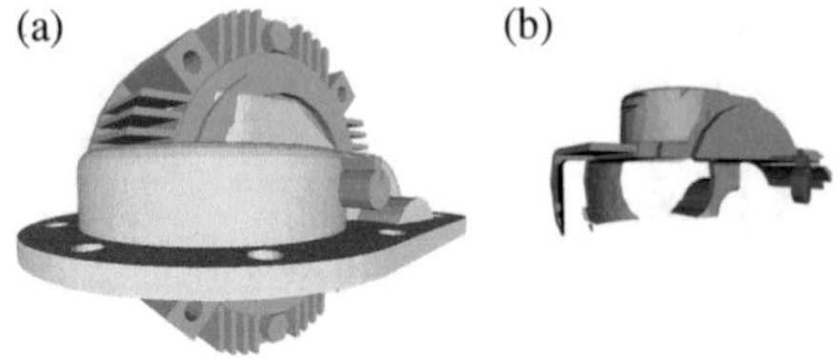

Fig. 3 Intersecting Casting10kf with Clutch2kf. (**a**) Visually overlaid. (**b**) Intersection

Table 2 presents the pairs of meshes used in the intersection experiments, the number of input triangles, the number of triangles in the resulting meshes and the uniform grid size.

Figure 3 shows one test, which took 0.2 s.

Arithmetic Filters and Other Optimizations To evaluate the effect of different optimizations, we profiled two key steps: the creation of the uniform grid and the detection of intersections between pairs of triangles. These experiments were performed with the Neptune and Neptune translated meshes using a uniform grid with first level resolution 64^3 and second level resolution 16^3. We evaluated various versions of the algorithm, and observed that the biggest impacts on speed were caused by using a good allocator, by using an interval arithmetic filter for the rational computations, and by coding the rational arithmetic expressions to minimize memory allocations.

The Importance of the Uniform Grid This accelerates the detection of pairs of triangles that intersect. To evaluate this idea, we compared it against an implementation using the CGAL method for intersecting dD Iso-oriented Boxes. Both algorithms are exact and employ arithmetic filters with interval arithmetic. Indeed, this CGAL algorithm is employed by LibiGL to accelerate the triangle-triangle intersection detection step of its mesh intersection method.

The CGAL method is sequential, and employs a hybrid approach composed of a sweep-line and a streaming algorithm to detect intersections between pairs of Axis Aligned Bounding Boxes. Thus, pairwise intersections of triangles can be detected by filtering the pairs of intersecting bounding-boxes, and then testing the triangles for intersection. Since the CGAL exact kernel was not thread-safe, even the triangle-triangle intersection tests were performed sequentially. Since our uniform grid was designed to be parallel, we evaluated it using 32 threads.

Table 3 presents these comparative experiments, performed on six pairs of meshes. The number of intersections detected is not necessarily the same in the

Table 3 Comparing the times (in seconds) for detecting pairwise intersections of triangles using CGAL (sequential) versus using a uniform grid (parallel)

CGAL							
		# faces ($\times 10^3$)		# int.[a]	Int.tests[b]	Time (s)	
M_0	M_1	M_0	M_1	($\times 10^3$)	($\times 10^3$)	Pre.proc.[c]	Inter.[d]
Camel	Armadillo	69	331	3	14	0.32	0.01
Armadillo	Armadillo	331	331	4611	5043	1.27	259.23
Kitten	RedC.Box[e]	274	1402	3	13	2.33	0.01
226633	461112	2452	805	23	128	7.18	0.08
Ramesses	Ram.Tran.[f]	1653	1653	36	237	12.38	0.17
Neptune	Nept.Tran.[g]	4008	4008	78	647	36.24	0.47
Uniform grid							
		# faces ($\times 10^3$)		# int.[a]	Int.tests[b]	Time (s)	
M_0	M_1	M_0	M_1	($\times 10^3$)	($\times 10^3$)	Pre.proc.[c]	Inter.[d]
Camel	Armadillo	69	331	3	33	0.06	0.02
Armadillo	Armadillo	331	331	50	5351	0.25	63.80
Kitten	RedC.Box[e]	274	1402	3	27	0.08	0.02
226633	461112	2452	805	23	307	0.16	0.05
Ramesses	Ram.Tran.[f]	1653	1653	36	866	0.16	0.10
Neptune	Nept.Tran.[g]	4008	4008	78	5087	0.27	0.35

[a]Number of intersections detected
[b]Number of intersection tests performed
[c]Pre-processing time
[d]Time spent testing pairs of triangles for intersection
[e]Red circular box
[f]Ramesses translated
[g]Neptune translated

two algorithms because our algorithm implements Simulation of Simplicity. E.g., co-planar triangles never intersect.

CGAL is better at culling pairs of non-intersecting bounding-boxes and so performs fewer intersection tests. However, since the uniform grid is lightweight and parallelizes well, its pre-processing time is much smaller (up to 134 times faster, which is much more than the degree of parallelism), and this difference is never recaptured. Indeed, except for the intersection of the Armadillo with itself, even if CGAL took 0 s to detect the intersections the total time spent by the uniform grid method would still be smaller.

The only situation where the intersection detection time was much larger than the pre-processing time was in the intersection of the Armadillo mesh with itself. In this situation the uniform grid was still faster than CGAL for two reasons: first, the number of intersection tests performed by the two methods was similar. Second, the intersection computation done by the uniform grid method is performed in parallel.

The worst performance for both methods happened during the intersection of the Armadillo mesh with itself. There are many coincidences (co-planar triangles being tested for intersection, triangles intersecting other triangles on the edges, etc). These coincidences lead to arithmetic filter failures (because the result of some of the orientation predicates is 0 and, thus, the intervals representing these results are likely to have different signs for their bounds), which lead to exact computations with rationals. Furthermore, coincidences lead to the use of SoS predicates (which we have not optimized yet) when using the uniform grid.

Comparing 3D-EPUG-OVERLAY *to Other Methods* We compared 3D-EPUG-OVERLAY against other three algorithms using the pairs of meshes presented in Table 2. The resulting running times (in seconds, excluding I/O) are presented in Table 4. Since the CGAL exact intersection algorithm deals with Nef Polyhedra, we also included the time it spent converting the triangulating meshes to this representation and to convert the result back to a triangular mesh (it often takes more time to convert the dataset than to compute the intersection). Both times are reported to let the user choose.

We can see that 3D-EPUG-OVERLAY was up to 101 times faster than LibiGL. The only test cases where the times spent by LibiGL were similar to the times spent by 3D-EPUG-OVERLAY were during the computation of the intersections of a mesh with itself (even in these test cases 3D-EPUG-OVERLAY was still faster than LibiGL). In this situation, the intersecting triangles from the two meshes are never in general position, and thus the computation has to frequently trigger the SoS version of the predicates, which we haven't not optimized yet. In the future, we intend to optimize this.

However, LibiGL also repairs meshes (by resolving self-intersections) during the intersection computation, which 3D-EPUG-OVERLAY does not attempt.

Because of the overhead of Nef Polyhedra and since it is a sequential algorithm, CGAL was always the slowest. When computing the intersections, 3D-EPUG-OVERLAY was up to 1284 times faster than CGAL. The difference is much higher if the time CGAL spends converting the triangular mesh to Nef Polyhedra is taken

Table 4 Times, in seconds, spent by different methods for intersecting pairs of meshes

M_0	M_1	Time (s)				
				CGAL		
		3D-Epug	LibiGL	Convert[a]	Intersect[b]	QuickCSG
Casting10kf	Clutch2kf	0.2	1.3	4.2	1.1	**0.1**∗
Armadillo52kf	Dinausor40kf	0.1	3.0	38.0	21.5	0.1
Horse40kf	Cow76kf	0.1	3.2	51.1	24.2	0.1
Camel69kf	Armadillo52kf	0.1	3.2	54.3	25.7	0.1
Camel	Camel	13.9	18.0	62.7	230.6	**0.9**∗
Camel	Armadillo	0.2	11.7	189.9	80.0	0.3
Armadillo	Armadillo	67.0	88.1	339.7	1198.2	**4.1**∗
461112	461115	0.8	58.9	753.2	473.2	1.1
Kitten	RedCircBox	0.3	28.6	819.8	329.6	1.1
Bimba	Vase	0.6	58.0	971.7	455.7	1.1
226633	461112	0.9	96.0	1723.7	905.5	**2.2**∗
Ramesses	Ram.Tran.[c]	1.3	93.0	1558.8	946.1	**2.4**∗
Ramesses	Ram.Rot.[d]	2.1	122.0	1577.3	989.8	2.4
Neptune	Ramesses	1.2	118.1	3535.5	1535.6	4.1
Neptune	Nept.Tran.[e]	2.7	220.2	5390.7	2726.2	6.1
68380Tet.[f]	914686Tet.[g]	51.3	–	–	–	–
Armad.Tet.[h]	Arm.Tet.Tran.[i]	263.3	–	–	–	–
518092Tetra	461112Tetra	136.6	–	–	–	–

QuickCSG reported errors during the intersections whose times are flagged with *. The tetrahedral mesh tests (last three rows) used only 3D-EPUG-OVERLAY

[a]Time converting the meshes to CGAL Nef Polyhedra
[b]Time intersecting the Nef Polyhedra
[c]Ramesses translated
[d]Ramesses rotated
[e]Neptune translated
[f]68380Tetra
[g]914686Tetra
[h]ArmadilloTetra
[i]ArmadilloTetra translated

into consideration: intersecting meshes with 3D-EPUG-OVERLAY was up to 4241 times faster than using CGAL to convert and intersect the meshes.

While 3D-EPUG-OVERLAY was faster than QuickCSG in most of the test cases (mainly the largest ones), in others QuickCSG was up to 20% faster than 3D-EPUG-OVERLAY. The relatively small performance difference between 3D-EPUG-OVERLAY and an inexact method (that was specifically designed to be very fast) indicates that 3D-EPUG-OVERLAY presents good performance allied with exact results. Besides reporting errors during the experiments detached with a * in Table 4, QuickCSG also failed in some situations where errors were not reported (this will be detailed later).

Finally, we also performed experiments with tetra-meshes. Each tetrahedron in these meshes is considered to be a different object and, thus, the output of 3D-EPUG-OVERLAY is a mesh where each object represents the intersection of two tetrahedra (from the two input meshes). These meshes are particularly hard to process because of their internal structure, which generates many triangle-triangle intersections. For example, during the intersection of the *Neptune* with the *Neptune translated* datasets (two meshes without internal structure), there are 78 thousand pairs of intersecting triangles and the resulting mesh contains 3 million triangles. On the other hand, in the intersection of *518092_tetra* (a mesh with 6 million triangles and 3 million tetrahedra) with *461112_tetra* (a mesh with 8 million triangles and 4 million tetrahedra) there are 5 million pairs of intersecting triangles and the output contains 23 million triangles.

To the best of our knowledge, LibiGL, CGAL and QuickCSG were not designed to handle meshes with multi-material and, thus, we couldn't compare the running time of 3D-EPUG-OVERLAY against them in these test cases.

We also evaluated the peak memory usage of each algorithm. 3D-EPUG-OVERLAY was: almost always smaller than LibiGL, with the difference increasing as the datasets became larger; smaller than QuickCSG in every case where QuickCSG returned the correct answer; and much smaller than CGAL. A typical result was the intersection of Neptune (4M triangles) with Ramesses (1.7M triangles): 3D-EPUG-OVERLAY used 2.6 GB, LibiGL used 6.7 GB, and CGAL 84 GB. The largest example that 3D-EPUG-OVERLAY processed, 518092Tetra (6M triangles) with 461112Tetra (8.5M triangles) used 43 GB. de Magalhães [7] contains detailed results.

Correctness Evaluation 3D-EPUG-OVERLAY was developed on a solid foundation (i.e., all computation is exact and special cases are properly handled using Simulation of Simplicity) in order to ensure correctness. However, perhaps its implementation has errors? Therefore, we performed extensive experiments comparing it against LibiGL (as a reference solution). We employed the Metro tool, Cignoni et al. [5], to compute the Hausdorff distances between the meshes being compared. Metro is widely employed, for example, to evaluate mesh simplification algorithms by comparing their results with the original meshes.

Let $e(p, S)$ be the minimum Euclidean distance between the point p and the surface S. Cignoni et al. [5] defines the one sided distance $E(S_1, S_2)$ between two surfaces S_1 and S_2 as: $E(S_1, S_2) = \max_{p \in S_1} e(p, S_2)$. The Hausdorff distance between two surfaces S_1 and S_2 is the maximum between $E(S_1, S_2)$ and $E(S_2, S_1)$. The Metro implementation employs an approximation strategy that samples points on the surface of the meshes in order to estimate the Hausdorff distance. In all experiments we employed the default parameters (where 10 points are sampled per face).

Since Metro is not exact (all the computation is performed using double variables), we use the distance between meshes only as evidence that our implementation is correct. In every test, the difference between 3D-EPUG-OVERLAY and LibiGL was reported as 0. In some situations the difference between LibiGL and

CGAL was a small number (maximum 0.0007% of the diagonal of the bounding-box). We guess this is because the exact results are stored using floating-point variables, and different strategies are used to round the vertices to floats and write them to the text file.

QuickCSG, on the other hand, generated errors much larger than CGAL: in the worst case, the difference between QuickCSG output and LibiGL was 0.13% of the diagonal of the bounding-box). de Magalhães [7] contains detailed results.

Visual Inspection We also visually inspected the results using MeshLab. Even though small changes in the coordinates of the vertices cannot be easily identified by visual inspection (and even the program employed for displaying the meshes may have roundoff errors), topological errors (such as triangles with reversed orientation, self-intersections, etc) often stand out.

Even when QuickCSG did not report a failure, results were frequently inconsistent, with open meshes, spurious triangles or inconsistent orientations.

Figure 4 shows the intersection of Bimba and the Vase. The first part is the complete overlay mesh, as computed by 3D-EPUG-OVERLAY. The second is a detail of an error-prone output region, computed correctly by 3D-EPUG-OVERLAY. The third part shows the same region computed by QuickCSG. Note the errors along the edges.

Figure 5a presents a zoom in the output of QuickCSG for the intersection of the Ramesses dataset with Ramesses Translated: some triangles are oriented incorrectly. These errors may be created either by floating-point errors or because QuickCSG doesn't handle the coincidences.

To mitigate this later problem, QuickCSG provides options where the user can apply a random perturbation in the input dataset. In contrast to the symbolic perturbations of Simulation of Simplicity, these numerical perturbations are not guaranteed to work, and the user has to choose a maximum range. A too-small

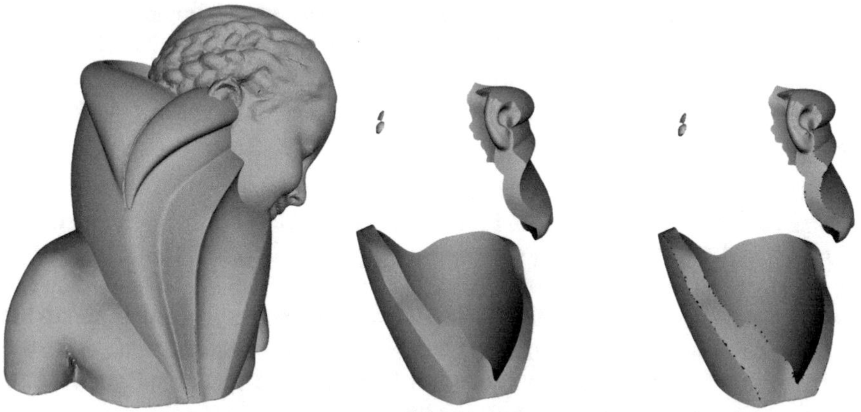

Fig. 4 Intersection of the Bimba and Vase meshes computed by 3D-EPUG-OVERLAY and QuickCSG, showing only 3D-EPUG-OVERLAY computing a region correctly

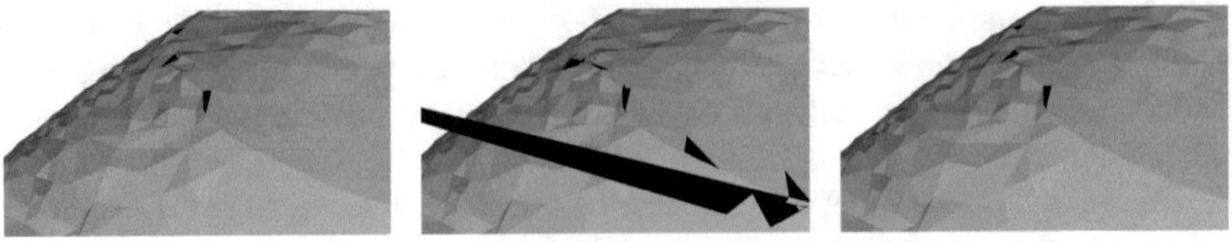

Fig. 5 Detail of the intersection of Ramesses with Ramesses translated generated by QuickCSG using different ranges for the numerical perturbation: (a) no perturbation, (b) 10^{-1}, and (c) 10^{-6}

range may not eliminate all errors while a too-big range may modify the mesh too much. Figure 5 displays the results from perturbations with maximum range 10^{-1}, and 10^{-6}. None of these perturbations removed all errors and the bigger perturbation (10^{-1}) even added undesirable artifacts to the output. Similar problems in QuickCSG have been reported by Zhou et al. [42].

Rotation Invariance We also validated 3D-EPUG-OVERLAY by verifying that its result does not change when the input meshes are rotated. I.e., a pair of meshes were rotated around the same point, intersected, and the resulting mesh was rotated back. To ensure exactness, we chose a rotation angle with rational sine and cosine. We evaluated all the pairs in Table 2. For each pair, we performed a rotation around the x axis and, then, a rotation around the y axis (the origin was defined as the center of the joint bounding-box of the two meshes). We chose rotation angle θ such that $\sin\theta = 400/10004$ and $\cos\theta = 9996/10004$. $\theta \approx 2.29$ degrees.

In all the experiments Metro reported that the resulting meshes were equal (i.e., the Hausdorff distance was 0.000000) to the corresponding ones obtained without rotation.

In addition, we intersected each mesh from Table 2 with a rotated version of itself. This is a notoriously difficult case for CAD systems because the large number of intersections and small triangles. Each mesh M was rotated around the center of its bounding-box using the above θ, and intersected with its original version, using both LibiGL and 3D-EPUG-OVERLAY. In every experiment the Hausdorff distance between the two outputs was 0.000000. That is, we can quickly process cases that can crash CAD systems.

Limitations Even though the computations are performed exactly, common file formats for 3D objects such as OFF represent data using floating-point numbers. Converting 3D-EPUG-OVERLAY's rational output into floats may introduce errors since most rationals cannot be represented exactly. Possible solutions include avoiding the conversion (i.e., always employing multiple-precision rationals in the representations), or using heuristics such [42] to try to choose floats for each coordinate so that the approximate output will not only be similar to the exact one, but also it will not present topological errors.

A limitation of symbolic perturbation is that the results are consistent considering the perturbed dataset, not necessarily considering the original one [10]. Thus, if the perturbation in the mesh resulting from the intersection is ignored, the unperturbed mesh may contain degeneracies such as triangles with area 0 or polyhedra with volume 0 (these polyhedra would have infinitesimal volume if the perturbation was not ignored). More details are in [7].

Summary 3D-EPUG-OVERLAY is an algorithm and implementation to intersect a pair of 3D triangular meshes. It is simultaneously the fastest, free from roundoff errors, handles geometric degeneracies, parallelizes well, and is economical of memory. The source code, albeit research quality, is freely available for nonprofit research and education at https://github.com/sallesviana/MeshIntersection. We have extensively tested it for errors; we encourage others to test it. It is a suitable subroutine for larger systems such as 3D GIS or CAD systems. Computing other kinds of overlays, such as union, difference, and exclusive-or, would require modifying only the classification step. We expect that 3D-EPUG-OVERLAY could easily process datasets that are orders of magnitude larger, with hundreds of millions of triangles. Finally, 3D-EPUG-OVERLAY has not nearly been fully optimized, and could be made much faster.

References

1. V. Akman, W.R. Franklin, M. Kankanhalli, C. Narayanaswami, Geometric computing and the uniform grid data technique. Comput. Aided Des. **21**(7), 410–420 (1989)
2. A. Belussi, S. Migliorini, M. Negri, G. Pelagatti, Snap rounding with restore: an algorithm for producing robust geometric datasets. ACM Trans. Spatial Algoritm. Syst. **2**(1), 1:1–1:36 (2016)
3. G. Bernstein, D. Fussell, Fast, exact, linear booleans. Eurographics Symp. Geom. Process. **28**(5), 1269–1278 (2009)
4. CGAL, Computational Geometry Algorithms Library. https://www.cgal.org. Retrieved Sept 2018
5. P. Cignoni, C. Rocchini, R. Scopigno, Metro: measuring error on simplified surfaces. Comput. Graphics Forum **17**(2), 167–174 (1998)
6. M. de Berg, D. Halperin, M. Overmars, An intersection-sensitive algorithm for snap rounding. Comput. Geom. **36**(3), 159–165 (2007)
7. S.V.G. de Magalhães, Exact and parallel intersection of 3D triangular meshes, Ph.D. thesis, Rensselaer Polytechnic Institute, 2017
8. S.V.G. de Magalhães, W.R. Franklin, M.V.A. Andrade, W. Li, An efficient algorithm for computing the exact overlay of triangulations, in *25th Fall Workshop on Computational Geometry*, U. Buffalo, New York, USA, 23–24 Oct 2015 (2015). Extended abstract
9. M. Douze, J.-S. Franco, B. Raffin, QuickCSG: arbitrary and faster boolean combinations of n solids, Ph.D. thesis, Inria-Research Centre, Grenoble–Rhône-Alpes, France, 2015
10. H. Edelsbrunner, E.P. Mücke, Simulation of simplicity: a technique to cope with degenerate cases in geometric algorithms. ACM TOG **9**(1), 66–104 (1990)
11. F. Feito, C. Ogayar, R. Segura, M. Rivero, Fast and accurate evaluation of regularized boolean operations on triangulated solids. Comput. Aided Des. **45**(3), 705–716 (2013)
12. W.R. Franklin, Efficient polyhedron intersection and union, in *Proceedings of Graphics Interface*, pp. 73–80, Toronto (1982)
13. W.R. Franklin, Adaptive grids for geometric operations. Cartographica **21**(2–3), 161–167, Summer – Autumn (1984). Monograph 32–33

14. W.R. Franklin, Polygon properties calculated from the vertex neighborhoods, in *Proceedings of 3rd Annual ACM Symposium on Computational Geometry*, pp. 110–118 (1987)
15. W.R. Franklin, S.V.G. Magalhães, Parallel intersection detection in massive sets of cubes, in *Proceedings of BigSpatial' 17: 6th ACM SIGSPATIAL Workshop on Analytics for Big Geospatial Data*, Los Angeles Area, CA, USA, 7–10 Nov 2017 (2017)
16. W.R. Franklin, N. Chandrasekhar, M. Kankanhalli, M. Seshan, V. Akman, Efficiency of uniform grids for intersection detection on serial and parallel machines, in *New Trends in Computer Graphics (Proc. Computer Graphics International'88)*, ed. by N. Magnenat-Thalmann, D. Thalmann, pp. 288–297 (Springer, Berlin, 1988)
17. W.R. Franklin, C. Narayanaswami, M. Kankanhalli, D. Sun, M.-C. Zhou, P. Y. Wu, Uniform grids: a technique for intersection detection on serial and parallel machines, in *Proceedings of Auto Carto 9: Ninth International Symposium on Computer-Assisted Cartography*, pp. 100–109, Baltimore, Maryland, 2–7 April 1989 (1989)
18. P.J. Frey, P. George, *Mesh Generation: Application to Finite Elements, Second Edition* (ISTE Ltd./Wiley, London/Hoboken, 2010)
19. C. Geuzaine, J.-F. Remacle, Gmsh: a 3-d finite element mesh generator with built-in pre-and post-processing facilities. Int. J. Numer. Methods Eng. **79**(11), 1309–1331 (2009)
20. S. Ghemawat, P. Menage, TCMalloc: thread-caching malloc (15 Nov 2015), http://goog-perftools.sourceforge.net/doc/tcmalloc.html. Retrieved on 13 Nov 2016
21. P. Hachenberger, L. Kettner, K. Mehlhorn, Boolean operations on 3d selective nef complexes: data structure, algorithms, optimized implementation and experiments. Comupt. Geom. **38**(1), 64–99 (2007)
22. D. Hedin, W.R. Franklin, NearptD: a parallel implementation of exact nearest neighbor search using a uniform grid, in *Canadian Conference on Computational Geometry*, Vancouver Canada (Aug. 2016)
23. J. Hershberger, Stable snap rounding. Comput. Geom. **46**(4), 403–416 (2013)
24. J.D. Hobby, Practical segment intersection with finite precision output. Comput. Geom. **13**(4), 199–214 (1999)
25. A. Jacobson, D. Panozzo, et al., *libigl: A Simple C++ Geometry Processing Library* (2016), http://libigl.github.io/libigl/. Retrieved on 18 Oct 2017
26. M. Kankanhalli, W.R. Franklin, Area and perimeter computation of the union of a set of iso-rectangles in parallel. J. Parallel Distrib. Comput. **27**(2), 107–117 (1995)
27. L. Kettner, K. Mehlhorn, S. Pion, S. Schirra, C. Yap, Classroom examples of robustness problems in geometric computations. Comput. Geom. Theory Appl. **40**(1), 61–78 (2008)
28. C. Leconte, H. Barki, F. Dupont, Exact and Efficient Booleans for Polyhedra. Technical Report RR-LIRIS-2010-018, LIRIS UMR 5205 CNRS/INSA de Lyon/Université Claude Bernard Lyon 1/Université Lumière Lyon 2/École Centrale de Lyon (Oct. 2010). Retrieved on 19 Oct 2017
29. C. Li, *Exact geometric computation: theory and applications*, Ph.D. thesis, Department of Computer Science, Courant Institute - New York University, January 2001
30. S.V.G. Magalhães, M.V.A. Andrade, W.R. Franklin, W. Li, Fast exact parallel map overlay using a two-level uniform grid, in *4th ACM SIGSPATIAL International Workshop on Analytics for Big Geospatial Data (BigSpatial)*, Bellevue WA USA, 3 Nov 2015
31. S.V.G. Magalhães, M.V.A. Andrade, W.R. Franklin, W. Li, PinMesh – fast and exact 3D point location queries using a uniform grid. Comput. Graph. J. **58**, 1–11 (2016). Special issue on Shape Modeling International 2016 (online 17 May). Awarded a reproducibility stamp, http://www.reproducibilitystamp.com/
32. S.V.G. Magalhães, M.V.A. Andrade, W.R. Franklin, W. Li, M.G. Gruppi, Exact intersection of 3D geometric models, in *Geoinfo 2016, XVII Brazilian Symposium on GeoInformatics*, Campos do Jordão, SP, Brazil (Nov. 2016)
33. D.J. Meagher, Geometric modelling using octree encoding. Comput. Graphics Image Process. **19**, 129–147 (1982)
34. K. Mehlhorn, R. Osbild, M. Sagraloff, Reliable and efficient computational geometry via controlled perturbation, in *ICALP (1)*, ed. by M. Bugliesi, B. Preneel, V. Sassone, I. Wegener. Lecture Notes in Computer Science, vol. 4051, pp. 299–310 (Springer, Berlin, 2006)

35. G. Mei, J.C. Tipper, Simple and robust boolean operations for triangulated surfaces. CoRR, abs/1308.4434 (2013)
36. Oslandia, IGN, *SFCGAL*, 2017, http://www.sfcgal.org/. Retrieved on 19 Oct 2017
37. D. Pavić, M. Campen, L. Kobbelt, Hybrid booleans. Comput. Graph. Forum **29**(1), 75–87 (2010)
38. S. Pion, A. Fabri, A generic lazy evaluation scheme for exact geometric computations. Sci. Comput. Program. **76**(4), 307–323 (2011)
39. J.R. Shewchuk, Adaptive precision floating-point arithmetic and fast robust geometric predicates. Discret. Comput. Geom. **18**(3), 305–363 (1997)
40. C.K. Yap, Symbolic treatment of geometric degeneracies, in *System Modelling and Optimization: Proceedings of 13th IFIP Conference*, ed. by M. Iri, K. Yajima, pp. 348–358 (Springer, Berlin, 1988)
41. J. Yongbin, W. Liguan, B. Lin, C. Jianhong, Boolean operations on polygonal meshes using obb trees, in *ESIAT 2009*, vol. 1, pp. 619–622 (IEEE, Piscataway, 2009)
42. Q. Zhou, E. Grinspun, D. Zorin, A. Jacobson, Mesh arrangements for solid geometry. ACM Trans. Graph. **35**(4), 39:1–39:15 (2016)

Performance Comparison and Workload Analysis of Mesh Untangling and Smoothing Algorithms

Domingo Benitez, J. M. Escobar, R. Montenegro, and E. Rodriguez

Abstract This paper compares methods for simultaneous mesh untangling and quality improvement that are based on repositioning the vertices. The execution times of these algorithms vary widely, usually with a trade-off between different parameters. Thus, computer performance and workloads are used to make comparisons. A range of algorithms in terms of quality metric, approach and formulation of the objective function, and optimization solver are considered. Among them, two new objective function formulations are proposed. Triangle and tetrahedral meshes and three processors architectures are also used in this study. We found that the execution time of vertex repositioning algorithms is more directly proportional to a new workload measure called *mesh element evaluations* than other workload measures such as mesh size or objective function evaluations. The comparisons are employed to propose a performance model for sequential algorithms. Using this model, the workload required by each mesh vertex is studied. Finally, the effects of processor architecture on performance are also analyzed.

1 Introduction

Mesh optimizing techniques reduce the total time to solution and improve the accuracy of results of PDE solvers. Processing a mesh can spend up to 25% of the overall running time of a PDE-based application [5]. When a mesh element is inverted, standard finite element simulation algorithms cannot numerically solve the PDE, although some methods are being investigated to solve a PDE on a tangled mesh [20]. Thus, researchers and practitioners recommend untangling the mesh prior to analysis using commercial packages.

D. Benitez (✉) · J. M. Escobar · R. Montenegro · E. Rodriguez
SIANI Institute & DIS Department, University of Las Palmas de Gran Canaria, Las Palmas de Gran Canaria, Spain
e-mail: domingo.benitez@ulpgc.es

X. Roca, A. Loseille (eds.), *27th International Meshing Roundtable*, Lecture Notes in Computational Science and Engineering 127,
https://doi.org/10.1007/978-3-030-13992-6_21

Vertex repositioning algorithms (VrPA) have been adopted by a vast majority of mesh optimization applications [6, 7, 17], including hex meshes [11, 14]. VrPA algorithms improve the quality of a mesh by moving its free vertices. They can be posed as numerical techniques in which the following parameters are considered [6]: objective function approach (A) and formulation (f), element quality metric (q), minimization method (NM) and convergence or termination criteria (TC). The execution times of these numerical algorithms vary widely, usually with a trade-off between parameters.

There are few studies that compare the performance of mesh optimization algorithms. Diachin et al. compared the performance of two VrPA algorithms [6]. One of them employed an inexact Newton method and the all-vertex approach to numerically optimize the global objective function. The other algorithm used a coordinate descent method and the single-vertex approach.

Sastry and Shontz compared the performance of several optimization methods for mesh quality improvement [15]. They considered all-vertex and single-vertex approaches in combination with gradient and Hessian-based optimization solvers. In their paper, it is showed that performance can vary significantly, depending on the choice of mesh quality metric.

These previous studies considered valid input meshes, smoothing algorithms, one objective function formulation, and one processor architecture. In our work, up to 34 algorithms that simultaneously untangle and smooth meshes composed of two different element topologies and three processor architectures are considered. Additionally, three mesh quality metrics and five objective function formulations, two of them new, are investigated (Sect. 2). A single generalized version of the sequential VrPA algorithm is used (Sect. 3), and its implementation details are explained in Sect. 4.

One of our goals is to determine when one of these methods for mesh untangling and smoothing and why one processor is preferable to the others (Sect. 5). Preference includes the execution time, success in untangling meshes and the quality of the optimum mesh produced. To gain insights into the causes of performance variability, the workload of VrPA algorithms is analyzed (Sects. 6 and 7). Another goal is to model the performance of sequential VrPA algorithms. Thus, we propose a performance model for mesh optimization and its accuracy is studied in Sect. 8. We have also investigated the causes of the variability in time of VrPA methods (Sect. 9).

2 Free Parameters of the Study

The vertex repositioning problem has been formulated by other authors [6]. There are many choices for each free parameter one could make in a study of VrPA algorithms. In this paper, we limited the options to those shown in Table 1. Each combination of choices will be called *VrPA configuration* and denoted: $\langle A\rangle$-$\langle f\rangle$-$\langle q\rangle$-$\langle NM\rangle$-$\langle TC\rangle$, for instance, "Lo-D1-hS-SD-TC2".

Table 1 Free VrPA parameters and their choices that are considered in this paper

Parameter	Options		RW
Objective function approach (A):	**Gl**: All-vertex (K: Global function)	$n = N_M$: free elements[a] of mesh	[6]
$K = \sum_{i=1}^{n} f(q_i)$ n: total free elements[a]	**Lo**: Single-vertex (K: Local function)	$n = N_v$: free elements[a] of local patch	[6]
Objective function formulation: $f(q_i)$	**D1**: Distortion 1	$f(q_i) = q_i^{-1}$	[3]
	D2: Distortion 2	$f(q_i) = q_i^{-2}$	[3]
q_i: quality of ith element $q_{min} = min(q_i)_{i \in \{1...n\}}$	**log1**: Logarithmic barrier 1	$f(q_i) = n^{-1} q_{min}^{-1} -$ $\mu\, log(\tau\, q_{min}^{-1} - q_i^{-1})$	New, see Sect. 2.1
$h(z) = \frac{1}{2}\left(z + \sqrt{z^2 + 4\delta^2}\right)$	**log2**: Logarithmic barrier 2	$f(q_i) = n^{-1} q_{min} +$ $\mu\, log(q_i - q_{min})$	[16]
δ, μ, τ = constants	**inv**: Regularized barrier	$f(q_i) = q_i^{-1} + \frac{1}{h\left(q_{min}^{-1} - q_i^{-1}\right)}$	New, see Sect. 2.1
Element quality metric: q_i S_i: Jacobian matrix	**hS**: Regularized mean-ratio	$q_i = \frac{d\, [h(\sigma_i)]^{2/d}}{\lVert S_i \rVert_F^2}$	[7]
$\lVert\ \rVert_F$: Frobenius norm $h(z) = \frac{1}{2}\left(z + \sqrt{z^2 + 4\delta^2}\right)$	**MQ**: Hybrid quality metric	$q_i = \frac{\lambda\ vol}{1+e^{a\ \lambda\ vol}} + \frac{A}{1+e^{-b\ A}}$	[16]
$\sigma_i = determinant(S_i)$ $triangle: d = 2, s = 3$	vol = element volume l_j= element edge lengths	$A = \frac{vol}{L^{d/2}} \quad L = \sum_{j=1}^{s} l_j^2$	
$tetrahedron{:}d = 3, s = 6$ $a, b, \delta, \lambda = constants$	**TU**: Untangle quality metric	$q_i = 2\left(-\sigma_i + \sqrt{\sigma_i^2 + \delta^2}\right)^{-1}$	[3]
Numerical minimization method (NM)	**CG**: Conjugate Gradient	Polack-Ribiere, analytical derivatives	[3]
	SD: Steepest Descent	Analytical derivatives	[3]
Termination criteria (TC)	**TC1** (untangled mesh)	$true = (Q_{min} > 0)$	[3]
Q_i: mean-ratio quality value of the ith element $Q_{min} = min(Q_i)_{i \in \{1...N_M\}}$	**TC2** (optimum mesh) $\overline{Q}$: average mean-ratio value of mesh Δ: maximum variation between outer iterations	$true = (Q_{min} > 0\ and$ $\Delta\overline{Q} < 10^{-3}\ and$ $\Delta Q_{min} < 10^{-3})$	

RW denotes related work
[a] *Free element*: mesh element with at least one free vertex

2.1 *Novel Objective Function Formulations*

Table 1 includes two new objective functions for mesh untangling and quality improvement called *Logarithmic distortion-based barrier* (Eq. (1), **log1**) and *Reg-*

ularized distortion-based barrier (Eq. (2), **inv**), where q_i is the quality metric function, n is the number of free elements involved in forming the objective function and μ, δ and τ are constants.

$$K = \frac{1}{q_{min}} - \mu \sum_{i=1}^{n} log\left(\frac{\tau}{q_{min}} - \frac{1}{q_i}\right) \qquad q_{min} = min(q_i)_{i\in\{1...n\}} \tag{1}$$

$$K = \sum_{i=1}^{n} \frac{1}{q_i} + \frac{1}{h\left(\frac{1}{q_{min}} - \frac{1}{q_i}\right)} \qquad h(z) = \frac{1}{2}\left(z + \sqrt{z^2 + 4\delta^2}\right) \tag{2}$$

These barrier objective functions are defined for differentiable quality metrics whose maximum and minimum values are the qualities of ideal and degenerate elements, respectively. This formulation assumes that the quality of an element should be maximized in order to obtain the ideal element. Thus, our new barrier functions are used in vertex repositioning methods that minimize the objective function.

2.2 Calculation of the Constant Value for δ

Our experience has shown that the constant values involved in the calculation of quality metrics can determine whether a mesh untangling algorithm is successful in producing meshes without inverted elements. δ in Eq. (2) and Table 1 represents a constant value that depends on the mesh in all-vertex approach ($n = N_M$) or submesh in single-vertex approach ($n = N_v$):

$$\delta = \max\{\ 10^{-3}\,\bar{\sigma},\ \mathrm{Re}(\ 10\,\sqrt{\epsilon(\epsilon - \sigma_{min})}\ )\ \} \tag{3}$$

$$\bar{\sigma} = \frac{1}{n}\sum_{i=1}^{n} |\sigma_i| \qquad \sigma_{min} = \min\{\sigma_i\}_{i\in\{1,\ldots,n\}} \qquad \epsilon = 10^{11} DBL\_EPS$$

where σ_i is defined in Table 1 and *DBL_EPS* is upper bound on the relative error due to rounding in floating-point arithmetic. δ is always recalculated before a different vertex is optimized by a single-vertex method. For all-vertex methods, δ is recalculated before a different mesh iteration begins.

This calculation of δ was applied to the known metric called *untangle quality metric* and denoted **TU** [3] (see Table 1). For the **TU** metric and the regularized mean-ratio quality metric (**hS**), this strategy is key to successfully producing a mesh without inverted elements when the input mesh is tangled.

3 Sequential VrPA Algorithm

Many mesh optimization applications employ a VrPA that is similar to Algorithm 1 [6, 7, 17]. It consists of a variable number of mesh sweeps. In each of them, every vertex is processed and can be repositioned by the numerical solver. The vertices that lie on the mesh surface are treated as fixed and are not updated.

The most time-consuming operation called *VertexRepositioning* moves free vertices (V) of an input mesh (M). It iterates an *inner loop* (lines 9–24) while an extreme of the objective function (K) is being reached by a numerical method (*NM*). K is constructed after determining the above mentioned VrPA parameters: A, f and q (lines 17–20). *LogicFunction* uses a termination criterion (TC) to stop the algorithm (line 30). The *outer loop* (lines 30–37) is iterated in the *Main* procedure while *LogicFunction* is not true. In each outer iteration, the spatial coordinates of all free vertices (X_V) are updated, and so a mesh sweep is implemented. *GlobalMeasures* provides the average and worst mean-ratio quality metric of the mesh [6]. At the end of the algorithm, an optimized mesh is obtained.

4 Experimental Setup

Algorithm 1 was used to compare the above mentioned VrPA configurations. The following paragraphs describe details of the implementation.

Software Framework We developed complete programs that include double-precision floating-point data structures and *PAPI* functions for hardware performance monitoring [4]. The source code includes some C++ classes and methods from the *Mesquite* framework [3]; all of them were modified to evaluate the computer performance and workload of vertex repositioning algorithms. The *Mesquite* method **TU** was also modified as previously explained in Sect. 2.2. We created new C++ classes and methods to support **hS** and **MQ** quality metrics, **log1**, **log2** and **inv** objective function formulations and **TC2** termination criteria (see Table 1). We used *gcc* 4.8.4 with `-O2` flag on *Linux* systems. For each VrPA configuration, we repeated the execution of the programs several times, such that the 95% confidence interval was lower than 1%.

Benchmark Meshes Algorithm 1 was applied on the unstructured, fixed-sized meshes shown in Fig. 1 whose characteristics are in Table 2. All the mesh sizes were always fixed. The 2D mesh was obtained by using *Gmsh* tool [8], taking a square, meshing with triangles and displacing selected nodes of the boundary. This type of tangled mesh can be found in some problems with evolving domains [9]. All 3D meshes were obtained from a tool, called *The Meccano Method*, for adaptive tetrahedral mesh generation that tangles the mesh in one of its intermediate stages [12].

Figure 2 shows some convergence plots that represent the worst mean-ratio quality metric and the number of inverted elements versus the number of mesh iterations. They were obtained when the Lo-D1-hS-SD-TC2 configuration was used to optimize two of the meshes. The rest of "...-TC2" configurations and meshes

Algorithm 1 Sequential mesh vertex repositioning algorithm

```
▷ Input: file with information of M mesh
#define: approach (A), formulation (f), quality metric (q), numerical optimization method (NM)
#define termination criteria: TC = LogicFunction(Q_min, ΔQ̄, ΔQ_min)
#define constants: τ = 10^-6 (maximum or minimum increase of the objective function), N_mII = 150 (maximum number of inner iterations), N_mOI = 100 (maximum number of outer iterations)
N_e ← 0, N_f ← 0          ▷ Global variables: element evaluations (N_e), objective function evaluations (N_f)
procedure VERTEXREPOSITIONING(W, X, n)
▷Inputs : W(free vertices),X(their coordinates),n(number of elements)
    ▷ Initiation : K = 0, ΔK = 0, m = 0 (inner loop index)
    while (ΔK ≤ τ(minimizing) or ΔK ≥ τ(maximizing)) and m ≤ N_mII do          ▷ Inner loop
        X̂ ← X          ▷ Returned spatial coordinates (X̂) of vertices (W)
        ▷ Initiation : P ← 0          ▷ Moving directions: P = {p_v}, v ∈ W
        for i = 1, ..., n do          ▷ n: number of free elements
            for each free vertex v of i^th free element do
                p_v += NM(f'(q_i),v)          ▷ f': derivatives used in NM
                N_e += 1          ▷ Number of mesh element evaluations
            end for
        end for
        X ← X̂ + P          ▷ Tentative positions of free vertices
        K_t ← 0          ▷ Initial value of objective function
        for i = 1, ..., n do
            K_t += f(q_i)          ▷ MESH ELEMENT EVALUATION
            N_e += 1          ▷ Number of mesh element evaluations
        end for
        ΔK ← K_t − K
        K ← K_t          ▷ Final value of objective function
        N_f += 1          ▷ Number of evaluations of the objective function and its derivative
        m += 1          ▷ Number of inner iterations
    end while
    return X̂          ▷ Output: updated coordinates of free vertices
end procedure
procedure MAIN( )
    ▷ Read the vertex and element information of M mesh
    Q_min ← GLOBALMEASURES(M)          ▷ Minimum quality of input mesh
    ▷ Initiation : ΔQ̄ = 10^6, ΔQ_min = 10^6, k = 0 (loop index)
    while TC ≠ true and k ≤ N_mOI do          ▷ Mesh/Outer loop
        if A = Gl then          ▷ Gl : all-vertex approach
            X_V ← VERTEXREPOSITIONING(V, X_V, N_M)
        else          ▷ Lo : single-vertex approach
            for each free vertex v ∈ M do
                x_v ← VERTEXREPOSITIONING(v, x_v, N_v)
            end for
        end if
        (Q_min, ΔQ̄, ΔQ_min) ← GLOBALMEASURES(M)
        k += 1          ▷ Number of mesh/outer iterations
    end while
    ▷ Output: file with information of optimized M mesh
end procedure
```

reported in this paper exhibit similar convergence behaviors. Two consecutive stages can be identified in each optimization problem. Firstly, an untangling stage in which the inverted elements decrease over time to zero. Secondly, a smoothing stage where the worst mean-ratio quality metric increases from zero to a stable value, when TC2 is met. The convergence behaviors of "…-TC1" configurations exhibit only the untangling stage.

Processor Cores Numerical experiments were conducted on three computers with different processor models. One of them is Intel Xeon E5645 that integrates 6 Westmere-EP cores whose clock speed is 2.4 GHz and are connected to 48 GB of DDR3/1600 MHz. Another processor is Intel Xeon E5-2670 that integrates 8

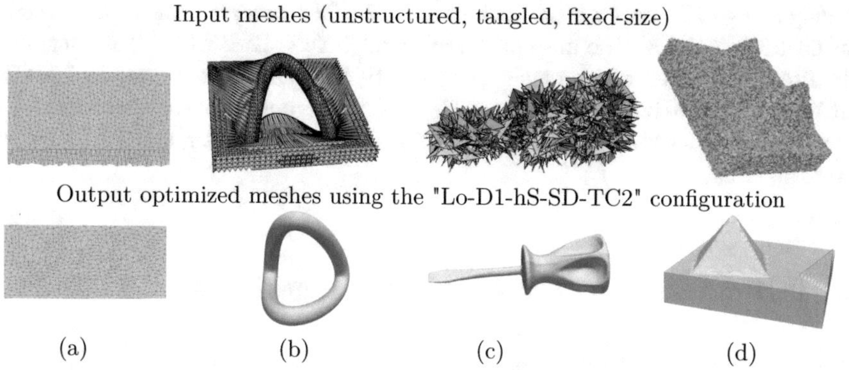

Fig. 1 Input and output meshes for four optimization problems solved with the same VrPA algorithm. (**a**) Square (2D). (**b**) Toroid (3D). (**c**) Screwdriver (3D). (**d**) Egypt (3D)

Table 2 Characteristics of input meshes, which have inverted elements: $Q_{min} = 0$

Mesh characteristic	Square	Toroid	Screwdriver	Egypt
Total vertices	3,314,499	9176	39,617	1,724,456
Free vertices (they can be moved)	3,309,498	3992	21,131	1,616,442
Fixed vertices (they are not moved)	5001	5184	18,486	108,014
Element type: triangle (2D), tetrahedron (3D)	2D	3D	3D	3D
Total free elements (N_M)	6,620,936	35,920	168,834	10,013,858
Inverted/tangled elements (%)	0.1%	38.2%	49.4%	46.2%
Average number of elements of local patch (N_v)	6.00	21.27	21.51	23.96
Average edge length	0.02	1.49	8.96	14.30
Standard deviation of the edge length	0.02	1.86	6.22	5.79
Average mean-ratio quality metric ($\overline{Q}$)	0.95	0.17	0.13	0.23
Standard deviation of the mean-ratio metric	0.05	0.31	0.21	0.27

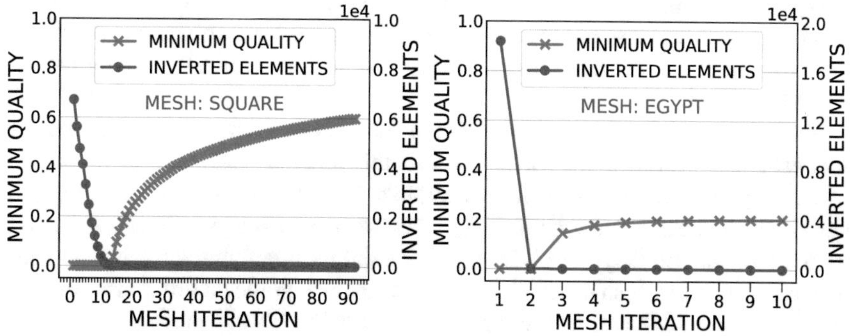

Fig. 2 Convergence plots obtained when the Lo-D1-hS-SD-TC2 configuration was employed to optimize the Square (left) and Egypt (right) meshes

Sandy Bridge-EP cores whose clock speed is 2.6 GHz and are connected to 32 GB of DDR3/1333 MHz. The third processor is Intel Xeon E5-2690V4 that integrates 14 Broadwell cores whose clock speed is 2.6 GHz and are connected to 256 GB of DDR4/2400 MHz. During the experiments, the compute nodes were not shared among other user-level workloads. Additionally, multithreading and Turbo Boost were disabled.

5 Performance Comparison of VrPA Algorithms

The execution time of Algorithm 1 varies widely, usually with a trade-off between different VrPA parameters. This can be seen in Figs. 3 and 4 that show the results of a performance evaluation using 136 configurations and two processors (E5645, E5-2670). Those configurations that not appear in these figures produce tangled meshes.

Results are grouped by benchmark mesh. Each graph distinguishes the convergence criterion established and the processor used. The goal of the TC1 criterion is to produce a mesh with no inverted elements. The other criterion, TC2, is met when the mesh is optimum, i.e., the increase in both the worst and average mean-ratio quality metrics after two successive mesh iterations are below a certain threshold. These graphs include the average and minimum mean-ratio quality metrics of output meshes. Note that there is more variability in the execution time than in the quality metrics.

Table 3 shows the fastest VrPA configuration for each mesh when the TC1 convergence criterion was established. The main results of this partial analysis are as follows: (1) the minimum quality metrics (Q_{min}) of output meshes vary significantly; (2) global approaches (Gl$-\ldots$) frequently achieve larger Q_{min} than local approaches (Lo$-\ldots$); (3) the average quality metrics ($\overline{Q}$) of output meshes exhibit much less variability than the minimum values; (4) the fastest algorithm always uses Gl approach, hS metric and SD solver; (5) the fastest configurations on all processors are the same; (6) the solver with superior performance is not always the same; however, the performance behavior of the SD solver is frequently superior to the CG solver.

Table 4 shows the fastest configuration for each mesh when the TC2 criterion was established. The main conclusions of this another analysis are: (1)Q_{min} is much larger and exhibits less variability than for mesh untangling (TC1); (2) the algorithms that achieve the largest Q_{min} always use local approach (Lo$-\ldots$) and barrier formulation (log1, log2, inv); (3) $\overline{Q}$ exhibits slightly larger values and less variability than for mesh untangling; (4) there are no significant differences in Q_{min} and $\overline{Q}$ between SD and CG solvers, except for the Toroid mesh whose results obtained with CG solver are superior; (5) the fastest algorithm always uses the SD solver and D1 formulation; additionally, the hS quality metric takes more times

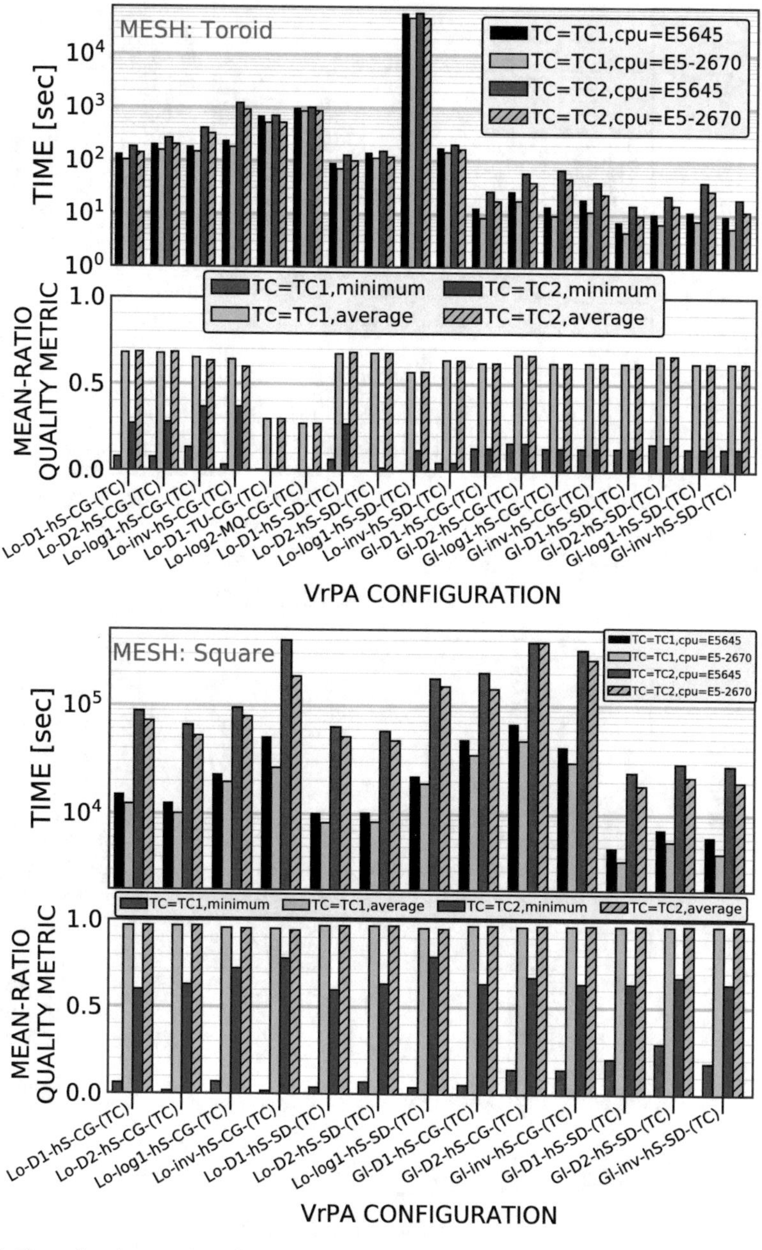

Fig. 3 Execution times and quality metrics for the Square and Toroid meshes

the first place in the rank ordering of performance than the other formulations and quality metrics; (6) the fastest configurations on both processors are again coincident; (7) the ratio of the time required by the configuration with the largest

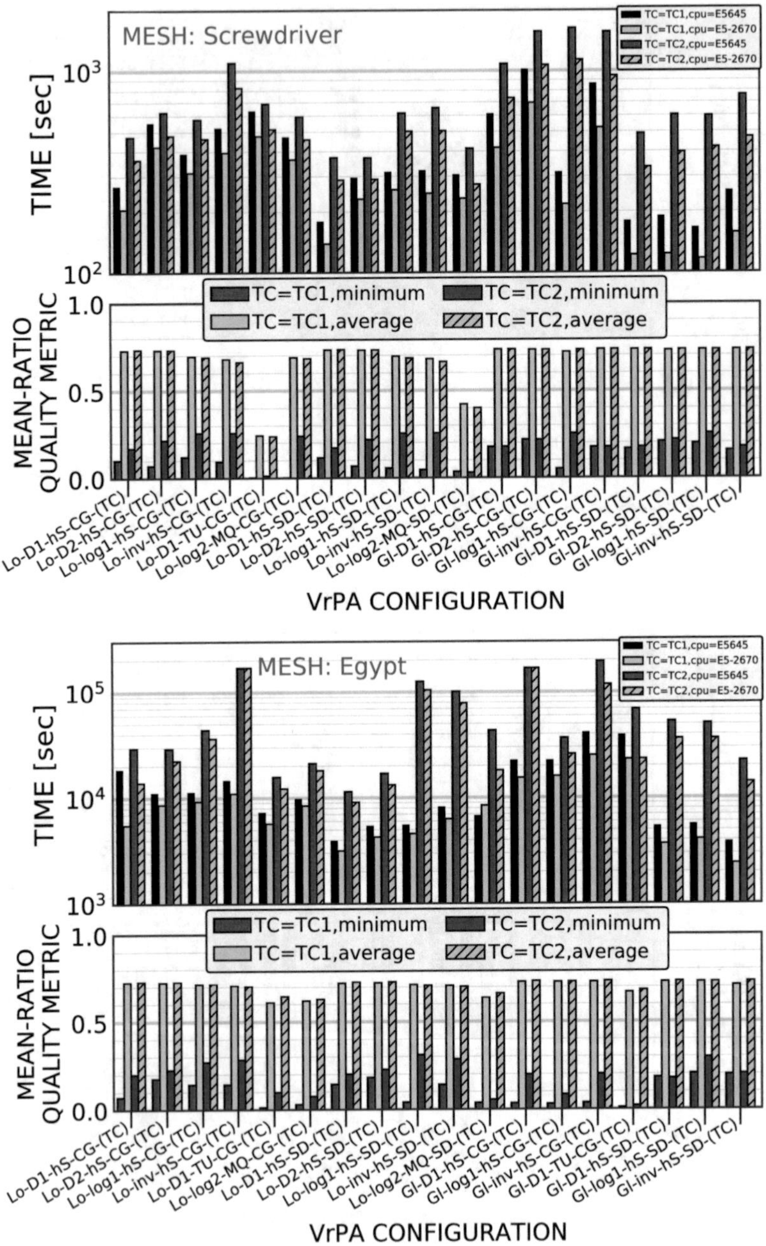

Fig. 4 Execution times and quality metrics for the Screwdriver and Egypt meshes

Q_{min} to the lowest time is larger than the ratio of respective Q_{min} except for the Screwdriver mesh; this means that the highest performance configurations may be the best choices for obtaining approximate solutions in the smallest amount of time.

Table 3 VrPA configurations with lowest runtimes for the TC1 convergence criterion

Benchmark mesh	Fastest VrPA configuration					
	Configuration	CPU time [s] E5645	CPU time [s] E5-2670	N_e	$\frac{largest\ CPU\ time}{CPU\ time}$ E5645	$\frac{largest\ CPU\ time}{CPU\ time}$ E5-2670
Square	Gl-D1-hS-SD-TC1	4.96×10^3	3.87×10^3	4.42×10^9	13.8	12.8
Toroid	Gl-D1-hS-SD-TC1	7.4	4.61	8.73×10^6	8320.9	10,691.2
Screwdriver	Gl-log1-hS-SD-TC1	1.66×10^2	1.17×10^2	1.84×10^8	6.1	6.0
Egypt	Gl-inv-hS-SD-TC1	3.86×10^3	2.41×10^3	3.38×10^9	10.8	10.4

N_e is the number of mesh element evaluations that were required by each configuration

Table 4 VrPA configurations that meet the TC2 convergence criterion and achieve the lowest CPU time or the largest Q_{min} (worst mean-ratio quality metric)

Mesh	Comparison criterion	Best configuration	CPU time (sec) E5645	CPU time (sec) E5-2670	Q_{min}	$\frac{time\ largest\ Q_{min}}{lowest\ time}$ E5645	$\frac{time\ largest\ Q_{min}}{lowest\ time}$ E5-2670
Square	Lowest t_{CPU}	Gl-D1-hS-SD-TC2	2.44×10^4	1.84×10^4	0.633	7.3	8.2
	Largest Q_{min}	Lo-log1-hS-SD-TC2	1.78×10^5	1.51×10^5	0.791		
Toroid	Lowest t_{CPU}	Gl-D1-hS-SD-TC2	1.47×10^1	9.56	0.13	82	95
	Largest Q_{min}	Lo-inv-hS-CG-TC2	1.21×10^3	9.12×10^2	0.369		
Screwdriver	Lowest t_{CPU}	Lo-log2-MQ-SD-TC2	3.04×10^2	2.73×10^2	0.03	3.6	3.0
	Largest Q_{min}	Lo-inv-hS-CG-TC2	1.1×10^3	8.27×10^2	0.257		
Egypt	Lowest t_{CPU}	Lo-D1-hS-SD-TC2	1.15×10^4	9.11×10^3	0.201	11	12
	Largest Q_{min}	Lo-log1-hS-SD-TC2	1.26×10^5	1.05×10^5	0.311		

6 Global Workload Analysis of VrPA Algorithms

Some authors use the number of elements (mesh size) as a workload measure to estimate the execution time of VrPA algorithms [15, 17]. However, this can be accurately done only when the configuration is fixed and the total numbers of inner and outer iterations of the algorithm are both fixed. In this case, the execution time is directly proportional to the size of the mesh.

For the algorithms where the VrPA parameters are free and the number of iterations is variable and based on convergence criteria, this proportionality is not evident as can be seen in Fig. 5a. This figure shows a graph of mesh size versus execution time that was derived from the results of the above-described experiments for mesh untangling. Each point represents one of 68 VrPA configurations that successfully untangle a fixed-size test mesh (see the "...-TC1" configurations in Figs. 3 and 4). The correlation coefficient between time and mesh size taken in the linear scale is $r = 0.44$.

The number of evaluations of the objective function and its derivative (N_f) has also been used as a workload measure in numerical algorithms [18]. Line 23 in Algorithm 1 was employed to count the number of evaluations of the objective function and its derivative. Figure 5b shows a graph of N_f versus time for the same

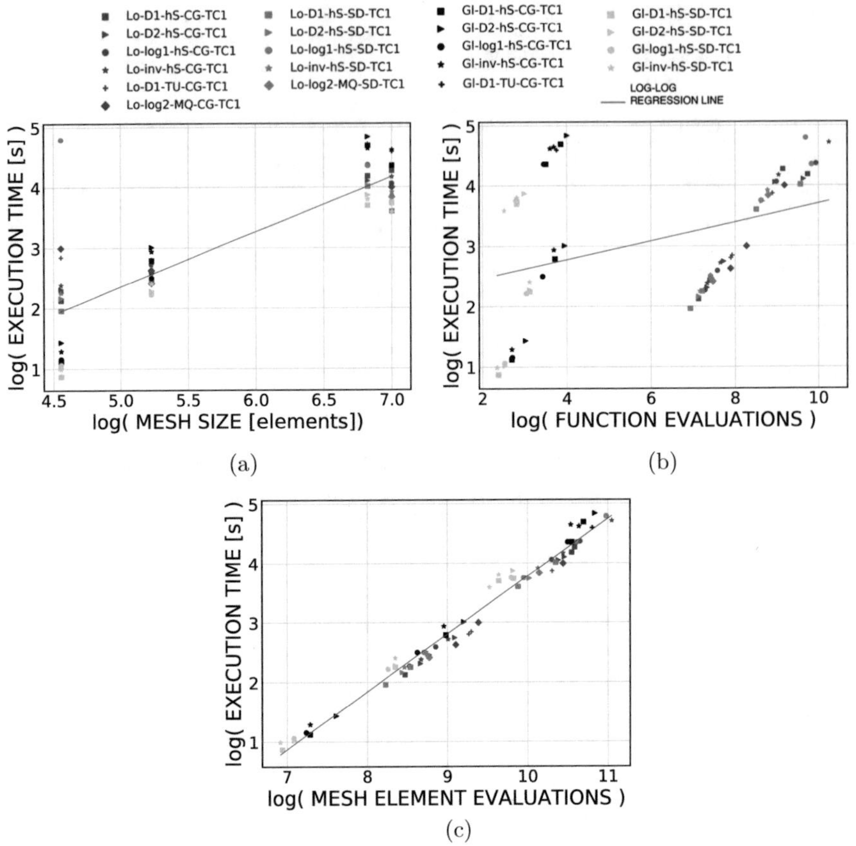

Fig. 5 Scalability of the algorithms for mesh untangling (convergence criterion: TC1) using different workload measures, the E5645 processor and the four benchmark meshes. r is the correlation coefficient taken in the linear scale. (**a**) Mesh size vs. time: $r = 0.44$. (**b**) Function evaluations vs. time: $r = 0.45$. (**c**) Element evaluations vs. time: $r = 0.93$

"...-TC1" configurations. In this case, the correlation coefficient between time and N_f taken in the linear scale is $r = 0.45$.

N_e in Algorithm 1 is called *number of mesh element evaluations* and measures the number of evaluations of the element quality metric and its derivative. This measure involves computing the separable but not independent parts of objective function evaluations. Although not exactly the same definition, the mesh element evaluation is slightly similar to the *concurrent function evaluation step* defined in [18] for identifying parallelism opportunities in finite difference gradients.

Figure 5c shows a graph of N_e versus execution time for the "...-TC1" configurations. In this case, the correlation coefficient between time and N_e taken in the linear scale is $r = 0.93$. Therefore, execution time is more directly proportional to the number of mesh element evaluations than the mesh size or the number of objective function evaluations. It is important to note that N_e is not very intrusive and depends not only on the problem size but also on the numbers of inner and outer iterations required to meet the convergence criteria. Thus, we will use N_e as a workload measure for VrPA algorithms in a new performance model that is proposed below.

Some of the fastest configurations that were evaluated in Sect. 5 achieve the highest performance because they need fewer element evaluations than the others (see N_e in Table 3). This was the case for all of our test meshes except Toroid when the TC1 criterion was established. To untangle the Toroid mesh, Gl-inv-hS-SD-TC1 configuration needs less N_e (8.23×10^6) than the fastest configuration (8.73×10^6). However, its time per element evaluation is sufficiently larger than the fastest configuration such that the time to convergence is not the lowest.

In summary, the performance of VrPA algorithms depends on the balance between two factors: the global workload measured in number of element evaluations and the time per element evaluation. The first factor is independent of the computer hardware; it depends on the algorithm and its implementation, the selected numerical accuracy of data structures, and the method chosen by the compiler to implement arithmetic operations. Moreover, the number of evaluations also depends on the characteristics of the input mesh such as the quality of elements or the number both of free vertices and of elements of each patch. The second factor is affected also by all these algorithmic, software and mesh aspects in addition to the computer hardware.

7 Analysis of the Workload Required by a Free Vertex

For a deeper understanding of VrPA algorithms, this section analyzes the workload required by vertices. Figure 6 shows the average and standard deviation of the number of element evaluations (N_e) that were needed by each free vertex in every outer iteration for a selection of configurations. Note that vertices are sorted by number of element evaluations from largest to smallest.

As it can be seen, the average workload per vertex of algorithms that use local approach (Lo$-\ldots$) is unbalanced, i.e., there is a large range of workloads (see the black lines in Figs. 6a,b and d). The maximum of the range can be one order of magnitude larger than the minimum. The variance of the number of element evaluations per vertex over different mesh iterations is also variable. It can be as large as the mean value (see Fig. 6b), or close to zero (see Figs. 6a,d).

The algorithms that use a global approach (Gl$-\ldots$) exhibit very different workload behaviors. The average and variance of the number of element evaluations per free vertex and mesh iteration are both constants (see Fig. 6c). This is due to that each free vertex is repositioned after the displacement directions of all vertices have been obtained. A common way to do this is to evaluate all elements in each evaluation of the single objective function and its derivative (see lines 12–15,18–20 in Algorithm 1).

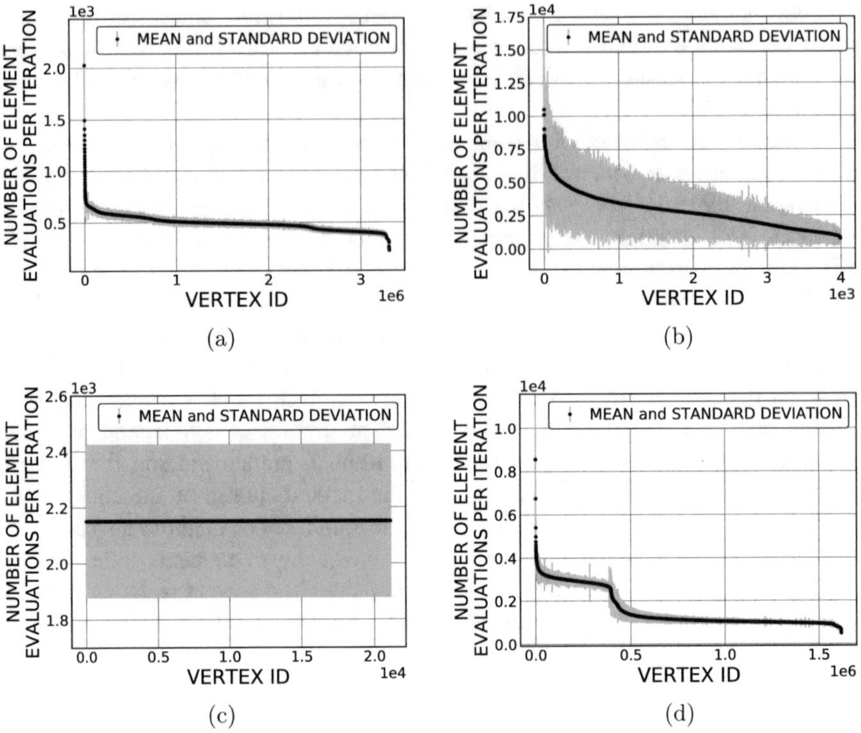

Fig. 6 Mean and standard deviation of the number of element evaluations per free vertex and mesh iteration. These results were obtained using the TC1 termination criterion for mesh untangling, the E5645 processor, a selected group of VrPA configurations and all the benchmark meshes. (**a**) Mesh: Square, VrPA: Lo-D2-hS-SD-TC1, 14 mesh iterations, 5.1×10^{-7} s/evaluation. (**b**) Mesh: Toroid, VrPA: Lo-D1-hS-CG-TC1, 14 mesh iterations, 4.2×10^{-7} s/evaluation. (**c**) Mesh: Screwdriver, Gl-log1-hS-SD-TC1, 4 mesh iterations, 9.5×10^{-7} s/evaluation. (**d**) Mesh: Egypt, VrPA: Lo-D1-hS-SD-TC1, 3 mesh iterations, 5.2×10^{-7} s/evaluation

In contrast, a local algorithm uses many independent objective functions, each of them determines the displacement direction of a single vertex. The number of element evaluations required for every objective function is variable. Thus, each free vertex causes a different computer workload. As the vertices are repositioned in series and the execution time was demonstrated in Sect. 6 that is highly correlated with the number of element evaluations, each vertex contributes to the total execution time differently.

8 Sequential Performance Model

Performance models combine methods that provide expectations on performance and instrumentation tools that find parameters with empirical measurements [1]. Taking the findings of Sect. 6, we use a simple one-parameter model to understand the performance of sequential VrPA algorithms,

$$t_{CPU} = \alpha \, N_e \tag{4}$$

where t_{CPU} denotes the execution time, N_e denotes the number of mesh element evaluations and α denotes the model parameter that represents the time per element evaluation. Equation (4) assumes that the computation time is much larger than the total input/output time. In this way, the time to optimize a mesh is directly proportional to the number of element evaluations.

This model may justify previous experimental observations where more element evaluations cause usually larger runtimes. However, there are VrPA configurations with fewer element evaluations than others that require more runtime (see Fig. 5c). This effect can be justified by our model.

Equation (4) calculates the time by multiplying two factors. A VrPA configuration can need larger number of element evaluations than other: $N_{e,1} > N_{e,2}$. However, the second algorithm can evaluate each element in more time than the first algorithm: $\alpha_1 < \alpha_2$. This is what happens to some configurations.

Taking the execution times and the respective numbers of element evaluations for the TC2 convergence criterion, we calculated the time per element evaluation for each configuration using Eq. (4). These results for the Screwdriver mesh are shown in Fig. 7a. We found similar results when the other benchmark meshes were optimized. As it can be seen, each configuration causes a different time per element evaluation. Combining the results shown in Figs. 5c and 7a, it can be demonstrated that the time per element evaluation is not correlated with the number of element evaluations (E5645 processor: $r = -0.04$). Therefore, the smallest product of the two factors of Eq. (4) determines what algorithm achieves the best performance.

8.1 Model Application and Accuracy

To check the accuracy of our one-parameter model, we used the 68 configurations of Sect. 5 that meet the TC1 convergence criterion. Note that we let all the VrPA parameters and input meshes vary freely except the termination criterion. Randomly chosen, half of the VrPA configurations were used to derive α. We calculated for each configuration the ratio of execution time to the number of element evaluations: $\alpha = t_{CPU}/N_e$. After averaging the results, $\overline{\alpha}$ was 6.7×10^{-7} and 4.9×10^{-7} [s/element] when E5645 and E5-2670 processor cores were used, respectively. The relative errors in estimating the execution times were obtained with the other half of configurations, $\overline{\alpha}$ and Eq. (4). The average relative errors were 0.27 and 0.34 for the E5645 and E5-2670 processor cores, respectively.

We extended this accuracy analysis by setting free only one of the five VrPA parameters (see Table 1). The input mesh was considered as a sixth parameter. For this study, we analyzed a total of 136 configurations. From these configurations, we took groups that have one free and five fixed parameters. Due to the possible

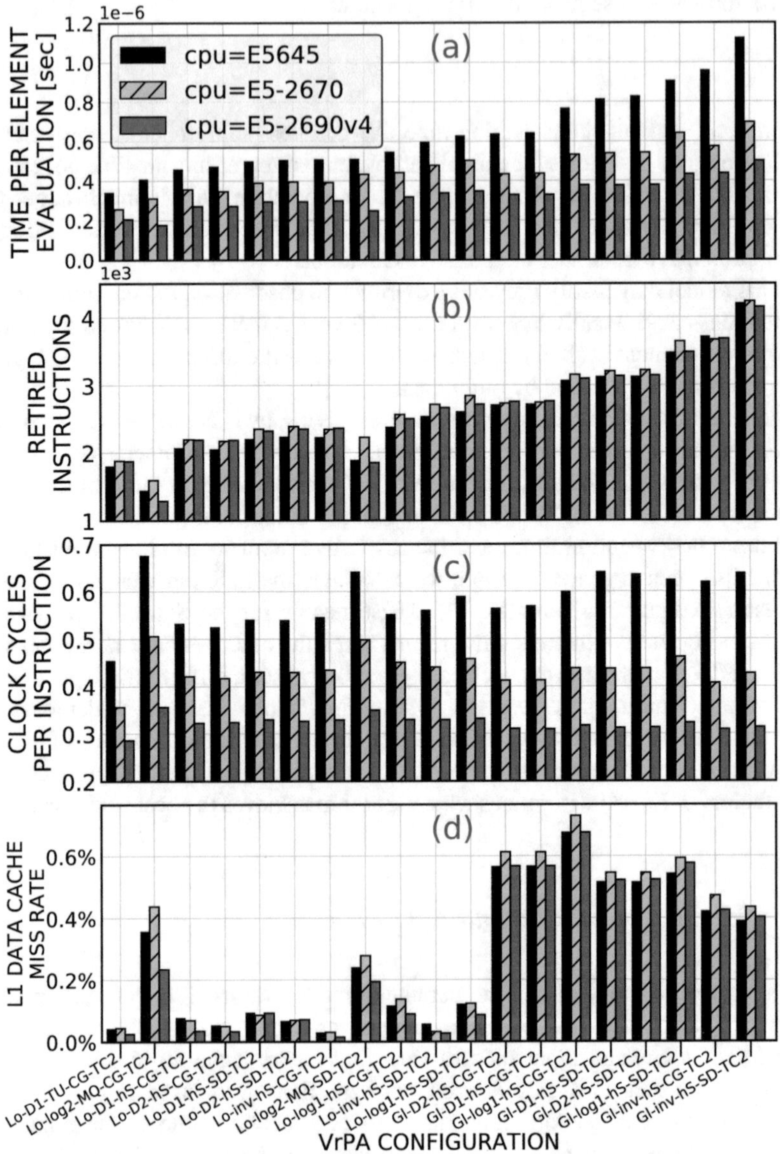

Fig. 7 Performance metrics per mesh element evaluation: (**a**) execution time. (**b**) Total retired instructions. (**c**) clock cycles per instruction (CPI) (**d**) miss rate of the Level-1 data cache Mesh: Screwdriver

choices of each parameter, the number of configurations included in every group was between two and five. For each group, we obtained the mean time per element evaluation ($\overline{\alpha}$). Then, we compared the time provided by our model (Eq. (4)) with the real

Table 5 Mean and maximum relative errors in estimating the execution times of VrPA algorithms on E5645 and E5-2670 processors

	CPU : E5645		CPU : E5-2670	
Free VrPA parameter	Mean	Max.	Mean	Max.
Approach (*A*)	0.28	0.61	0.19	0.61
Formulation (f)	0.07	0.12	0.06	0.17
Quality metric (q)	0.05	0.10	0.05	0.11
Numerical method (*NM*)	0.08	0.17	0.08	0.16
Convergence criteria (*TC*)	0.01	0.02	0.01	0.04
Mesh	0.07	0.20	0.09	0.41

execution time for each configuration in every group. The mean and maximum relative errors of our model using the above-mentioned processors are shown in Table 5.

The errors are caused by the variability in the values of α that are obtained for the configurations that constitute each group. The largest variability about the mean occurs in the groups that combine all-vertex and single-vertex algorithms where the rest of VrPA parameters are the same. Our model fits best when it is applied to a single VrPA algorithm and mesh. In this case, the same α can be used to justify accurately the execution time for different convergence criteria. Moreover, our results indicate that the model parameter (α) depends on the processor architecture. Thus, α must be recalculated if the processor changes.

9 Architectural Analysis

As it can be seen in Figs. 3 and 4, the E5-2670 processor core outperforms the E5645 core. Our performance model (Eq. (4)) indicates that it is due to smaller time per element evaluation (α) as the number of element evaluations (N_e) of a VrPA algorithm is the same on both processor cores.

Figure 7a shows the effects of three processor microarchitectures on the time per element evaluation. The main conclusions drawn from the results presented in this figure are the following: (1) the E5645 core spends on average 1.4 and 2.1 times more execution time per element evaluation than the E5-2670 and E5-2690V4 cores, respectively; (2) global methods (Gl$-\ldots$) always require more execution time per element evaluation than local methods (Lo$-\ldots$); (3) the configurations that employ the SD solver also spend more time in each element evaluation than the configurations that use CG solver if the rest of VrPA parameters are the same.

We have also investigated the causes of this variability in execution time per element evaluation at the hardware level. Our programs were instrumented using the PAPI programming interface [19] and informative performance counters of Intel cores [10]. In this investigation, we also used the classic CPU performance equation that relates the execution time to instruction count, clock cycles-per-instruction (CPI) and clock rate [13],

$$t_{CPU} = \frac{InstructionCount\ CPI}{ClockRate} \tag{5}$$

CPI is a performance metric that mainly depends on the hardware microarchitecture, instruction-set architecture, compiler and programming language [13]. In our experiments, instruction-set architecture, compiler and programming language remained the same.

Figure 7b–d show measurements of three performance metrics per element evaluation for the Screwdriver mesh and VrPA configurations that meet the TC2 convergence criterion. These metrics are: instruction count (retired instructions), CPI and miss rate of the L1 data cache. The configurations have been sorted by time per element evaluation from the smallest to the largest numerical value. So, the correlation between time per element evaluation and the measurements provided by the hardware counters can be perceived.

The main conclusions of the analysis of these performance metrics per element evaluation are: (1) longer time per element evaluation is frequently caused by a larger number of retired instructions per evaluation when the same microarchitecture is analyzed (see Fig. 7b); note that global methods (Gl$-\ldots$) execute more instructions than local methods (Lo$-\ldots$); (2) when the same VrPA configuration is evaluated on the three processors, the number of retired instructions per evaluation is approximately the same, which is to be expected because we are using the same program, compiler and instruction-set; (3) for each configuration, the CPI of the E5-2690V4 core is always lower than the CPI of the E5-2670 core whose CPI is lower than the CPI of the E5645 core (see Fig. 7c); this is to be expected from newer generations of microarchitectures given that fewer number of stall cycles per instruction is frequently a hardware design goal for commodity processors; (4) because of the ratio of clock rates of these three processors is lower than 1.08 and the instruction counts are similar, most of the performance advantage comes from a much lower CPI for the newer processors; on average, the ratio between the CPIs of E5-2670 and E5645 is 1.22, and the ratio between the CPIs of E5-2690V4 and E5645 is 2.3; (5) global configurations always cause larger miss rates of the L1 data cache than local configurations (see Fig. 7d), although the impact on the CPI is not very significant.

10 Conclusions

We have studied the performance and workload of vertex repositioning algorithms (VrPA) for simultaneous mesh untangling and smoothing. The influence on performance and workload of a large number of VrPA based on five parameters using triangle and tetrahedral meshes has been investigated. Among the choices for possible VrPA parameters, two new objective function formulations have been proposed. The largest worst mean-ratio quality metric for each benchmark mesh was obtained using one of these formulations. Additionally, a performance model for sequential VrPA algorithms has been proposed. This model involves a new workload measure called *number of mesh element evaluations* that is independent of the processor if numerical precision, compiler, and program are the same. We have shown that the execution times of VrPA methods are more proportional to the number of element evaluations than mesh size or the number of objective function evaluations. The other factor of the model, the time per element evaluation, is more affected by the processor and objective function approach than the objective function formulation, the quality metric, numerical solver, convergence criteria or mesh. The performance of VrPA methods has been compared using three processor cores with different microarchitectures. Most of the advantage of newer processors come from smaller values of the CPI performance metric.

This paper has also shown that the workload per vertex of a local optimization algorithm is very unbalanced. Using this finding, we have devised a new mesh partitioning approach [2]. Our study methodology may be applied to meshes with elements that are different from triangles and tetrahedra. For instance, some vertex repositioning methods improve the quality of hex-meshes after solving an optimization problem [11, 14]. They employ iterative solvers to minimize different objective functions composed of terms that measure the element qualities. However, the validity of our findings across these other methods has to be examined.

Acknowledgements This work has been supported by Spanish Government, "Secretaría de Estado de Universidades e Investigación", "Ministerio de Economía y Competitividad" and FEDER, grant contract: CTM2014-55014-C3-1-R. One of the computers used in this work was provided by the "Instituto Tecnológico y de Energías Renovables, S.A.". We thank to anonymous reviewers for their valuable comments and suggestions on this manuscript.

References

1. K. Barker, N. Chrisochoides, Practical performance model for optimizing dynamic load balancing of adaptive applications, in *Proceedings of* 19th *IPDPS*, pp. 28.a–28.b (2005)
2. D. Benítez, J.M. Escobar, R. Montenegro, E. Rodríguez, Parallel performance model for vertex repositioning algorithms and application to mesh partitioning, in *Proceedings of* 27th *International Meshing Roundtable* (2018)
3. M. Brewer, L. Diachin, P. Knupp, T. Leurent, D. Melander, The mesquite mesh quality improvement toolkit, in *Proceedings of* 12th *International Meshing Roundtable*, pp. 239–250 (2003)
4. S. Browne, J. Dongarra, N. Garner, K. London, P. Mucci, A scalable cross-platform infrastructure for application performance tuning using hardware counters, in *Proceedings of Supercomputing*, Article 42 (IEEE Computer Society, Washington, 2000)
5. Y. Che, L. Zhang, C. Xu, Y. Wang, W. Liu, Z. Wang, Optimization of a parallel CFD code and its performance evaluation on Tianhe1A. Comput. Inform. **33**(6), 1377–1399 (2015)
6. L. Diachin, P. Knupp, T. Munson, S. Shontz, A comparison of inexact newton and coordinate descent mesh optimization techniques, in *Proceedings of* 13th *International Meshing Roundtable*, pp. 243–254 (2004)
7. J.M. Escobar, E. Rodríguez, R. Montenegro, G. Montero, J.M. González-Yuste, Simultaneous untangling and smoothing of tetrahedral meshes. Comput. Methods Appl. Mech. Eng. **192**, 2775–2787 (2003)
8. C. Geuzaine, J.F. Remacle, GMSH: a three-dimensional finite element mesh generator with built-in pre- and post-processing facilities. Int. J. Numer. Methods Eng. **79**(11), 1309–1331 (2009)

9. P. Knupp, Updating meshes on deforming domains: an application of the target-matrix paradigm. Commun. Numer. Methods Eng. **24**, 467–476 (2007)
10. D. Levinthal, *Performance Analysis Guide for Intel Core i7 Processor and Intel Xeon 5500 Processors* (Intel, Santa Clara, 2014)
11. M. Livesu, A. Sheffer, N. Vining, M. Tarini, Practical hex-mesh optimization via edge-cone rectification. ACM Trans. Graph. **34**(4), Article 141 (2015)
12. R. Montenegro, J.M. Cascón, J.M. Escobar, E. Rodríguez, G. Montero, An automatic strategy for adaptive tetrahedral mesh generation. Appl. Numer. Math. **59**(9), 2203–2217 (2009)
13. D.A. Patterson, J.L. *Hennessy, Computer Organization and Design: The Hardware/Software Interface*, ARM edn. (Morgan Kaufmann Publishers Inc., Burlington, 2017)
14. E. Ruiz-Girones, X. Roca, J. Sarrate, R. Montenegro, J.M. Escobar, Simultaneous untangling and smoothing of quadrilateral and hexahedral meshes using an object-oriented framework. Adv. Eng. Softw. **80**, 12–24 (2015)
15. S.P. Sastry, S.M. Shontz, Performance characterization of nonlinear optimization methods for mesh quality improvement. Eng. Comput. **28**, 269–286 (2012)
16. S.P. Sastry, S.M. Shontz, A parallel log-barrier method for mesh quality improvement and untangling. Eng. Comput. **30**(4), 503–515 (2014)
17. S.P. Sastry, S.M. Shontz, S.A. Vavasis, A log-barrier method for mesh quality improvement and untangling. Eng. Comput. **30**(3), 315–329 (2014)
18. R.B. Schnabel, Concurrent function evaluations in local and global optimization. CU-CS-345-86. Computer Science Technical Reports, 332 (University of Colorado, Boulder, 1986)
19. D. Terpstra, H. Jagode, H. You, J. Dongarra, Collecting performance data with PAPI-C, in *Tools for High Performance Computing 2009*, pp. 157–173 (Springer, Berlin, 2010)
20. C.S. Verman, K. Suresh, Towards FEA over tangled quads. Proc. Eng. **82**, 187–199 (2014)

Accurate Manycore-Accelerated Manifold Surface Remesh Kernels

Hoby Rakotoarivelo and Franck Ledoux

Abstract In this work, we devise surface remesh kernels suitable for massively multithreaded machines. They fulfill the locality constraints induced by these hardware, while preserving accuracy and effectiveness. To achieve that, our kernels rely on:

- a point projection based on geodesic computations,
- a mixed diffusion-optimization smoothing kernel,
- an optimal direction-preserving transport of metric tensors,
- a fine-grained parallelization dedicated to manycore architectures.

The validity of metric transport is proven. The impact of point projection as well as the accuracy of smoothing kernel are assessed by comparisons with efficient existing schemes, in terms of surface deformation and mesh quality. Kernels compliance are shown by representative examples involving surface approximation or numerical solution field guided adaptations. Finally, their scaling are highlighted by conclusive profiles on recent dual-socket multicore and dual-memory manycore machines.

1 Context, Issues and Features Overview

In this work, we focus on surface mesh adaptation kernels suitable for massively multithreaded architectures without sacrificing effectiveness or accuracy, in context of adaptive simulation involving both a numerical solver and a 3D remesher. Given an initial uniform triangular mesh and a budget of points, we aim at providing a mesh which minimizes the surface approximation error or the numerical solution

H. Rakotoarivelo (✉)
Centre de mathématiques et leurs applications (CMLA), ENS Cachan, Cachan, France
e-mail: hoby.rakotoarivelo@ens-cachan.fr

F. Ledoux
CEA, DAM, DIF, Bruyères-le-Châtel, France
e-mail: franck.ledoux@cea.fr

X. Roca, A. Loseille (eds.), *27th International Meshing Roundtable*,
Lecture Notes in Computational Science and Engineering 127,
https://doi.org/10.1007/978-3-030-13992-6_22

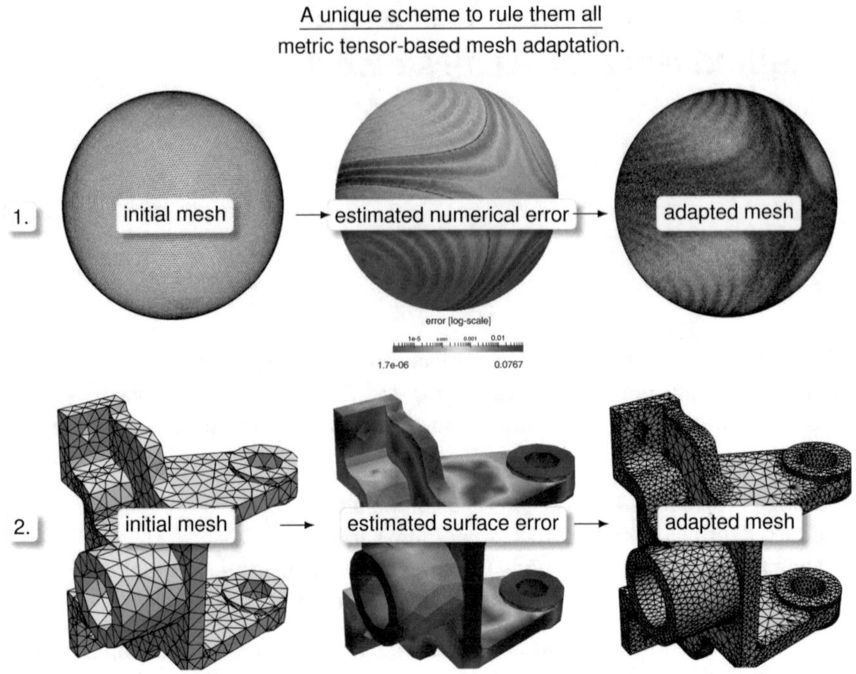

Fig. 1 Kinds of adaptation supported. If the mesh is intended to be adapted to the error of a numerical solution u (defined on each point), then the metric field may be derived from local Hessian matrices of u. If it is intended to be adapted to the error of the surface itself, then the metric field may be derived from local curvature tensors. Finally, both errors may be simultaneously supported using metric field intersection

interpolation error with respect to the given budget of points (as depicted in Fig. 1), while achieving good cell aspect ratios in both isotropic and anisotropic context.

Related Works Many robust open-source libraries for surface-volume remeshing exists such as tetgen, CGAL, MMG3D, MMGS, GMSH among others. The issue is that the involved kernels do not expose sufficient locality required by these architectures. Indeed, each operation on a mesh point or cell should involve a small, static and bounded vicinity. This locality constraint is far from being trivial, and most of the state-of-the-art kernels rely a dynamic point or cell vicinity. For instance,

- cavity-based remeshing kernels (surfacic Delaunay or hybrid cavity) enable to significantly improve mesh quality [1, 2]. They aim at removing bad cells when adding or removing points. However those cells are not statically known.
- atlas-based remeshing kernels provide well-sampled surface meshes, with a quality comparable to a variational scheme [3]. It consist of on-the-fly embedding the surface into a plane, and then to remesh the embedded region using 2D kernels. The issue is that it often require growing a non-predefined region at each time [4].

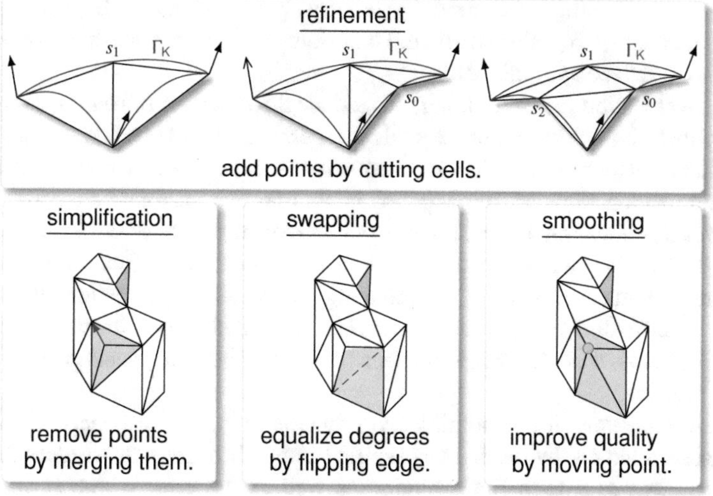

Fig. 2 Our static remesh kernels. Here, refinement and simplification aim at resampling the surface, whereas swapping and smoothing aim at regularizing it. They actually involve a cell, a couple of cells or the direct vicinity of a point

- numerical schemes are more stable on quasi-structured meshes. Relaxing mesh point degrees is tedious though, due to numerous local minima. To address this issue, related heuristics (such as 5-6-7 or puzzle solving) rely on a dynamic edge flip-refine-collapse sequence [4, 5]. Therefore, impacted cells cannot be inferred.

In fact, kernel locality and effectiveness are two antagonistic constraints. To achieve locality, one have to resort to basic kernels (no dynamic cavity, no local embedding, no dynamic sequence of operations). To achieve effectiveness though, one have to resort to dynamic kernels to quickly achieve mesh convergence, in terms of surface approximation, mesh quality, or solution interpolation of a finite element simulation.

Contributions In this work, we aim to conciliate both constraints. To achieve maximal locality, we resort to basic kernels depicted in Fig. 2, and we avoid any of the dynamic features described above. To achieve effectiveness, we rely on advanced geometric features instead. They include :

- Point Projection (for refinement, simplification and smoothing). In our case, the real surface is only known on mesh points, and we do not have an atlas for local parametrization. Thus we have to find a way to accurately put points on the real surface after cutting an edge, merging two points or moving a point. If the point lies on the mesh then it is not a problem since we may use a BEZIER curve, a PN-triangle or a local quadric surface in that case. However, if it does not lie on the mesh, then the situation is not clear. In that case, one have to project it on the mesh before projecting it to the surface, with a potential loss of accuracy. Here, we provide a point projection scheme which relies on geodesic curve computations using a differential geometry operator (Sect. 2). It ensures that the loss of accuracy remains small, and contributes to speedup mesh convergence in terms of surface approximation.

- Point Relocation (for smoothing). To reduce their number of rounds, refinement, simplification and swapping kernels are performed without taking care of cell qualities. Hence, improving mesh quality is entirely entrusted to smoothing kernel, which aims at moving points to improve surrounding cells quality. However, relocating a point involves an unavoidable surface deformation. Here, we provide a point relocation scheme for smoothing kernel, which aims at improving mesh quality while reducing the surface degradation. It relies on both a diffusion filter which is applied first, and a non-linear optimization routine invoked on failure cases (Sect. 3). It contributes to speedup mesh quality convergence.
- Metric Transport (for smoothing and refinement). In our case, desired edge size are prescribed on the mesh using metric tensors. It enables us to adapt the mesh to the error of a numerical solution as well as the error of the surface itself in anisotropic context (see Fig. 1). For any point, its metric tensor encodes the edge length at any direction incident to it. Hence, when a point p is created or relocated, then its metric tensor has to be interpolated from its neighbors. However, repeated metric interpolation would involve a loss of anisotropy. To reduce that, one may move all involved metric tensors at p before averaging them (if necessary). However, moving a metric tensor on a surface may deviate its directions if no precautions are taken. Here, we provide a safe way to move them on the surface without altering their principal directions (Sect. 4). It is based on the notion of parallel transport in differential geometry, and contributes to speedup mesh convergence in terms of surface approximation and numerical solution interpolation.

Hence, the locality constraint induced by the hardware is respected. Indeed, our kernels involve a small and static vicinity for surface resampling, cells regularization or degree equalization (Fig. 2). Besides, their effectiveness is preserved thanks to the three geometric features introduced above. The remaining question is about how to efficiently port them on our target hardware:

- Fine-Grained Parallelization (for all kernels). Manycore machines consist of numerous underclocked cores with a very limited cache-memory per core. Besides, multicore machines have less but faster cores. The latter have multiple cache and memory levels, leading to unequal data access latencies (NUMA). Hence, kernels must expose a huge amount of parallelism and a high rate of data reuse to scale on both machines. However, remesh kernels are not trivially parallelizable in high performance computing context. Indeed they are data-driven since task dependencies evolve at runtime, and cannot be statically predicted. They are also data-intensive since most of their instructions are data accesses but often on different data. Hence, usual optimizations (such as cache tiling, static load balancing, prefetching) will not work well on them. Here, we devise a multithreaded scheme enabling to ease both data-driven and data-intensive issues (Sect. 5). It is based on our work in [6, 7], but extended in 3D with ridges support.

2 Accurate Point Projection Based on the Exponential Map

As stated before, we need to find a way to accurately put points on the real surface during refinement, simplification or smoothing. Hence, if the point lies on the mesh, then it may be directly projected to this surface through a local parametrization scheme. The main issue is on finding an effective way to project the point when it does not lie on mesh. First, we show how we locally recover the real surface, and then we will describe how to accurately project the point on this real surface.

Reconstruction In fact, the real surface Γ is only known on mesh points and needs to be locally recovered. It may be done by many ways, however achieving both accuracy and regularity is quite difficult. For instance, one may recover it in the vicinity of each point using a quadric surface obtained by a least squares method. Since Γ is approximated by a smooth surface, then it achieves regularity. However, it only provides a quadratic reconstruction of Γ. Besides, remeshers and shaders such as Inria's MMGS or the DIRECT-X 11 tesselation engine often resort to a cubic reconstruction through PN-triangles [8]. Despite their compacity, they do not ensure point tangent plane unicity across patch boundaries. Here, we aim at conciliating both constraints. For that, we resort to a per-cell local parametrization with complete G1-continuity. It is achieved through a GREGORY PATCH construction with twist points BLENDING like in [9] but with special care for ridges and corners (Fig. 3).

Point Projection In fine, the projection of a point p on the ideal surface may be directly obtained as long as its tangent plane $T_p\Gamma$ is supported by a mesh cell. However, there are situations where p does not lie on the mesh, and cannot be directly parametrized. During smoothing for instance, the projected point q has to be computed such that the length of underlying geodesic curve of $[pq]$ directed by a

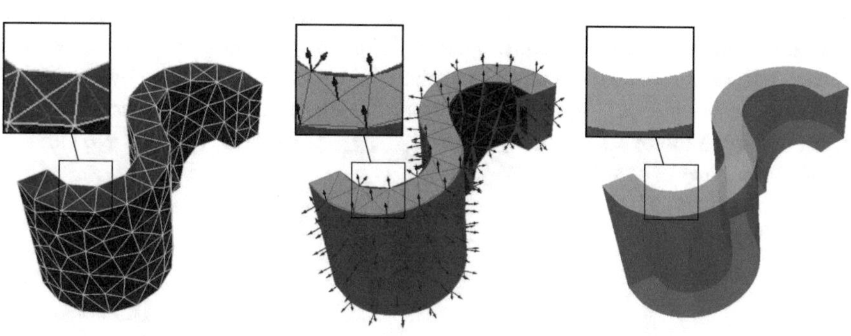

Fig. 3 Surface reconstruction using quartic patches with G^1-continuity

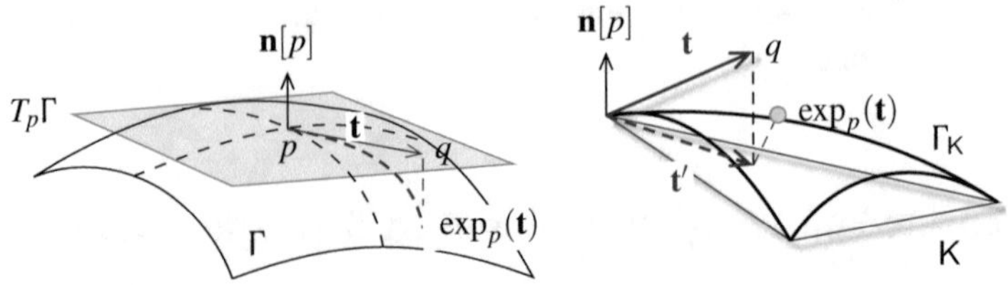

(a) view in the vicinity of p (b) view on cell pointed by **t**.

Fig. 4 Discrete approximation of exponential map in the vicinity of a point. (**a**) View in the vicinity of p. (**b**) View on cell pointed by **t**

displacement vector $\alpha\mathbf{t}$ must be equal to $\|\alpha\mathbf{t}\|$. Such an operator exists in continuous context, it is the *exponential map* described in Definition 1.

Definition 1 (Exponential Map) Let p be a point of a manifold Γ, and R a region of its tangent plane. Its exponential map $\exp_p : \mathsf{R} \to \Gamma$ maps any tangent vector $\mathbf{t}$ of R to the geodesic curve segment γ starting from p with initial speed $\frac{\mathrm{d}\gamma}{\mathrm{d}s}(0) = \mathbf{t}$ and length $\ell(\gamma) = \|\mathbf{t}\|$.

Intuitively, $\exp_p(\mathbf{t})$ may be seen as an operator giving the optimal projection of q for a specified tangent vector $\mathbf{t} = \overrightarrow{pq}$ as shown in Fig. 4. It may be obtained through a reparametrization of γ by arc length. For that, we have to compute curve length $\mathrm{s}(t) = \int_0^t \|\frac{\mathrm{d}\gamma}{\mathrm{d}t}(x)\|\,\mathrm{d}x$ with $\frac{\mathrm{d}\gamma}{\mathrm{d}t}(0) = \mathbf{t}$, then express $\mathbf{t}$ and thus γ in terms of s. However, we do not have a parametrization of $\frac{\mathrm{d}\gamma}{\mathrm{d}t}$ in terms of t, and thus no explicit expression of $\exp_p$. Therefore, we use the following heuristic to approximate $\exp_p : T_p\Gamma \to \Gamma$.

Algorithm The first step is to find which cell K incident to p is pointed by $\mathbf{t}$. Indeed, $\exp_p(\mathbf{t})$ will lie in the quartic patch Γ_{K} related to K. Then the difficulty lies in the choice of a point $\tilde{q} \in \mathsf{K}$ such that the length of underlying curve equals to $\|\mathbf{t}\|$. In other words, we have to find a vector $\mathbf{t}' = [p\tilde{q}]$ from $\mathbf{t}$ such that $\exp_p(\mathbf{t}) = \Gamma_{\mathsf{K}}[\tilde{q}]$. Here, this issue is resolved by a single step linear search. To find an initial value of $\tilde{q}$, the idea is to perform a rotation of $\mathbf{t}$ on the tangent plane $T_{\mathsf{K}}\Gamma$ of K as shown in Fig. 4b. Then we compute the projection $\Gamma_{\mathsf{K}}[\tilde{q}]$ of $\tilde{q}$ on Γ_{K}, as well as its normal vector. Afterward, the idea is to approximate the geodesic segment γ by a B-spline curve spanned by $[p, \Gamma_{\mathsf{K}}[\tilde{q}]]$ and to compute its length. It enables us to adjust the norm of $\mathbf{t}'$ by a factor s such that $\|\mathbf{t}'\| = \|\mathbf{t}\|^2 \ell[\gamma]^{-1}$. By recomputing $\Gamma_{\mathsf{K}}[\tilde{q}]$ and γ, we actually have $\ell(\gamma) \approx \|\mathbf{t}\|$.

3 Accurate Mixed Diffusion-Descent Smoothing Kernel

So far, we have an accurate way to project any point q lying either on the mesh or on the tangent plane of another point p, onto the real surface. Hence, we are able to project the resulting point on the real surface after an cutting an edge, merging two points or relocating a point. Here, we aim to specifically show how it may be used to devise an effective smoothing kernel. In our case, it is the only kernel entrusted to improve mesh quality, and thus is really important. For each point, it aims at improving surrounding cells qualities with respect to the metric tensor field.

Issues Existing kernels may be classified in two categories: laplacian-based and non-linear optimization ones. The former is simple and enables both geometry denoising [10] as well as cell shape regularization. However, it tends to shrink the surface since it is not a low-pass filter [11], and does not strictly improve cell quality since it is a heuristic. The latter enables advanced features such as volume-preserving denoising [12, 13] or cell quality improvement on problematic cases [14]. However, it is clearly more expensive. Based on these facts, we devised a mixed kernel like [15, 16]: a diffusion filter is primarily used for point relocation, followed by a cell quality maximization in failure cases. However, we use a different scheme for diffusion as well as for optimal step computation.

- Diffusion. For each point, we aim at equalizing its incident cell patches, while reducing the unavoidable surface deformation. Here the idea is to move the point p toward the projected weighted center of mass of its vicinity, as shown in (1). First, we compute the displacement direction $\mathbf{t}$ toward the center of mass of the region $\mathsf{C} = \bigcup_{j=1}^{n} \Gamma_j$. Then, we compute the geodesic segment curve γ starting from p with initial speed $\mathbf{t}$. Finally, the new point position is given by $\gamma(1)$.

$$p = \exp_p \left[\alpha \, \frac{\int_{\mathsf{C}} \log_{p[t]}[x]\, \rho[x]\, \mathrm{d}x}{\int_{\mathsf{C}} \rho[x]\, \mathrm{d}x} \right], \text{ with } \begin{cases} \mathsf{C} : \text{continuous vicinity of } p, \\ \rho : \text{density function}, \\ \alpha : \text{scaling factor}. \end{cases} \quad (1)$$

  Here, $\rho[x] = \sqrt{\det[J_x^{\mathsf{T}} g_x J_x]}$ aims to weight the displacement of p with respect to the metric tensor g_x of each surrounding point x, and J_x is the Jacobian matrix of the surface parametrization at point x. To avoid useless computations, we exclude the case in which the center of mass is too far from p. Here, the displacement is accepted if the quality of the worst cell is enhanced and if the deviation of incident cells from $T_p\Gamma$ does not exceed an angular threshold θ_{max}. Otherwise, we reduce the scaling factor α, which is initially set to 1, in a dichotomic way.
- Optimization. Here, the goal is to enforce worst cell quality improvement in the vicinity of p. Let $f_i[p]$ be the distorsion[1] of an incident cell K_i according to the current position of p. The goal is to solve $\min \max_i \; f_i[p]$ with the constraint that

[1]The distorsion of a cell is just the inverse of its quality.

p must lie in C. For that, the idea is to move p gradually on the surface according to a displacement step α toward a direction $-\nabla f_k[p]$ until convergence or if a round threshold is achieved. The position of p is then updated as follow:

$$p^{[t+1]} = \exp_{p^{[t]}}[-\alpha \nabla f_k(p^{[t]})],\ \alpha \in [0, 1],\ \text{with} \begin{cases} f_j \ : \text{distorsion of cell } \mathsf{K}_j \\ f_k = \max_j f_j[p^{[t]}] \\ \alpha : \text{displacement step} \end{cases} \tag{2}$$

Here, minimizing $f_k[p]$ may increase the distorsion of other cells. Hence, we must take them into account in the choice of initial step α. Let $j = \arg\min_{i \neq k} f_i[p]$. By solving $f_j[p - \alpha \nabla f_k(p)] = f_k[p - \alpha \nabla f_k(p)]$, and by considering the first order Taylor-Young expansion of f_k, we have:

$$\alpha = \frac{f_j[p] - f_k[p]}{\|\nabla f_k[p]\| - \langle \nabla f_k[p], \nabla f_j[p] \rangle}. \tag{3}$$

The next step is determined by a linear search verifying WOLFE conditions [17]. They guarantee that f_k decreases significantly and that α is large enough to converge rapidly. Note that the convergence rate is closely related to the initial point position (or seed). Since this routine is called after moving the point toward its center of mass, then p is relatively close to its optimal position. Thus, a few number of rounds is expected in practice (3–5). Finally, if the point lies on a ridge, then we relocate it on the midpoint of the curve segment related to its neighboring ridges.

4 Accurate Direction-Preserving Transport of Metrics

Up to now, we have an effective kernel designed to improve mesh quality by moving points. Note that when a point is relocated, its geometric data (normal and metric tensor) need to be recomputed or interpolated from its neighbors, since they have changed. However, repeated metric tensor interpolation would involve a loss of anisotropy due to diffusion effects. It is at least the case when a simple linear scheme is used. Besides, moving a metric tensor along a curve may deviate its directions. To ease this deviation, we resort to a parallel transport scheme, as described in Definition 2, but extended to metric tensors. Intuitively, it generalizes the notion of translation on manifold surfaces. Note that it is already used in MMGS in a heuristically way. Here, our goal is to formally prove that a direction-preserving transport of metric tensors may be simply achieved through tangent vectors parallel transport.

Definition 2 (Parallel Transport) Let ∇ be an affine connection[2] related to a manifold Γ. A vector field $\mathbf{v}^{[t]}$ on the tangent bundle of Γ along a curve $\gamma : I \to \Gamma$ is said parallel with respect to this connection, if $\nabla_{\dot{\gamma}[t]}\mathbf{v}^{[t]}$ vanishes for all $t \in I$. Hence, a tangent vector $\mathbf{v}$ is parallel transported from p to q on γ if the vector field $\mathbf{v}^{[t]}$ induced by its displacement is parallel.

Definition 3 (Levi-Civita Connection) There exists a unique affine connection ∇ on a Riemannian manifold (Γ, g) such that:

- it is torsion-free: there is no tangent planes rotation along a geodesic γ when they are parallel transported along γ through ∇.
- it is compatible with g: the induced dot product at any point p of Γ is preserved by parallel transport of p along γ: this displacement is actually an isometry.

For each curve $\gamma : I \to \Gamma$, a metric $g_{\gamma[t]}$ is compatible with the Levi-Civita connection ∇ of the surface (see Definition 3), if for any $t \in I$ and for each vector fields $\mathbf{u}, \mathbf{v}$ of the tangent bundle of Γ, we have:

$$(\nabla_{\dot{\gamma}[t]} g_{\gamma[t]})(\mathbf{u}, \mathbf{v}) = \partial_{\dot{\gamma}[t]}(g_{\gamma[t]}(\mathbf{u}, \mathbf{v})) - g_{\gamma[t]}(\nabla_{\dot{\gamma}[t]}\mathbf{u}, \mathbf{v}) - g_{\gamma[t]}(\mathbf{u}, \nabla_{\dot{\gamma}[t]}\mathbf{v}) = 0. \tag{4}$$

In fact, the metric tensor g_p related to a point p is a symmetric bilinear form corresponding to a symmetric matrix M in a local basis of its tangent plane $T_p\Gamma$. In particular, there exists a local basis P of $T_p\Gamma$ such that M is congruent to a diagonal matrix D in P. In other words, $g_p(\mathbf{v}_i, \mathbf{v}_j) = 0$ for any $\mathbf{v}_i, \mathbf{v}_j \in \mathsf{P}$. Here, we aim to show that the eigenvalues λ_i of D are invariant by parallel transport of column vectors $\mathbf{v}_i$ of P along a curve γ spanned by $[pq]$. For a given $t \in I$, let $\mathbf{w}^{[t]}$ be the transported tangent vector $\mathbf{w}$ at $\gamma(t)$, D the eigenvalue diagonal matrix at $p = \gamma(0)$, and $\mathsf{P}^{[t]} = (\mathbf{v}_1^{[t]}, \mathbf{v}_2^{[t]})$. Thus, we have:

$$\begin{aligned}
\forall \mathbf{u}^{[0]} \in T_{\gamma[0]}\Gamma : g_{\gamma[t]}(\mathbf{u}^{[t]}, \mathbf{v}_2^{[t]}) &= \langle \mathbf{u}^{[t]}, \mathsf{M}_{\gamma[t]}\mathbf{v}_2^{[t]}\rangle = \langle \mathbf{u}^{[t]}, \textstyle\sum_{i=1}^{2} \lambda_i^{[t]} \mathbf{v}_i^{[t]} \mathbf{v}_i^{[t],\mathsf{T}} \mathbf{v}_2^{[t]}\rangle, \\
&= \langle \mathbf{u}^{[t]}, \lambda_1^{[t]} \mathbf{v}_1^{[t]} (\mathbf{v}_i^{[t],\mathsf{T}} \mathbf{v}_2^{[t]}) + \lambda_2 (\mathbf{v}_2^{[t]} \mathbf{v}_i^{[t],\mathsf{T}}) \mathbf{v}_2^{[t]}\rangle, \\
&= \langle \mathbf{u}^{[t]}, \lambda_2^{[t]} \|\mathbf{v}_2^{[t]}\| \, \mathbf{v}_2^{[t]}\rangle = \lambda_2^{[t]} \langle \mathbf{u}^{[t]}, \mathbf{v}_2^{[t]}\rangle,
\end{aligned}$$

At the same time $\nabla_{\dot{\gamma}[t]}\mathbf{u} = \nabla_{\dot{\gamma}[t]}\mathbf{v}_i = \mathbf{0}, \ \forall t \in I,$

thus $g_{\gamma[t]}(\nabla_{\dot{\gamma}[t]}\mathbf{u}, \mathbf{v}_i) + g_{\gamma[t]}(\mathbf{u}, \nabla_{\dot{\gamma}[t]}\mathbf{v}_i) = 0,$

hence $\nabla_{\dot{\gamma}[t]} g_{\gamma[t]}(\mathbf{u}, \mathbf{v}_i) = \partial_{\dot{\gamma}[t]} g_{\gamma[t]}(\mathbf{u}^{[t]}, \mathbf{v}_i^{[t]}),$

$$\text{therefore } \nabla_{\dot{\gamma}[t]} g_{\gamma[t]}(\mathbf{u}, \mathbf{v}_i) = 0 \Leftrightarrow \frac{g_{\gamma[t]}(\mathbf{u}^{[t]}, \mathbf{v}_i^{[t]}) - g_{\gamma[0]}(\mathbf{u}^{[0]}, \mathbf{v}_i^{[0]})}{\int_0^t \|\dot{\gamma}(s)\| \mathrm{d}s} = 0.$$

[2] An affine connection generalizes the notion of derivative for vector fields on a manifold.

$$\begin{aligned} \text{by the way } \langle \mathbf{u}, \mathbf{v}_i \rangle \text{ constant on} \gamma &\Leftrightarrow \frac{\langle \mathbf{u}^{[0]}, \mathbf{v}_i^{[0]} \rangle}{\int_0^t \|\dot{\gamma}(s)\| ds} (\lambda_i^{[t]} - \lambda_i^{[0]}) = 0, \\ &\Leftrightarrow \lambda_i^{[t]} - \lambda_i^{[0]} = 0, \\ &\Leftrightarrow \mathsf{M}_{\gamma[t]} = \textstyle\sum_{i=1}^{2} \lambda_i^{[0]} \mathbf{v}_i^{[t]} \mathbf{v}_i^{[t],\mathsf{T}}, \\ &\Leftrightarrow \mathsf{M}_{\gamma[t]} = \mathsf{P}^{[t],\mathsf{T}} \mathsf{D}\, \mathsf{P}^{[t]}, \\ \text{therefore } \nabla_{\dot{\gamma}[t]} g_{\gamma[t]}(\mathbf{u}, \mathbf{v}) = 0 &\Leftrightarrow \begin{cases} \nabla_{\dot{\gamma}[t]} \mathbf{u} = \nabla_{\dot{\gamma}[t]} \mathbf{v} = \mathbf{0} \\ g_{\gamma[t]}(\mathbf{u}, \mathbf{v}) = \langle \mathbf{u},\ \mathsf{P}^{[t],\mathsf{T}} \mathsf{D}\, \mathsf{P}^{[t]} \mathbf{v} \rangle \end{cases} . \end{aligned} \tag{5}$$

Algorithm 1 Parallel transport of a metric tensor

let $\tilde{p} = p$.
for each timestep $t \in I$ **do**
  set $\tilde{q} = \gamma[t]$, and extract Jacobians and normals at $\tilde{p}$ and $\tilde{q}$.
  extract a local basis $\mathsf{P} = (\mathbf{v}_1, \mathbf{v}_2)$ of $T_{\tilde{p}}\Gamma$ such that $g_{\tilde{p}} = J_{\tilde{p}}\, \mathsf{P}^\mathsf{T}\, \mathsf{D}\, \mathsf{P}\, J_{\tilde{p}}^\mathsf{T}$.
  determine the point s as follow: {SCHILD'S LADDER.}
  - let r be the endpoint of $\mathbf{v}_1$,
  - compute a segment $[r\tilde{q}]$ and let m be its midpoint,
  - compute a segment $[ps]$ such that $\|ps\| = 2\|pm\|$,

  compute $\mathbf{v}_1^{[t]} = \overrightarrow{\tilde{q}s}$, $\mathbf{v}_2^{[t]} = \mathbf{v}_1^{[t]} \times \mathbf{n}(\tilde{q})$, and $\mathsf{P}^{[t]} = (\mathbf{v}_1^{[t]}, \mathbf{v}_2^{[t]})$.
  store $g_{\tilde{p}}^{[t]} = J_{\tilde{q}}\, \mathsf{P}^{[t],\mathsf{T}}\, \mathsf{D}\, \mathsf{P}^{[t]}\, J_{\tilde{q}}^\mathsf{T}$.
  update $\tilde{p} = \tilde{q}$ and $\mathbf{n}(\tilde{p}) = \mathbf{n}(\tilde{q})$ and return $g_p^{[1]}$.
end for

According to (5), a parallel transport of a metric tensor g_p along a curve γ is equivalent to a parallel transport of its eigenvectors $\mathbf{v}_i$ through the Levi-Civita connection ∇ related to the intrinsic metric of the surface (also called the first fondamental form). In fact, since the directions of the metric tensor must be orthogonal at each point of $\gamma[t]$, then it is enough to transport a unique direction $\mathbf{v}_1$ and to find $\mathbf{v}_2$ such that $\langle \mathbf{v}_1^{[t]}, \mathbf{v}_2^{[t]} \rangle = 0$ for any $t \in I$. Here, it is achieved through the routine depicted in Algorithm 1.

5 Fine-Grained Lock-Free Parallelization

At this point, we have four effective kernels for surface resampling, mesh regularization or degree equalization. They respect the locality constraint induced by our target hardware, since they only involve a small static vicinity at each time, (Fig. 2). Besides, they do not rely on a dynamic sequence of operations to achieve effectiveness. Instead, they rely on the three geometric features described

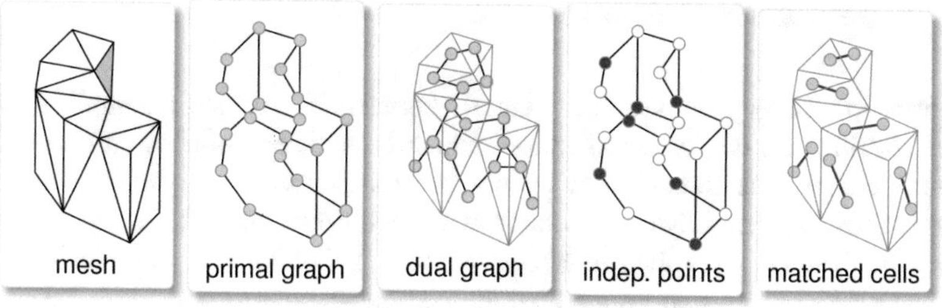

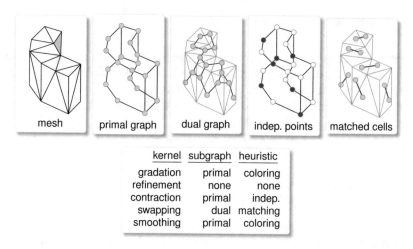

kernel	subgraph	heuristic
gradation	primal	coloring
refinement	none	none
contraction	primal	indep.
swapping	dual	matching
smoothing	primal	coloring

Fig. 5 Task graphs and related heuristics

in Sects. 2, 3, and 4. Hence, the remaining concern is about how to parallelize them efficiently, despite the fact that they are data-driven and data-intensive as explained in Sect. 1.

As stated in Sect. 1, a huge amount of parallelism is required for kernels scaling on these either low frequency and limited memory per core, or deep memory hardware (multicore, manycore). For that, we clearly advocate the use of massively multithreaded schemes instead of usual MPI-based ones. Here, we devise a fine-grained lock-free scheme based on our work in [6, 7], but with additional ridges adjacency graph support. For each kernel, data dependencies are formulated into a graph, and tasks are extracted through multithreaded maximal stable set, graph coloring and matching heuristics as shown in Fig. 5, except for refinement. Incidence graphs updates are done through a dual-step scheme: related incidence lists are grown asynchronously using cheap atomic primitives (15–30 CPU cycles), then obsolete references are removed in a single sweep.

To ease their irregularity, we structure kernels into bulk-synchronous tasks [18]. Hence, each kernel is organized into sweeps, and each sweep consists of local computation, shared-data writes and a thread barrier.To ease data indirections penalties, we use a locality-aware reduction[3] scheme. It enables coalescent data writes in shared memory, and is further explained in [7].

[3] A reduction is an aggregation of local data.

6 Numerical Evaluation

In short, we have proposed four remesh kernels for surface resampling, mesh regularization and degree equalization, which conciliates both locality and effectiveness constraints. Indeed, they do not involve neither a dynamic cell or point vicinity growing, nor a dynamic sequence of operations. Instead, they rely on three geometric features described in Sects. 2, 3 and 4. For sake of completeness, we also have proposed a fine-grained parallelization scheme to port these kernels on multicore and manycore architectures. At this point, we aim to numerically assess the effectiveness of each devised feature. For that purpose, the devised kernels are written in C++14 and OPENMP4 as a library called trigen. It is being integrated into a new version of GMDS [19], and will be open-source (Fig. 6).

Projection To assess the effectiveness of our point projection scheme, we evaluate it on both mesh simplification and smoothing cases in terms of surface approximation error as depicted in Fig. 7. Indeed, a major issue on both kernels relies on reducing the unavoidable surface degradation while performing point relocation or removal. Here, we aim at assessing that our exponential-map based kernels are as accurate as those of the state-of-the-art.

- First, the surface error induced by our simplification kernel compared to the quadric error metric [20] is given. Note that both algorithms are quite identical, the difference mainly rely on how do we put the remaining point after merging two points. Here, normals deviation threshold is set to $\theta_{\max} = 10°$ and no edge swap is allowed. The pointwise Hausdorff distance is computed through METRO plugin [21] with roughly 400,000 sampled points. The given error distribution shows that it is better reduced and equidistributed on strong curvature areas and features curves with our scheme unlike quadric error metric.
- The surface error induced by our smoothing kernel compared to Taubin [22] and HC-laplacian [23] (using Meshlab) is given in case of a non trivial smooth manifold. Note that the two other schemes are optimized to reduce shrinkage effects induced by an usual laplacian, and contributes to reduce the surface

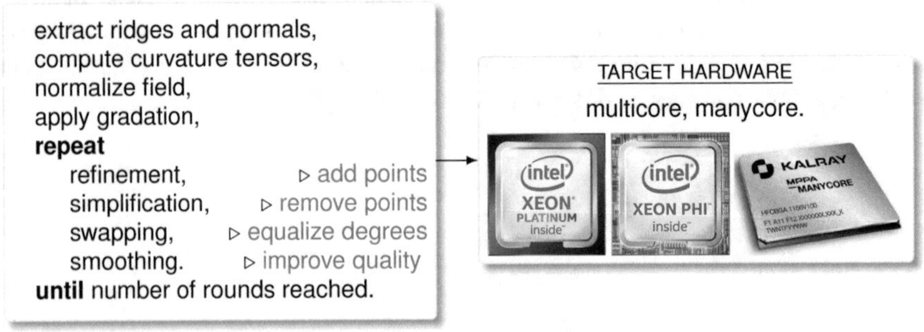

Fig. 6 Algorithm and kind of hardware involved in our numerical evaluations

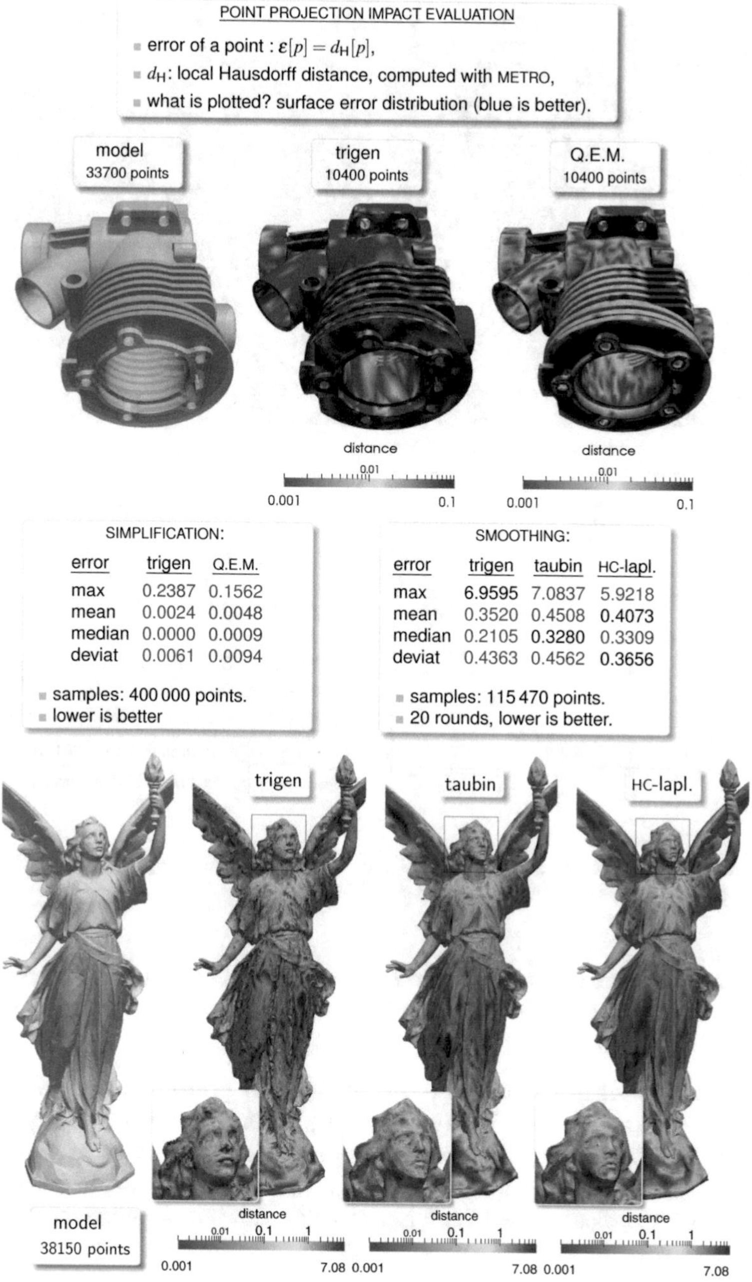

Fig. 7 Impact of the point projection scheme on kernels. Surface degradation is a major issue in mesh simplification and smoothing. Here, the goal is to assess the accuracy of our exponential-map based kernels compared to the state-of-the-art

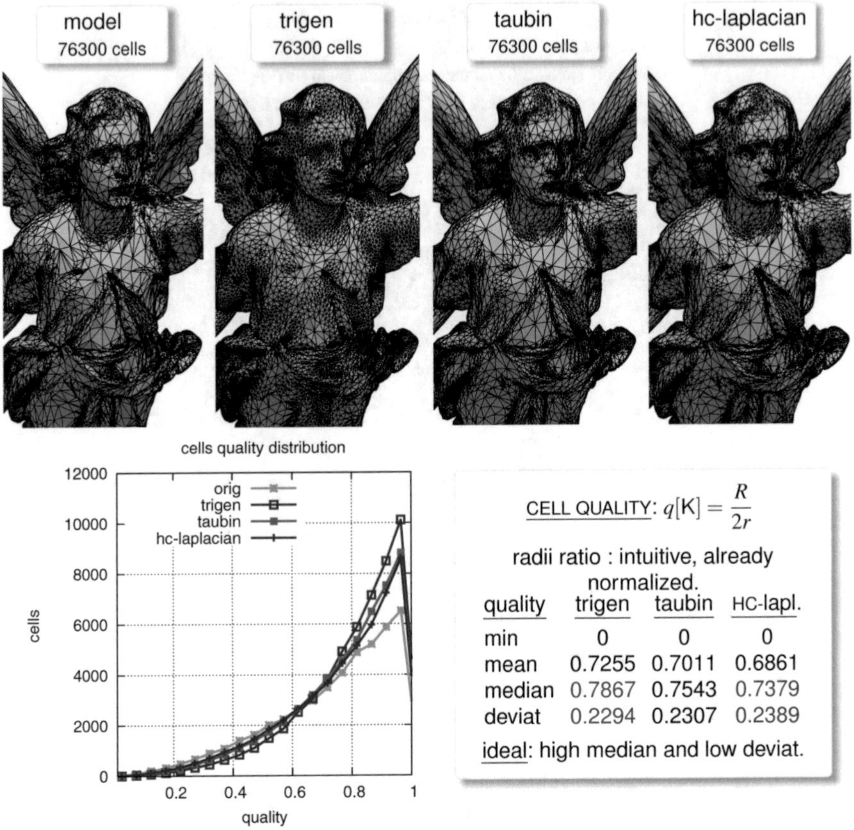

CELL QUALITY: $q[\mathsf{K}] = \frac{R}{2r}$

radii ratio : intuitive, already normalized.

quality	trigen	taubin	HC-lapl.
min	0	0	0
mean	0.7255	0.7011	0.6861
median	0.7867	0.7543	0.7379
deviat	0.2294	0.2307	0.2389

ideal: high median and low deviat.

Fig. 8 Evaluation of our smoothing kernel. Here, we aim to assess if a good mesh quality may actually be achieved while minimizing the surface deformation

deformation. The diameter is set to $h_{\mathsf{max}} = 10\%$ of bounding-box diagonal and normals deviation threshold is set to $\theta_{\mathsf{max}} = 7°$. Here, we can see that the surface is well preserved with our kernel.

Smoothing Recall that in our case, smoothing aims primarily at improving cell quality. A comparison with Taubin and HC-laplacian in terms of mesh quality is given in Fig. 8. It is exactly the same testcase as in Fig. 7 with the same parameters. We have already assessed that the surface error is reduced with our smoothing kernel. Here, the distribution of cell aspect ratios shows that a good mesh quality is also achieved.

Convergence This time, we aim at assessing mesh error and quality convergence when all kernels are combined. Precisely, we aim to evaluate how both surface error and mesh quality evolve when the number of points is increased or decreased. For that, local curvature guided adaptations are first considered in Fig. 9 through

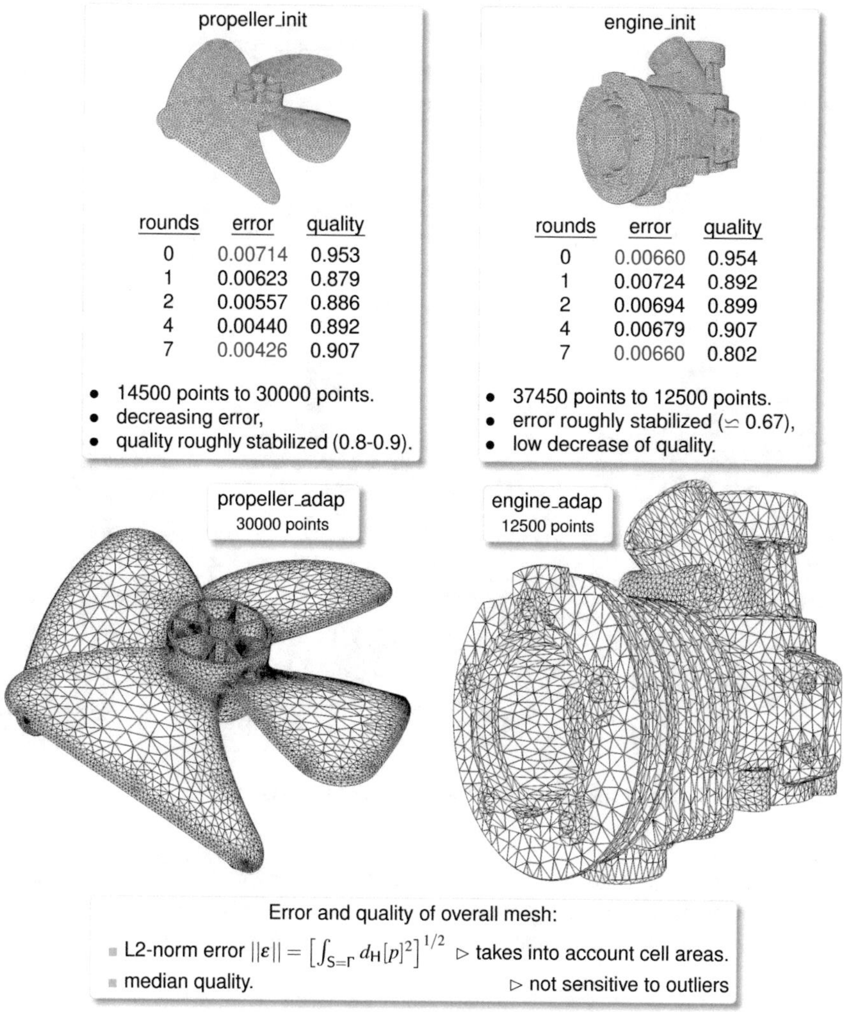
propeller_init
rounds error quality
0 0.00714 0.953
1 0.00623 0.879
2 0.00557 0.886
4 0.00440 0.892
7 0.00426 0.907
• 14500 points to 30000 points.
• decreasing error,
• quality roughly stabilized (0.8-0.9).
engine_init
rounds error quality
0 0.00660 0.954
1 0.00724 0.892
2 0.00694 0.899
4 0.00679 0.907
7 0.00660 0.802
• 37450 points to 12500 points.
• error roughly stabilized (≃ 0.67),
• low decrease of quality.
propeller_adap
30000 points
engine_adap
12500 points
Error and quality of overall mesh:
L2-norm error $||\varepsilon|| = \left[\int_{S=\Gamma} d_H[p]^2\right]^{1/2}$ ▷ takes into account cell areas.
median quality. ▷ not sensitive to outliers

Fig. 9 Convergence in error and quality when all kernels are being combined

isotropic model enrichment (PROPELLER, $n_{\max} = 2n$) and anisotropic coarsening (ENGINE, $n_{\max} = \frac{n}{3}$). Here, curvature tensors are computed using MEYER's scheme [24], and a simple L^∞-norm error estimate is used for metric field extraction. A gradation process based on [25] is used with a H-shock threshold set to 1.5. For PROPELLER, we can clearly see that points are mostly located on feature curves and high curvature areas, and that edge sizes are well-graded. For ENGINE, a surface degradation is necessarily introduced. However, feature curves are well-preserved, and cells are correctly stretched along minimal curvature direction. The surface approximation error as well as mesh quality evolutions are also depicted

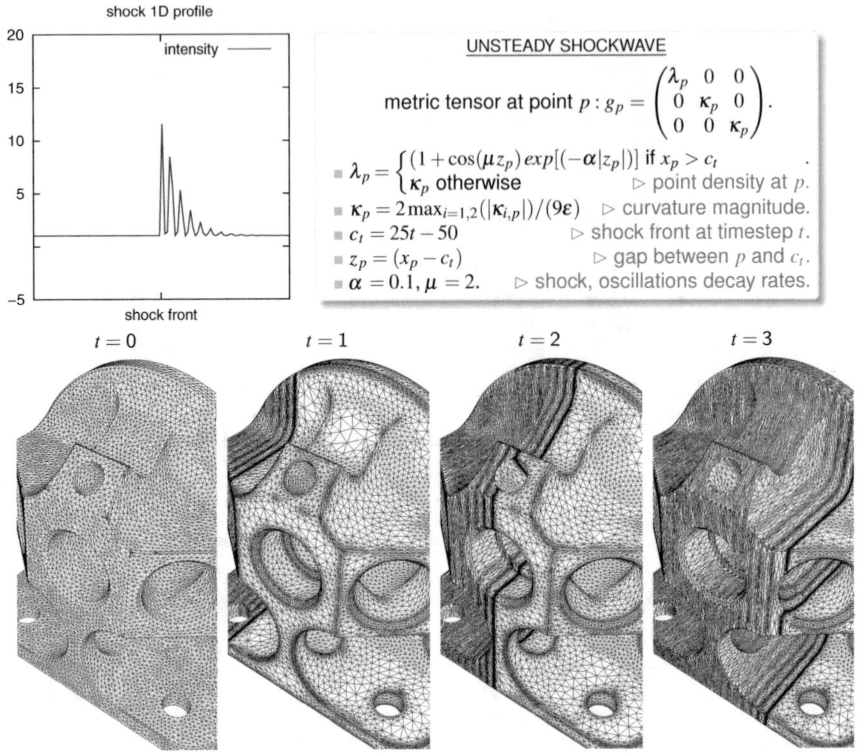

Fig. 10 Anisotropic adaptation to a user-defined size field using all kernels

throughout remeshing rounds. For PROPELLER, mesh quality is relatively preserved whereas error decreases along rounds. For ENGINE, the surface approximation is relatively well-preserved despite a number of points reduced by a factor three.

Finally, user-defined density field guided adaptation is shown in Fig. 10. Here, we can see that scales are finely captured (shock front and small flows), and a correct cell stretching is achieved.

Profiling For our benchmarks, we used three Intel-based machines: two NUMA dual-socket 32/48-core and a dual-memory 68-core processors (Table 1). The latter involves 4 VPU per core as well as on-chip MCDRAM at 300 GB/S and usual DDR4 at 68 GB/S. Here, trigen's code has been compiled through Intel compiler (version 17) with O3 and qopt-prefetch=5 optimization flags, including auto-vectorization and software prefetching capabilities. To enable specific features, we set march=native flag while compiling on HSW-SKL, and xmic-avx512 on KNL. Thread-core binding is done in a round-robin way by setting KMP_AFFINITY to scatter on normal mode (1 per core on HSW-SKL), and to compact on hyperthreading enabled (our default mode on KNL with 4 HT per core as recommended by Intel). Finally, two testcases were considered: (1) a curvature-based piecewise manifold isotropic adaptation of

Table 1 Benchmark machines features

Code	Sockets	NUMA	Cores	GHz	Threads	GB/core	Full reference
HSW	2	4	32	2.5	64	4.0	Xeon Haswell E5-2698 v3
SKL	2	2	48	2.7	96	7.9	Xeon Skylake Platinum 8168
KNL	1	4	72	1.4	288	1.5	Xeon-Phi Knights Landing 7250

- KNL: low frequency, low per thread memory ▷ 1.5 GHz, 375 MB per thread
- KNL: high memory-access latencies ▷ 30 ns for MCDRAM, 28 ns for DDR4.
- KNL: enable hyperthreading to hide latency. ▷ 4 threads per core.
- HSW and SKL: no hyperthreading to save GB per thread. ▷ 4 to 7.9 GB per thread.

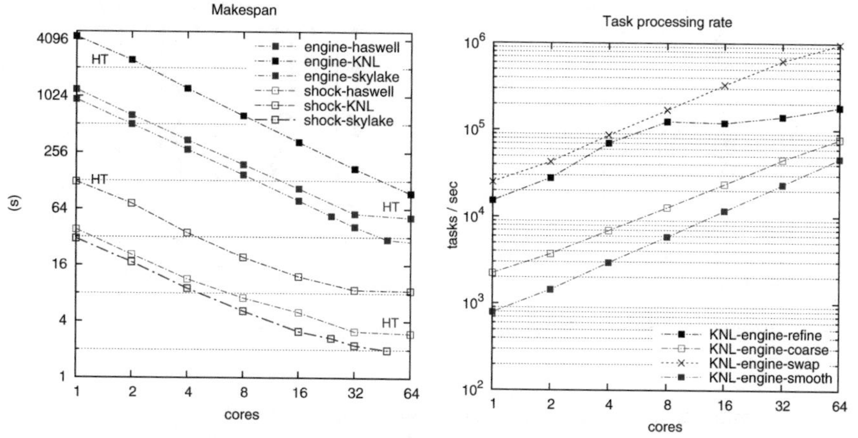

Fig. 11 Restitution time on all architectures and kernels processing rate on KNL. Note that a KNL core is more like a GPU core rather than a XEON one. It does not really make sense to compare single core performance of KNL with HSW or SKL

the engine model in [3] (engine) with 1,826,000 points and 3,652,058 cells, (2) the anisotropic hessian-based shock adaptation (shock) in Fig. 1 with 1,014,890 points, 2,029,772 cells and a target resolution $n = 250{,}000$ for the latter.

Scaling Strong scaling profile for four adaptive rounds on all architectures are given in Fig. 11, as well as kernel processing rate on KNL. A good scaling is achieved and a similar profile is observed on all architectures. However kernels are 3–4 times slower on KNL compared to HSW-SKL which is normal considering its per-core frequency and cache size limitations. The engine case is more CPU-expensive due to ridge-specific processing. A quasi-linear kernel processing rate scaling is achieved on KNL, except for refinement which requires no graph but involves more thread synchronization. Restitution time distribution per step on KNL is given in Fig. 12. Recall that in our context, each kernel is structured into bulk-synchronous sweeps. The time spent on each sweep is then represented here, and overheads related to parallelization are depicted in red (task extraction and synchronization involved in mesh topology updates). These overheads are negligible and scale roughly at

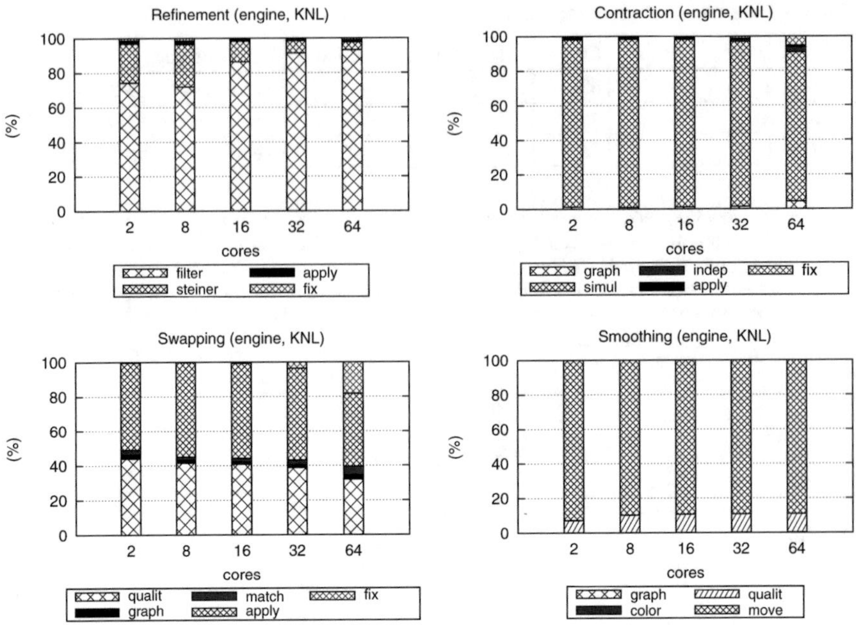

Fig. 12 Kernels time distribution per step on KNL; overheads are depicted in red

the same rate as other steps. However, an exceptional overhead is observed on relaxation kernel at 64 cores (256 threads). Indeed, memory indirections involved by this low arithmetic-intensity kernel induces high cache contentions and RAM access penalties. This behavior is not observed on larger per-core cache/RAM machines (SKL,HSW). Note that in HPC, we clearly advocate comparison of algorithms instead of software tools. Indeed, involved schemes are not comparable if they are not implemented exactly in the same way (with the same level of code optimization). Thus no performance comparison is included here.

References

1. J.-D. Boissonnat, K.-L. Shi, J. Tournois, M. Yvinec, Anisotropic delaunay meshes of surfaces. ACM Trans. Graph. **34**(2), 10 (2015)
2. A. Loseille, V. Menier, Serial and parallel mesh modification through a unique cavity-based primitive, in *IMR-22* (2014), pp. 541–558
3. B. Lévy, N. Bonneel, Variational anisotropic surface meshing with voronoi parallel linear enumeration, in *IMR-21* (2013), pp. 349–366
4. V. Surazhsky, C. Gotsman, Explicit surface remeshing, in *Eurographics*, SGP'03 (2003), pp. 20–30
5. V. Vidal, G. Lavoué, F. Dupont, Low budget and high fidelity relaxed 567-remeshing. Comput. Graph. **47**, 16–23 (2015)

6. H. Rakotoarivelo, F. Ledoux, F. Pommereau, Fine-grained parallel scheme for anisotropic mesh adaptation, in *IMR-25* (2016), pp. 123–135
7. H. Rakotoarivelo, F. Ledoux, F. Pommereau, N. Le-Goff, Scalable fine-grained metric-based remeshing algorithm for manycore-NUMA architectures, in *EuroPar'23* (2017), pp. 594–606
8. A. Vlachos, J. Peters, C. Boyd, J. Mitchell, Curved PN triangles, in *ACM I3D* (2001), pp. 159–166
9. D.J. Walton, D.S. Meek, A triangular G^1 patch from boundary curves. Comput. Aided Des. **28**(2) , 113–123 (1996)
10. G. Taubin, A signal processing approach to fair surface design, in *SIGGRAPH '95* (1995), pp. 351–358
11. Y. Ohtake, A. Belyaev, I. Bogaevski, Polyhedral surface smoothing with simultaneous mesh regularization, in *GMP* (2000), pp. 229–237
12. A. Kuprat, A. Khamayseh, D. George, L. Larkey, Volume conserving smoothing for piecewise linear curves, surfaces, and triple lines. J. Comput. Phys. **172**(1), 99–118 (2001)
13. X. Jiao, Volume and feature preservation in surface mesh optimization, in *IMR-15* (2006), pp. 359–373
14. D. Aubram, Optimization-based smoothing algorithm for triangle meshes over arbitrarily shaped domains. Technical report (2014)
15. S. Canann, J. Tristano, M. Staten, An approach to combined laplacian and optimization-based smoothing for triangular, quadrilateral, and quad-dominant meshes, in *IMR-7* (1998), pp. 479–494
16. L. Freitag, On combining laplacian and optimization-based mesh smoothing techniques. MeshTrends **220**, 37–43 (1999)
17. P. Wolfe, Convergence conditions for ascent methods. II: Some corrections. SIAM Rev. **13**(2), 185–188 (1971)
18. L. Valiant, A bridging-model for multicore computing. J. Comput. Syst. Sci. **77**, 154–166 (2011)
19. F. Ledoux, J.-C. Weill, Y. Bertrand, Gmds: a generic mesh data structure. Technical report, IMR-17 (2008)
20. M. Garland, P. Heckbert, Surface simplification using quadric error metrics, in *SIGGRAPH'97* (1997), pp. 209–216
21. P. Cignoni, C. Rocchini, R. Scopigno, Metro: measuring error on simplified surfaces. Technical report, CNRS (1996)
22. G. Taubin, Curve and surface smoothing without shrinkage, in *ICCV '95* (1995), pp. 852–857
23. J. Vollmer, R. Mencl, H. Müller. Improved laplacian smoothing of noisy surface meshes, in *Eurographics* (1999), pp. 131–138
24. M. Meyer, M. Desbrun, P. Schröder, A. Barr, Discrete differential-geometry operators for triangulated 2-manifolds, in *Visualization and Mathematics III* (Springer, Berlin, 2003), pp. 35–57
25. F. Alauzet, Size gradation control of anisotropic meshes. Finite Elem. Anal. Des. **46**(1), 181–202 (2010)

Parallel Performance Model for Vertex Repositioning Algorithms and Application to Mesh Partitioning

D. Benitez, J. M. Escobar, R. Montenegro, and E. Rodriguez

Abstract Many mesh optimization applications are based on vertex repositioning algorithms (VrPA). Since the time required for VrPA programs may be large and there is concurrency in processing mesh elements, parallelism has been used to improve performance. In this paper, we propose a performance model for parallel VrPA algorithms that are implemented on memory-distributed computers. This model is validated on two parallel computers and used in a quantitative analysis of performance scalability, load balancing and synchronization and communication overheads. We show that load imbalance and synchronization between boundary partitions are the major causes of the parallel bottlenecks. In order to diminish load imbalance, a new approach to mesh partitioning is proposed. This strategy reduces the imbalance in mesh element evaluations caused by multilevel k-way partitioning algorithms and consequently, improves the performance of parallel VrPA algorithms.

1 Introduction

There are several areas of research involving parallel processing of meshes. For example, many mesh processing techniques have been developed to generate meshes in parallel [5]. The sizes and shapes of generated elements affect the efficiency and accuracy of computational applications. Thus, other parallel algorithms are used for mesh optimization [8]. Additionally, parallel mesh warping algorithms have been developed which employ optimization methods for use in computational simulations with deforming domains [15].

A few performance models for parallel meshing algorithms have been developed. Such models can enable us to understand, fine-tune and predict the performance

D. Benitez (✉) · J. M. Escobar · R. Montenegro · E. Rodriguez
SIANI Institute and DIS Department, University of Las Palmas de Gran Canaria, Las Palmas, Spain
e-mail: domingo.benitez@ulpgc.es

X. Roca, A. Loseille (eds.), *27th International Meshing Roundtable*, Lecture Notes in Computational Science and Engineering 127,
https://doi.org/10.1007/978-3-030-13992-6_23

of applications. Barker and Chrisochoides applied an analytical model for load balancing to mesh generation asynchronous applications [1]. Sarje et al. used a performance model to propose a mesh partitioning that improves the load balancing of an ocean modeling code [16]. Mathis and Kerbyson presented a parametric model to predict the parallel performance of a partial differential equation solver on unstructured meshes [13].

Vertex repositioning algorithms (VrPA) have been adopted by a vast majority of mesh optimization applications [6–8, 17, 18], but no performance model for distributed-memory computers has been proposed yet. In another paper, we have proposed a performance model for sequential VrPA algorithms [2]. Using this model, we propose in this paper a performance model for loosely synchronous algorithms executed on distributed-memory computers (Sects. 4 and 5). The parallel model was applied to several VrPA algorithms for mesh untangling and smoothing and the results in prediction accuracy are shown in Sect. 6. Additionally, the parallel model is used in Sect. 7 to study the performance scalability, load balancing and synchronization and communication overheads of VrPA. Based on the parallel model and the results of its validation, a new approach to mesh partitioning that reduces load imbalance is proposed in Sect. 8. Before contributions are explained, the following two sections describe a generalized version of parallel VrPA algorithms and implementation details.

2 Generalized Parallel Algorithm

VrPA algorithms have been used to untangle a mesh and/or improve its quality by moving the free vertices [8, 17, 18]. They can be posed as numerical techniques in which the following parameters are considered [6]: objective function approach (A) and formulation (f), element quality metric (q), minimization method (NM) and convergence or termination criteria (TC). There are many choices for each free parameter one could make in a study of VrPA algorithms. In this paper, we limited the options to those shown in Table 1. Each combination of choices will be called *VrPA configuration* and denoted: $\langle A\rangle$-$\langle f\rangle$-$\langle q\rangle$-$\langle NM\rangle$-$\langle TC\rangle$, for instance, "Lo-D2-hS-CG-TC2".

Algorithm 1 shows a generalized VrPA for distributed-memory computers that is similar to others [8, 17]. The input is a set of nC files with information of vertices and elements of mesh partitions: P_i, $i \in \{1, \ldots, nC\}$. Every partition includes spatial coordinates of vertices and information of element edges. A partitioning tool is used to obtain these files from the file with information of a mesh, M.

The M mesh is frequently partitioned by assigning each vertex to one partition [8, 17]. In this way, the number of send and receive messages between parallel processes is minimized. Each partition is required to additionally include information of all vertices of elements where at least one vertex is assigned to that partition. The boundary of a partition is constituted by shared elements, each of them is formed by vertices assigned to that partition and at least to another partition.

Table 1 Free VrPA parameters and their choices that are considered in this paper

Parameter	Options		RW
Objective function approach (A):	**Gl**: All-vertex (K: Global function)	$n = N_M$: free elements[a] of mesh	[6]
$K = \sum_{i=1}^{n} f(q_i)$ n: total free elements[a]	**Lo**: Single-vertex (K: Local function)	$n = N_v$: free elements[a] of local patch	[6]
Objective function formulation: $f(q_i)$	**D1**: Distortion 1	$f(q_i) = q_i^{-1}$	[4]
q_i: quality of ith element	**D2**: Distortion 2	$f(q_i) = q_i^{-2}$	[4]
$q_{min} = min(q_i)_{i \in \{1...n\}}$	**log1**: Logarithmic barrier 1	$f(q_i) = n^{-1} q_{min}^{-1} - \mu\, log(\tau\, q_{min}^{-1} - q_i^{-1})$	[2]
$h(z) = \frac{1}{2}\left(z + \sqrt{z^2 + 4\delta^2}\right)$ $\delta, \mu, \tau =$ constants	**inv**: Regularized barrier	$f(q_i) = q_i^{-1} + \frac{1}{h\left(q_{min}^{-1} - q_i^{-1}\right)}$	[2]
Element quality metric: q_i	**hS**: Regularized mean ratio	$q_i = \frac{d\,[h(\sigma_i)]^{2/d}}{\Vert S_i \Vert_F^2}$	[7]
S_i: Jacobian matrix $\Vert\ \Vert_F$: Frobenius norm	$h(z) = \frac{1}{2}\left(z + \sqrt{z^2 + 4\delta^2}\right)$ $\sigma_i = determinant(S_i)$	$triangle : d = 2, s = 3$ $tetrahedron : d = 3, s = 6$ $a, b, \delta, \lambda = constants$	
Numerical minimization method (NM)	**CG**: Conjugate Gradient	Polack-Ribiere, analytical derivatives	[4]
	SD: Steepest Descent	Analytical derivatives	[4]
Termination criteria (TC) Q_i: mean-ratio quality value of the ith element $Q_{min} = min(Q_i)_{i \in \{1...N_M\}}$	**TC2** (optimum mesh) $\overline{Q}$: average mean-ratio value of mesh Δ: maximum variation between outer iterations	$true = (\ Q_{min} > 0\ and$ $\Delta\overline{Q} < 10^{-3}\ and$ $\Delta Q_{min} < 10^{-3}\ )$	

RW denotes related work
[a]*Free element*: mesh element with at least one free vertex

In each mesh partition, vertices are classified as interior, non-ghost boundary (or simply, boundary), ghost or fixed. Interior vertices form elements whose all vertices belong to that partition. Boundary vertices form shared elements where at least one vertex belongs to another partition. Ghost vertices are these vertices that belong to other partitions. Thus, ghost vertices are replicated in shared partitions.

Each partition is assigned to a different parallel process that optimizes interior and boundary vertices but not ghost vertices. The numerical processing is divided into three parallel phases. The first phase is implemented in lines 26–32. It is used only once to prepare the processing of vertices laying on the partition boundaries in phase 3.

When a boundary vertex is being repositioned in phase 3, the numerical method needs the coordinates of all connected vertices that should remain fixed. Computational dependency appears between the adjacent boundary and ghost vertices

Algorithm 1 Parallel mesh vertex repositioning algorithm

$\triangleright$ Input: files with information of P_i partitions, $P_i \leftarrow$ *Partition(M)*, $i \in \{1, \ldots, nC\}$
#define: approach (A), formulation (f), quality metric (q), numerical minimization method (NM)
#define termination criteria: $TC = LogicFunction(Q_{min}, \Delta\overline{Q}, \Delta Q_{min})$
#define constants: $\tau = 10^{-6}$ (maximum or minimum increase of the objective function), $N_{mII} = 150$ (maximum number of inner iterations), $N_{mOI} = 100$ (maximum number of outer iterations)
$N_{e,i} \leftarrow 0$ $\triangleright$ element evaluations for partitions $P_i, i \in \{1, \ldots, nC\}$
procedure VERTEXREPOSITIONING(W, X, n)
$\triangleright Inputs:$ W(free vertices),X(their coordinates),n(number of elements)
$\triangleright$ $Initiation:$ $K = 0,\ \Delta K = 0,\ m = 0$ (inner loop index)
while ($\Delta K \leq \tau$(minimizing) or $\Delta K \geq \tau$(maximizing)) and $m \leq N_{mII}$ **do** $\triangleright$ Inner loop
$\hat{X} \leftarrow X$ $\triangleright$ Returned spatial coordinates ($\hat{X}$) of vertices (W)
$\triangleright$ $Initiation:$ $P \leftarrow 0$ $\triangleright$ Moving directions: $P = \{p_v\}, v \in W$
for $i = 1, \ldots, n$ **do** $\triangleright$ n: number of free elements
for *each free vertex* v *of* i^{th} *free element* **do**
p_v += NM($f'(q_i)$,v) $\triangleright$ f': derivatives used in NM
$N_e += 1$ $\triangleright$ Number of mesh element evaluations
$X \leftarrow \hat{X} + P$ $\triangleright$ Tentative positions of free vertices
$K_t \leftarrow 0$ $\triangleright$ Initial value of objective function
for $i = 1, \ldots, n$ **do**
K_t += $f(q_i)$ $\triangleright$ MESH ELEMENT EVALUATION
$N_e += 1$ $\triangleright$ Number of mesh element evaluations
$\Delta K \leftarrow K_t - K$
$K \leftarrow K_t$ $\triangleright$ Final value of objective function
$m += 1$ $\triangleright$ Number of inner iterations
return $\hat{X}$ $\triangleright$ Output: updated coordinates of free vertices
procedure MAIN( )
for $P_i \in M$ **in parallel do** $\triangleright$ Parallel phase 1: begin
$\triangleright$ *Read the vertex and element information of* P_i *partition*
$I_i \leftarrow$ BOUNDARYCOLORING(P_i) $\triangleright$ $I_i = \{I_{ij}\}_{j \in \{1..nF_i\}, i \in \{1..nC\}}$
`MPI_Send-MPI_Receive` information of boundary/ghost vertices
$\triangleright$ *Store the order of partition free boundary/ghost vertices*
$Q_{min,i} \leftarrow$ GLOBALMEASURES(P_i) $\triangleright$ Initial partition quality
Synchronization `MPI_Allreduce` $\triangleright$ Parallel phase 1: end
for $P_i \in M$ **in parallel do**
$\triangleright$ $Initiation:$ $\Delta\overline{Q_i} = 10^6,\ \Delta Q_{min,i} = 10^6,\ k_i = 0$ (loop index)
while $TC \neq true$ *and* $k_i \leq N_{mOI}$ **do** $\triangleright$ Mesh/Outer loop
if $A = Gl$ **then** $\triangleright$ Par. pha. 2 - Interior processing: begin
$X_V \leftarrow$ VERTEXREPOSITIONING(V, X_V, N_M)
else $\triangleright$ $A = Lo$ (single-vertex)
for *each free interior vertex* $v \in P_i$ **do**
$x_v \leftarrow$ VERTEXREPOSITIONING(v, x_v, N_v)
Synchronization `MPI_Barrier` $\triangleright$ Parallel phase 2: end
for *each boundary independent-set* $I_{ij} \in P_i$ **do** $\triangleright$ Parallel phase 3: begin
for *each free boundary vertex of partition* $v \in I_{ij}$ **do**
$x_v \leftarrow$ VERTEXREPOSITIONING(v, x_v, N_v)
`MPI_Send-MPI_Receive` *coordinates of vertices* x_v
Synchronization `MPI_Barrier` $\triangleright$ All boundary vertices
$(Q_{min,i}, \Delta\overline{Q_i}, \Delta Q_{min,i}) \leftarrow$ GLOBALMEASURES(P_i)
$k_i += 1$ $\triangleright$ Number of mesh/outer iterations
Synchronization `MPI_Allreduce` $\triangleright$ Parallel phase 3: end
`MPI_Send-MPI_Receive` $N_{e,i}$
$\triangleright$ Output: files with information of optimized P_i partitions

because one vertex begins to be processed after another has been repositioned. Thus, vertices of shared elements cannot be optimized in parallel.

BoundaryColoring divides the boundary of a partition (P_i) into nF independent sets (I_{ij}), also called *colors* (line 28) [3]. After that, the order of processing and interchange of boundary and ghost vertices is established. The resulting orderings are interchanged among shared partitions (line 29).

Finally, a list with the order of boundary and ghost vertices is created in line 30. This list determines the order in which these vertices are optimized in the parallel process, or received from other processes in phase 3. Interior vertices do not need to be reordered because all adjacent vertices are assigned to the same process. Using the function `MPI_Allreduce` at the end of phase 1, a synchronization barrier ensures that all partitions have completed these steps before continuing computation (line 32).

Mainly, the algorithm spends most of the time in a variable number of concurrent partition sweeps (lines 35–50). In each $partition\ sweep$, also called $outer$ or $mesh\ iteration$, all interior and boundary vertices are optimized separately in parallel phases 2 (lines 36...41) and 3 (lines 42...49), respectively, adjusting the spatial coordinates of all free vertices (X_V). The vertices that lie on the mesh surface are treated as fixed and are not updated.

Interior vertices of every partition are not dependent on vertices of other partitions and are sequentially optimized by the same parallel process. In this way, the interior vertices of all partitions are optimized concurrently. A single synchronization phase between partitions is established to ensure that all interior vertices are completely repositioned (line 41).

For partition boundaries, some independent sets of free vertices from different partitions (I_{ij}) are optimized in parallel. After an independent set has been optimized, the interchange of updated coordinates is implemented using message send/receive functions (line 45). These computation-synchronization-communication phases are repeated until all boundary vertices have been optimized (lines 42...45).

When all boundary vertices have been updated (line 46), the minimum quality metrics of all partitions are calculated and distributed (line 47...49) and the partition sweep finishes. $GlobalMeasures$ provides the average and minimum mean-ratio quality metric of the mesh [6]. At this moment, a new partition sweep may begin if convergence conditions are not met by every partition (line 35). $LogicFunction$ uses termination criteria (TC) to stop the algorithm. The *outer loop* is iterated in $Main$ procedure while $LogicFunction$ is not true. After a variable number of partition sweeps, the output of our parallel algorithm provides optimized mesh partitions (line 51).

The most time-consuming operation is a sequential procedure called $VertexRepositioning$ (lines 37, 40, 44), which moves free vertices (V) of an input mesh partition (P_i). Thus, the parallel performance is based on the performance of the underlying serial numerical method. $VertexRepositioning$ iterates an *inner loop* while an extreme of the objective function (K) is being reached by using an optimization solver (NM). K is constructed with A, f and q as it is shown in Table 1.

3 Experimental Setup

Software Framework We developed programs for simultaneous mesh untangling and smoothing that implement Algorithm 1. They include double-precision floating-point data structures and functions from *MPI* and the *Mesquite* C++ library [4], which is specialized in mesh smoothing. *Mesquite* was extended to support **hS** quality metric, **log1** and **inv** objective function formulations and **TC2** termination criteria (see Table 1). TC2 is met when the output mesh has no inverted elements and is optimum, i.e., the minimum and average values of the mean-ratio quality metric in two successive mesh iterations do not change significantly. We used *OpenMPI* 1.6.5 and *gcc* 4.8.4 with `-O2` flag on *Linux* systems. A pure sequential version was selected for baseline runs. For each VrPA configuration, we repeated the execution of the sequential and parallel programs several times, such that the 95% confidence interval was lower than 1%.

Benchmark Meshes Algorithm 1 was applied on the unstructured, fixed-sized meshes shown in Fig. 1 whose characteristics are in Table 2. The 2D mesh was obtained by using *Gmsh* tool [9], taking a square, meshing with triangles and displacing selected nodes of the boundary. This type of tangled mesh can be found in some problems with evolving domains [12]. All 3D meshes were obtained from a tool for adaptive tetrahedral mesh generation that tangles the mesh [14]. All the mesh sizes were always fixed, and we used *Metis* 5.1.0 for mesh partitioning [11].

Platforms Numerical experiments were conducted on two cluster computers called *Cluster1* and *Cluster2* that are in two different locations. *Cluster1* is a Bull computer with 28 compute nodes that are organized in 7 BullxR424E2 servers. They are interconnected with Infiniband QDR 4X (32 Gbit/s). Each node integrates two Intel Xeon E5645 (6 cores each, 2.4 GHz), and 48 GB of DDR3/1333. So, up to 336 cores were used in parallel. The storage system is a RAID-5 disk array consisting of 7200 RPM SATA2 disk drives. All compute nodes share a common file system through

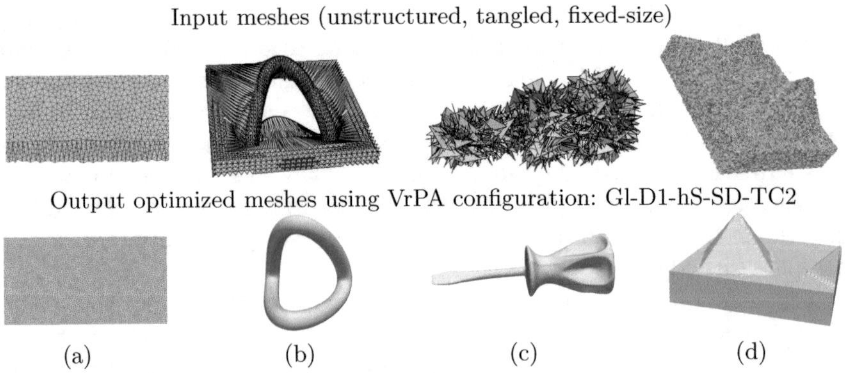

Fig. 1 Input and output meshes for four optimization problems solved with the same VrPA algorithm. (**a**) Square (2D). (**b**) Toroid (3D). (**c**) Screwdr (3D). (**d**) Egypt (3D)

Table 2 Characteristics of input meshes. All meshes have inverted elements: Q_{min}=0

Mesh characteristic	Square	Toroid	Screwdriver	Egypt
Total vertices	3,314,499	9176	39,617	1,724,456
Free vertices (they can be moved)	3,309,498	3992	21,131	1,616,442
Fixed vertices (they are not moved)	5001	5184	18486	108014
Element type: triangle (2D), tetrahedron (3D)	2D	3D	3D	3D
Total free elements (N_M)	6620936	35920	168834	10013858
Inverted/Tangled elements (%)	0.1%	38.2%	49.4%	46.2%
Average mean-ratio quality metric ($\overline{Q}$)	0.95	0.17	0.13	0.23
Standard deviation of the mean-ratio metric	0.05	0.31	0.21	0.27

NFS over a gigabit Ethernet LAN. *Cluster2* is a Fujitsu computer that has the same type of network, storage and file system as *Cluster1* but only four compute nodes (Primergy CX250) with 16 E5-2670@2.6GHz cores and 32 GB of DDR3/1600 per node. We activated multiples of 12 or 16 cores to completely occupy the compute nodes. During the experiments, the compute nodes were not shared among other user-level workloads. Additionally, multithreading and Turbo Boost were disabled.

4 Sequential Performance Model

In another paper, we have proposed a performance model for sequential VrPA algorithms that tries to justify their execution times [2],

$$t_{CPU}^{Smodel} = \alpha\ N_e \tag{1}$$

with t_{CPU}^{Smodel} the execution time, N_e the number of mesh element evaluations and α the model parameter that represents the time per element evaluation. Equation 1 assumes that computation time is much larger than total input/output time.

N_e takes into account multiple evaluations of an element quality metric and its derivative (see lines 15 and 20 in Algorithm 1). Although not exactly equal, N_e is similar to the *concurrent function evaluation steps* defined in [19]. We have demonstrated that the execution time of VrPA algorithms is more proportional to N_e than other workload measures such as mesh size or objective function evaluations (see [2] for a more extended discussion). It is important to note that N_e depends not only on the problem size but also on the number of inner and outer iterations required to meet the convergence criteria. However, N_e is independent of the computer hardware; it depends on the algorithm and its implementation, the selected numerical accuracy of data structures, and the method chosen by the compiler to implement arithmetic operations. We use the number of element evaluations as workload measure in the new parallel performance model that is presented below.

The other factor, the time per element evaluation (α), is more affected by the processor and objective function approach than the objective function formulation, the quality metric, numerical solver, convergence criteria or mesh [2]. Thus, to precisely determine execution times with Eq. 1, α must be recalculated when the processor and algorithm configuration change.

In summary, this simple performance model determines that the time required by sequential VrPA algorithms to optimize a mesh is directly proportional to the number of element evaluations.

5 Parallel Performance Model

In this section, we describe a new model to justify the parallel runtimes of Algorithm 1 for a selected VrPA configuration on a determined distributed-memory computer. This model uses the time per mesh element evaluation (α), which is obtained from the sequential execution, denoted *Sreal*, of the same configuration using a pure sequential version of Algorithm 1. Then, selecting one VrPA configuration and employing Eq. 1,

$$\alpha = \frac{t_{CPU}^{Sreal}}{N_e} \tag{2}$$

We assume that this model parameter is constant for all parallel experiments that use the same VrPA configuration, mesh and cluster computer.

Since there is an MPI barrier between the repositioning of interior and partition boundary vertices (line 41), Eq. 3 models the parallel execution time that is divided into two components, one for optimizing interior vertices and the other for partition boundary vertices,

$$t_{CPU}^{Pmodel} = t_{interior}^{Pmodel} + t_{boundary}^{Pmodel} \tag{3}$$

In this case, we have assumed that the execution time for the mesh partitioning phase and parallel phase 1 are negligible with respect to parallel phases 2 and 3. The parallel time for interior vertices is expressed as a sum,

$$t_{interior}^{Pmodel} = t_{scalable,interior}^{Pmodel} + t_{imbalance,interior}^{Pmodel} \tag{4}$$

where the first term denotes the *scalable interior parallelism*. If the workload was evenly distributed among nC partitions, the total workload for optimizing interior vertices in all partitions ($N_{e,interior}^{Pmodel}$) would be divided by nC,

$$t_{scalable,interior}^{Pmodel} = \alpha \frac{N_{e,interior}^{Pmodel}}{nC} \tag{5}$$

The second component of Eq. 4, called *interior imbalance*, measures the additional time required by the most loaded partition when the workload for processing interior vertices is not evenly distributed. It is given by Eq. 6, where $N_{e,interior,max}^{Pmodel}$ is the maximum number of interior element evaluations of a partition,

$$t_{imbalance,interior}^{Pmodel} = \alpha \left(N_{e,interior,max}^{Pmodel} - \frac{N_{e,interior}^{Pmodel}}{nC} \right) \quad (6)$$

The values of $N_{e,interior}^{Pmodel}$ and $N_{e,interior,max}^{Pmodel}$ for Eqs. 5 and 6 are measured at the end of the parallel execution. The time needed to optimize all partition boundary vertices has four terms,

$$t_{boundary}^{Pmodel} = t_{scalable,boundary}^{Pmodel} + t_{imbalance,boundary}^{Pmodel} + t_{synchro,boundary}^{Pmodel} + t_{comm,boundary}^{Pmodel} \quad (7)$$

Scalable boundary parallelism (Eq. 8) assumes that the workload of boundary vertices ($N_{e,boundary}^{Pmodel}$) are evenly distributed among nC partitions.

$$t_{scalable,boundary}^{Pmodel} = \alpha \frac{N_{e,boundary}^{Pmodel}}{nC} \quad (8)$$

Boundary imbalance (Eq. 9) measures the additional time needed by the most loaded partitions when workloads of the nF independent sets are not evenly distributed,

$$t_{imbalance,boundary}^{Pmodel} = \alpha \left(\sum_{j=1}^{nF} N_{e,boundary,j,max}^{Pmodel} - \frac{N_{e,boundary}^{Pmodel}}{nC} \right) \quad (9)$$

where $N_{e,boundary,j,max}^{Pmodel}$ is the maximum number of element evaluations of the jth independent-set of a partition. $N_{e,boundary}^{Pmodel}$ and the accumulated value ($\sum N_{e,boundary,j,max}^{Pmodel}$) in Eqs. 8 and 9 are obtained at the end of parallel execution for a given number of partitions (nC).

Synchronization (Eq. 10) assumes that all partitions cannot optimize boundary vertices concurrently in phase 3. It is due to the vertex dependence imposed by the processing and interchange order of boundary and ghost vertices that is determined in phase 1. In this term of the model, the scalable workload of boundary processing is factored with $(nC - nC')/nC'$, where nC' is the number of partitions that actually are optimizing vertices concurrently in phase 3 ($0 < nC' \leq nC$). As fewer opportunities for parallelism are available in boundary phase, nC' will reduce and the modeled effect causes an increase in execution time. nC' is obtained by averaging the number of partitions that finish a vertex reposition between another partition terminates two consecutive repositioning of boundary vertices.

$$t_{synchro,boundary}^{Pmodel} = \frac{\alpha\, N_{e,boundary}^{Pmodel}}{nC} \frac{nC - nC'}{nC'} \quad (10)$$

Equation 11 measures the MPI communication overhead of boundary processing using a two-parameter model for SMP nodes working in the short regime [10]. This equation has two terms times the number of outer iterations (k). The first term represents the *communication latency*, which is modeled as the network latency (LAT) times the number of data block communications ($2\ \beta\ nF$) during a single outer iteration. β denotes the total number of edges of a new graph that represents which partitions share boundary elements. An edge represents the boundary between two partitions. Neighboring partitions are represented by adjacent vertices. An MPI communication is performed through each edge of this graph after processing an independent set. nF denotes the average number of independent sets per partition that is obtained in phase 1.

$$t_{comm,boundary}^{Pmodel} = k \left(LAT\ 2\ \beta\ nF\ + \frac{32\ (\gamma - 1)\ nV_{boundary}}{BW} \right) \tag{11}$$

The other term is *data transmission* time, where BW denotes the data rate that each process can achieve in sending or receiving a message. The effective rate is dependent on transmitted data size. However, we assume this parameter is constant because the variability of message sizes is small and computing time significantly exceeds communication time. Each vertex has a data size of 32 bytes to send/receive spatial coordinates and global ID. γ denotes the average number of partitions that share the same vertex, and $nV_{boundary}$ the total number of free boundary vertices of all partitions. Using code instrumentation, we measured LAT and BW in each MPI process and their average values were used to determine the parameters in our performance model. The rest of parameters, β, γ and $nV_{boundary}$, are obtained from the partitions of the input mesh at the end of partitioning phase.

6 Validation of the Parallel Model

Algorithm 1 was applied to four mesh optimization problems using different VrPA configurations but fixing the TC2 convergence criteria. TC2 is met when the mesh is untangled and smoothed. In order to demonstrate the applicability of our parallel model to a variety of VrPA configurations, a different one was selected for each mesh. The values used for the α model parameter were calculated with Eq. 2 and are shown in Table 3.

Figures 2 and 3 depict results that were obtained from *Clusters 1* and *2*, respectively. The resulting execution times (t_{CPU}^{Preal}) are compared to the predictions of our parallel model (t_{CPU}^{Pmodel}). In these tests, the numbers of partitions, MPI processes and CPU cores had the same value. We include results obtained using partitions that activated all cores of different subsets of compute nodes. Thus, each bar diagram shows execution times for numbers of cores that are multiple of 12 (*Cluster1*) or 16 (*Cluster2*). For each mesh optimization problem, note in Table 3 that the minimum qualities of output meshes are similar.

Table 3 Performance of the baseline sequential experiments and average values of the minimum mean-ratio quality metric ($\overline{Q_{min}}$) of the output meshes of parallel experiments whose results are shown in Figs. 2 and 3

		Sequential experiments				Parallel experiments	
		CPU time [sec]		α [$\mu sec/element$]		$\overline{Q_{min}}$	
Mesh	VrPA configuration	Cluster1	Cluster2	Cluster1	Cluster2	Cluster1	Cluster2
Square	Lo-D2-hS-SD-TC2	5.8×10^4	5×10^4	0.5	0.4	0.633 ± 0.001	0.633 ± 0.001
Toroid	Gl-inv-hS-SD-TC2	19.5	11.0	1.1	0.7	0.333 ± 0.094	0.318 ± 0.108
Screwdriver	Gl-log1-hS-CG-TC2	1.6×10^3	1.0×10^3	0.8	0.6	0.255 ± 0.002	0.256 ± 0.001
Egypt	Lo-D1-hS-SD-TC2	1.1×10^4	9.0×10^3	0.5	0.4	0.201 ± 0.002	0.202 ± 0.002

α denotes the time per mesh element evaluation that was employed when the parallel model was applied

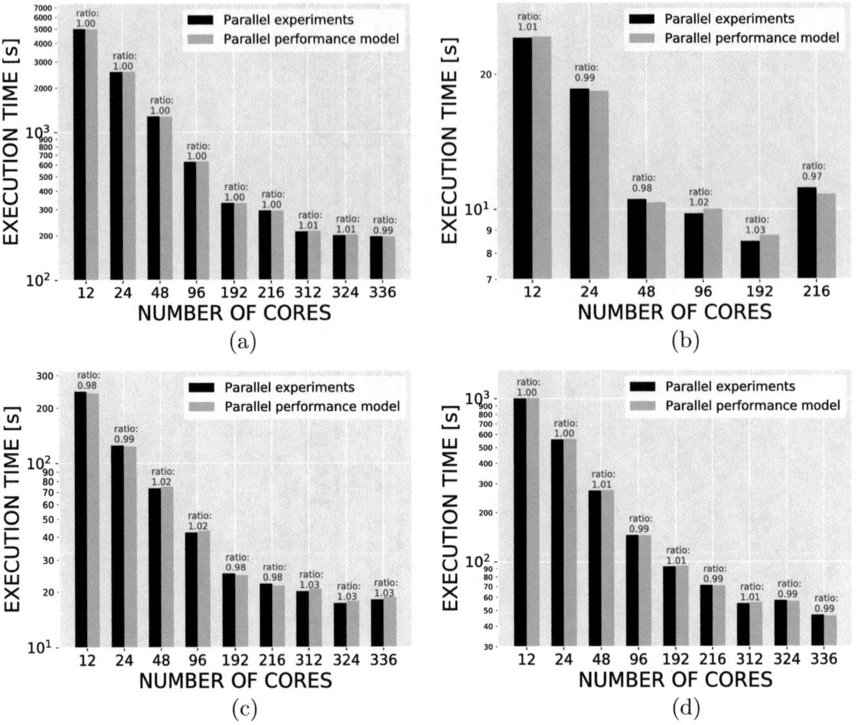

Fig. 2 Results of the parallel experiments and performance model using *Cluster1*. The ratio of the performance model to experiment execution time is shown in blue. Table 3 shows the VrPA configurations that were applied and the average values of the minimum mean-ratio quality metric of output meshes. Performance of the sequential executions of respective configurations is also shown in Table 3. (**a**) Square. (**b**) Toroid. (**c**) Screwdriver. (**d**) Egypt

On average, the mean relative errors of our parallel model in the estimation of the times obtained from *Cluster1* and *Cluster2* were 0.027 and 0.031, respectively. This discrepancy can be explained by the inaccuracy introduced when nC' and $\sum_j N^{Pmodel}_{e,boundary,j,max}$ were obtained. Another source of inaccuracy is introduced by α that may be slightly different between parallel and sequential processing.

7 Parallel Performance Analysis

Figure 4 shows stacked column graphs for the times provided by our parallel model when *Cluster1* was used to run Algorithm 1. Every single column corresponds to a determined VrPA configuration, benchmark mesh and number of partitions. It is divided into six sections, which are grouped into four categories: *scalable parallelism*, *imbalance*, *synchronization* and *communication*.

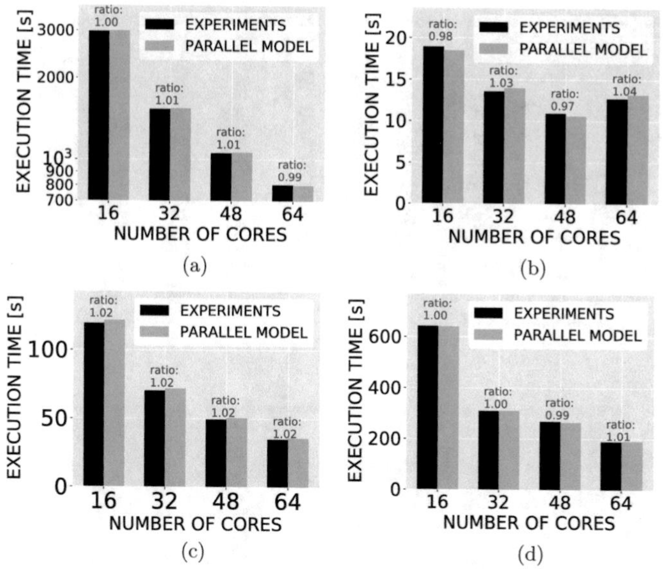

Fig. 3 Results of the parallel experiments and performance model using *Cluster2*. The ratio of the performance model to experiment execution time is shown in blue. Table 3 shows the VrPA configurations that were applied and the average values of the minimum mean-ratio quality metric of output meshes. Performance of the sequential executions of respective configurations is also shown in Table 3. (**a**) Square. (**b**) Toroid. (**c**) Screwdriver. (**d**) Egypt

Scalable parallelism includes runtimes for optimizing interior and boundary vertices if the sequential workloads were evenly distributed over all mesh partitions (Eqs. 5 and 8). These times are represented in Fig. 4 by the two bottom columns denoted as *Inter-Scaling* and *Boun-Scaling*, respectively.

As the number of partitions increases in strong scaling when a mesh is optimized, the time devoted to this category reduces. Equations 5 and 8 predict that it is due to the fact that we are solving fixed-size problems and the element evaluations in each partition reduce. Note in Fig. 4 that the fraction of time in scalable parallelism also reduces, which means that overheads are more relevant. However, the fraction of time in boundary optimization tends to increase because the ratio of boundary to interior element evaluations increases when the number of partitions increases.

Another increasing trend is observed when problems of different sizes are compared for a given number of cores. For example, using 324 cores in *Cluster1*, note that the fraction of time in scalable parallelism is 25% for Screwdriver mesh, 68% for Egypt mesh and 78% for the Square mesh. Since for 324 cores Screwdriver requires fewer element evaluations than Egypt and Egypt fewer element evaluations than Square ($1.8 \times 10^9, 2.5 \times 10^{10}, 1.1 \times 10^{11}$, respectively), the workload distributed among partitions is lower for Screwdriver than for Egypt, which is lower than for Square. Thus, VrPA algorithms cannot compensate for the parallel overheads when Screwdriver is optimized as much as when Egypt or Square are optimized. In general, this performance category is associated with parallel efficiency, which

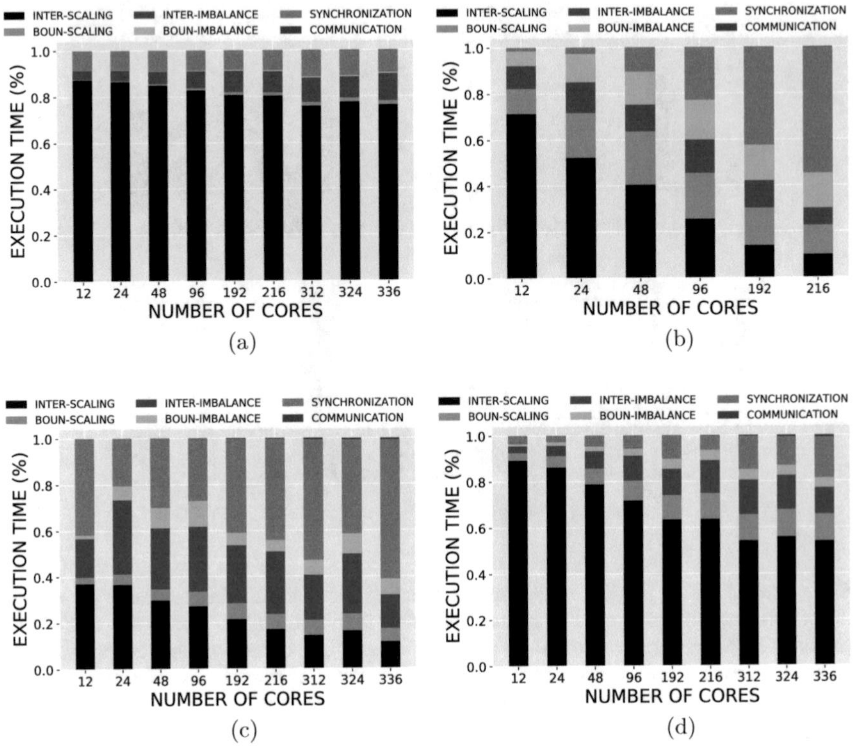

Fig. 4 Time breakdowns provided by the parallel performance model for *Cluster*1. (**a**) Mesh: Square, VrPA configuration: Lo-D2-hS-SD-TC2. (**b**) Mesh: Toroid, VrPA configuration: Gl-inv-hS-SD-TC2. (**c**) Mesh: Screwdriver, VrPA configuration: Gl-log1-hS-CG-TC2. (**d**) Mesh: Egypt, VrPA configuration: Lo-D1-hS-SD-TC2

depends mainly on the fraction of time occupied by mesh element evaluations perfectly balanced.

Load imbalance (Eqs. 6 and 9) is another category that includes the execution times due to processor overload during vertex repositioning when the element evaluations are not well balanced (see *Inter-Imbalance* and *Boun-Imbalance* in Fig. 4). Although the load imbalance cost decreases as the number of partitions for a given problem increase because the elements evaluations per partition decrease, its percentage relevance tends to be larger. It is due to the less homogeneous distribution of workload that is assigned by the mesh partitioning tool. Note that this tool distributes vertices and elements among partitions but it does not know in advance how many mesh elements evaluations will be completed. For our largest problems, Square and Egypt, this overhead category is the major cause of the parallel bottlenecks. For example, using 324 cores in *Cluster1*, load imbalance is responsible for 11% and 19% of the total runtime, respectively.

Synchronization (Eq. 10) includes the overheads caused by the independent sets of partition boundary vertices that have to be processed in the order determined in parallel phase 1. Note in Fig. 4 that, as the number of partitions (nC) increases, the percentage relevance of this category tends to be larger in all of our optimization problems. It is due to the number of processes that concurrently reposition partition boundary vertices (nC'), which increases less than the number of partitions. This percentage relevance is also affected by the increasing ratio of boundary to interior element evaluations, which is larger as described above when the number of partitions increases.

Another effect of synchronization overhead can be observed when problems of different sizes are compared for a given number of partitions. Taking any number of partitions, the percentage relevance of this category is larger for Toroid and Screwdriver than Square and Egypt. It is due to that nC' tends to reduce when the number of boundary element evaluations reduces. So, the factor of our model $(nC - nC')/nC'$ is larger. The modeled effect is concordant with fewer opportunities for parallelism when the concurrent boundary element evaluations reduce. Moreover, although the number of boundary elements evaluations is smaller in Toroid and Screwdriver than Square and Egypt for a given number of partitions, the ratio of boundary to interior element evaluations is larger in Toroid and Screwdriver than Square and Egypt. Thus, the percentage relevance of synchronization is also larger.

Communication is a category that considers the overhead caused by the transmission of updated coordinates of partition boundary vertices (Eq. 11). This overhead increases with the number of partitions because it depends on the numbers of boundary vertices and independent sets. However, its percentage relevance is the lowest, from 1% to 2% when 336 cores are used (see Fig. 4). Therefore, VrPA algorithms do not suffer significantly from the MPI communication overhead in our experiments. This is due to the dependence of communication time on the numbers of partition boundary vertices and independent sets, in contrast to the optimization time of boundary vertices and other overhead categories that are dependent on the concurrent mesh element evaluations.

8 Application to Load Balancing

Mesh optimization algorithms for distributed-memory computers use a previous phase of mesh partitioning to balance and distribute vertices among parallel processes [17]. As stated in Sect. 3, we used a Metis program (`gpmetis`) based on the multilevel k-way graph partitioning algorithm with edge cut minimization [11]. This program requires as input a file storing a mesh. Part of this file contains information relevant for vertices.

The results of previous section show that load imbalance is a significant parallel overhead. To reduce this overhead, we propose to include in the input file of the partitioning program a weight associated with each vertex. This weight coincides with the number of element evaluations that are needed by the vertex in a previous outer iteration of the parallel execution of the VrPA algorithm. Thus, the first outer iteration is repeated twice, one for weight calculation and the other for mesh

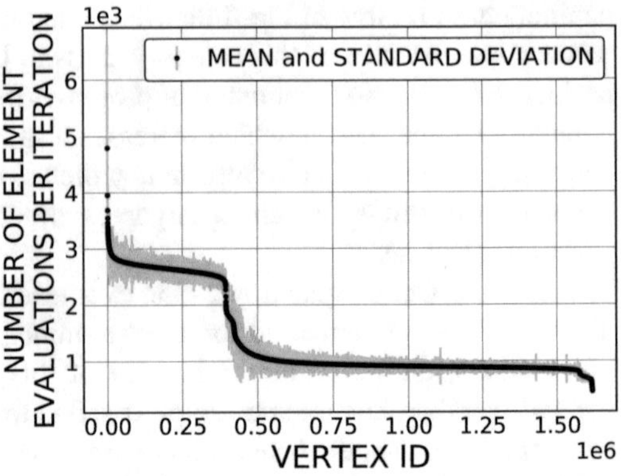

Fig. 5 Element evaluations per vertex and mesh iteration of a VrPA algorithm (configuration: Lo-D1-hS-SD-TC2) when it was employed to untangle and smooth the Egypt mesh using one compute node of *Cluster1* and 16 mesh partitions. Mesh vertices are sorted by element evaluations from largest to smallest

optimization. Without vertex weights, the partitioning program balances vertices. With our proposal, this program balances element evaluations, e.g., the sum of evaluations of the vertices assigned to each parallel process is approximately the same across the partitions.

8.1 Hypothesis

Figure 5 shows the mesh element evaluations that were needed on average in every outer iteration by each free vertex of Egypt mesh when Lo-D1-hS-SD-TC2 configuration was used. Note that vertices are sorted by element evaluations from largest to smallest. This figure shows that there is a large range of workloads per vertex (black line). Given that the optimization of a free vertex in the parallel algorithm is a serial process and using Eq. 1 with constant α, this workload variability means that each vertex requires a runtime that can range in a large interval.

Equations 5 and 8 show that the main workload of a partition (P_i) is due to the element evaluations of all assigned vertices. Equations 6 and 9 show that load imbalance is caused by the difference in element evaluations between the most loaded partition and the average partition. Thus, we might expect that load balancing would improve when mesh partitioning uses the sum of workloads assigned to partitions rather than the sum of vertices.

8.2 Mesh Partitioning by Workload Decomposition

Our hypothesis was tested in a new experiment by comparing the performance of parallel VrPA algorithms that use meshes partitioned both with and without workload information. The new proposal of mesh partitioning needs to know the workload of every vertex. This information cannot be obtained from the input mesh.

Thus, a previous mesh iteration of the parallel optimization is used to derive the weight that represents the workload of a vertex. For this previous stage, the mesh is partitioned without workload information using `gpmetis` program. The second stage of our method consists in repartitioning the input mesh, including the weights.

8.3 Results

The reduction in load imbalance is significant as can be seen in Fig. 6. This figure shows both the maximum and minimum numbers of element evaluations per mesh partition normalized to the mean number of evaluations. Consequently, the execution times decreased in this parallel experiment. Our proposal achieved average speedups of 1.28X and 1.13X when Screwdriver and Egypt meshes were optimized, respectively. The extra times of both the previous mesh iteration and another mesh partitioning were added to the evaluation of our proposal.

Performance improvement is not as high for Egypt mesh as it is for Screwdriver mesh because the significance of load imbalance is smaller (see Fig. 4). Thus, our mesh repartitioning strategy achieves larger performance improvement when the relevance of load imbalance is greater.

Since the main cost of our proposal is the previous optimization stage, larger benefits will be achieved when its execution time is much smaller than the total time to convergence. A circumstance where our proposal has a beneficial effect on performance occurs when the variance of the number of elements evaluations per vertex in successive mesh iterations is low (see Fig. 5). In this case, a single mesh iteration is sufficient to derive the relative workloads of vertices that are valid for the rest of the iterations.

9 Conclusions

We have proposed a performance model for vertex repositioning algorithms on distributed-memory computers. This model is based on a workload measure called *number of mesh element evaluations*. The parallel model has been shown to be accurate with low average errors across a range of configurations in terms of the number of parallel processes, processor microarchitecture, mesh geometry, and algorithm configuration utilized. Furthermore, the parallel model was used to quantitatively understand the performance scalability, load balancing and synchronization and communication overheads. The results in this paper have shown that imbalance in the number of element evaluations and synchronization between boundary partitions are the major causes of the parallel bottlenecks. Finally, we have proposed a new approach to mesh partitioning that uses the number of mesh element evaluations to distribute vertices among parallel processes. This mesh partitioning strategy has been shown to reduce the imbalance in element evaluations caused

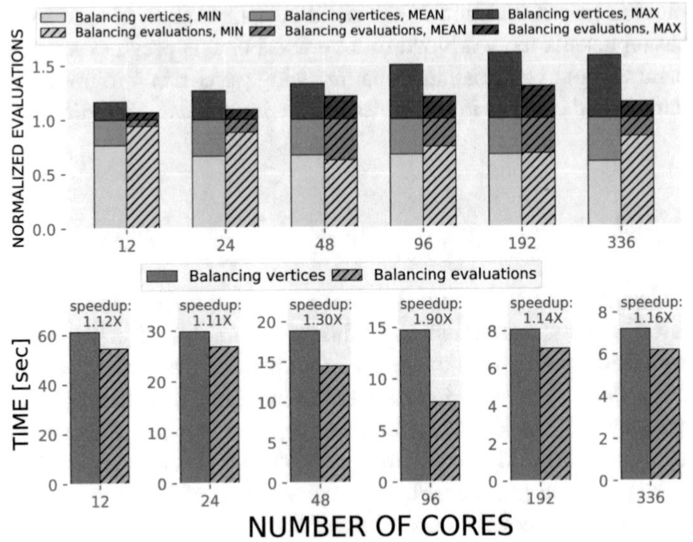

Mesh: Screwdriver. VrPA configuration: Lo-D1-hS-CG-TC2

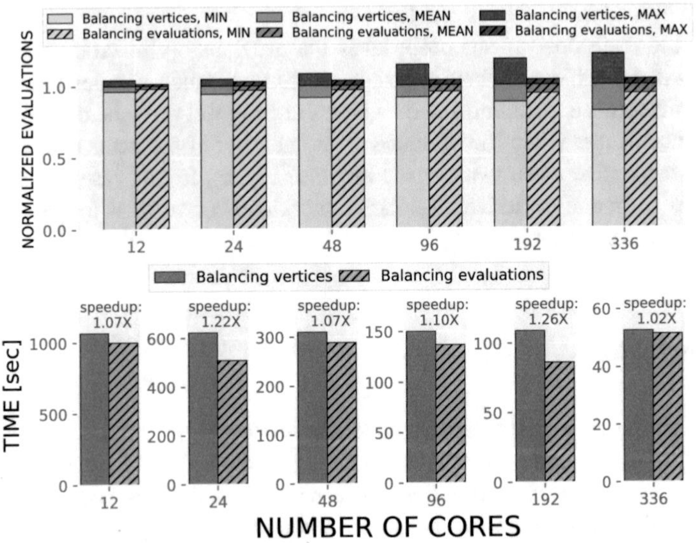

Fig. 6 Comparison of mesh partitioning strategies that balance either vertices or mesh element evaluations. MIN, MEAN and MAX denote the minimum, mean and maximum numbers of element evaluations per mesh partition, respectively. The element evaluations for each number of partitions were divided by the mean value. Speedup denotes the ratio of execution times

by multilevel k-way partitioning algorithms. Consequently, our mesh repartitioning proposal improves the parallel performance of VrPA algorithms. As future work, we will study if performance improvement may be achieved using our strategy in other known load balancing techniques. Additionally, we will also investigate how to reduce the synchronization overhead of distributed-memory algorithms for repositioning mesh vertices.

Acknowledgements This work has been supported by Spanish Government, "Secretaría de Estado de Universidades e Investigación", "Ministerio de Economía y Competitividad" and FEDER, grant contract: CTM2014-55014-C3-1-R. *Cluster2* (TeideHPC) was provided by the "Instituto Tecnológico y de Energías Renovables, S.A.". We thank to anonymous reviewers for their valuable comments and suggestions on this manuscript.

References

1. K. Barker, N. Chrisochoides, Practical performance model for optimizing dynamic load balancing of adaptive applications, in *Proceedings of the 19th IPDPS*, 28.a-28.b (2005)
2. D. Benitez, J.M. Escobar, R. Montenegro, E. Rodriguez, Performance comparison and workload analysis of mesh untangling and smoothing algorithms, in *Proceedings of the 27th International Meshing Roundtable* (2018)
3. D. Bozdag, A. Gebremedhin, F. Manne, E. Boman, U. Catalyurek, A framework for scalable greedy coloring on distributed memory parallel computers. J. Parallel Distrib. Comput. **68**(4), 515–535 (2008)
4. M. Brewer, L. Diachin, P. Knupp, T. Leurent, D. Melander, The mesquite mesh quality improvement toolkit, in *Proceedings of the 12th International Meshing Roundtable* (2003), pp. 239–250
5. N. Chrisochoides, A survey of parallel mesh generation methods. Technical Report. SC-2005-09. Brown University (2005)
6. L. Diachin, P. Knupp, T. Munson, S. Shontz, A comparison of inexact Newton and coordinate descent mesh optimization techniques, in *Proceedings of the 13th International Meshing Roundtable* (2004), pp. 243–254
7. J.M. Escobar, E. Rodríguez, R. Montenegro, G. Montero, J.M. González-Yuste, Simultaneous untangling and smoothing of tetrahedral meshes. Comput. Methods Appl. Mech. Eng. **192**, 2775–2787 (2003)
8. L. Freitag, M.T. Jones, P.E. Plassmann, A parallel algorithm for mesh smoothing. SIAM J. Sci. Comput. **20**(6), 2023–2040 (1999)
9. C. Geuzaine, J.F. Remacle, Gmsh: a three-dimensional finite element mesh generator with built-in pre- and post-processing facilities. Int. J. Numer. Methods Eng. **79**(11), 1309–1331 (2009)
10. W. Gropp, L.N. Olson, P. Samfass, Modeling MPI communication performance on SMP nodes, in *Proceedings of the 23rd European MPI Users Group Meeting* (2016)
11. G. Karypis, METIS (version 5.1.0) – A Software Package for Partitioning Unstructured Graphs, Partitioning Meshes, and Computing Fill-reducing Orderings of Sparse Matrices. University of Minnesota (2013)
12. P. Knupp, Updating meshes on deforming domains: an application of the target-matrix paradigm. Commun. Numer. Method Eng. **24**, 467–476 (2007)
13. M. Mathis, D. Kerbyson, A general performance model of structured and unstructured mesh particle transport computations. J. Supercomput. **34**, 181–199 (2005)

14. R. Montenegro, J.M. Cascón, J.M. Escobar, E. Rodríguez, G. Montero, An automatic strategy for adaptive tetrahedral mesh generation. Appl. Numer. Math. **59**(9), 2203–2217 (2009)
15. T. Panitanarak, S.M. Shontz, A parallel log barrier-based mesh warping algorithm for distributed memory machines. Eng. Comput. **34**, 59–76 (2018)
16. A. Sarje, S. Song, D. Jacobsen, K. Huck, J. Hollingsworth, A. Malony, S. Williams, L. Oliker, Parallel performance optimizations on unstructured mesh-based simulations. Procedia Comput. Sci. **51**, 2016–2025 (2015)
17. S.P. Sastry, S.M. Shontz, A parallel log-barrier method for mesh quality improvement and untangling. Eng. Comput. **30**(4), 503–515 (2014)
18. S.P. Sastry, S.M. Shontz, S.A. Vavasis, A log-barrier method for mesh quality improvement and untangling. Eng. Comput. **30**(3), 315–329 (2014)
19. R.B. Schnabel, Concurrent function evaluations in local and global optimization. CU-CS-345-86. Computer Science Technical Report. 332. University of Colorado, Boulder (1986)

Discrete Mesh Optimization on GPU

Daniel Zint and Roberto Grosso

Abstract We present an algorithm called *discrete mesh optimization* (DMO), a greedy approach to topology-consistent mesh quality improvement. The method requires a quality metric for all element types that appear in a given mesh. It is easily adaptable to any mesh and metric as it does not rely on differentiable functions. We give examples for triangle, quadrilateral, and tetrahedral meshes and for various metrics. The method improves quality iteratively by finding the optimal position for each vertex on a discretized domain. We show that DMO outperforms other state of the art methods in terms of convergence and runtime.

1 Introduction

Simulations based on finite elements require meshes with high quality. Element shape has a strong impact on convergence [3, 32]. In cases like fluid simulations [1] anisotropy or a locally varying element size influence stability and accuracy. Such attributes of a mesh can be described with a quality metric. A vast range of smoothing methods considers purely element shape [7, 8, 11, 17, 19–22, 24, 26, 37, 41, 42, 44, 45]. Changing the quality metric within these methods requires major changes in the algorithm structure.

We present *discrete mesh optimization* (DMO), a greedy approach to topology-consistent mesh smoothing. The exhaustive search efficiently exploits the parallel computing power of GPUs. Full utilization is achieved by applying a coloring scheme to update several independent vertices at once. Due to the versatility of DMO it optimizes meshes of any kind for any quality metric without relying on differentiable functions. DMO aims to maximize the global mesh quality by iteratively finding the optimal position for each vertex with respect to its one-ring neighborhood. The optimal position is found by evaluating the quality metric on

---

D. Zint (✉) · R. Grosso
Computer Graphics Chair, Universität Erlangen-Nürnberg, Erlangen, Germany
e-mail: daniel.zint@fau.de; roberto.grosso@fau.de

X. Roca, A. Loseille (eds.), *27th International Meshing Roundtable*,
Lecture Notes in Computational Science and Engineering 127,
https://doi.org/10.1007/978-3-030-13992-6_24

a coarse grid of candidate positions. The best candidate is chosen as new vertex position. The process is repeated several times while the grid spacing is reduced after every iteration. This results in a runtime comparable to Laplacian-based smoothing while creating high quality results. For example, with our setup DMO converges within 50 iterations on a mesh with 37,907 interior vertices taking around 7 ms. Besides the GPU version, DMO was also implemented on CPU. Even though DMO was developed to benefit from parallelization, it still outperforms other smoothing methods on CPU due to its fast convergence.

The next section introduces the *max-min* problem for mesh optimization. Section 3 gives an overview of smoothing methods. DMO is presented in Sect. 4. Examples and a performance analysis are stated in Sect. 5. Conclusions are given in Sect. 6.

2 The Max-Min Problem for Mesh Optimization

For each mesh element e_{k} a quality $q_{\mathrm{k}}^{(e)}$ is obtained by evaluating a quality metric $q^{(e)}(e_{\mathrm{k}})$. The quality metric should be strictly quasiconcave, continuous, and decaying from its unique global maximum. This holds for commonly used quality metrics. Other functions that do not fulfill this criteria do not guarantee optimal quality improvement.

The formulation of the mesh optimization problem is independent of the particular quality function $q^{(e)}$. Assume a mesh M in $\mathbb{R}^n$ which consists of n-dimensional elements. They do not have to be of one type, hybrid meshes are also possible.

For finite element simulations the minimal element quality is of importance [39]. Therefore, the quality $q_{\mathrm{i}}^{(v)}$ of a vertex v_{i} that is positioned at x is defined as the minimal quality of its incident elements $e_{\mathrm{k}} \in N_{\mathrm{e}}(v_{\mathrm{i}})$,

$$q_{\mathrm{i}}^{(v)}(x) = \min_{e_{\mathrm{k}} \in N_{\mathrm{e}}(v_{\mathrm{i}})} q_{\mathrm{k}}^{(e)}(x)\,, \tag{1}$$

where element quality $q_{\mathrm{k}}^{(e)}$ depends merely on the vertex position x as the one-ring neighborhood of v_{i} is kept fixed. If the element quality function $q_{\mathrm{k}}^{(e)}$ is quasiconcave, continuous, and decaying from its maximum, it follows that the vertex quality function $q_{\mathrm{i}}^{(v)}$ defined in (1) has a unique global maximum and is quasiconcave, [36].

From (1) follows the local optimization problem for finding the maximum quality $q_{\mathrm{i,\,max}}^{(v)}$ for a vertex v_{i},

$$q_{\mathrm{i,\,max}}^{(v)} = \max_{x} q_{\mathrm{i}}^{(v)}(x) = \max_{x} \min_{e_{\mathrm{k}} \in N_{\mathrm{e}}(v_{\mathrm{i}})} q_{\mathrm{k}}^{(e)}(x)\,. \tag{2}$$

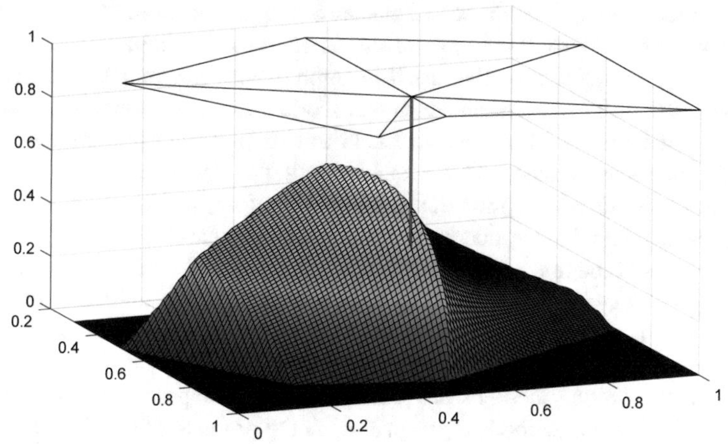

Fig. 1 Vertex quality function $q_{\mathrm{i}}^{(v)}$ for some one-ring neighborhood. Negative values were projected to zero. The red line points to the current quality of the vertex for the shown triangle fan

The vertex quality function is nonlinear and nonsmooth (Fig. 1) and thus cannot be solved by any optimization method that requires derivatives. Because of the properties of the vertex quality function an efficient exhaustive search as proposed in our algorithm in Sect. 4.1 will certainly lead to the optimum of (2). The optimal position x^* for v_{i} is given as

$$x^* = \arg\max_x q_{\mathrm{i}}^{(v)}(x) = \arg\max_x \min_{e_{\mathrm{k}} \in N_{\mathrm{e}}(v_{\mathrm{i}})} q_{\mathrm{k}}^{(e)}(x) . \tag{3}$$

In the next section we summarize different smoothing methods intending to solve the *max-min* problem of mesh optimization. They mostly differ in their strategy of approximating Eq. (3).

3 Related Work

Smoothing methods can be divided into three main groups, Laplacian-, physics-, and optimization-based. We will briefly discuss the three groups and give an overview of their advantages and disadvantages.

The classical Laplace-smoothing [15] is known to be fast but unstable in case of concave domains. A wide range of methods were proposed that modify the classical Laplace smoothing to overcome this problem [7, 8, 11, 17, 22, 24, 26, 28, 45]. Limitation still exists, for example angle-based smoothing [45] cannot deal with a strongly varying element size, Fig. 4. Another representative is the "Smart" Laplacian Smoothing of Freitag [17] which only performs smoothing when mesh quality is increased. Using "Smart" Laplacian Smoothing without further processing steps does not lead to satisfying results as it does not improve mesh quality considerably. Freitag proposed to use it in combination with an optimization-based method. This will be discussed later in this section. Laplacian-smoothers share the advantage of being fast but they also lead to suboptimal results. Many of these methods cannot guarantee that the quality will not decrease. DMO overcomes this problem by directly optimizing quality. Therefore, it will not converge to a suboptimal result and a non-decreasing quality can be guaranteed.

Physics-based methods consider the mesh as a physical model. Some examples are spring-mass systems [9, 14], truss networks [35], or elasticity models [4, 13, 38]. Just like Laplace-based methods they do not provide any guarantee of mesh improvement.

Optimization-based methods are named after their approach for optimizing a quality metric. Some methods try to overcome the problem of nondifferentiability by replacing Eq. (2) with a smooth function [18, 27, 31, 43, 44]. They run into the same problems as Laplacian-based methods.

Another approach is taken by the previously mentioned method of Freitag [17, 19–21]. While in [17, 19] the optimization is done with an analogue of the steepest descent method for smooth functions, later versions use the simplex algorithm to solve a linear programming problem [20, 21]. To keep runtime low it combines Laplacian-based smoothing with an optimization-based approach. Thus, the method does not converge to the optimal solution in general. DMO avoids that by optimizing all vertices the same way.

A derivative-free approach is done by Park and Shontz in [33]. They use pattern search in combination with backtracking line search to find a better vertex position. The convergence is suboptimal as it depends on search directions.

Rangarajan and Lew introduced the directional vertex relaxation (DVR) algorithm [37]. It solves the optimization problem by breaking it down to one dimension. This is achieved by providing a smoothing direction which can be chosen either randomly or by using previous knowledge. Within this one dimension the optimal solution can be found analytically. The major concern about this method is its randomness of relaxation directions as it leads to an inefficient smoothing with slow convergence rates. In contrast, DMO is not restricted to search directions and therefore yields faster convergence than DVR or the method of Park and Shontz.

4 Discrete Mesh Optimization

DVR solves the *max-min* problem in Eq. (2) analytically for only one direction. DMO uses a different strategy for solving the mesh optimization problem. We evaluate the vertex quality function with a greedy algorithm on a discretized domain. No derivatives are necessary which allows easy replacement of the quality metric.

In order to demonstrate the strength of DMO we use the standard mean ratio quality metric for triangles $q_{\mathrm{m_{tri}}}^{(e)}$ [2, 5, 6, 12, 20, 37],

$$q_{\mathrm{m_{tri}}}^{(e)} = 4\sqrt{3}\frac{A}{\sum_{i=1}^{3} l_i^2}, \tag{4}$$

where A is the signed area of the triangle and l_i is the length of its incident edges. The metric can be replaced by any other. For further informations on metrics see [29, 30, 39]. Some examples are given in Sect. 5.3.

In Sect. 4.1 our method is described. Section 4.2 explains a way of combining different quality metrics, here, performing density adaption while preserving mesh quality.

4.1 Algorithm

The method starts assigning each vertex to one of the sets S_m and S_f. Vertices that should be optimized are in set S_m, fixed vertices in S_f. In the given examples boundary vertices are fixed, see Algorithm 1. For each vertex the *argmax-min* problem of Eq. (3) is solved on a discretized domain using a greedy approach. Vertices that are not adjacent can be smoothed in parallel. Thus, graph coloring is applied once as preprocessing step to create subsets of vertices for parallel optimization [25].

Solving the discretized *argmax-min* problem is done iteratively with a uniform grid, see Algorithm 2. The grid is positioned around the vertex that should be optimized. The grid size is defined by the axis aligned bounding box for the one-ring neighborhood and a scaling factor ω. In our case, using half the size of the bounding box, $\omega = 0.5$, as an initial scaling factor led to convenient results.

Algorithm 1 Discrete mesh optimization

```
function SMOOTHMESH(M, n_iterations)
    Vertex-set S_m, S_f
    for all v ∈ M do
        if v is boundary vertex then
            S_f ← S_f ∪ v
        else
            S_m ← S_m ∪ v
    Color-set S_c = COLORMESH(S_m, M)          ▷ Each color is a subset of S_m
    for i = 0 to n_iterations do               ▷ Perform smoothing n_iterations times
        for all C ∈ S_c do                     ▷ iterate over colors
            for all v ∈ C do                   ▷ Iterate over vertices in C
                OPTIMIZEVERTEXPOSITION(v, N_e(v), 8, 3)   ▷ see Algorithm 2
    return
```

Algorithm 2 Discrete optimization of vertex position

function OPTIMIZEVERTEXPOSITION(v, $N_e(v)$, n, n_{greedy})
    $\omega \leftarrow 0.5$ ▷ Scaling-factor for grid
    for $counter = 0$ **to** n_{greedy} **do** ▷ Do n_{greedy} iterations of the algorithm
        $(x_{min}, y_{min}, dx, dy) =$ GETGRID(v, $N_e(v)$, n, ω) ▷ Get geometric grid specifications
        Create *quality_grid*[n][n]
        for $i, j = 0$ **to** n **do**
            $v' \leftarrow (x_{min} + i \cdot dx,\ y_{min} + j \cdot dy)$ ▷ Position v' at grid-point (i,j)
            quality_grid[i][j] $\leftarrow$ VERTEXQUALITY (v', $N_e(v)$)
        $q_{max} \leftarrow -\infty,\ i_{max} \leftarrow 0,\ j_{max} \leftarrow 0$
        for $i, j = 0$ **to** n **do**
            if *quality_grid*[i][j] $> q_{max}$ **then** ▷ Get argmax of *quality_grid*
                $q_{max} \leftarrow$ *quality_grid*[i][j]
                $i_{max} \leftarrow i,\ j_{max} \leftarrow j$
        if $q_{max} >$ VERTEXQUALITY(v, $N_e(v)$) **then** ▷ Check if initial position is better
            $v \leftarrow (x_{min} + i_{max} \cdot dx,\ y_{min} + j_{max} \cdot dy)$ ▷ Move vertex to new position
        $\omega \leftarrow \omega \cdot 2/(n-1)$ ▷ Reduce scaling factor
    return

Each grid-point is considered as candidate position where Eq. (1) is evaluated. The vertex is repositioned at the best candidate if this increases its quality, Fig. 2. After each iteration step the scaling factor ω is reduced such that the new grid size is twice the old grid spacing,

$$\omega \leftarrow \omega \cdot 2/(n-1), \tag{5}$$

where n is the number of grid points in one dimension, Fig. 3. Furthermore, the grid is repositioned around the current best vertex position. The process of candidate evaluation and mesh downscaling is repeated iteratively until the desired level of precision is reached.

We know that the vertex quality function stated in Eq. (1) is decaying from its unique maximum. Thus, we state that our exhaustive greedy approach provides the optimal solution for the vertex within the one-ring in a discrete sense. Accuracy of

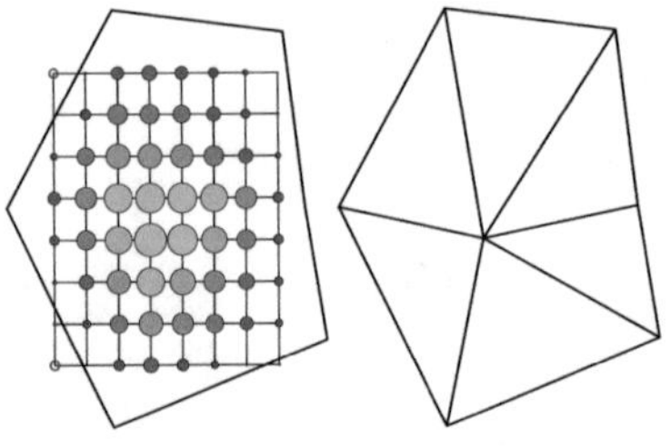

Fig. 2 *Left*: Uniform grid with quality metric evaluation for each candidate. A large green point represents good, a small red one bad quality. Positioning the vertex at one of the blue circles would result in triangle flipping. *Right*: the resulting fan after positioning the vertex at the best candidate

Fig. 3 Two iterations of the greedy method for finding the optimal vertex position. The dots represent the best candidate of the corresponding grid. The grid of the next iteration (red) is scaled down and placed at the best candidate of the previous grid (black)

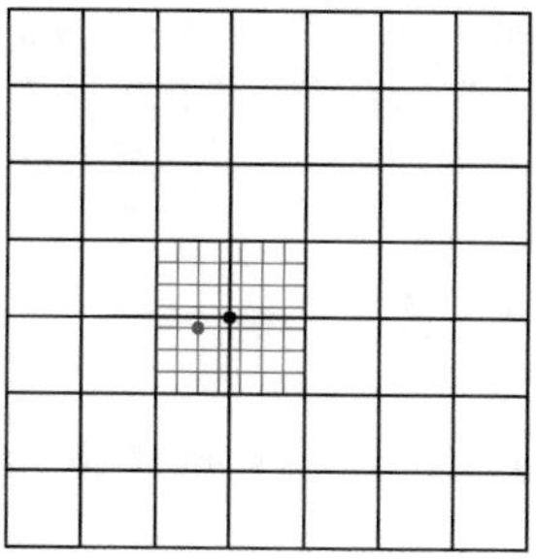

the result can be increased up to floating point precision by adapting the number of iterations in the greedy algorithm. We found three iterations combined with $n = 8$ grid points in each dimension to be sufficient.

Working on a uniform grid enables efficient parallelization of the greedy algorithm. Using one block of 32 threads on a grid with 64 points leads to an optimal utilization of the GPU if enough vertices are processed in parallel.

4.2 Mesh Density Adaptive Optimization

We introduce a way of combining several quality metrics. As an example, we fuse the mean ratio metric $q^{(e)}_{\mathrm{m_{tri}}}$ of Eq. (4), and the density metric $q^{(v)}_{\mathrm{d}}$, defined as the inverse of the squared distance between the position x_k of a vertex v_k and its optimal position x^*_k (in terms of density) relative to its one-ring neighborhood,

$$q^{(v)}_{\mathrm{d,\,k}}(x_\mathrm{k}) = \frac{1}{\|x_\mathrm{k} - x^*_\mathrm{k}\|^2}. \tag{6}$$

This metric tends to infinity the closer we get to the optimum and towards zero the further we move away. Computing the optimal position x^*_k is based on the spring-model for parametrization [16],

$$x^*_\mathrm{k} = \frac{\sum_{\mathrm{j} \in N(\mathrm{k})} D_\mathrm{j} x_\mathrm{j}}{\sum_{\mathrm{j} \in N(\mathrm{k})} D_\mathrm{j}}, \quad D_\mathrm{j} = \frac{1}{h(x_\mathrm{j})}, \tag{7}$$

with $N(\mathrm{k})$ the indices of the one-ring neighborhood of v_k and $h(v_\mathrm{j})$ the size function at position x_j [34].

Assume we want to adapt mesh density while ensuring a minimal element quality $\hat{q}^{(e)}_{\mathrm{m_{tri}}}$. We combine the two metrics to a new one as follows

$$q^{(v)}_{\mathrm{md,\,k}}(x) = \begin{cases} q^{(v)}_{\mathrm{m_{tri}}}(x) & \text{if} \;\; q^{(v)}_{\mathrm{m_{tri}}}(x) \le \hat{q}^{(e)}_{\mathrm{m_{tri}}} \\ \hat{q}^{(e)}_{\mathrm{m_{tri}}} + q^{(v)}_{\mathrm{d,\,k}}(x) & \text{otherwise} \end{cases}, \tag{8}$$

and solve the same *argmax-min* problem as before. An application of this metric is demonstrated in Sect. 5.3.

5 Results

DMO is compared to other methods in terms of convergence in Sect. 5.1 and performance in Sect. 5.2. Here, performance denotes the throughput of smoothed vertices per second. Examples for several mesh types and quality metrics are given in Sect. 5.3.

5.1 Convergence

We compare DMO with Laplace smoothing, angle-based smoothing, and DVR by analyzing convergence of minimal element quality. One iteration denotes that smoothing was applied to all vertices of S_m once. DMO and DVR optimize both for the mean ratio metric $q^{(e)}_{\mathrm{m_{tri}}}$ of Eq. (4). Figure 4a shows convergence on the triangle mesh "*east*". The input mesh has already satisfactory quality. While DMO and

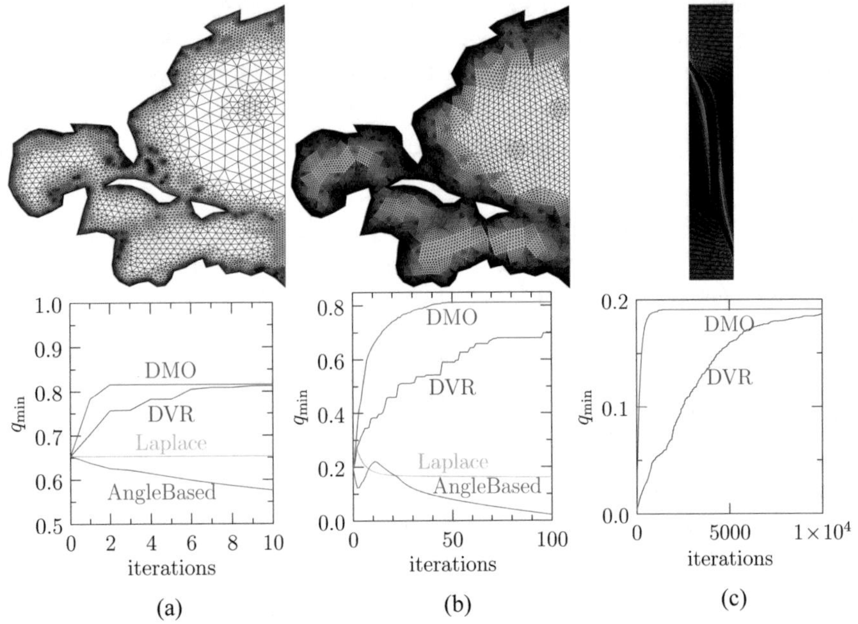

Fig. 4 Improvement of minimal quality measured with the mean ratio metric. (**a**) "*east*", $n_m = 8249$. (**b**) "*east*", $n_m = 37{,}907$. (**c**) "*shashkov*", $n_m = 3969$

DVR increase the minimal element quality, Laplace smoothing keeps it unchanged. Angle-based smoothing even decreases the quality. The bad behavior of angle-based smoothing is caused by the strongly varying element size within the mesh. DMO and DVR converge towards the same result. Nevertheless, DMO reached its optimum already after two iterations whereas DVR required ten iterations.

A finer and block structured version of mesh "*east*" is shown in Fig. 4b. Here, Laplace smoothing is unstable. This follows from the topology which is not perfectly adapted to the geometry anymore. Angle-based smoothing decreases quality. DVR and DMO improve the mesh quality. However, DVR takes around 900 iterations to reach the same minimal quality that is achieved by DMO in 60 iterations. Thus, DMO converges 15 times faster than DVR.

The last example considers an anisotropic mesh for which it is not reasonable to optimize its elements for roundness. Nevertheless, it can be used to test convergence of smoothing methods in extreme scenarios. Depending on a search direction, especially on a random one, leads to very slow convergence. Figure 4c shows an excerpt of the mesh "*shashkov*" which was taken from the example files of the Mesquite Toolkit [10]. DMO converged within 1400 iterations, whereas DVR was not converged after the stated 10,000 iterations.

5.2 *Performance*

We compare the performance of DMO with Laplace smoothing and DVR on triangle grids. Performance is measured as smoothed vertices per second. Preprocessing steps like mesh loading or coloring were excluded from measurements. Each method performed 1000 iterations on each mesh, ignoring convergence as we are only interested in runtime per iteration. The performance of DMO was achieved on a Nvidia GTX 1070. The other methods were profiled on an Intel i7-6700K CPU with 4.00 GHz and four cores. Note that the comparison was done using our own implementation of Laplace smoothing and DVR. DMO was performed using eight grid points in each dimension and three iterations of the greedy algorithm. Several tests were run on different meshes, Fig. 5. DMO and Laplace smoothing reach

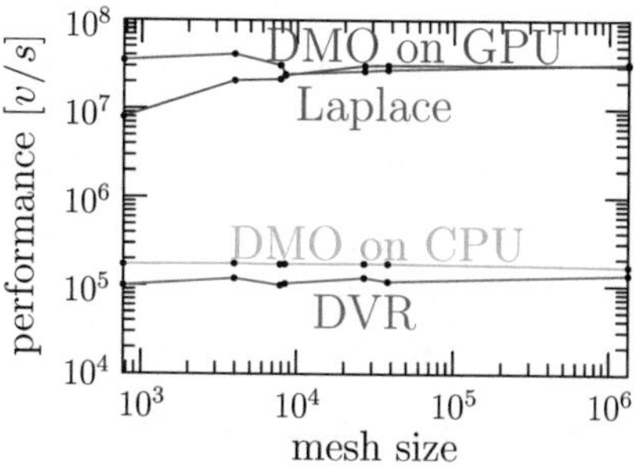

Fig. 5 Performance of smoothing methods for different mesh sizes. The performance is measured in smoothed vertices per second

around $2 \cdot 10^7$, DVR and DMO on CPU 10^5 vertices per second. The performance of DMO increases for larger meshes because the GPU is not fully utilized in case of small meshes.

On a mesh with 1,333,540 vertices in S_m and an initial minimal mean ratio quality of 0.033 DMO required four iterations and an execution time of 400 ms on GPU, including about 200 ms for graph coloring and copying data to the GPU, to converge towards a quality of 0.34. On CPU DMO ran for 31 s to execute the same number of iterations and reach the same quality. DVR required 43 iterations which took about 435 s to converge towards the same quality. Thus, in this example DMO is 14 times faster on CPU than DVR. On GPU it runs more than 1000 times faster.

Another mesh is given with 8429 vertices in S_m and an initial minimal mean ratio quality of 0.65. DMO converged within four iterations. On GPU it terminated in 4 ms, on CPU in 100 ms. DVR converged within ten iterations and ran for 780 ms. All methods reached a minimal quality of 0.82. Here, DMO is 7.8 times faster on CPU than DVR and on GPU it is 195 times faster.

5.3 *Different Quality Metrics*

All results stated up to now were computed using the mean ratio metric for triangles, Eq. (4). In this subsection we apply different metrics on various mesh types showing the flexibility of our method.

First, we smooth the block structured triangular mesh "*east*" (Fig. 6a) with respect to rectangularity [23],

$$q_{\text{rect}}^{(e)} = \max_{\text{i}=1,2,3} q^{\text{i}}, \tag{9}$$

$$q^{\text{i}} = \left(1 - \frac{|\frac{\pi}{2} - \theta_{\text{i}}|}{\frac{\pi}{2}}\right) \cdot \left(1 - \frac{|e_{\text{ij}} - e_{\text{ik}}|}{\max(e_{\text{ij}}, e_{\text{ik}})}\right). \tag{10}$$

Here, q^{i} denotes the local quality of one vertex within the triangle, θ_{i} is the interior angle at this vertex, e_{ij} and e_{ik} are the incident edges. The right-angled quality criterion in [23] contains another factor that aligns the triangles according to a cross field. We skipped this term as our purpose is just to show the flexibility of DMO. Figure 6c clearly states a rapid convergence of DMO to an optimal mesh. The resulting mesh is shown in Fig. 6b.

We also tested DMO with several other triangle shape quality metrics such as

- Max angle [3]: $\max_{\text{i}=1,2,3} \theta_{\text{i}} \cdot \frac{3}{\pi}$
- Min angle [19]: $1.5(1 - \frac{\min_{\text{i}=1,2,3} \theta_{\text{i}}}{\pi})$
- Radius ratio [39]: $2\frac{r}{R}$, where r is the incricle and R the circumcircle.

Figure 6f states the convergence plots for these metrics, the resulting meshes for *radius ratio* and *max angle* are given in Fig. 6d and e.

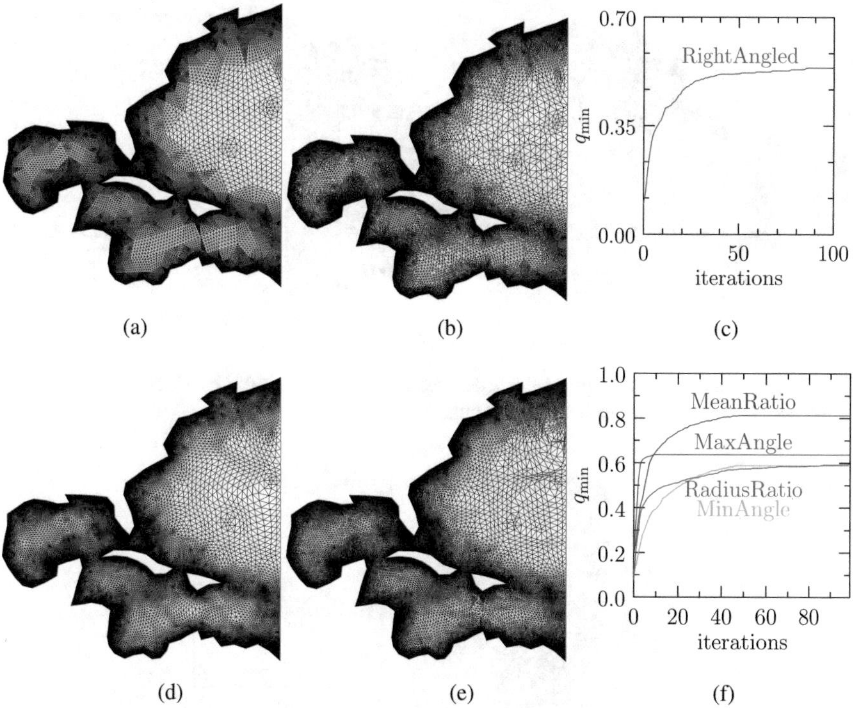

Fig. 6 Block structured triangle mesh "*east*" optimized with different quality metrics. (**a**) Initial block structured "*east*". (**b**) Right-angled quality. (**c**) Convergence for right-angled quality. (**d**) Radius ratio metric. (**e**) Max angle metric. (**f**) Convergence for several shape metrics

In the following example we apply the mean ratio metric for quad meshes,

$$q^{(e)}_{\mathrm{m_{quad}}} = 2\frac{A}{\sum_{i=1}^{4} l_i^2}. \tag{11}$$

The mesh in Fig. 7a is a block structured quad version of "*east*". The plot in Fig. 7c shows fast convergence and significant improvement in terms of the given quality metric. Figure 7b shows the resulting mesh.

Adding a third dimension to DMO enables optimization for volume meshes. Figure 8 shows a tetrahedral mesh where the mean ratio metric for tetrahedrons [20, 30, 37] was applied. The mesh quality was improved significantly within ten iterations.

Finally, we want to show that our method is capable of adapting density as described in Sect. 4.2. We use the distance to some point (x_0, y_0) as size function,

$$h_1(x, y) = \sqrt{(x_0 - x)^2 + (y_0 - y)^2} \tag{12}$$

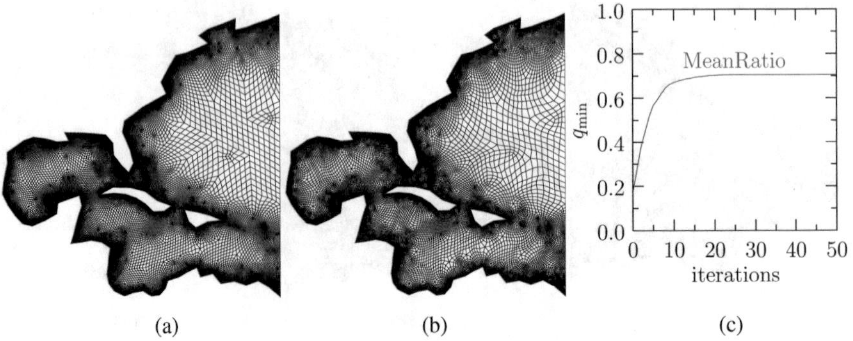

Fig. 7 Quad mesh "*east*" optimized with mean ratio metric. (**a**) "*east*" initial. (**b**) "*east*" smoothed. (**c**) "*east*" as a block structured quad mesh

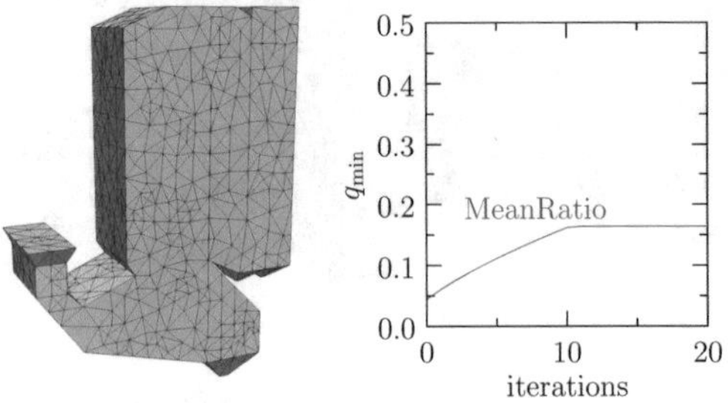

Fig. 8 Tetrahedral volume mesh "*tire*" from [40]

$$h_2(x, y) = \frac{1}{h_1(x, y)}. \tag{13}$$

Figure 9b shows the application of h_1 and Fig. 9c of h_2 on the triangle mesh "*bahamas*", Fig. 9a. In both cases the minimal mean ratio $\hat{q}^{(e)}_{\mathrm{m_{tri}}} = 0.7$ was given.

5.4 Robustness Against Smoothing Order

We give examples which indicate robustness of DMO against the order in which vertices are processed. We change the smoothing order by shuffling the vertex indices before applying DMO. We measure the quality of the mesh by lexicographically ordering the elements by their quality. Figure 10 shows the results of two meshes which already have an initial good quality. The smoothing was done 50 times with different orders and for each time a black line was added to the plot. These lines mostly overlap which corresponds to equal overall mesh quality.

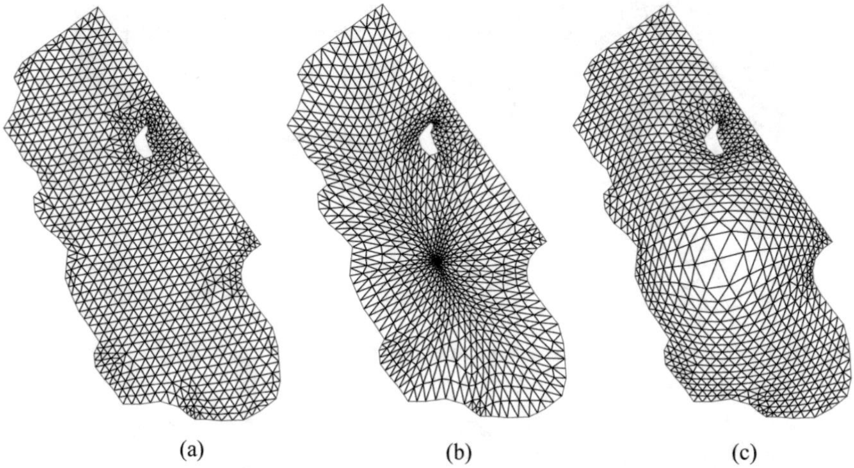

Fig. 9 Density adaption on mesh "*bahamas*". (**a**) Mesh "*bahamas*". (**b**) h_1 applied, $\hat{q}^{(e)}_{\mathrm{m_{tri}}} = 0.7$. (**c**) h_2 applied, $\hat{q}^{(e)}_{\mathrm{m_{tri}}} = 0.7$

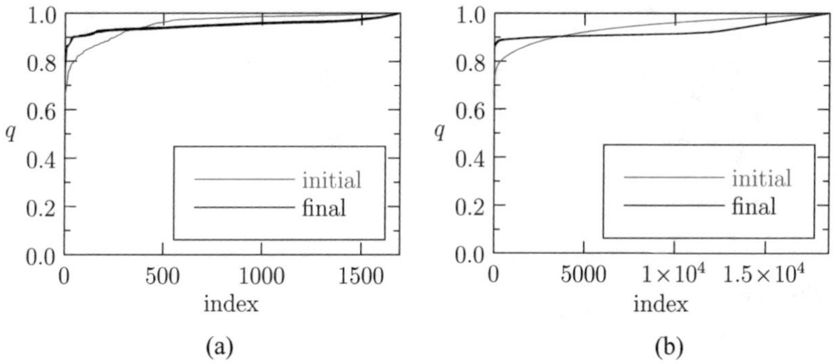

Fig. 10 Quality with different smoothing orders on good initial meshes. (**a**) "*bahamas*". (**b**) "*east*"

Figure 11 shows the same plot for a Delaunay triangulation of 100 randomly distributed points. Here, a larger deviation in quality can be observed. This is caused by the extremely low quality of the input mesh. Due to the large movements during smoothing the order in which vertices are processed has a strong influence on the output mesh. Figure 12 gives two examples. Both meshes have approximately the same minimal element quality but look completely different. This behavior could only be observed in such extreme cases as a Delaunay triangulation of randomly distributed points. If an input mesh has a reasonable quality, the result of our optimization algorithm is expected to be independent from the order in which the vertices are processed.

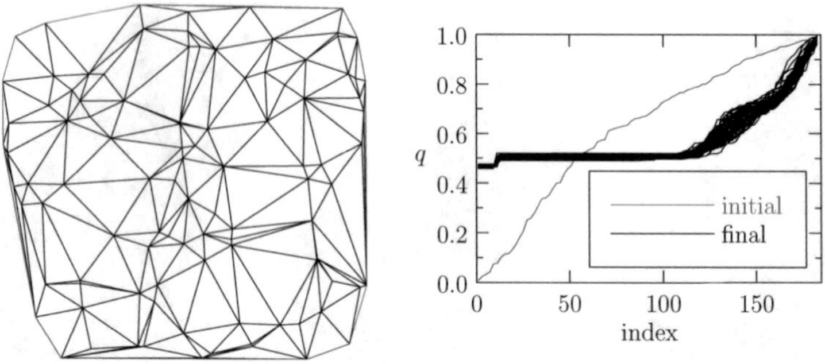

Fig. 11 Quality with different smoothing orders on a Delaunay triangulation (*left*)

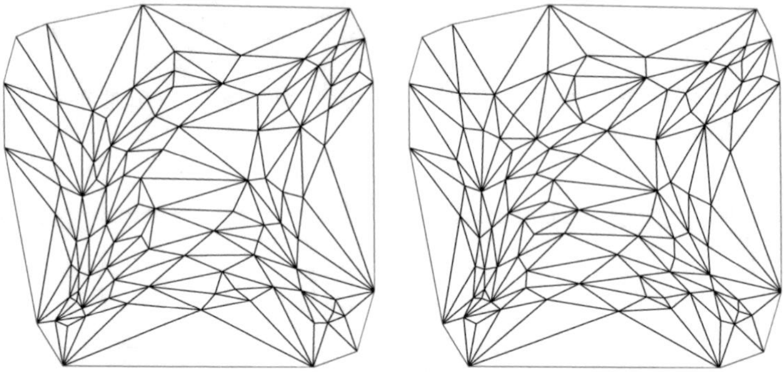

Fig. 12 Results for different smoothing orders

6 Conclusion

We presented a greedy approach to mesh optimization with fast convergence and adaptability to any kind of mesh or quality metric. This method can take full advantage of the parallelism of a consumer GPU. The minimal quality of a mesh is either improved or at least not decreased within each iteration. The interchangeability of the quality metric enables various cases of practical relevance, e.g. adaptive mesh smoothing.

In future work we will apply DMO to block structured ocean meshes. Creating a block structured mesh from an unstructured mesh results in a loss of minor features along the boundary. We aim to reposition the boundary of a block structured mesh to restore minor features while preserving high mesh quality.

References

1. V. Aizinger, C. Dawson, A discontinuous galerkin method for two-dimensional flow and transport in shallow water. Adv. Water Res. **25**(1), 67–84 (2002)
2. N. Amenta, M. Bern, D. Eppstein, Optimal point placement for mesh smoothing. J. Algorithms **30**(2), 302–322 (1999)
3. I. Babuška, A.K. Aziz, On the angle condition in the finite element method. SIAM J. Numer. Anal. **13**(2), 214–226 (1976)
4. T.J. Baker, Mesh movement and metamorphosis. Eng. Comput. **18**(3), 188–198 (2002)
5. R.E. Bank, R.K. Smith, Mesh smoothing using a posteriori error estimates. SIAM J. Numer. Anal. **34**(3), 979–997 (1997)
6. R.E. Bank, *A Software Package for Solving Elliptic Partial Differential Equations–Users Guide 7.0*. Frontiers in Applied Mathematics, vol. 15 (SIAM, Philadelphia, 1998)
7. T.D. Blacker, M.B. Stephenson, Paving: a new approach to automated quadrilateral mesh generation. Int. J. Numer. Methods Eng. **32**(4), 811–847 (1991)
8. T.D. Blacker, M.B. Stephenson, S. Canann, Analysis automation with paving: a new quadrilateral meshing technique. Adv. Eng. Softw. Work. **13**(5–6), 332–337 (1991)
9. F.J. Blom, Considerations on the spring analogy. Int. J. Numer. Methods Fluids **32**(6), 647–668 (2000)
10. M.L. Brewer, L.F. Diachin, P.M. Knupp, T. Leurent, D.J. Melander, The mesquite mesh quality improvement toolkit, in *IMR* (2003)
11. S.A. Canann, Y.-C. Liu, A.V. Mobley, Automatic 3d surface meshing to address today's industrial needs. Finite Elem. Anal. Des. **25**(1–2), 185–198 (1997)
12. S.A. Canann, J.R. Tristano, M.L. Staten et al., An approach to combined laplacian and optimization-based smoothing for triangular, quadrilateral, and quad-dominant meshes, in *IMR* (1998), pp. 479–494. Citeseer
13. V.F. De Almeida, Domain deformation mapping: application to variational mesh generation. SIAM J. Sci. Comput. **20**(4), 1252–1275 (1999)
14. Ch. Farhat, C. Degand, B. Koobus, M. Lesoinne, Torsional springs for two-dimensional dynamic unstructured fluid meshes. Comput. Methods Appl. Mech. Eng. **163**(1–4), 231–245 (1998)
15. D.A. Field, Laplacian smoothing and delaunay triangulations. Int. J. Numer. Methods Biomed. Eng. **4**(6), 709–712 (1988)
16. M.S. Floater, Parametrization and smooth approximation of surface triangulations. Comput. Aided Geom. Des. **14**(3), 231–250 (1997)
17. L.A. Freitag, On combining laplacian and optimization-based mesh smoothing techniques. ASME Applied mechanics division-publications-amd, vol. 220 (1997), pp. 37–44
18. L.A. Freitag, P.M. Knupp, Tetrahedral mesh improvement via optimization of the element condition number. Int. J. Numer. Methods Eng. **53**(6), 1377–1391 (2002)
19. L. Freitag, P. Plassmann, M Jones, An efficient parallel algorithm for mesh smoothing. Technical report, Argonne National Laboratory, IL (1995)
20. L. Freitag, M. Jones, P. Plassmann, A parallel algorithm for mesh smoothing. SIAM J. Sci. Comput. **20**(6), 2023–2040 (1999)
21. L.A. Freitag, P. Plassmann et al., Local optimization-based simplicial mesh untangling and improvement. Int. J. Numer. Methods Eng. **49**(1), 109–125 (2000)
22. P.-L. George, H. Borouchaki, *Delaunay Triangulation and Meshing: Application to Finite Elements* (Hermés Science, Paris, 1998)
23. C. Georgiadis, P.-A. Beaufort, J. Lambrechts, J.-F. Remacle, High quality mesh generation using cross and asterisk fields: application on coastal domains. arXiv preprint arXiv:1706.02236 (2017)
24. L.R. Herrmann, Laplacian-isoparametric grid generation scheme. J. Eng. Mech. Div. **102**(5), 749–907 (1976)
25. T.R. Jensen, B. Toft, *Graph Coloring Problems*, vol. 39 (Wiley, New York, 2011)

26. R.E. Jones, Qmesh: a self-organizing mesh generation program. Technical report, Sandia Laboratories, Albuquerque, NM (1974)
27. J. Kim, A multiobjective mesh optimization algorithm for improving the solution accuracy of pde computations. Int. J. Comput. Methods **13**(01), 1650002 (2016)
28. P.M. Knupp, Winslow smoothing on two-dimensional unstructured meshes. Eng. Comput. **15**(3), 263–268 (1999)
29. P.M. Knupp, Achieving finite element mesh quality via optimization of the Jacobian matrix norm and associated quantities. Part II – a framework for volume mesh optimization and the condition number of the Jacobian matrix. Int. J. Numer. Methods Eng. **48**(8), 1165–1185 (2000)
30. P.M. Knupp, Algebraic mesh quality metrics. SIAM J. Sci. Comput. **23**(1), 193–218 (2001)
31. P. Knupp, Updating meshes on deforming domains: an application of the target-matrix paradigm. Int. J. Numer. Methods Biomed. Eng. **24**(6), 467–476 (2008)
32. M. Křížek, On the maximum angle condition for linear tetrahedral elements. SIAM J. Numer. Anal. **29**(2), 513–520 (1992)
33. J. Park, S.M. Shontz, Two derivative-free optimization algorithms for mesh quality improvement. Procedia Comput. Sci. **1**(1), 387–396 (2010)
34. P.-O. Persson, Mesh size functions for implicit geometries and pde-based gradient limiting. Eng. Comput. **22**(2), 95–109 (2006)
35. P.-O. Persson, G. Strang, A simple mesh generator in matlab. SIAM Rev. **46**(2), 329–345 (2004)
36. R. Rangarajan, On the resolution of certain discrete univariate max–min problems. Comput. Optim. Appl. **68**(1), 163–192 (2017)
37. R. Rangarajan, A.J. Lew, Provably robust directional vertex relaxation for geometric mesh optimization. SIAM J. Sci. Comput. **39**(6), A2438–A2471 (2017)
38. M. Rumpf, A variational approach to optimal meshes. Numer. Math. **72**(4), 523–540 (1996)
39. J. Shewchuk, What is a good linear finite element? interpolation, conditioning, anisotropy, and quality measures. Preprint. University of California at Berkeley, 73:137 (2002)
40. Shewchuk. Stellar: A tetrahedral mesh improvement program, 05-23-2018. Available from: https://people.eecs.berkeley.edu/~jrs/stellar/input_meshes.zip
41. S.M. Shontz, S.A. Vavasis, A mesh warping algorithm based on weighted laplacian smoothing, in *IMR* (2003), pp. 147–158
42. H. Xu, T.S. Newman, 2D FE quad mesh smoothing via angle-based optimization, in *International Conference on Computational Science* (Springer, Berlin, 2005), pp. 9–16
43. K. Xu, X. Gao, G. Chen, Hexahedral mesh quality improvement via edge-angle optimization. Comput. Graph. **70**, 17–27 (2018)
44. P.D. Zavattieri, E.A. Dari, G.C. Buscaglia, Optimization strategies in unstructured mesh generation. Int. J. Numer. Methods Eng. **39**(12), 2055–2071 (1996)
45. T. Zhou, K. Shimada, An angle-based approach to two-dimensional mesh smoothing, in *IMR* (2000), pp. 373–384

Mesh Morphing for Turbomachinery Applications Using Radial Basis Functions

Ismail Bello and Shahrokh Shahpar

Abstract This article introduces an implementation of mesh morphing using radial basis functions (RBF) in a proprietary meshing framework ABIHex. RBF Morphing is a meshless method for mesh morphing that derives its origin from optimisation and machine learning. The method is used to perturb an arbitrarily complex mesh given a sample of displacements at known locations. Using this method, a number of test cases shown in the literature were replicated, and the method was applied to solve some industrial applications for gas turbines. It was found that the RBF mesh morphing method is robust in the face of sparse data, and provides a generic tool for mesh deformation provided the numerical conditions for stability of the method are met in the initial problem setup. A novel application is presented where manufacturing variation data is input to the mesh morphing program to produce a suitable volume mesh from manufactured geometry in order to perform follow on fluid flow simulations.

1 Introduction

There are many applications in CFD which require deforming the computational domain or mesh for various reasons without changing its topology [1]. In fluid-structure interaction computations, one may study the flow field over an immersed moving boundary, or in coupled fluid-structure simulations common when studying aeroelastic phenomena [2]. Other applications also include design optimisation where we wish to optimise geometry subject to an objective function [3]. In either case, this involves deforming the computational domain by some means in order to iterate and or search a design space for an optimum solution, or to emulate a moving domain in a numerical simulation.

I. Bello · S. Shahpar (✉)
Rolls-Royce PLC CFD Methods Group, Derby, UK
e-mail: ismail.bello@rolls-royce.com; shahrokh.shahpar@rolls-royce.com

X. Roca, A. Loseille (eds.), *27th International Meshing Roundtable*,
Lecture Notes in Computational Science and Engineering 127,
https://doi.org/10.1007/978-3-030-13992-6_25

Traditionally, this is where parametric geometry shines where there is a set of parameters which uniquely define the computational domain, and morphing is achieved by variation of these parameters to generate new geometry. The new domain can then be re-meshed as seen fit by the analyst.

In the absence of parametric geometry, numerous mesh-based methods are available such as LBWARP, FEMWARP, and simplex methods [4]. A comprehensive comparison of the performance of these methods has been presented by Staten et al in [4] where the authors recommended the simplex linear, simplex natural neighbour methods as fast solutions for a mesh morphing routine.

These methods can be labour intensive and may incur a large computational cost, due to their dependence on element type and iterative nature. On the hand, Creating a parametric representation of the target geometry often requires domain specific knowledge and does not apply to arbitrary geometry. As such, a method which is mesh-agnostic, and offers additional flexibility without needing a parametrically defined domain is very desirable.

Radial basis functions (RBF) mesh morphing is well known alternative to mesh-based morphing methods. The method finds its origins as a generic interpolant for scattered data, and is used heavily in optimisation and machine learning applications [3, 5]. In this method, the problem of updating geometry by specifying the movement of an arbitrary set of vertices is cast as an interpolation problem. This can be stated informally as follows: given a vector field sampled over a domain, can we construct a new vector field globally over the domain which is exact at the sample points, and is smoothly varying elsewhere?

Applications of RBF morphing to numerical simulation problems have been demonstrated by many authors. In particular, we mention, Rendall et al. [5] in investigating fluid-structure interactions, and Seiger et al. [3] in design optimisation. In this work, the application of these techniques will be demonstrated for preparing a representative computational domain using scanned geometry from manufacturing data for follow on CFD simulations.

The mechanics associated with this method are outlined along with features of a prototype implementation that may be used for generic mesh deformation problems in the context of CFD in Roll-Royce. This was implemented within an in-house tool ABIHex, a framework for automatic blocking hexahedral mesh generation. The prototype implementations along with said facilities are integrated into ABIHex to make it easier to use for a wide range of problems. At the time of writing this paper, this decision was made to facilitate using mesh morphing in an automated fashion in a larger simulation loop. There are commercial tools which provide this functionality, in particular, ANSYS-RBFMorph [6]. Additional investigation is also under way to use RBFMorph in the same context as the ABIHex implementation.

2 RBF Interpolants in Mesh Morphing

2.1 Construction of RBF Interpolant

In this section, a brief introduction to the construction of an RBF interpolant is presented based primarily on the work of Boer et al. [7], Sieger et al. [8] and Schaback [9]. The RBF interpolant is defined in k dimensions as follows:

Let Ω be a subset of $\mathbb{R}^k$, and let $\boldsymbol{x}_i$ be a node in a mesh, given Ω_b, a subset of Ω where a displacement field $V_b : \Omega_b \to \mathbb{R}^k$ is defined for every $\boldsymbol{x}_{bi}$ in Ω_b, Furthermore, we make a choice of basis for k-variate scalar valued polynomials of order $Q-1$ denoted as P^k_{Q-1}.

For such a choice of basis $B \in \text{span}\left(P^k_{Q-1}\right)$ with elements $\pi_j : \mathbb{R}^k \to \mathbb{R}$, set a compatible radial basis function $\phi : \mathbb{R}^+ \to \mathbb{R}$, then the RBF interpolant generated by V_b is a function $s : \mathbb{R}^k \to \mathbb{R}^k$ given in Eq. (1):

$$s\left(\boldsymbol{x}\right) = \sum_{i=1}^{|\Omega_b|} \boldsymbol{\alpha}_i \phi\left(|\boldsymbol{x} - \boldsymbol{x}_{bi}|\right) + \sum_{j=1}^{l} \boldsymbol{q}_j \pi_j\left(\boldsymbol{x}\right) \tag{1}$$

Where, $\boldsymbol{\alpha}_i$ and $\boldsymbol{q}_j$ are weight vectors for the radial basis functions and the polynomial basis respectively. These are to be determined by solving the following system (a notational convenience is used here to denote $f(a)$ for all $a \in A$ as $f(A)$)

$$s\left(\Omega_b\right) = V_b\left(\Omega_b\right) \tag{2a}$$

$$\sum_{j=1}^{|\Omega_b|} \boldsymbol{\alpha}_i p\left(\boldsymbol{x}_{bi}\right) = \boldsymbol{0} \tag{2b}$$

Where, $p(\boldsymbol{x})$ is any polynomial in P^k_{Q-1}. In general, the choice of radial basis function determines the maximal degree of the polynomial [9] and thus the number of terms in the sum l in the polynomial term above is given as $l = \dim\left(P^k_{Q-1}\right) := \binom{Q-1+k}{k}$. However, we restrict ourselves to the kernels discussed in [7] for which $Q = 2$.

Given a solution to the system in Eqs. (2a) and (2b) for $\boldsymbol{\alpha}_i$ and $\boldsymbol{q}_j$, the deformation of the domain Ω is given by mapping $\boldsymbol{x} \longmapsto \boldsymbol{x} + s(\boldsymbol{x}) \quad \forall\, \boldsymbol{x} \in \Omega$.

The size of the linear system to be solved is $(|\Omega_b| + l) \times (|\Omega_b| + l)$, and a simple choice of basis set $\{\pi_1(\boldsymbol{x}), \pi_2(\boldsymbol{x}), \pi_3(\boldsymbol{x}), \pi_4(\boldsymbol{x})\}$ as $\{1, x, y, z\}$, the problem in Eqs.

(2a) and (2b) can be written in matrix form as ($N = |\Omega_b|$):

$$\begin{pmatrix} \phi(|x_{b1}-x_{b1}|) & \cdots & \phi(|x_{b1}-x_{bN}|) & 1 & x_{b1} & y_{b1} & z_{b1} \\ \vdots & \ddots & \vdots & \vdots & \ddots & & \vdots \\ \phi(|x_{bN}-x_{b1}|) & \cdots & \phi(|x_{bN}-x_{bN}|) & 1 & x_{bN} & y_{bN} & z_{bN} \\ 1 & \ldots & 1 & 0 & 0 & 0 & 0 \\ x_{b1} & \ddots & x_{bN} & 0 & 0 & 0 & 0 \\ y_{b1} & & y_{bN} & 0 & 0 & 0 & 0 \\ z_{b1} & \cdots & z_{bN} & 0 & 0 & 0 & 0 \end{pmatrix} \begin{pmatrix} \alpha_{x1}\ \alpha_{y1}\ \alpha_{z1} \\ \vdots \\ \alpha_{xN}\ \alpha_{yN}\ \alpha_{zN} \\ q_{x1}\ q_{y1}\ q_{z1} \\ q_{x2}\ q_{y2}\ q_{z2} \\ q_{x3}\ q_{y3}\ q_{z3} \\ q_{x4}\ q_{y4}\ q_{z4} \end{pmatrix} = \begin{pmatrix} d_{x1}\ d_{y1}\ d_{z1} \\ \vdots \\ d_{xN}\ d_{yN}\ d_{zN} \\ \mathbf{0}_{4\times 4} \end{pmatrix} \tag{3}$$

The block structure of the system above can be observed, allowing us to make the following definitions:

Let Φ be a linear map which consumes r vectors and produces an $r \times |\Omega_b|$ matrix.

$$\Phi : \mathbb{R}^k \times \mathbb{R}^k \times \cdots \times \mathbb{R}^k \to \mathbb{R}^{rN}$$

$$\Phi(\boldsymbol{x}_1, \boldsymbol{x}_2, \ldots \boldsymbol{x}_r) = \begin{pmatrix} \phi(|\boldsymbol{x}_1 - \boldsymbol{x}_{b1}|) & \cdots & \phi(|\boldsymbol{x}_1 - \boldsymbol{x}_{bN}|) \\ \vdots & \ddots & \vdots \\ \phi(|\boldsymbol{x}_r - \boldsymbol{x}_{b1}|) & \cdots & \phi(|\boldsymbol{x}_{rN} - \boldsymbol{x}_{bN}|) \end{pmatrix} \tag{4}$$

Furthermore, let Π be a similarly defined map which takes as input, r vectors and outputs an $r \times l$ matrix by simply applying the m basis of the polynomial term.

$$\Pi : \mathbb{R}^k \times \mathbb{R}^k \times \cdots \times \mathbb{R}^k \to \mathbb{R}^{rl}$$

$$\Pi(\boldsymbol{x}_1, \boldsymbol{x}_2, \ldots \boldsymbol{x}_r) = \begin{pmatrix} \pi_1(\boldsymbol{x}_1) & \ldots & \pi_l(\boldsymbol{x}_1) \\ \vdots & \ddots & \vdots \\ \pi_1(\boldsymbol{x}_r) & \ldots & \pi_l(\boldsymbol{x}_r) \end{pmatrix} \tag{5}$$

Then we may rewrite the solution and evaluation steps in Eqs. (1), (2a) and (2b) more compactly as:

$$\begin{pmatrix} \Phi\left(\boldsymbol{x}_{b1},\ldots\boldsymbol{x}_{bN}\right) & \Pi\left(\boldsymbol{x}_{b1},\ldots\boldsymbol{x}_{bN}\right) \\ \Pi(\boldsymbol{x}_{b1},\ldots\boldsymbol{x}_{bN})^T & \boldsymbol{0}_{k\times k} \end{pmatrix} W = \begin{pmatrix} V_b(\boldsymbol{x}_{b1})^T \\ \vdots \\ V_b(\boldsymbol{x}_{bN})^T \\ \boldsymbol{0}_{k\times k} \end{pmatrix} \tag{6a}$$

$$s\left(\boldsymbol{x}\right) = \left(\Phi\left(\boldsymbol{x}\right)\ \Pi\left(\boldsymbol{x}\right)\right) \bullet W \tag{6b}$$

Where, $W = (\boldsymbol{\alpha}^T\ \boldsymbol{q}^T)^T$ is a matrix containing the weights of the interpolant. One may have observed that no restriction is placed on Ω_b. Furthermore, no information about connectivity is needed in this process. This ultimately makes this method a much simpler method than traditional mesh-based methods of morphing. In general, the matrix equation (3) is dense. However, depending on the choice of kernel, it may be sparse which can be solved efficiently [8].

2.2 RBF Kernel Types

There are two types of kernel that are discussed in the literature; these are kernels with either compact or global support.

- Kernels with compact support satisfy the following requirements:
  Given a function $f : \mathbb{R}^+ \to \mathbb{R}$, and a radius r, a kernel can be defined as:

$$\phi(x) = \begin{cases} f(x) & x \leq r \\ 0 & \text{otherwise} \end{cases} \tag{7}$$

Typically, one may wish to use a kernel with compact support to localise the effect of displacing a node within a ball of radius r around that node. A collection of these kernels were constructed by Wendland [10], a feature which is typically achieved in traditional methods by using node connectivity data. In this event, better results are usually achieved by using as large a support radius as possible to avoid extremely localised deformations. In addition, smoother deformations can be obtained when $\phi(x)$ is in a high smoothness class.

- Kernels with global support are defined for all real positive values, for example the Gaussian functione^{-x^2}. These have the benefit of global support and as such problem specific choice of radii is eliminated.

  Additionally, kernels with global support naturally apply when we wish to generate an interpolant that vanishes far-field of the sample points of V_b, or when the domain Ω is infinite. A comprehensive discussion on different kernel types and the value of Q for those kernels is given in [9].

2.3 Solution and Evaluation of Morphing Problems

Solvability of the system in Eq. (3) is a big problem. In general, the matrix $\Phi(\Omega_b)$ is symmetric and conditionally positive definite. The condition for positive definiteness is dependent on the choice of the kernel. [9]

For example, the Gaussian kernel which is defined non-zero globally generates a conditionally positive definite system for $Q = 0$. This transforms the system in Eq. (3) to:

$$\Phi\left(\Omega_b\right) W = V_b(\Omega_b)^T \tag{8}$$

In contrast the Wendland kernels are a family of kernels with compact support that generate a positive definite system up to a maximal dimension of the underlying domain [9]. The simplest one of these kernels is the C^0 Wendland kernel given by:

$$\phi(x) = \begin{cases} \left(1 - \frac{x}{r}\right)^2 & x \leq r \\ 0 & \text{otherwise} \end{cases} \tag{9}$$

As such, a suitable solver must be applicable for general square systems and performs well for large systems. In this work, the implementation uses one of two solvers depending on the size of the system. LU decomposition with full pivoting is used when the system is small. Otherwise, LU decomposition with partial pivoting is used. This decision was based on benchmarks for the linear algebra library Eigen [11].

For very aggressive deformations, the system may result in degenerate cells or even affect the solvability of the linear system in Eq. (3). Often this can be remedied by scaling the displacements and incrementally solving and evaluating in a finite number of steps [3].

For example, to solve the system in Eq. (3) in n steps, we repeat the following computations steps n times:

$$\begin{pmatrix} \Phi\left(\Omega_b\right) & \Pi\left(\Omega_b\right) \\ \Pi(\Omega_b)^T & 0 \end{pmatrix} W_n = \begin{pmatrix} V_{b,n}(\Omega_b)^T \\ 0 \end{pmatrix} \tag{10a}$$

$$\text{Where,} \quad V_{b,n}\left(\boldsymbol{x}\right) := \frac{V_b\left(\boldsymbol{x}\right)}{n} \tag{10b}$$

$$s_n\left(\boldsymbol{x}\right) = \left(\Phi\left(\boldsymbol{x}\right)\ \Pi\left(\boldsymbol{x}\right)\right)\ W_n \tag{10c}$$

$$\forall \boldsymbol{x} \in \Omega, \quad \boldsymbol{x} \to \boldsymbol{x} + s_n\left(\boldsymbol{x}\right) \tag{10d}$$

By performing successive solve-evaluate-update computations n times we recover the original constraints encoded in V_b. However, since this is a linear scaling, it only allows us to incrementally solve smaller displacements and avoid large weights due to large displacements. For simple underlying fields V_b with small displacements, the solution in a single step is often enough to generate a stable result, otherwise there is a trade-off to be made in the number of steps, kernel type and kernel radius where applicable.

Practically, it may seem that a large number of steps may yield more stable results; however, this is not always the case. For a large number of steps the value of the displacements may become so small that the right hand side of Eq. (10a) becomes almost zero for a certain number of points. Therefore a low number of steps is preferred when using successive updates to avoid such numerical issues. Alternatively, a zonal interpolant can be constructed by solving the system on disjoint subsets of the mesh; this was demonstrated by Wang et al [12]. The authors presented a novel mesh morphing method using RBF interpolants and Delaunay Graph mapping. The sub-domains circumvent the need to construct and solve large linear systems and offer other performance benefits such as fast node identification.

Mesh morphing will invariably alter the quality of the cells in the mesh. A good morphing will avoid introducing degenerate cells, and always produce a valid mesh without negative volumes. The RBF method performs well in this regard, however, its performance is heavily dependent on the kernel type and shape and/or kernel radius [9]. Boer et al [7] provide a study investigating the effect of the morphing parameters on the mesh quality along with some recommendations for kernel radii and types depending on the problem.

3 Prototype Implementation

The prototype implementation comprises of a number of facilities for algebraically manipulating primitives, the basic building blocks required to fully define the morphing problem, as well as typical IO operations needed in the program. The domain model is described as follows:

1. The target domain for morphing Ω is represented as an unstructured grid with identical structure to a set of nodes in $\mathbb{R}^3$ representing node coordinates.
2. The sampled displacement field $V_b : \Omega_b \to \mathbb{R}^3$ is represented as a set $V_b \in \Omega_b \times \mathbb{R}^3$ where $\mathbb{R}^3$ here represents displacements. This allows the user to leverage the set structure to define V_b using algebraic operations of unions, intersections, and differences.
3. A kernel is represented as any binary function $\phi : \mathbb{R}_+ \times \mathbb{R}_+ \to \mathbb{R}$. Where the second argument of the kernel is a parameter which controls the shape/radius of the kernel.

All facilities for interacting with the types in this representation are exposed in out in-house code ABIHex through a lua scripting interface. Given a fully defined morphing problem, the displacement field is then used to construct the linear system in Eq. (3), and an appropriate solver is used to compute the interpolant weights.

Applying the computed morphing onto the original mesh is done either as a direct matrix multiplication [as in Eq. (6b)] if the number of nodes in the domain is small. Alternatively, an in-place update is performed by evaluating the deformations pointwise.

Algebraic manipulation of mesh node sets and elements of V_b can also pose a performance bottle neck when these sets become large. In our implementation, standard tree-based data structures are used to represent mesh subsets and constraints. This means search and insertion operations scale logarithmically with the size of the inputs. During this work it was found that the algebraic manipulation of node sets in problem construction increased the overall computational time significantly when the number of constraints approaches **100k** nodes. A fast alternative node identification method is presented by Wang et al. using Delaunay graph mapping [12] which could eliminate this issue by using a graph data structure and comparison operation for node identification.

3.1 *Simple Mesh Morphing Problems*

As discussed, setting up a morphing problem is done by specifying constraints and kernel parameters. For example, consider torsion of a cube containing only hexahedral elements. To examine the quality of the mesh before and after deformation, we will chiefly make use of element skewness as implemented in paraview [13].

Element skew is defined as the maximum angle between pairs of the element's principal axes $\{\boldsymbol{e_1}, \boldsymbol{e_2}, \boldsymbol{e_3}\}$. For a unit cube, skewness (see Eq. 11) is **0** and a desirable range for hexahedral cells is between **0** and **0.5** (Figs. 1 and 2).

$$skew = \max\left(|\boldsymbol{e}_1 \circ \boldsymbol{e}_2|, |\boldsymbol{e}_2 \circ \boldsymbol{e}_3|, |\boldsymbol{e}_3 \circ \boldsymbol{e}_1|\right) \tag{11}$$

A similar test case to that found in [7, 12] was also replicated using our implementation as shown in Fig. 3.

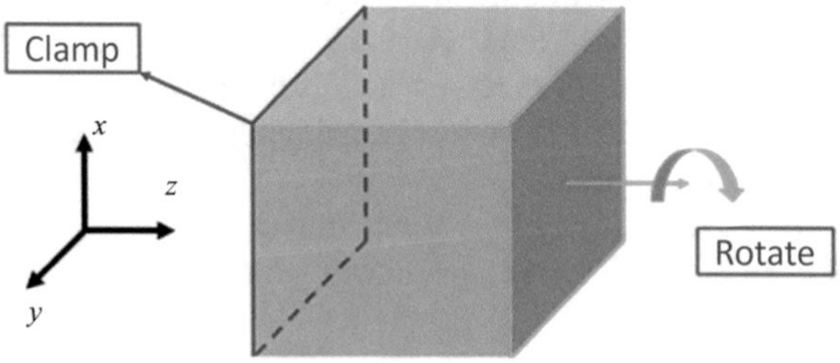

Fig. 1 Simple Torsion of a cube. The cube above covers the interval $[-1, 1] \times [-1, 1] \times [0, 2]$. The domain $\Omega_b$in this case is the union of the two surfaces where the displacement is known i.e $\Omega_b = \{x \in \mathbb{R}^3 | \, z = 0 \vee z = 2\}$

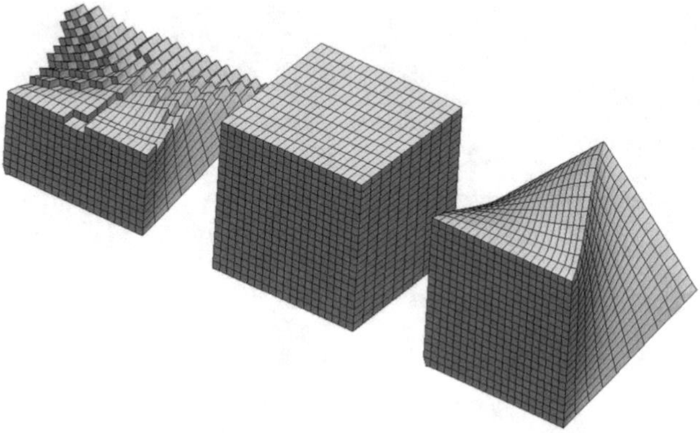

Fig. 2 Results of twisting the 2-by-2-by-2 cube

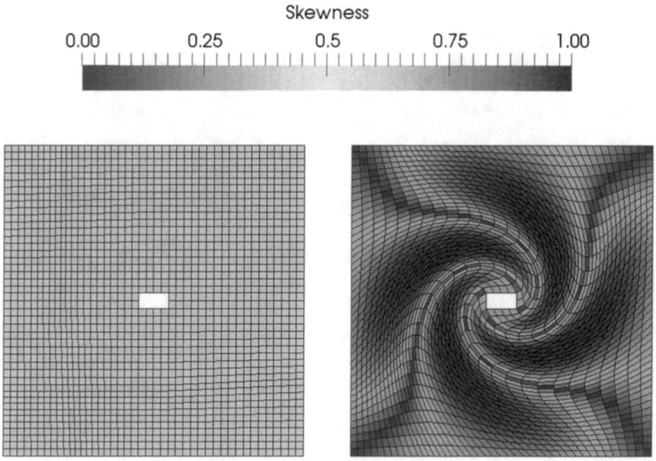

Fig. 3 Replicated test case from Boer et al showing skewness of the cells and the initial (right) and final mesh (left). The kernel used in this particular case was the thin plate spline kernel

Using an identical deformation to that shown in Figs. 3 and 4 shows the effect of choosing different kernel types and parameters on the resulting deformation.

In Fig. 4, we observe the choice of kernel and radius affects the strength of propagation of the deformation over the domain. Using a smaller radius confines the propagation closer to the points of application while a smoother kernel propagates the deformation farther away from the points of application. Which result is more desirable is left to the user to decide by examining element quality and the nature of the application. In general, a larger radius and a smoother kernel yield better results as this avoids sharp transitions near the points of application of the deformation and thereby less deterioration of element quality.

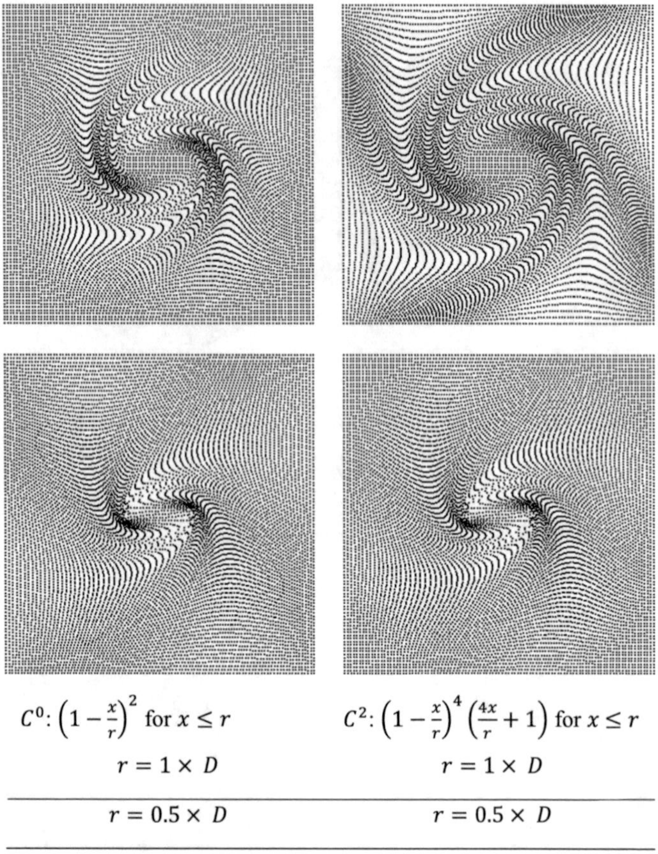

Fig. 4 Effect of kernel choice and shape on deformation for two of the Wendland polynomial Kernels with compact support. The value D above denotes the size of the domain

Due to the meshless nature of the method, no special treatment is given to more complex meshes. This can be seen in Fig. 5, where an identical deformation to that shown in Fig. 3 was applied to a different domain where high mesh clustering is present. In this test case, it is also seen that the method is able to propagate the displacement naturally throughout the domain.

In Fig. 5, we observe the introduction of clustering to the domain produces poor quality cells as the interpolant deforms high aspect ratio cells. However, this is a contrived case to demonstrate features in the mesh that are sensitive to morphing. In practice such an aggressive deformation is rarely applied near features of interest in the mesh. If so, the user is likely to consider re-meshing.

For simple underlying displacement fields, the method does well in reproducing the specified displacements in a single step computation. This is seen in Fig. 6 where the initial domain, a unit cube has a displacement field defined at the boundary nodes which project it onto an enclosing sphere.

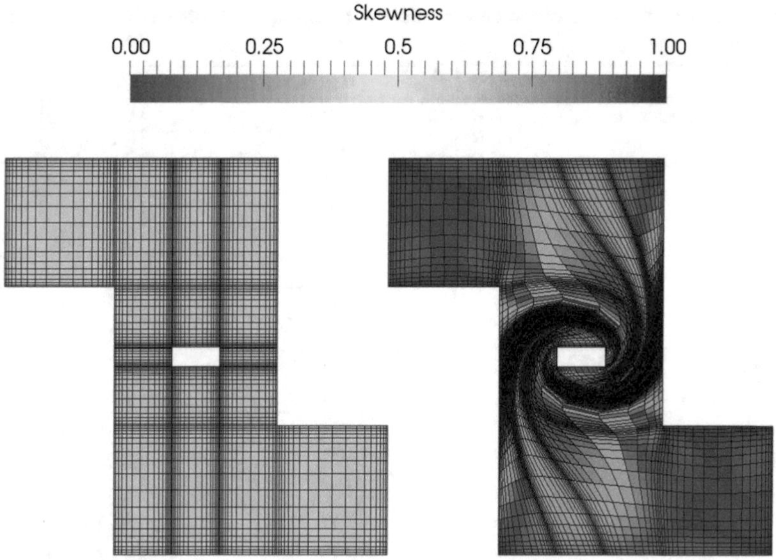

Fig. 5 Modified test case adapted from Boer et al with clustering using the thin plate spline kerne initial mesh (left) and deformed mesh (right)

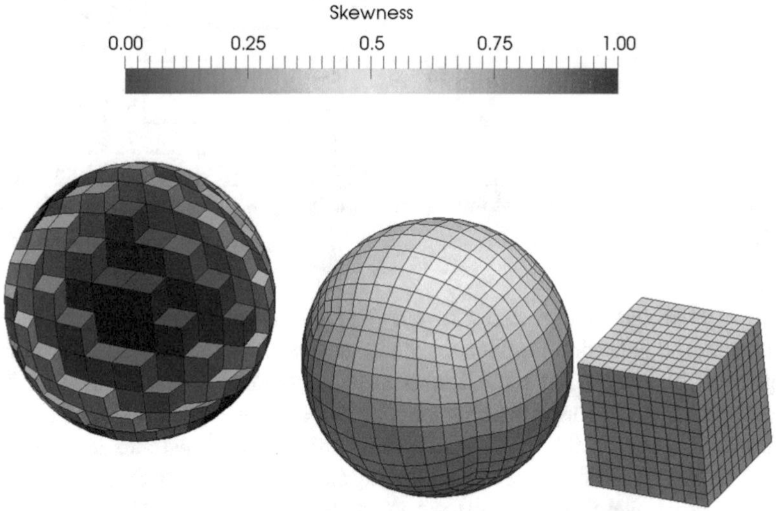

Fig. 6 Morphing a cube to a sphere using the C^2 smooth Wendland kernel of unit radius. The quality of the cells in the volume mesh degrade outwards as we approach the surface due to the aggressive changes to the surface geometry near the corners of the cube

To illustrate the solution difficulty highlighted, using the test case shown in Fig. 6, Table 1 shows the condition numbers of the RBF system for different problem sizes.

Table 1 change in matrix condition number due to change in problem size and number of constraints

Kernel type	#Nodes	Number of constraints	Condition number
Thin plate spline: $x^2 \log(x)$	10^3	100	9.18×10^3
		200	2.51×10^4
		300	4.30×10^4
		400	6.39×10^4
	50^3	100	2.82×10^4
		200	2.01×10^5
		300	3.65×10^5
		400	5.34×10^5
C0 Wendland: $(1 - x/r)\,\hat{}\,2$	10^3	100	5.18×10^1
		200	1.18×10^2
		300	8.21×10^1
		400	3.00×10^2
	50^3	100	1.60×10^2
		200	2.75×10^2
		300	3.57×10^2
		400	4.26×10^2
C2 Wendland: $(1 - x/r)\,\hat{}\,4\,(1 + 4x/r)$	10^3	100	1.18×10^2
		200	8.21×10^1
		300	3.00×10^2
		400	1.07×10^2
	50^3	100	1.56×10^3
		200	5.18×10^3
		300	7.25×10^3
		400	1.73×10^4

The data in table one was generated from the test case in Fig. 6 where a cube is morphed into a sphere

From table one above one can see that for a relatively small number of constraints, the condition number for the system of equations which determine the weights of the interpolant grow consistently by about an order of magnitude. In general globally supported kernels such as the thin plate spline suffer from this more than the compactly supported kernels. This further illustrates the need for a careful choice of RBF kernel in order to minimise loss in accuracy. As a general rule of thumb, it was found that using compactly supported kerenels with large radii resulted in better performance when the mesh is a a bounded region.

As discussed earlier, possible applications of mesh morphing include exploring smooth variations in the geometry of a domain. This is naturally achieved when the domain has a parametric representation. In addition to this, the deformations applied to the geometry must be achievable by variation of these parameters. One such application in gas turbine design is variations in twist over the length of a blade-like geometry. This is easily achieved using our implementation by simply defining

Fig. 7 Example of applying RBF morphing to achieve similar variations in geometry in the absence of parametric representations on a turbine nozzle guide vane (NGV). From left to right, Original Geometry,C^0, C^2, C^4, C^6 polynomial wendland kernels

the constraints required by the morphing process. The choice of kernel affects the nature of the variation along the blade length as illustrated in Fig. 7.

4 Application to Outlet Guide Vane Geometry Matching

In this test case, the application of mesh morphing in altering the mesh of an outlet guide vane (OGV) of a modern high bypass gas turbine engine is investigated. Traditionally, a block structured grid of each blade is produced from a parametric representation of the component using the in-house meshing code Padram [14]. Great care is taken to produce high quality cells in every block to adequately resolve the flow during fluid flow simulations.

A 48-blade assembly of a real OGV assembly was laser scanned to produce a stereolithography (STL) representation of the assembly surface, and it was found that variations in the geometry were introduced by various external factors. To understand the differences introduced by these variations on the dynamics of the flow around the OGVs, an accurate representation of the manufactured assembly must be generated and meshed in order to perform new CFD simulations. This is critical in order to understand local effects introduced by these variations, as well as system-level effects such as changes in the gas turbine performance.

These variations are not easily represented by changes in the parametric representation of the blades, and as such form an ideal candidate for our test case.

For each blade, the displacement field required to morph the mesh generated from the parametric representation can be computed by calculating the signed shortest distance field between the STL scan and the mesh, this was done using in-house code built for this purpose called Point2Surface (P2S) [15]. P2S computes the Hausdorff distance between the two surfaces and its associated direction. This distance computation is only required on the outlet guide vane surface, the inner boundary surface in Fig. 8. For all other surfaces, we require these to be fixed. In other words, the morphing process is purely needed to adapt the volume mesh local to the OGV surface naturally and smoothly.

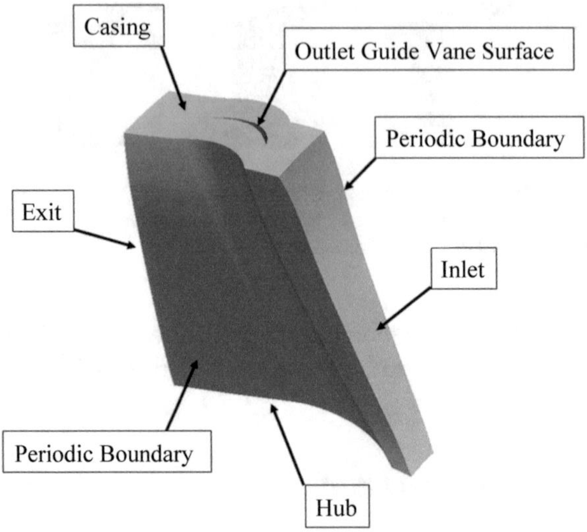

Fig. 8 Outlet Guide Vane (OGV) boundary surfaces defined for a single passage. Each OGV mesh has ~ **1.2** million nodes and **1.17** million cells

A number of challenges were encountered in this application, but practical solutions were found which allowed the use of the method to produce valid meshes for all blades.

The number of nodes in the constrained subset of the mesh was ~**19**k Nodes. Solving a linear system of this size can be a big numerical challenge. In order for this to scale, it was found that a good match could be achieved by randomly down-sampling the input displacement field. This is possible because the underlying displacement we wish to interpolate is not chaotic outside the sample set. This allows the size of the linear system to be reduced to within practical sizes. In this work, **4000** samples were drawn from the aforementioned set, **2000** on the OGV surface and **2000** from the union of the bounding surfaces for each blade volume mesh.

The update phase of the morphing can also present a challenge when applying the interpolant to a large set of nodes. In the OGV test case, each OGV comprises of ~**1.2M** nodes which must be updated when the weights of the interpolant have been computed. To mitigate this, we avoid application of the interpolant as a matrix but rather as a function evaluation for every point in the domain, i.e Eq. (6b) is evaluated at a single point to perform an in-place update thereby eliminating the need to evaluate and store the evaluation matrix. From Eq. (6b), the size of an evaluation matrix for a set of M points with an interplant contructed from N constraints is $M \times (N + 4)$.

A similar study on the effect of sampling on interpolation accuracy is also presented by Powell [16], the author concludes by numerical experiments that the

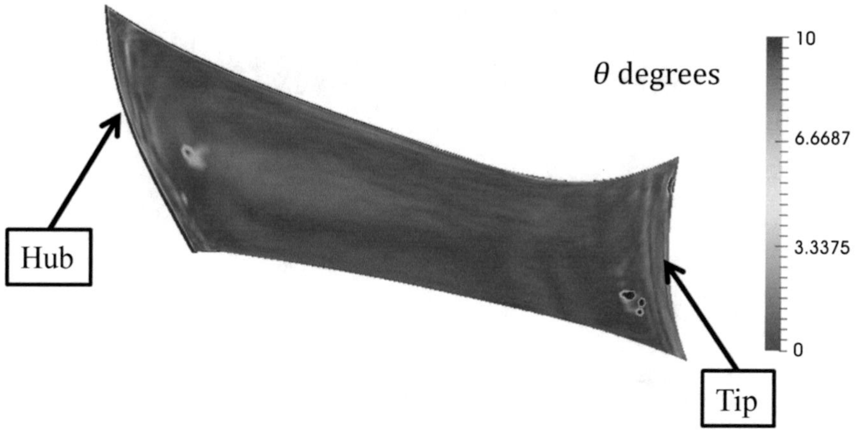

Fig. 9 Typical angular variation between the computed distance field vector and the surface normal (OGV not shown to scale)

accuracy of the interpolant is in-fact bounded by an amount that is independent of the number of samples.

Local features in the underlying displacement field presented an issue in avoiding negative volumes. Three kinds of features were found to affect the validity of the result. High displacements applied to cells with high curvature area. Particularly near the leading and trailing edge. This is currently being investigated as an improvement to the sampling routines and distance computations.

Because the Hausdorff distance is not necessarily normal to the original surface, angular variations are found where certain features are present on the surfaces. Where locally the distance field deviated largely with the surface normal, the morphing process was found to be prone to inversions, i.e negative volume cells.

These variations were usually due to local features in the scanned component such as surface imperfections and fillet radii. In order to avoid this, an additional filtration of the input data was done to exclude such measurements from the sampled data. In other words, given a threshold angle, all distance vectors that deviate from the surface normal by greater than that value are excluded from the morphing constraints in order to not capture such features (Fig. 9).

The results of applying these additional steps to the computation are shown in Fig. 10. The mode of the displacement field after morphing is seen to drop by ~**40%**. Table 2 shows a summary of the problem setup and runtimes.

The runtimes shown in Table 1 include IO operations, solution and update, however we can treat IO as an additive constant factor to the times since the size of the data has not changed between the two test cases. It is worthy of note that the size of the mesh and the size of the RBF matrix both contribute to the performance of our implementation. The size of the mesh affects the update phase where all nodes need to be displaced by evaluating the interpolant at those points (Fig. 11).

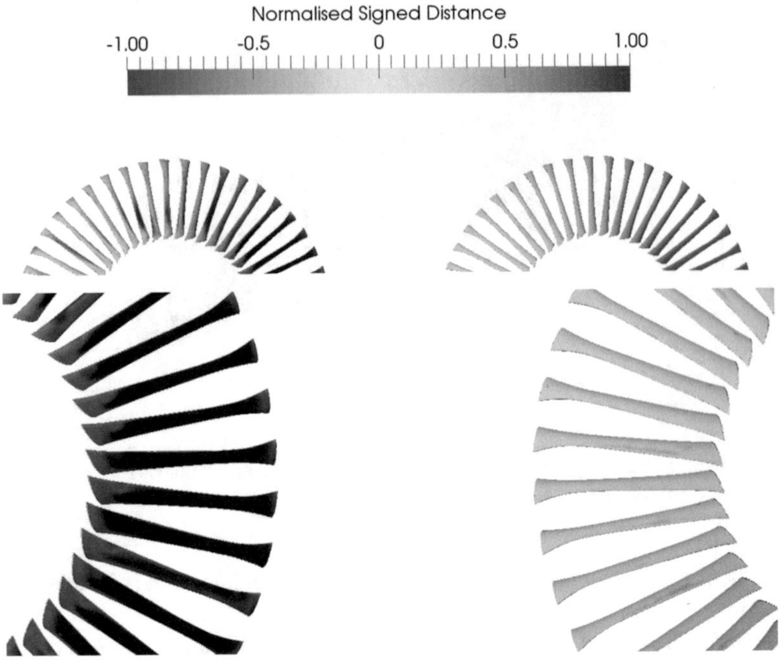

Fig. 10 Visual inspection of signed distance field before and after morphing

Table 2 Size and runtime of two test cases for the OGV mesh with sampling

Nodes (Millions)	Cells (Millions)	Size of RBF Matrix	Max CPU Time (s)	Min CPU Time (s)
1.22	1.17	4004 × 4004	50.04	43.0
1.22	1.17	8004 × 8004	331.7	166.3

Please note the time quoted above includes IO, solution and update. The data shown covers all 48 individual blades which were morphed independently

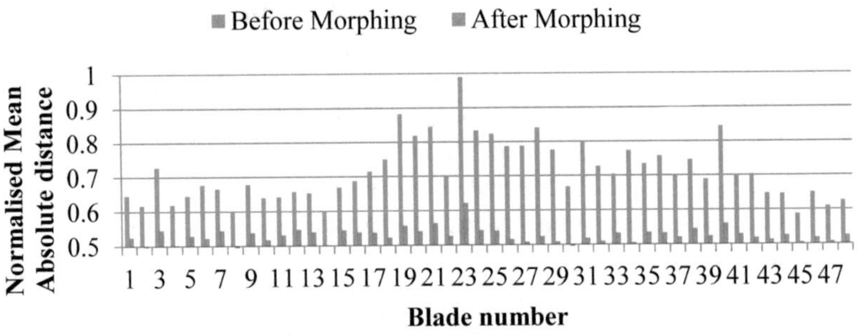

Fig. 11 Normalised mean absolute distance over OGV surface before and after morphing

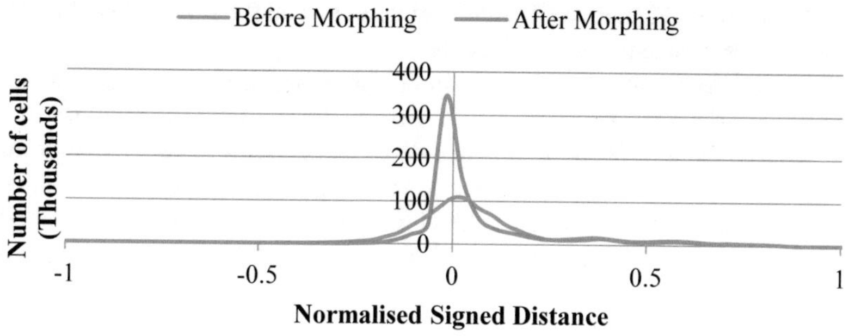

Fig. 12 Distribution of normalised signed distance field values over the entire OGV assembly before and after morphing

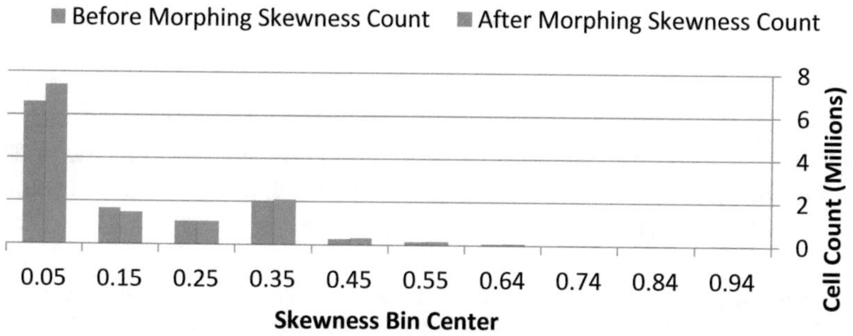

Fig. 13 Distribution of cell skewness before and after morphing

The residual distances observed after morphing can be interpreted as the effect of the random sampling and filtration. This is seen in Fig. 12 where the distribution of the displacement field computation is shown. Little variation is observed at the higher displacement values as these points were not included in the constraint set used to define the morphing problem due to the filtration of the candidate sample points.

As seen in Fig. 13, for the complete OGV assembly, little change has been introduced to the meshes after morphing. As such we are able to make a new set of OGV meshes for follow on simulation without introducing significant deterioration to element quality.

5 Conclusion

In this report the utility and robustness of the RBF morphing method was demonstrated and applied to solve an industrial problem. It was found that although the method is very sensitive to kernel type and scale (which affect the solvability of the system), the compact supported Wendland kernels, and the thin plate spline kernel were a suitable choice for this application.

A high quality mesh typically used in industrial applications can contain a very large number of nodes. This can lead to performance bottlenecks when solving for the weights of the interpolant in practice when deforming a large number of nodes. In this work, it was demonstrated that down-sampling strategies can be employed to alleviate this concern and still achieve reasonable results quickly provided the underlying data is not chaotic and the down-sampled data captures all features of interest in the underlying displacement field.

Furthermore, it was also demonstrated that the method is able to cope with deformations that may not be easily reproduced by using a parametric representation. It was also shown that for complex deformations, the method can produce degenerate cells particularly near areas with high curvature area. By additional filtration of the data to exclude such locations in the domain, it was shown that a valid mesh could be obtained.

For gas turbine applications, the leading and trailing edges are extremely important locations in the geometry of the computational domain. As such even though a valid mesh was obtained by allowing the interpolant to fit the nodes at these locations, a more explicit matching of these locations is needed to ensure the results of any subsequent simulation is representative.

Acknowledgement The authors would like to thank Roll-Royce PLC for their support and permission to publish the work. Personal thanks also to the members of Rolls-Royce 3D Geometry and Meshing Group for their help support with software tools and guidance, Alejandro Moreno, Ralf Tilch and Alan Le Moigne amongst others.

References

1. D.N. Domenico, C. Groth, A. Wade, T. Berg, M. Biancolini, Fluid structure interaction analysis: vortex shedding induced vibrations. Procedia Structural Integrity **8**, 422–432 (2018)
2. T. Rendall, C.B. Allen, Fluid-structure interpolation and mesh motion using radial basis functions. International Journal for Numerical Methods in Engineering **74**, 1519–1559 (2008)
3. D. Seiger, S. Menzel, B. Mario, RBF morphing techniques for simulation-based design optimisation, 2012. [Online]. http://www.honda-ri.de/pubs/pdf/924.pdf. Accessed 18 May 2018
4. M.L. Staten, S.J. Owen, S.M. Shontz, A.G. Salinger, T.S. Coffey, A comparison of mesh morphing methods for 3D shape optimisation, in *Proceedings of the 20th International Meshing Roundtable, IMR 2011*, Paris, 2011
5. R. Schaback, W. Holger, Kernel techniques: from machine learning to meshless methods. Acta Numerica **15**, 543–639 (2006)
6. Ansys Inc, RBF Morph, Ansys, [Online]. http://www.rbf-morph.com/. Accessed 28 July 2018
7. A.D. Boer, M.V.D. Schoot, H. Bijl, Mesh deformation based on radial basis function interpolation. Computers & Structures **85**(11–14), 784–795 (2007)
8. D. Sieger, S. Menzel, M. Botsch, High quality mesh morphing using triharmonic radial basis functions, in *International Meshing Roundtable*, (2012), pp. 1–15
9. R. Schaback, A practical guide to radial basis functions, 2018

10. H. Wendland, Piecewise polynomial, positive definite and compactly supported radial basis function of minimal degree. Advances in computational Mathematics **4**(1), 389–396 (1995)
11. Eigen Tuxfamily, Eigen: Benchmark Of Dense Decompositions, Eigen Tuxfamily, 18 May 2018. [Online]. http://eigen.tuxfamily.org/dox/group__DenseDecompositionBenchmark.html. Accessed 18 May 2018
12. Y. Wang, N. Qin, N. Zhao, Delaunay graph and radial basis function for fast quality mesh deformation. Journal of Computational Physics **294**, 149–172 (2015)
13. P. Knupp, P.P. Pebay, D. Thompson, *The Verdict Library Reference Manual* (Kitware, New York, 2007)
14. S. Shahpar, L. Lapworth, PADRAM: parametric design and rapid meshing system for turbomachinery optimisation, in *ASME Turbo Expo 2003, collocated with the 2003 International Joint Power Generation Conference*, Atlanta, Georgia, 2003
15. R. Tilch, P2S user guide, Rolls-Royce internal Document, Derby, 2018
16. M. Powell, Radial basis function methods for interpolation to functions of many variables, in *Fifth Hellenic-European Conference on Computer Mathematicas and its Applications*, Athens, 2002
17. H. Wendland, *Scattered Data Approximation* (Cambridge Unversity Press, Cambridge, 2004)

Editorial Policy

1. Volumes in the following three categories will be published in LNCSE:

i) Research monographs
ii) Tutorials
iii) Conference proceedings

Those considering a book which might be suitable for the series are strongly advised to contact the publisher or the series editors at an early stage.

2. Categories i) and ii). Tutorials are lecture notes typically arising via summer schools or similar events, which are used to teach graduate students. These categories will be emphasized by Lecture Notes in Computational Science and Engineering. **Submissions by interdisciplinary teams of authors are encouraged**. The goal is to report new developments – quickly, informally, and in a way that will make them accessible to non-specialists. In the evaluation of submissions timeliness of the work is an important criterion. Texts should be well-rounded, well-written and reasonably self-contained. In most cases the work will contain results of others as well as those of the author(s). In each case the author(s) should provide sufficient motivation, examples, and applications. In this respect, Ph.D. theses will usually be deemed unsuitable for the Lecture Notes series. Proposals for volumes in these categories should be submitted either to one of the series editors or to Springer-Verlag, Heidelberg, and will be refereed. A provisional judgement on the acceptability of a project can be based on partial information about the work: a detailed outline describing the contents of each chapter, the estimated length, a bibliography, and one or two sample chapters – or a first draft. A final decision whether to accept will rest on an evaluation of the completed work which should include

- at least 100 pages of text;
- a table of contents;
- an informative introduction perhaps with some historical remarks which should be accessible to readers unfamiliar with the topic treated;
- a subject index.

3. Category iii). Conference proceedings will be considered for publication provided that they are both of exceptional interest and devoted to a single topic. One (or more) expert participants will act as the scientific editor(s) of the volume. They select the papers which are suitable for inclusion and have them individually refereed as for a journal. Papers not closely related to the central topic are to be excluded. Organizers should contact the Editor for CSE at Springer at the planning stage, see *Addresses* below.

In exceptional cases some other multi-author-volumes may be considered in this category.

4. Only works in English will be considered. For evaluation purposes, manuscripts may be submitted in print or electronic form, in the latter case, preferably as pdf- or zipped ps-files. Authors are requested to use the LaTeX style files available from Springer at http://www.springer.com/gp/authors-editors/book-authors-editors/manuscript-preparation/5636 (Click on LaTeX Template → monographs or contributed books).

For categories ii) and iii) we strongly recommend that all contributions in a volume be written in the same LaTeX version, preferably LaTeX2e. Electronic material can be included if appropriate. Please contact the publisher.

Careful preparation of the manuscripts will help keep production time short besides ensuring satisfactory appearance of the finished book in print and online.

5. The following terms and conditions hold. Categories i), ii) and iii):

Authors receive 50 free copies of their book. No royalty is paid.
Volume editors receive a total of 50 free copies of their volume to be shared with authors, but no royalties.

Authors and volume editors are entitled to a discount of 40 % on the price of Springer books purchased for their personal use, if ordering directly from Springer.

6. Springer secures the copyright for each volume.

Addresses:

Timothy J. Barth
NASA Ames Research Center
NAS Division
Moffett Field, CA 94035, USA
barth@nas.nasa.gov

Michael Griebel
Institut für Numerische Simulation
der Universität Bonn
Wegelerstr. 6
53115 Bonn, Germany
griebel@ins.uni-bonn.de

David E. Keyes
Mathematical and Computer Sciences
and Engineering
King Abdullah University of Science
and Technology
P.O. Box 55455
Jeddah 21534, Saudi Arabia
david.keyes@kaust.edu.sa

and

Department of Applied Physics
and Applied Mathematics
Columbia University
500 W. 120 th Street
New York, NY 10027, USA
kd2112@columbia.edu

Risto M. Nieminen
Department of Applied Physics
Aalto University School of Science
and Technology
00076 Aalto, Finland
risto.nieminen@aalto.fi

Dirk Roose
Department of Computer Science
Katholieke Universiteit Leuven
Celestijnenlaan 200A
3001 Leuven-Heverlee, Belgium
dirk.roose@cs.kuleuven.be

Tamar Schlick
Department of Chemistry
and Courant Institute
of Mathematical Sciences
New York University
251 Mercer Street
New York, NY 10012, USA
schlick@nyu.edu

Editor for Computational Science
and Engineering at Springer:
Martin Peters
Springer-Verlag
Mathematics Editorial IV
Tiergartenstrasse 17
69121 Heidelberg, Germany
martin.peters@springer.com

Lecture Notes in Computational Science and Engineering

1. D. Funaro, *Spectral Elements for Transport-Dominated Equations.*
2. H.P. Langtangen, *Computational Partial Differential Equations.* Numerical Methods and Diffpack Programming.
3. W. Hackbusch, G. Wittum (eds.), *Multigrid Methods V.*
4. P. Deuflhard, J. Hermans, B. Leimkuhler, A.E. Mark, S. Reich, R.D. Skeel (eds.), *Computational Molecular Dynamics: Challenges, Methods, Ideas.*
5. D. Kröner, M. Ohlberger, C. Rohde (eds.), *An Introduction to Recent Developments in Theory and Numerics for Conservation Laws.*
6. S. Turek, *Efficient Solvers for Incompressible Flow Problems.* An Algorithmic and Computational Approach.
7. R. von Schwerin, *Multi **B**ody **S**ystem **SIM**ulation.* Numerical Methods, Algorithms, and Software.
8. H.-J. Bungartz, F. Durst, C. Zenger (eds.), *High Performance Scientific and Engineering Computing.*
9. T.J. Barth, H. Deconinck (eds.), *High-Order Methods for Computational Physics.*
10. H.P. Langtangen, A.M. Bruaset, E. Quak (eds.), *Advances in Software Tools for Scientific Computing.*
11. B. Cockburn, G.E. Karniadakis, C.-W. Shu (eds.), *Discontinuous Galerkin Methods.* Theory, Computation and Applications.
12. U. van Rienen, *Numerical Methods in Computational Electrodynamics.* Linear Systems in Practical Applications.
13. B. Engquist, L. Johnsson, M. Hammill, F. Short (eds.), *Simulation and Visualization on the Grid.*
14. E. Dick, K. Riemslagh, J. Vierendeels (eds.), *Multigrid Methods VI.*
15. A. Frommer, T. Lippert, B. Medeke, K. Schilling (eds.), *Numerical Challenges in Lattice Quantum Chromodynamics.*
16. J. Lang, *Adaptive Multilevel Solution of Nonlinear Parabolic PDE Systems.* Theory, Algorithm, and Applications.
17. B.I. Wohlmuth, *Discretization Methods and Iterative Solvers Based on Domain Decomposition.*
18. U. van Rienen, M. Günther, D. Hecht (eds.), *Scientific Computing in Electrical Engineering.*
19. I. Babuška, P.G. Ciarlet, T. Miyoshi (eds.), *Mathematical Modeling and Numerical Simulation in Continuum Mechanics.*
20. T.J. Barth, T. Chan, R. Haimes (eds.), *Multiscale and Multiresolution Methods.* Theory and Applications.
21. M. Breuer, F. Durst, C. Zenger (eds.), *High Performance Scientific and Engineering Computing.*
22. K. Urban, *Wavelets in Numerical Simulation.* Problem Adapted Construction and Applications.
23. L.F. Pavarino, A. Toselli (eds.), *Recent Developments in Domain Decomposition Methods.*

24. T. Schlick, H.H. Gan (eds.), *Computational Methods for Macromolecules: Challenges and Applications.*

25. T.J. Barth, H. Deconinck (eds.), *Error Estimation and Adaptive Discretization Methods in Computational Fluid Dynamics.*

26. M. Griebel, M.A. Schweitzer (eds.), *Meshfree Methods for Partial Differential Equations.*

27. S. Müller, *Adaptive Multiscale Schemes for Conservation Laws.*

28. C. Carstensen, S. Funken, W. Hackbusch, R.H.W. Hoppe, P. Monk (eds.), *Computational Electromagnetics.*

29. M.A. Schweitzer, *A Parallel Multilevel Partition of Unity Method for Elliptic Partial Differential Equations.*

30. T. Biegler, O. Ghattas, M. Heinkenschloss, B. van Bloemen Waanders (eds.), *Large-Scale PDE-Constrained Optimization.*

31. M. Ainsworth, P. Davies, D. Duncan, P. Martin, B. Rynne (eds.), *Topics in Computational Wave Propagation.* Direct and Inverse Problems.

32. H. Emmerich, B. Nestler, M. Schreckenberg (eds.), *Interface and Transport Dynamics.* Computational Modelling.

33. H.P. Langtangen, A. Tveito (eds.), *Advanced Topics in Computational Partial Differential Equations.* Numerical Methods and Diffpack Programming.

34. V. John, *Large Eddy Simulation of Turbulent Incompressible Flows.* Analytical and Numerical Results for a Class of LES Models.

35. E. Bänsch (ed.), *Challenges in Scientific Computing - CISC 2002.*

36. B.N. Khoromskij, G. Wittum, *Numerical Solution of Elliptic Differential Equations by Reduction to the Interface.*

37. A. Iske, *Multiresolution Methods in Scattered Data Modelling.*

38. S.-I. Niculescu, K. Gu (eds.), *Advances in Time-Delay Systems.*

39. S. Attinger, P. Koumoutsakos (eds.), *Multiscale Modelling and Simulation.*

40. R. Kornhuber, R. Hoppe, J. Périaux, O. Pironneau, O. Wildlund, J. Xu (eds.), *Domain Decomposition Methods in Science and Engineering.*

41. T. Plewa, T. Linde, V.G. Weirs (eds.), *Adaptive Mesh Refinement – Theory and Applications.*

42. A. Schmidt, K.G. Siebert, *Design of Adaptive Finite Element Software.* The Finite Element Toolbox ALBERTA.

43. M. Griebel, M.A. Schweitzer (eds.), *Meshfree Methods for Partial Differential Equations II.*

44. B. Engquist, P. Lötstedt, O. Runborg (eds.), *Multiscale Methods in Science and Engineering.*

45. P. Benner, V. Mehrmann, D.C. Sorensen (eds.), *Dimension Reduction of Large-Scale Systems.*

46. D. Kressner, *Numerical Methods for General and Structured Eigenvalue Problems.*

47. A. Boriçi, A. Frommer, B. Joó, A. Kennedy, B. Pendleton (eds.), *QCD and Numerical Analysis III.*

48. F. Graziani (ed.), *Computational Methods in Transport.*

49. B. Leimkuhler, C. Chipot, R. Elber, A. Laaksonen, A. Mark, T. Schlick, C. Schütte, R. Skeel (eds.), *New Algorithms for Macromolecular Simulation.*

50. M. Bücker, G. Corliss, P. Hovland, U. Naumann, B. Norris (eds.), *Automatic Differentiation: Applications, Theory, and Implementations.*

51. A.M. Bruaset, A. Tveito (eds.), *Numerical Solution of Partial Differential Equations on Parallel Computers.*

52. K.H. Hoffmann, A. Meyer (eds.), *Parallel Algorithms and Cluster Computing.*

53. H.-J. Bungartz, M. Schäfer (eds.), *Fluid-Structure Interaction.*

54. J. Behrens, *Adaptive Atmospheric Modeling.*

55. O. Widlund, D. Keyes (eds.), *Domain Decomposition Methods in Science and Engineering XVI.*

56. S. Kassinos, C. Langer, G. Iaccarino, P. Moin (eds.), *Complex Effects in Large Eddy Simulations.*

57. M. Griebel, M.A Schweitzer (eds.), *Meshfree Methods for Partial Differential Equations III.*

58. A.N. Gorban, B. Kégl, D.C. Wunsch, A. Zinovyev (eds.), *Principal Manifolds for Data Visualization and Dimension Reduction.*

59. H. Ammari (ed.), *Modeling and Computations in Electromagnetics: A Volume Dedicated to Jean-Claude Nédélec.*

60. U. Langer, M. Discacciati, D. Keyes, O. Widlund, W. Zulehner (eds.), *Domain Decomposition Methods in Science and Engineering XVII.*

61. T. Mathew, *Domain Decomposition Methods for the Numerical Solution of Partial Differential Equations.*

62. F. Graziani (ed.), *Computational Methods in Transport: Verification and Validation.*

63. M. Bebendorf, *Hierarchical Matrices.* A Means to Efficiently Solve Elliptic Boundary Value Problems.

64. C.H. Bischof, H.M. Bücker, P. Hovland, U. Naumann, J. Utke (eds.), *Advances in Automatic Differentiation.*

65. M. Griebel, M.A. Schweitzer (eds.), *Meshfree Methods for Partial Differential Equations IV.*

66. B. Engquist, P. Lötstedt, O. Runborg (eds.), *Multiscale Modeling and Simulation in Science.*

67. I.H. Tuncer, Ü. Gülcat, D.R. Emerson, K. Matsuno (eds.), *Parallel Computational Fluid Dynamics 2007.*

68. S. Yip, T. Diaz de la Rubia (eds.), *Scientific Modeling and Simulations.*

69. A. Hegarty, N. Kopteva, E. O'Riordan, M. Stynes (eds.), *BAIL* 2008 – *Boundary and Interior Layers.*

70. M. Bercovier, M.J. Gander, R. Kornhuber, O. Widlund (eds.), *Domain Decomposition Methods in Science and Engineering XVIII.*

71. B. Koren, C. Vuik (eds.), *Advanced Computational Methods in Science and Engineering.*

72. M. Peters (ed.), *Computational Fluid Dynamics for Sport Simulation.*

73. H.-J. Bungartz, M. Mehl, M. Schäfer (eds.), *Fluid Structure Interaction II - Modelling, Simulation, Optimization.*

74. D. Tromeur-Dervout, G. Brenner, D.R. Emerson, J. Erhel (eds.), *Parallel Computational Fluid Dynamics 2008.*

75. A.N. Gorban, D. Roose (eds.), *Coping with Complexity: Model Reduction and Data Analysis.*

76. J.S. Hesthaven, E.M. Rønquist (eds.), *Spectral and High Order Methods for Partial Differential Equations.*

77. M. Holtz, *Sparse Grid Quadrature in High Dimensions with Applications in Finance and Insurance.*

78. Y. Huang, R. Kornhuber, O.Widlund, J. Xu (eds.), *Domain Decomposition Methods in Science and Engineering XIX.*

79. M. Griebel, M.A. Schweitzer (eds.), *Meshfree Methods for Partial Differential Equations V.*

80. P.H. Lauritzen, C. Jablonowski, M.A. Taylor, R.D. Nair (eds.), *Numerical Techniques for Global Atmospheric Models.*

81. C. Clavero, J.L. Gracia, F.J. Lisbona (eds.), *BAIL 2010 – Boundary and Interior Layers, Computational and Asymptotic Methods.*

82. B. Engquist, O. Runborg, Y.R. Tsai (eds.), *Numerical Analysis and Multiscale Computations.*

83. I.G. Graham, T.Y. Hou, O. Lakkis, R. Scheichl (eds.), *Numerical Analysis of Multiscale Problems.*

84. A. Logg, K.-A. Mardal, G. Wells (eds.), *Automated Solution of Differential Equations by the Finite Element Method.*

85. J. Blowey, M. Jensen (eds.), *Frontiers in Numerical Analysis - Durham 2010.*

86. O. Kolditz, U.-J. Gorke, H. Shao, W. Wang (eds.), *Thermo-Hydro-Mechanical-Chemical Processes in Fractured Porous Media - Benchmarks and Examples.*

87. S. Forth, P. Hovland, E. Phipps, J. Utke, A. Walther (eds.), *Recent Advances in Algorithmic Differentiation.*

88. J. Garcke, M. Griebel (eds.), *Sparse Grids and Applications.*

89. M. Griebel, M.A. Schweitzer (eds.), *Meshfree Methods for Partial Differential Equations VI.*

90. C. Pechstein, *Finite and Boundary Element Tearing and Interconnecting Solvers for Multiscale Problems.*

91. R. Bank, M. Holst, O. Widlund, J. Xu (eds.), *Domain Decomposition Methods in Science and Engineering XX.*

92. H. Bijl, D. Lucor, S. Mishra, C. Schwab (eds.), *Uncertainty Quantification in Computational Fluid Dynamics.*

93. M. Bader, H.-J. Bungartz, T. Weinzierl (eds.), *Advanced Computing.*

94. M. Ehrhardt, T. Koprucki (eds.), *Advanced Mathematical Models and Numerical Techniques for Multi-Band Effective Mass Approximations.*

95. M. Azaïez, H. El Fekih, J.S. Hesthaven (eds.), *Spectral and High Order Methods for Partial Differential Equations ICOSAHOM 2012.*

96. F. Graziani, M.P. Desjarlais, R. Redmer, S.B. Trickey (eds.), *Frontiers and Challenges in Warm Dense Matter.*

97. J. Garcke, D. Pflüger (eds.), *Sparse Grids and Applications – Munich 2012.*

98. J. Erhel, M. Gander, L. Halpern, G. Pichot, T. Sassi, O. Widlund (eds.), *Domain Decomposition Methods in Science and Engineering XXI.*

99. R. Abgrall, H. Beaugendre, P.M. Congedo, C. Dobrzynski, V. Perrier, M. Ricchiuto (eds.), *High Order Nonlinear Numerical Methods for Evolutionary PDEs - HONOM 2013.*

100. M. Griebel, M.A. Schweitzer (eds.), *Meshfree Methods for Partial Differential Equations VII.*

101. R. Hoppe (ed.), *Optimization with PDE Constraints - OPTPDE 2014.*

102. S. Dahlke, W. Dahmen, M. Griebel, W. Hackbusch, K. Ritter, R. Schneider, C. Schwab, H. Yserentant (eds.), *Extraction of Quantifiable Information from Complex Systems.*

103. A. Abdulle, S. Deparis, D. Kressner, F. Nobile, M. Picasso (eds.), *Numerical Mathematics and Advanced Applications - ENUMATH 2013.*

104. T. Dickopf, M.J. Gander, L. Halpern, R. Krause, L.F. Pavarino (eds.), *Domain Decomposition Methods in Science and Engineering XXII.*

105. M. Mehl, M. Bischoff, M. Schäfer (eds.), *Recent Trends in Computational Engineering - CE2014.* Optimization, Uncertainty, Parallel Algorithms, Coupled and Complex Problems.

106. R.M. Kirby, M. Berzins, J.S. Hesthaven (eds.), *Spectral and High Order Methods for Partial Differential Equations - ICOSAHOM'14.*

107. B. Jüttler, B. Simeon (eds.), *Isogeometric Analysis and Applications 2014.*

108. P. Knobloch (ed.), *Boundary and Interior Layers, Computational and Asymptotic Methods – BAIL 2014.*

109. J. Garcke, D. Pflüger (eds.), *Sparse Grids and Applications – Stuttgart 2014.*

110. H. P. Langtangen, *Finite Difference Computing with Exponential Decay Models.*

111. A. Tveito, G.T. Lines, *Computing Characterizations of Drugs for Ion Channels and Receptors Using Markov Models.*

112. B. Karazösen, M. Manguoğlu, M. Tezer-Sezgin, S. Göktepe, Ö. Uğur (eds.), *Numerical Mathematics and Advanced Applications - ENUMATH 2015.*

113. H.-J. Bungartz, P. Neumann, W.E. Nagel (eds.), *Software for Exascale Computing - SPPEXA 2013-2015.*

114. G.R. Barrenechea, F. Brezzi, A. Cangiani, E.H. Georgoulis (eds.), *Building Bridges: Connections and Challenges in Modern Approaches to Numerical Partial Differential Equations.*

115. M. Griebel, M.A. Schweitzer (eds.), *Meshfree Methods for Partial Differential Equations VIII.*

116. C.-O. Lee, X.-C. Cai, D.E. Keyes, H.H. Kim, A. Klawonn, E.-J. Park, O.B. Widlund (eds.), *Domain Decomposition Methods in Science and Engineering XXIII.*

117. T. Sakurai, S.-L. Zhang, T. Imamura, Y. Yamamoto, Y. Kuramashi, T. Hoshi (eds.), *Eigenvalue Problems: Algorithms, Software and Applications in Petascale Computing.* EPASA 2015, Tsukuba, Japan, September 2015.

118. T. Richter (ed.), *Fluid-structure Interactions.* Models, Analysis and Finite Elements.

119. M.L. Bittencourt, N.A. Dumont, J.S. Hesthaven (eds.), *Spectral and High Order Methods for Partial Differential Equations ICOSAHOM 2016.* Selected Papers from the ICOSAHOM Conference, June 27-July 1, 2016, Rio de Janeiro, Brazil.

120. Z. Huang, M. Stynes, Z. Zhang (eds.), *Boundary and Interior Layers, Computational and Asymptotic Methods BAIL 2016.*

121. S.P.A. Bordas, E.N. Burman, M.G. Larson, M.A. Olshanskii (eds.), *Geometrically Unfitted Finite Element Methods and Applications.* Proceedings of the UCL Workshop 2016.

122. A. Gerisch, R. Penta, J. Lang (eds.), *Multiscale Models in Mechano and Tumor Biology*. Modeling, Homogenization, and Applications.

123. J. Garcke, D. Pflüger, C.G. Webster, G. Zhang (eds.), *Sparse Grids and Applications - Miami 2016*.

124. M. Schäfer, M. Behr, M. Mehl, B. Wohlmuth (eds.), *Recent Advances in Computational Engineering*. Proceedings of the 4th International Conference on Computational Engineering (ICCE 2017) in Darmstadt.

125. P.E. Bjørstad, S.C. Brenner, L. Halpern, R. Kornhuber, H.H. Kim, T. Rahman, O.B. Widlund (eds.), *Domain Decomposition Methods in Science and Engineering XXIV*. 24th International Conference on Domain Decomposition Methods, Svalbard, Norway, February 6–10, 2017.

126. F.A. Radu, K. Kumar, I. Berre, J.M. Nordbotten, I.S. Pop (eds.), *Numerical Mathematics and Advanced Applications – ENUMATH 2017*.

127. X. Roca, A. Loseille (eds.), 27th International Meshing Roundtable.

128. Th. Apel, U. Langer, A. Meyer, O. Steinbach (eds.), Advanced Finite Element Methods with Applications. Selected Papers from the 30th Chemnitz Finite Element Symposium 2017.

129. M. Griebel, M. A. Schweitzer (eds.), Meshfree Methods for Partial Differential Equations IX.

For further information on these books please have a look at our mathematics catalogue at the following URL: www.springer.com/series/3527

Monographs in Computational Science and Engineering

1. J. Sundnes, G.T. Lines, X. Cai, B.F. Nielsen, K.-A. Mardal, A. Tveito, *Computing the Electrical Activity in the Heart.*

For further information on this book, please have a look at our mathematics catalogue at the following URL: www.springer.com/series/7417

Texts in Computational Science and Engineering

1. H. P. Langtangen, *Computational Partial Differential Equations.* Numerical Methods and Diffpack Programming. 2nd Edition
2. A. Quarteroni, F. Saleri, P. Gervasio, *Scientific Computing with MATLAB and Octave.* 4th Edition
3. H. P. Langtangen, *Python Scripting for Computational Science.* 3rd Edition
4. H. Gardner, G. Manduchi, *Design Patterns for e-Science.*
5. M. Griebel, S. Knapek, G. Zumbusch, *Numerical Simulation in Molecular Dynamics.*
6. H. P. Langtangen, *A Primer on Scientific Programming with Python.* 5th Edition
7. A. Tveito, H. P. Langtangen, B. F. Nielsen, X. Cai, *Elements of Scientific Computing.*
8. B. Gustafsson, *Fundamentals of Scientific Computing.*
9. M. Bader, *Space-Filling Curves.*
10. M. Larson, F. Bengzon, *The Finite Element Method: Theory, Implementation and Applications.*
11. W. Gander, M. Gander, F. Kwok, *Scientific Computing: An Introduction using Maple and MATLAB.*
12. P. Deuflhard, S. Röblitz, *A Guide to Numerical Modelling in Systems Biology.*
13. M. H. Holmes, *Introduction to Scientific Computing and Data Analysis.*
14. S. Linge, H. P. Langtangen, *Programming for Computations* - A Gentle Introduction to Numerical Simulations with MATLAB /Octave.
15. S. Linge, H. P. Langtangen, *Programming for Computations* - A Gentle Introduction to Numerical Simulations with Python.
16. H.P. Langtangen, S. Linge, *Finite Difference Computing with PDEs - A* Modern Software Approach.
17. B. Gustafsson, *Scientific Computing from a Historical Perspective.*
18. J. A. Trangenstein, *Scientific Computing.* Volume I - Linear and Nonlinear Equations.

19. J. A. Trangenstein, *Scientific Computing*. Volume II - Eigenvalues and Optimization.

20. J. A. Trangenstein, *Scientific Computing*. Volume III - Approximation and Integration.

For further information on these books please have a look at our mathematics catalogue at the following URL: `www.springer.com/series/5151`

Printed in the United States
By Bookmasters